FUNGAL METABOLITES

FUNGAL
METABOLITES

W. B. TURNER

I.C.I., Pharmaceuticals Division,
Alderley Park, Macclesfield,
Cheshire,
England

1971

ACADEMIC PRESS

London and New York

ACADEMIC PRESS INC. (LONDON) LTD
Berkeley Square House,
Berkeley Square,
London, WIX 6BA

U.S. Edition published by
ACADEMIC PRESS INC.
111 Fifth Avenue,
New York, New York 10003

ISBN: O–12–704550–3
Library of Congress Catalog Card Number: 73–129797

Printed in Great Britain at
the Pitman Press, Bath

Preface

THE ORGANIC compounds produced by fungi have been studied for a variety of reasons by scientists from many disciplines. As a result of this diversity of interest, work on fungal products is reported in a wide range of scientific journals. In Chemical Abstracts, for example, entries dealing with the isolation of a fungal metabolite may be found in the sections headed "General Biochemistry", "Plant Biochemistry", "Microbial Biochemistry", "Fermentations", "Pharmacodynamics", "Pharmaceuticals", and "Organic Chemistry". This book attempts to draw together this wide-spread literature and to present a coherent picture of the subject, and also to provide a reference work which may be of use to those working in the field.

Most of the fungal products are "secondary metabolites", so called because they play no obvious role in the metabolism of the organisms which produce them, and we begin with a brief account of the relationship of primary and secondary metabolism. Next we consider that fungi are and some aspects of their cultivation. This discussion is not intended to provide a do-it-yourself manual; rather, it presents the sort of background information which I, as a chemist working with fungal products, find useful and interesting. The remainder of the book presents a comprehensive list of fungal products.

The compounds are classified according to biosynthetic origin rather than structural type. Such a classification reveals the relationship of compounds of different chemical types, and highlights the most important unifying aspects of fungal secondary metabolism—that in spite of the diversity of their chemical structures and their biological activities, the fungal metabolites are derived from some half-dozen basic biosynthetic pathways. In the discussion accompanying the lists, emphasis is placed on the biosynthetic origins and interrelationship of the compounds. The discussion also includes comments upon their biological activities and sometimes historical notes; for we should not forget that many fungal products are of considerable economic and social importance.

The lists of compounds are based on an index started many years ago by Dr. J. F. Grove at the Akers Laboratories of I.C.I., and continued by me since his resignation in 1963. I have also made extensive use of two previous comprehensive lists of fungal metabolites—"The Pfizer Handbook of Microbial Metabolites" by M. W. Miller (McGraw-Hill, 1961) and "List of Fungal

v

Products" by S. Shibata, S. Natori and S. Udagawa (University of Tokyo Press, 1964). The accuracy of the lists depends upon the thoroughness with which I have scanned the literature since 1963 and there are undoubtedly some omissions; I would be most grateful to hear of these.

I have been fortunate in having many colleagues willing and able to help and advise me during the writing of the book and it is a pleasure to acknowledge their contribution. Of my friends at Alderley Park I must make particular mention of the following: Dr. D. F. Jones, who read, and criticized in detail, the bulk of the manuscript and who made many suggestions which have been incorporated into the book; Dr. D. Broadbent, Dr. E. G. Jefferys and Mr. A. Borrow, for advice on Chapter 2; and Mr. D. C. Aldridge, for invaluable help in checking both the manuscript and the proofs. Dr. J. D. Bu'Lock of Manchester University, who first introduced me to fungal metabolites, has read and criticized many parts of the manuscript, especially those concerned with polyketides and fatty acids, and Dr. J. R. Hanson of Sussex University has helped me considerably with Chapter 6.

Finally, I am grateful to Pharmaceuticals Division for permission to undertake this venture, for patience during its progress, and for the excellent facilities which it provides.

Contents

Chapter 1
Primary and Secondary Metabolism

Chapter 2
Fungi, Their Cultivation and Their Secondary Metabolism

Chapter 3
Secondary Metabolites Derived without the Intervention of Acetate

Chapter 4
Secondary Metabolites Derived from Fatty Acids

Chapter 5
Polyketides

CONTENTS

Chapter 6
Terpenes and Steroids

Chapter 7
Secondary Metabolites Derived from Intermediates of the Tricarboxylic Acid Cycle

Chapter 8
Secondary Metabolites Derived from Amino-acids

Chapter 9
Miscellaneous Secondary Metabolites

Introduction

IN PREPARING the lists of fungal metabolites, the following policy has been adopted:

1. Classification is by biosynthetic origin.

2. The compounds whose biosynthesis has been studied by labelling experiments are distinguished by the letter L, with reference to the experiments.

3. Except for the following classes, the lists are intended to be comprehensive: the polyacetylenes, the depsides, the carotenoids, the ergot alkaloids, the *Amanita* toxins, and the siderochromes. These classes are exemplified and references to review articles are given.

CHAPTER 1

Primary and Secondary Metabolism

THE PRIMARY metabolism of an organism is the summation of an interrelated series of enzyme-catalysed chemical reactions (both degradative and synthetic) which provide the organism with its energy, its synthetic intermediates and its key macromolecules such as protein and DNA. On the other hand, secondary metabolism involves mainly synthetic processes whose end-products, the secondary metabolites, play no obvious role in the economy of the organism. Whereas primary metabolism is basically the same for all living systems, secondary metabolism is restricted to the lower forms of life and is species, often strain, specific.

We now have a broad understanding of the biosynthetic processes leading to secondary metabolites, and the compounds are most meaningfully classified in terms of these processes; this is the approach adopted in this book. Because the secondary metabolites are derived from the intermediates of primary metabolism, often by processes related to those leading to primary metabolites, it will be useful to give some account of primary metabolism. This will be accomplished by presenting here a broad review of primary metabolism drawing particular attention to the intermediates used for the formation of secondary metabolites, and by prefacing the chapters dealing with the fungal products with a more detailed account of the relevant processes (three biological reactions which are involved in the biosynthesis of several of the classes of secondary metabolite are discussed in more detail in this chapter). Throughout we shall be mainly concerned with the progress of carbon and, to a lesser extent, nitrogen along the various pathways. We shall often ignore the interplay of substrate, enzyme and cofactor which is necessary for the biochemical reactions to proceed; nor shall we discuss in detail the energy balance of the processes, though we note here that many primary processes release energy, usually in the form of adenosine triphosphate (ATP), and that all secondary processes require energy.

A. THE MAIN CARBON PATHWAYS

The major source of carbon and of energy for most heterotrophic organisms is glucose, usually supplied as such in laboratory cultures and derived from carbohydrate in nature. A few fungal metabolites are derived directly from glucose (see p. 29) but the carbon of glucose becomes available for most

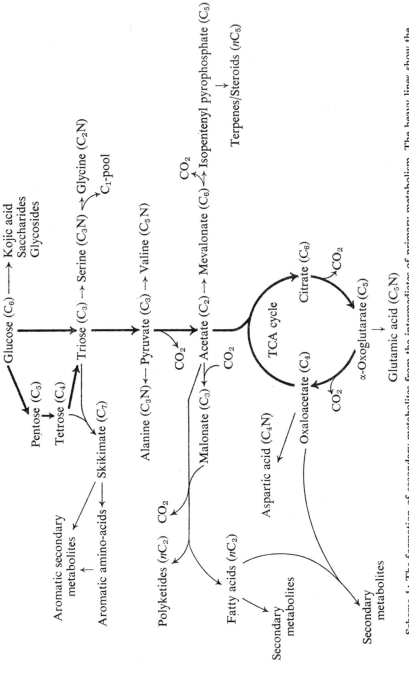

Scheme 1: The formation of secondary metabolites from the intermediates of primary metabolism. The heavy lines show the main pathways for the oxidation of glucose

biosynthetic processes in the form of the intermediates in a series of energy-releasing reactions (heavy lines in Scheme 1). If intermediates are not withdrawn for synthetic purposes, the overall result of the reaction sequence is the oxidation of glucose to carbon dioxide and water with the production of thirty-eight molecules of ATP. But for our purpose, it is the synthetic uses to which the intermediates are put which is of interest.

The breakdown of glucose begins with its conversion to a triose, either by the Embden–Meyerhof pathway or by the pentose phosphate cycle (see p. 29). The latter route also makes available pentoses, important in nucleotide biosynthesis, and a tetrose which can react with phosphoenolpyruvate to give shikimic acid (see p. 29). Shikimic acid is an intermediate for the aromatic amino-acids and also for many aromatic secondary metabolites, discussed in Chapter 3. The triose is also a precursor of serine which is converted to glycine with loss of a carbon atom which enters the C_1-pool (see p. 297).

Proceeding along the carbon pathway, the triose is converted first to pyruvate and then to acetyl coenzyme A (acetyl CoA) which is the most important single intermediate in fungal secondary metabolism. Carboxylation of acetyl CoA gives malonyl CoA (for an alternative derivation of malonate see ref. 4), and linear condensation of acetyl CoA with several molecules of malonyl CoA leads either to the polyketides, the most numerous secondary metabolites of fungi (Chapter 5) or to the fatty acids which can in turn give rise to secondary metabolites (Chapter 4). Alternatively, condensation of three molecules of acetyl CoA gives mevalonic acid, the key intermediate in terpene biosynthesis (Chapter 6). Finally, by condensation of acetyl CoA with oxaloacetate, carbon from glucose enters the tricarboxylic acid (TCA) cycle (p. 280) which serves not only to complete the oxidation of glucose but also as a source of the carbon skeletons of several amino-acids (Chapter 8) and of some secondary metabolites (Chapter 7).

B. THE CONTROL OF METABOLISM

An interrelated series of reactions such as that of Scheme 1 requires some control to prevent the overproduction of one intermediate at the expense of another. This is achieved either by inhibition of enzyme synthesis or inhibition of enzyme activity. There are two classes of enzymes: *constitutive enzymes* which are present in much the same amount under all circumstances, and *inducible enzymes* which are only synthesized in appreciable quantity in the presence of inducers, usually their substrates. The existence of inducible enzymes immediately provides an organism with some control over its metabolism since it prevents the unnecessary production of an enzyme.

But organisms have available two further control mechanisms: *end-product repression* of enzyme synthesis and *end-product* (*allosteric*) *inhibition* of enzyme activity. In end-product repression the last product of a biosynthetic sequence specifically represses the synthesis of all the enzymes catalysing the steps of the sequence. In allosteric inhibition, the end-product inhibits the activity of one of the enzymes of the sequence; this inhibition is thought to result from absorption of the end-product on to a receptor site (not that occupied by the substrate) of the enzyme, causing a conformational change in the enzyme so that it can no longer accommodate its substrate.

The enzymes of secondary metabolism must also be under metabolic control and this problem is beginning to receive attention (see, for example, p. 19).

C. GENERAL BIOLOGICAL REACTIONS

In addition to the biosynthetic pathways which form the basis for our classification of fungal metabolites, and which are discussed in the appropriate chapters, there are three biochemical processes which are used to modify the products of the major pathways. For convenience these are discussed here.

C.1. BIOLOGICAL OXIDATION AND REDUCTION

The oxidation of alcohols or the reduction of carbonyl groups, the introduction or reduction of double bonds, the introduction of oxygen atoms, and the oxidative cleavage of aromatic rings are common reactions in secondary metabolism, greatly increasing the number and variety of structures which are derivable by the basic synthetic processes. The enzymes catalysing

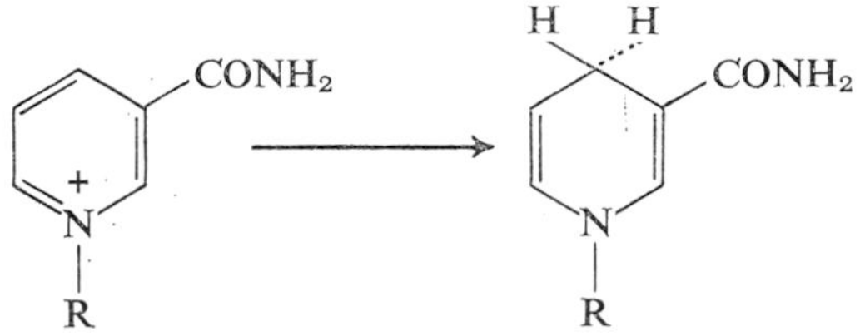

Scheme 2: The transfer of hydrogen to NAD or NADP

biological oxidations may be divided into two broad classes: the dehydrogenases, which catalyse the transfer of a hydrogen atom from the substrate to a receptor and thence to oxygen via the respiratory chain, and the oxygenases, which catalyse the introduction of oxygen into the substrate.

Dehydrogenase reactions are particularly important in primary metabolism, being involved in all the oxidative steps in the catabolism of glucose. The most common hydrogen receptors are nicotinamide adenine diphosphate (NAD, formerly DPN) and its phosphate ester NADP (formerly TPN) which are converted to the reduced forms (NADH and NADPH) with reduction of a pyridine ring (Scheme 2). The process is asymmetric, so that a deuterated substrate gives deuterated NADH which is optically active. The hydrogen is then transferred to other carrier(s) (indicated by X in Scheme 3), along a series of iron-containing pigments (the cytochromes) which constitute the respiratory chain, and finally to oxygen. The respiratory chain serves two purposes: the potential energy of each step is low so that the overall reaction proceeds more readily, and at various points along the

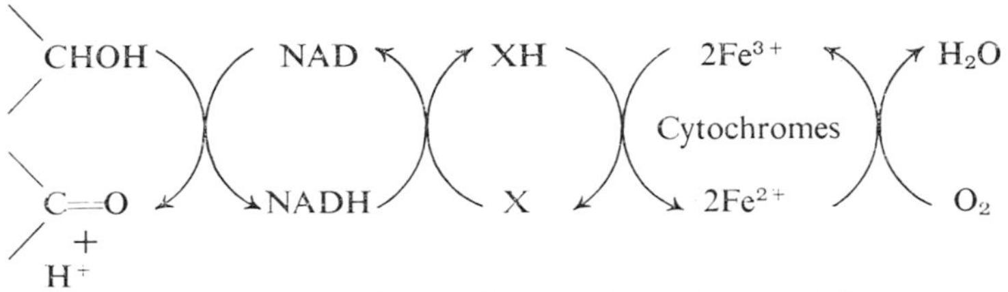

Scheme 3: Dehydrogenation and the respiratory chain

chain the oxidation–reduction process is coupled to the phosphorylation of ADP to ATP ("oxidative phosphorylation"), so that energy liberated during the oxidation of carbon substrates is stored and made available for synthetic processes. Reduction is carried out by the reverse process and therefore consumes energy with conversion of ATP to ADP.

The oxygenases probably play a larger part in secondary metabolism than in primary metabolism. They serve to activate oxygen and the most important for secondary metabolism are the "mixed function oxygenases", so called because they catalyse the introduction of one atom of an oxygen molecule into the substrate while the other is reduced to water, the hydrogen being derived ultimately from a carrier such as NADH:

$$RH + O_2 + NADH + H^+ \rightarrow ROH + H_2O + NAD$$

Presumably these are the enzymes which catalyse the introduction of the "extra" oxygen atoms into polyketide-derived compounds and into terpenes and steroids. Many organisms possess mixed function oxidases which will catalyse the oxidation of extraneous substances as well as of natural intermediates, an ability which is of considerable economic importance for the manufacture of certain steroid drugs. This lack of specificity of some oxidases has also led to confusion in the study of biosynthetic pathways since the ability to transform a supposed intermediate into a natural product has sometimes been taken to imply a natural role for the precursor (see also p. 21). In all the examples studied it has been found that hydroxylation of aliphatic substrates proceeds with retention of configuration (Scheme 4).

Another class of oxygenase reaction, less common in secondary metabolism, involves the introduction of both the atoms of a molecule of oxygen into the substrate, for example in prostaglandin biosynthesis (p. 71). The process is

Scheme 4: Retention of configuration during hydroxylation

often accompanied by cleavage of an aromatic ring, as in the biosynthesis of patulin and penicillic acid (p. 106).

C.2. THE ONE-CARBON (C_1) POOL AND BIOLOGICAL METHYLATION

One-carbon units are involved in the biosynthesis of many primary metabolites, especially in the construction of the purine and pyrimidine ring-systems of the nucleotides and in biological methylations. It is the latter aspect which is of greatest interest to us since the biosynthesis of many secondary metabolites, especially polyketides, involves methylation steps.

The key substance in the interrelationship of the one-carbon units is tetrahydrofolic acid (FH_4) (1), N-5 and N-10 of which can serve as acceptors of one-carbon units (Scheme 5). Serine and glycine are in equilibrium with N^5, N^{10}-methylene-FH_4 (see p. 297), and homocysteine and methionine are in equilibrium with N^5-methyl-FH_2 (see p. 298).

It is the methyl group of methionine which is transferred during most biological methylations and probably all methylations involved in fungal secondary metabolism. Methionine is activated for this purpose by formation of S-adenosylmethionine (2) whose methyl group can undergo nucleophilic attack.

The mechanism of the methyl transfer has been studied by Lederer and his colleagues using CD_3-methionine. In polyketide biosynthesis, both C- and O-methylations proceed with retention of all three hydrogen atoms

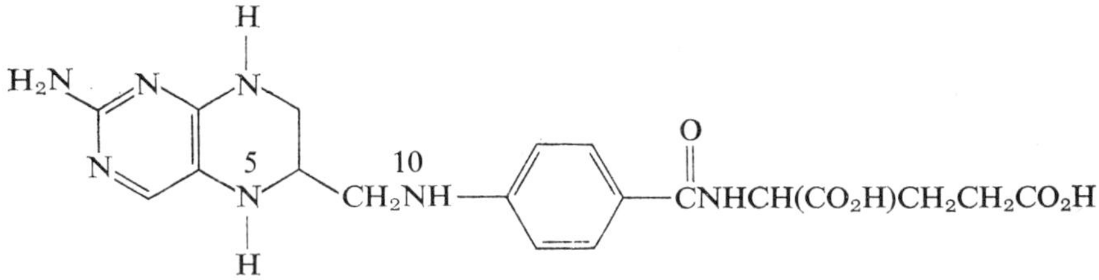

1

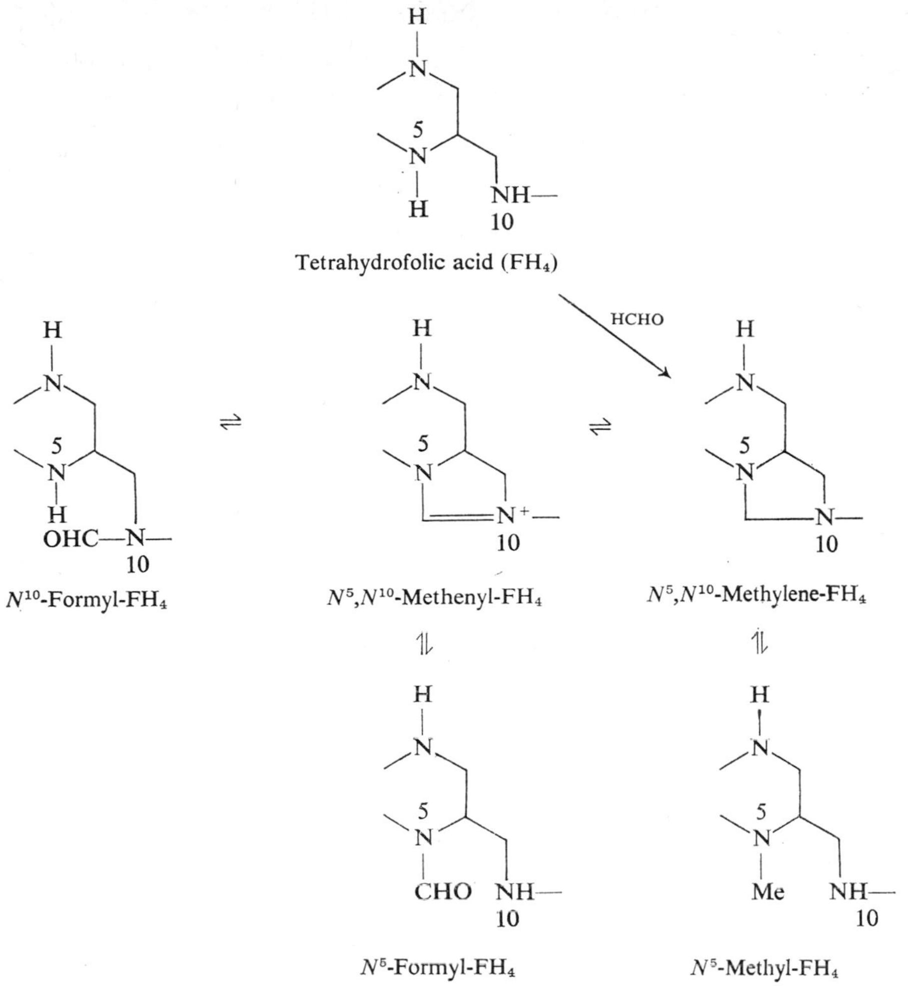

Tetrahydrofolic acid (FH₄)

N^{10}-Formyl-FH$_4$ N^5,N^{10}-Methenyl-FH$_4$ N^5,N^{10}-Methylene-FH$_4$

N^5-Formyl-FH$_4$ N^5-Methyl-FH$_4$

Scheme 5: The role of tetrahydrofolic acid (FH₄) in the C₁-pool

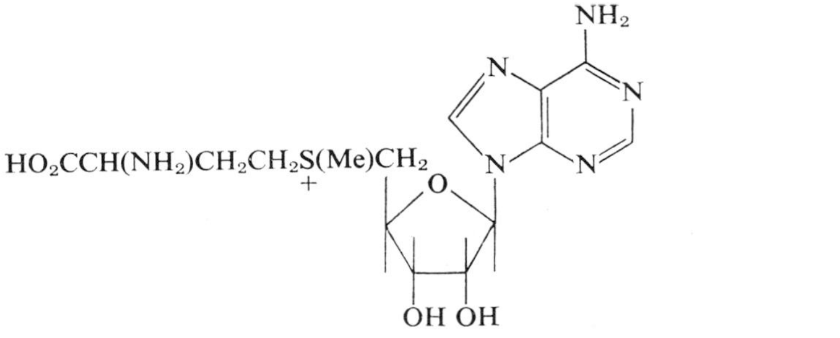

$$HO_2CCH(NH_2)CH_2CH_2\overset{+}{S}(Me)CH_2$$

2

in accord with a simple S 2 mechanism (Scheme 6) (cf. aurantiogliocladin, **p.** 102, mycophenolic acid, p. 116, and sclerotiorin, p. 174). In the methylation

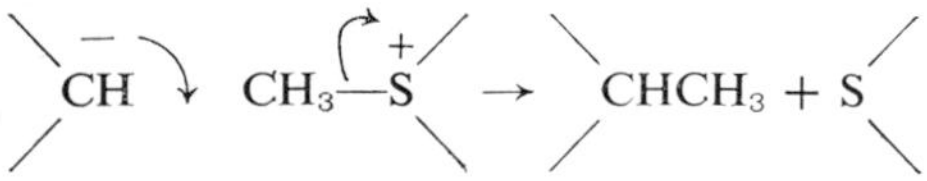

Scheme 6

of unsaturated compounds the first step appears to be the formation of a carbonium ion (**3** in Scheme 7) which can be stabilized in various ways.

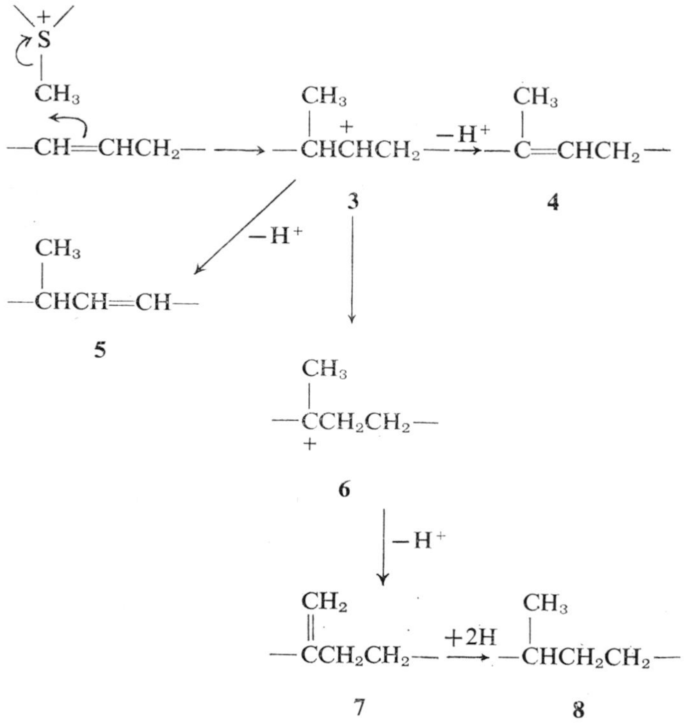

Scheme 7: Alternative reactions following the transfer of methyl from methionine to a double bond

Loss of a proton gives either of the unsaturated compounds (**4**) or (**5**), cf. for example α-smegmamycolic acid (**9**)[1]. Rearrangment of the carbonium ion

$$CH_3[CH_2]_nCH=CH[CH_2]_mCH=CHCH(*CH_3)[CH_2]_{17}CH(OH)CH(C_{22}H_{45})CO_2H$$

9

(3) to the ion (6) can lead to a methylene intermediate (7) which can be reduced to the methyl derivative (8). This process is involved in the biosynthesis of ergosterol (p. 254) and involves the loss of one of the hydrogen atoms of the methyl group of methionine. A hydrogen atom is also lost from the methyl group during the formation of tuberculostearic acid (10)[2].

$$Me[CH_2]_7CH(Me)[CH_2]_8CO_2H$$
$$10$$

C.3. BIOLOGICAL HALOGENATION

Several fungal secondary metabolites contain chlorine atoms, and in some cases the corresponding bromine derivatives are formed on media containing bromide in place of chloride. Chlorination is often not the last step in the biosynthesis and occurs in intermediates such as β-dicarbonyl compounds or phenols, which are readily halogenated *in vitro* by sources of cationic halogens.

Hager and his co-workers[3] have isolated from *Caldariomyces fumago*, which produces the chlorine-containing compound caldariomycin (see p. 50), a crystalline chloroperoxidase which catalyses the reaction:

$$RH + H_2O_2 + Cl^- + H^+ \rightarrow RCl + 2H_2O$$

The enzyme, which is a glycoprotein and has ferriprotoporphyrin-IX as prosthetic group, is relatively non-specific with regard to both substrate and halogen, and catalyses the introduction of chlorine or bromine into several β-dicarbonyl compounds and tyrosine. The detailed mechanism of the process has not been established but possibly involves oxidation of Cl^- to Cl^+.

REFERENCES

1. G. Jaureguiberry, M. Lenfant, B. C. Das and E. Lederer, *Tetrahedron*, 1966, Suppl. 8, Part I, 27.
2. G. Jaureguiberry, J. H. Law, J. A. McCloskey and E. Lederer, *Biochemistry*, 1965, **4**, 347.
3. D. R. Morris and L. P. Hager, *J. Biol. Chem.*, 1966, **241**, 1763; L. P. Hager, D. R. Morris, F. S. Brown and H. Eberwein, *J. Biol. Chem.*, 1966, **241**, 1769.
4. S. Gatenbeck and K. Mahlén, *Acta Chem. Scand.*, 1968, **22**, 1696.

CHAPTER 2

Fungi, Their Cultivation and Their Secondary Metabolism

A. FUNGI AND THEIR CLASSIFICATION

THE FUNGI, along with the algae and the bacteria, are members of the Thallophyta, a somewhat artificial division of the plant kingdom comprising organisms with no true roots, stems or leaves (some authorities prefer to regard fungi and other heterotrophic thallophytes, i.e. those which require a carbon source other than carbon dioxide, as neither plants nor animals but belonging to a separate kingdom). The vegetative body of the Thallophytes is the thallus; in the group as a whole the thallus is very simple and in many of the lower forms consists of a single cell.

The characteristic thallus of the fungi is the mycelium, which consists of an interwoven system of branching tubes called hyphae, whose walls enclose multinucleate protoplasm that continuously lays down new cell wall at the growing tip. In some classes (see below) the hyphae are divided at intervals by cross-walls (septa), while in others the hyphae are aseptate. The septa have central pores through which cytoplasm, and sometimes nuclei, can pass, so that even septate hyphae are not divided into discrete mononucleate cells. Such organisms are described as "coenocytic". Nutrients are absorbed over the whole surface of the mycelium, making for the efficient utilization of natural resources and also making possible the cultivation of fungi on artificial media which has led to their scientific and commercial importance.

We can distinguish two types of mycelium: vegetative mycelium which exists in the substrate and serves for the uptake of nutrients, and aerial mycelium which is often specialized for the formation of spores, either sexual or asexual, which serve for the preservation of the species during unfavourable conditions and for dispersal by carriers such as wind and water. Spore-formation is a complex process and it is sufficient for our purpose to note the difference between sexual and asexual spores. In the formation of sexual spores, cells (gametes) from two hyphae coalesce to form a fused cell (zygote) in which nuclear fusion and subsequent reductive division occurs (the reductive division is necessary to bring the number of chromosomes per cell back to the level of the gametes). The hyphae involved in sexual spore formation may belong to the same thallus (homothallism) or to two distinct thalli, usually denoted *plus* and *minus* (heterothallism). Asexual spores are formed from a single hypha and nuclear fusion is not involved; the process is initiated by the sealing off of the hyphal tip by a cross-wall.

The classification of the fungi rests upon the type of spores which are formed, together with the nature of the mycelium. There are four classes of fungi discussed below, which are further divided into orders (names ending in *-ales*), families (*-iaceae*), genera (defined by the first name of the fungi) and species (defined by the second name). The name of the fungus is usually followed by the authority on whose classification it is based, but we have dispensed with this throughout this book.

The Phycomycetes are the most primitive class of fungi, the lower forms being mainly aquatic with very simple thalli, often unicellular. If present, mycelium is aseptate, a feature which distinguishes the Phycomycetes from the other classes. The class includes parasites of plants (*Pythium, Phytophthera*) and of man (*Absidia, Mucor*). *Rhizopus* often forms a characteristic furry grey-white growth on bread, the appearance being due to the sporangiophores, which bear the asexual spores.

The Ascomycetes derive their name from the ascus, the sac-like vessel which contains the sexual spores. The class includes, at one extreme, the unicellular yeasts and, at the other extreme, species with large fruiting-bodies some of which, e.g. truffles, are edible. The mycelium is septate, and the most common asexual spores are conidia, borne on aerial hyphae called conidiophores. Most yeasts form loose aggregates of single cells, though some form true mycelium under certain circumstances so that it is not always easy to distinguish between them and other fungi. Asexual reproduction of yeasts is by "budding", in which a small outgrowth appears on the cell and increases in size until it is equal to the parent cell. Nuclear division occurs during this process and one of the nuclei is accommodated in the outgrowth which then separates to form a new cell. Many Ascomycetes are parasitic on plants, especially in the conidial stage, and some yeasts are responsible for human disease. An example of a plant pathogen is *Claviceps purpurea* (see the ergot alkaloids, p. 316).

The Basidiomycetes, the highest class of fungi, bear their spores on basidia which are often associated in large numbers in organized fruiting-bodies: the mushrooms, bracket fungi, puff-balls, etc. The mycelium is often perennial in soil or wood, forming sclerotia (firm masses of mycelium serving as a storage organ) or rhizomorphs (string-like structures of hyphae often having a toughened outer surface). Basidiomycetes often cause "fairy rings", resulting from the radial spread of mycelium; grass is discoloured at the perimeter where the mycelium is active but grows normally, or is even stimulated, within the ring where the mycelium is spent. Fruiting-bodies often appear at the outer edge of the ring. Basidiomycetes are also responsible for the rusts (caused by the order Uredinales) and smuts (caused by the Ustilaginales) of cereals.

We have seen that the assignment of an organism to the Ascomycetes or the Basidiomycetes depends upon the nature of its sexual spores. Those organisms for which no sexual (perfect) stage has been observed are grouped together as Fungi Imperfecti. Members of this class form asexual spores borne on conidiophores which usually occur on the surface of the mycelium but which may be formed in small fruiting-bodies called pycnidia. The most common genera are the *Penicillia* and *Aspergilli* which are often responsible for the grey-green growth on "mouldy" food (the term "mould" has no strict taxonomic meaning but is a colloquialism reserved mainly for the Fungi Imperfecti and for other organisms which do not form macroscopic fruiting-bodies). Many species allocated to the Fungi Imperfecti are now known to be conidial states of perfect fungi, mainly Ascomycetes, and are referred to, often indiscriminately, by the names of both the conidial and the perfect stages. Examples are *Fusarium moniliforme*, which is the conidial stage of *Gibberella fujikuroi*, and *Helminthosporium sativum*, the conidial stage of *Cochliobolus sativus*. In laboratory cultures it is normally the conidial stage which is grown, and it seems logical to retain the conidial name for this purpose.

The failure to obtain sexual spores may sometimes be due to heterothallism (see above) which requires both *plus* and *minus* mycelium to be present simultaneously, though many other factors are involved.

Most of the secondary metabolites discussed in this book are produced by members of the four classes of fungi discussed above but we shall also include the metabolites of the lichens, and also those metabolites of bacteria which are related to fungal metabolites. The lichens are symbiotic associations of fungi (Ascomycetes or Basidiomycetes) with unicellular algae, which contain chlorophyll and are therefore able to synthesize cell components from carbon dioxide. The fungal component (mycobiont) is thus able to draw on nutrients synthesized by the alga (phycobiont). The thallus of the lichens is unrelated in external appearance to either the fungi or the algae, and can range from a powdery crust on rocks or bark to large plants. The component organisms can often be separated and grown in laboratory culture but it is extremely difficult to reconstitute a lichen from its separated components. Many of the secondary metabolites of lichens have now been obtained from the isolated mycobiont so that they are true fungal metabolites.

The bacteria (Schizomycetes) are unicellular organisms which reproduce mainly by simple division of the cells. The order with which we shall be most concerned, because its members often produce compounds related to those from fungi, is the Actinomycetales. In these the cells form chains which branch and interweave to form a mycelium, giving the organisms a mould-like appearance. Reproduction is often by spores and some Actinomycetales, notably the family Streptomycetaceae, form conidiophores. The Actinomycetales are widespread in nature, especially in soils and composts, and species are parasitic on plants and animals. Many of the most important antibiotics are produced by Actinomycetales and as a result vast numbers have been isolated and screened for antibacterial and other biological activity.

B. THE PRODUCTION OF SECONDARY METABOLITES BY FUNGI

The fungi are able, in common with the higher plants and the bacteria, to produce secondary metabolites. In the case of the Basidiomycetes, the larger Ascomycetes and the lichens, the compounds may be obtained simply by extraction of the organism collected in the field. But the great advantage of the fungi as sources of secondary metabolites is their ability to produce the compounds on aqueous media. As a result, secondary metabolites of diverse type are conveniently available in the laboratory for chemical, biochemical and biological studies, and a few are manufactured on a commercial scale.

In some cases the same secondary metabolites have been obtained from fruiting-bodies and from aqueous culture of Basidiomycetes, though in most cases the compounds have so far only been obtained from one of the sources. The laboratory cultures of Basidiomycetes are, of course, mycelial; Basidiomycetes do not normally form fruiting-bodies under laboratory conditions and in some cases have resisted all attempts to induce them to do so. Lichens, too, are often difficult to grow *in vitro* and are always slow. For this reason, the biosynthesis of lichen products is studied with freshly collected thallus suspended in a suitable medium or, in favourable cases, with the isolated lichen fungi (though these, too, are often extremely slow growing).

In aqueous cultures, secondary metabolites accumulate both in the medium and in the mycelium. For related compounds, the distribution between medium and mycelium can often be correlated with water-solubility, though this apparent correlation may be a result of some other factor such as ease of transport across cell membranes.

C. THE LABORATORY CULTIVATION OF FUNGI

C.1. Culture Maintenance

Before embarking upon a series of experiments with a fungus it is necessary to ensure a continuing source of the organism. This involves three problems: keeping the organism alive, keeping it free from contamination by other organisms and maintaining the biochemical property which is being studied or used so that an experiment may be repeated with essentially the same result months, or even years, later.

The last of these requirements is the most difficult to fulfil, for fungi are extremely variable organisms—strains obtained from different sources, while appearing morphologically identical, will not necessarily behave in the same way biochemically. What is worse, variation can occur within a given strain, a property which manifests itself in the phenomenon of "sectoring". If a spore or a mycelial cell is placed on a nutrient agar plate (see below) the organism will grow radially, maintaining a roughly circular boundary. If at some point the mycelium undergoes a change, possibly a mutation, the variant progency will also grow radially outwards, forming a sector. If the mutation is associated with a visible change, e.g. production or absence of pigment, then the sector will be obvious. If, on the other hand, the change is associated with a biochemical function which does not lead to a visible change we are presented with an apparently uniform colony which in fact contains cells of different biochemical capabilities; so that if inocula are taken from different parts of the plate different results will be obtained. This can happen during repeated subculturing of an organism, and if a variant becomes dominant which lacks the biochemical property of interest then the property will be lost altogether. This tendency to change is universal among fungi, though some are more stable than others, so that, in the words of Foster,[5] "all investigations dealing with specific metabolic functions of a fungus sooner or later encounter physiological degeneration manifested by progressive loss of the function of particular interest". This, of course, is particularly true of the secondary metabolic processes.

In order to keep an organism alive at ambient temperature it must be transferred periodically to fresh medium. But to reduce the chance of strain variation, i.e. of subculturing from the "wrong" part of the parent culture, and of contamination by other organisms, the transfers must be as infrequent as possible. For this reason, and also to reduce the chance of variation within the culture, the growth of the stock culture should be kept to a minimum.

For most experiments fungi are grown on liquid media (see below) but for storage purposes a solid medium is more convenient. This can be prepared by the addition of agar to the warm medium, causing it to solidify at temperatures below about 40°C. An agar "slope", i.e. agar medium which has been allowed to solidify in a sloping test-tube or bottle, gives a relatively large surface on which the organism can grow while keeping storage space to a minimum. In order to keep the stock cultures moist the slopes are often kept under liquid paraffin, and to reduce growth cultures are stored at about 5°C. For longer storage, spores (if formed) may be freeze-dried or stored in dry sterile sand, or suspensions of spores or mycelium in a suitable medium may be frozen at a controlled rate in a liquid nitrogen-cooled container and stored at −196°C.

C.2. MEDIA

The media for growing fungi contain a carbon source (usually glucose though a host of other substrates may be used), a nitrogen source (usually ammonia or nitrate but often an amino-acid), phosphate, sulphate, magnesium, potassium and the trace elements: iron, manganese, zinc, molybdenum and copper. In addition to, and sometimes in place of, these chemically defined constituents, complex natural materials are often added to the medium; these include corn-steep liquor, yeast extract, vegetable juices and protein hydrolysates.

Often several different media will support the growth of an organism but not all lead to the production of a desired compound. In our screening programme at I.C.I. we use two media—Raulin–Thom and Czapek–Dox— which differ primarily in the nitrogen source (ammonia in the former and nitrate in the latter), and frequently observe the production of biological activity on one medium but not the other. A striking example of the effect of medium on the production of secondary metabolites is provided by *Penicillium baarnense* which produces mainly orsellinic acid on Czapek–Dox medium and barnol on Raulin–Thom medium.[6] The nitrogen source is the key factor in determining which compound is produced, and using replacement culture techniques it is possible to induce mycelium which is producing one of the compounds to switch to production of the other.

The differences in secondary metabolism observed on media containing different nitrogen sources may result from an indirect effect. For example, the change in pH of the medium during the course of a fermentation is partly dependent upon the nitrogen source, and pH is known in some cases to have a marked effect on secondary metabolism. Nor is it only the qualitative composition of the medium which is important—the ratio of the constituents determines the order in which each nutrient becomes exhausted and this, in turn, influences the course of a fermentation (see p. 18). Other factors which influence fermentations are discussed below.

C.3. TECHNIQUES FOR THE GROWTH OF FUNGI

C.3.a. *Surface Culture*

When a nutrient medium is inoculated with fungal spores or mycelium, the mycelium grows over the surface of the liquid to form what is variously referred to as a felt, a pad or a mat. This form of cultivation is the simplest and cheapest but suffers from several disadvantages.

Growth is inhomogeneous so that the felt contains mycelium at various stages of development and in a variety of environments—that at the surface of the felt is under more aerobic conditions than that at, or below, the surface of the medium, while that in contact with the medium is better provided with nutrients. Moreover, since nutrient taken up by the mycelium is only slowly replaced by diffusion from the lower parts of the medium, a gradient is set up within the medium so that the process is inefficient in terms of nutrient uptake and therefore relatively slow compared with the submerged techniques described below.

The composition of the gas phase also changes during fermentation and is likely to vary from flask to flask depending on the nature of the stoppers used to prevent contamination (this is true, too, of shaken cultures).

The main use of surface culture is in screening where the inhomogeneity of the mycelium becomes an advantage in that, if an organism has the ability to produce a given compound, some of the mycelium should be at the right stage and in the right environment for its production. The slowness of the process, too, becomes an advantage, increasing the chance of sampling at the right time.

For physiological or biochemical experiments or for the large-scale production of fungal products, submerged conditions, either shaken or stirred, are always to be preferred, though it is often extremely difficult to obtain the production of a desired metabolite, known to be produced in surface culture, in submerged cultures.

C.3.b. *Shaken Culture*

In this technique the medium is shaken after inoculation with spores or mycelium so that growth occurs in the body of the liquid. The advantages over surface cultures are that nutrient uptake is more efficient, giving more rapid growth and that growth is more homogeneous. Even in shaken or in stirred aerated conditions (see below) it may be difficult to obtain sufficiently homogenous growth for physiological experiments, especially in those organisms whose mycelium tends to form pellets rather than loose aggregates (conditions at the centre of a pellet will be very different from those at the surface).

C.3.c. *Stirred Aerated Culture*

This is a logical extension of the shaken culture technique and involves vigorous stirring of the medium with passage of air or oxygen. Because of the efficient agitation and aeration the process may be readily scaled up and is the most efficient method for the large-scale production of fungal metabolites. With the exception of the production of citric acid by *Aspergillus niger* all industrial processes are carried out under stirred aerated conditions and fermentors of up to 50,000 gallon capacity are now in use.

C,3,d *Continuous Culture*

By continuously introducing fresh medium into a stirred fermentation with simultaneous withdrawal of spent medium and cells, it is possible, in principle, to maintain a fermentation in a steady state indefinitely. This technique can be very valuable in laboratory studies of fermentations, for by suitable adjustment of the rate of flow (dilution rate) and composition of the medium it is possible to hold a fermentation at any desired stage while the effects of various parameters are studied.

For example, we have already noted that pH can influence secondary metabolism and that the pH of the medium changes during fermentation. Using the continuous culture technique the fermentation can be held at any stage while the pH is varied, so that the effects due to pH can be isolated from those produced by other variables. Another factor whose effect on fermentations is best studied by this technique is oxygen tension.

Continuous fermentation also offers obvious advantages for production processes. In practice two difficulties have prevented its widespread use: the difficulty of avoiding contamination over the long periods involved (this, surely, is not insurmountable) and the greater problem of maintaining the organism in the correct state for production of the desired product, i.e. of avoiding physiological degeneration (see p. 15). The method is, however, successfully used in modified form in the brewing industry and is likely to be used in the future for the production of some antibiotics.

D. THE PHASES OF FERMENTATION

There is an accumulating body of evidence, of which the experiments of Borrow *et al.*[7] and of Bu'Lock[8] may be cited, that growth and metabolism of a fungus in submerged conditions pass through distinct phases (in surface culture the greater inhomogeneity of the mycelium makes the phases more difficult to define). Initially, during what Borrow *et al.* refer to as the "balanced phase" and Bu'Lock as the "trophophase", the organism grows in an exponential manner with uptake of essential nutrients in a constant ratio. Production of secondary metabolites rarely occurs during this phase which ends with the exhaustion of one of the nutrients, usually nitrogen or phosphorus. At this point in nitrogen-limited fermentations, cell replication (though not necessarily increase in cell weight) ceases, a variety of metabolic changes occur and the production of secondary metabolites commences. The organism has then entered what Bu'Lock calls the "idiophase", i.e. the phase during which the species-specific secondary metabolites appear, which continues until the carbon source is exhausted and autolysis sets in. On the other hand, for nitrogen-limited fermentations Borrow *et al.* divided the idiophase into the "storage phase", during which cell weight continues to increase due to accumulation of fat and carbohydrate and production of secondary metabolites commences, and the "maintenance phase", during which dry weight is constant but uptake of glucose and production of secondary metabolites continues. In fermentations in which nutrients other than nitrogen are exhausted, Borrow *et al.* also distinguish a "transition phase" between the balanced phase and the storage phase, during which cell proliferation continues at a reduced rate. The various stages are shown schematically in Figure 2.1.

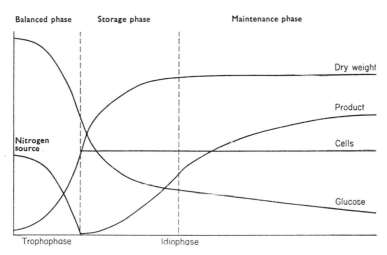

Figure 2.1: Schematic representation of a nitrogen-limited fermentation. The nomenclature of Borrow *et al.* is shown at the top of the figure and that of Bu'Lock at the bottom

Bu'Lock has suggested that the sequence of events leading to the production of secondary metabolites during the idiophase is (1) exhaustion of an essential nutrient leading to termination of cell replication and to (2) an accumulation of primary intermediates which (3) triggers either the induction of the enzymes necessary for secondary biosynthesis or the activation of enzymes formed during the trophophase; the formation of a secondary metabolite may in turn induce the synthesis of enzymes catalysing its further modification. In one case, Bu'Lock[9] has been able to distinguish between induction and activation of enzymes mediating secondary biosynthesis. *Penicillium urticae* produces 6-methysalicylic acid (6-MS) which is converted first to gentisyl derivatives and then to patulin (see p. 107). By addition of inhibitors of protein biosynthesis at various stages of the fermentations, Bu'Lock was able to show that 6-MS synthetase is a metabolically stable enzyme produced during the trophophase and activated during the idiophase but that conversion of 6-MS to gentisyl derivatives and to patulin is mediated by metabolically labile enzymes whose formation is induced during the idiophase. The mechanisms of the inhibition of 6-MS synthetase during the trophophase, and of the induction during the idiophase of the enzymes leading to patulin, remain obscure.

E. THE STUDY OF BIOSYNTHETIC PATHWAYS IN FUNGI

Fungi are convenient organisms for the study of both primary and secondary metabolism and have been used to establish the details of many biosynthetic processes of general interest as well as of those peculiar to fungi. Although the same techniques have been used in the study of both primary and secondary metabolic pathways, the emphasis has been different. Mutation experiments have been widely used in the study of primary metabolism where a block, due to the loss of an enzyme, on the pathway to an essential metabolite is easily detectable—growth is not possible unless the essential metabolite, or an intermediate after the block, is supplied in the medium. In contrast, the loss of the ability to produce a secondary metabolite is less easily detectable, except perhaps in the case of pigment production, and involves the growth and analysis of many mutant colonies; mutation experiments have therefore been used less in the study of secondary metabolism, but examples are provided by citrinin, the anthraquinones, sulochrin and related compounds and, especially, the tetracyclines.

Another difference in emphasis is the widespread use of cell-free systems or purified enzymes for the study of primary metabolism; this approach has been used successfully in the study of the biosynthesis of very few secondary metabolites. This difference is partly due to the domination of the secondary metabolite field by the organic chemist, but also arises from the extreme difficulty of obtaining cell-free systems capable of secondary biosynthesis.

The study of secondary biosynthesis has been dominated by the use of labelled precursors. The technique is exemplified throughout the book and it is necessary here only to mention two of the newer approaches and to draw attention to the pitfalls associated with this kind of work.

The problem can be attacked at two levels: the incorporation of small units, e.g. acetate or mevalonate, is used to define the general pathway involved, and the incorporation of larger molecules thought to be intermediates is used to establish the steps in the pathway. With either approach, when carbon-14 is used as the marker atom (which it is in most experiments) it is essential to establish that the product is radiochemically pure and that the chemical degradations used to locate the radioactive atoms in the product are sufficiently selective. Failure to meet these criteria has undoubtedly led to erroneous results and is the cause of some of the discrepancies between the results of different groups. Results obtained with precursors labelled with deuterium or, especially, tritium are sometimes confused by isotope effects, either during biosynthetic reactions or during degradation of the labelled products.

The use of chemical degradation to locate the isotope can sometimes be avoided by the use of alternative techniques which will probably be more widely used in future. In the first of these, deuterium-labelled precursors are used and the product is analysed by mass spectrometry (see patulin, p. 108); this requires a detailed knowledge of the fragmentation pattern of the product, and is subject to possible errors due to isotope effects. In the second, precursors labelled with carbon-13 are used and the product is analysed by n.m.r. spectroscopy (see sepedonin, p. 134, and fusaric acid, p. 315). For this technique to succeed the signals in the n.m.r. spectrum of the product must be well resolved and unambiguously assigned.

Experiments with supposed intermediates also pose a different, and more difficult problem. As Davis[10] has pointed out, "it is much easier to show that a compound *can serve as a precursor* of a cell constituent than to determine whether it *is a normal, obligatory intermediate* in the biosynthesis of that constituent from a carbohydrate". This problem has been discussed in detail by Davis[10] and by Bu'Lock,[11] and those intending to work in this field (and some who already are!) are strongly advised to read their arguments. The ambiguity arises in various ways, of which three seem to be particularly important in fungi. First, microorganisms possess non-specific enzyme

$$A \rightarrow B \rightarrow C$$
$$\updownarrow$$
$$B'$$

Scheme 1: The incorporation of B′ into C and its normal presence in the organism would suggest that it is an intermediate for C

systems which will accept a wide variety of substrates—a fact upon which the science of "microbiological transformations" depends. Secondly, although the demonstration that a precursor is normally present in the organism increases the likelihood of its being an intermediate, it may, in fact, be in equilibrium with a true intermediate (Scheme 1). Thirdly, the specificity of the enzymes of secondary biosynthesis is sometimes only relative, leading to what Bu'Lock calls a "metabolic grid" (Scheme 2). In this, the enzymes e_1 to e_4 will each catalyse the reaction of three substrates, though with differing efficiencies, with the result that there are six different pathways from A to C″. Such a system would explain many apparent inconsistencies in the literature.

Finally, it is becoming apparent in many cases that the search for intermediates is fruitless since several steps of a reaction sequence are carried out

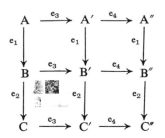

$$A \xrightarrow{\ e_3\ } A' \xrightarrow{\ e_4\ } A''$$

Scheme 2: A "metabolic grid" (for explanation see text)

on protein-bound intermediates (an outstanding example from primary metabolism is provided by fatty acid biosynthesis).

F. WHY SECONDARY METABOLITES ARE PRODUCED

We have defined secondary metabolites as compounds which play no obvious role in the economy of the organisms which produce them. Why, then, are they produced at all? This has been a topic of lively debate over the years and various suggestions have been put forward, of which three merit discussion here.

The first, criticized by Foster,[5] is that the secondary metabolites are reserve foodstuffs, accumulated during times of abundant carbohydrate and available at times when supplies of carbohydrate are limited. While it is true that most secondary metabolites are consumed when carbohydrate is exhausted, this suggestion does nothing to explain the diversity of the secondary metabolites and their uneven distribution between cell and environment. If they are produced as reserve sources of carbon, the range of structures produced is surprising since all organisms have the ability to lay down fat for this purpose.

In the case of the antibiotics it has been argued that their production confers upon the organism an advantage in its natural environment (for discussions of this point see, for example, Brian[12] or Pollock[13]). Whether or not this is true, the fact remains that most secondary metabolites have no antibiotic activity; moreover, the biosynthetic origins of those that do and those that do not are often so similar that the possession of antibiotic activity must be regarded as a fortuitous property of the product.

What is needed is an explanation for the production of *all* secondary metabolites, and in seeking such an explanation one must ask what this chemically diverse group has in common. One answer is that they are all produced from a few key intermediates of primary metabolism and this, in conjunction with the fact that, in fermentations at least, they are produced when a substrate other than carbon becomes limited, led Bu'Lock[14] to suggest that it is not the secondary metabolites which are important to the organism but the processes of secondary metabolism. In other words, secondary metabolism provides a pathway for the removal of intermediates which would otherwise accumulate and thus enables the primary processes leading to these intermediates to remain operational during times of stress. This view of secondary metabolism is an extension of the theory of secondary metabolites as shunt products (see Foster,[5] p. 164) formed in the presence of excess carbohydrate.

These arguments are attempts to explain teleologically the production of secondary metabolites, and suggest three ways in which their formation might benefit the producing organism. But it might be that the production of secondary metabolites is, after all, a mere chance event and that any advantages which they confer are fortuitous. What is more certain is that their production has provided many benefits, both intellectual and material, for mankind.

REFERENCES

The first five references are to general accounts of the subject matter of this chapter.

1. G. C. Ainsworth, "Ainsworth and Bisby's Dictionary of the Fungi", Commonwealth Mycological Institute, Kew, 5th Edn., 1961.
2. G. C. Ainsworth and A. S. Sussman (Editors), "The Fungi: An Advanced Treatise". Vol. I, The Fungal Cell, 1965; Vol. II, The Fungal Organism, 1966; Vol. III, The Fungal Population, 1968. Academic Press, New York and London.
3. G. Smith, "An Introduction to Industrial Mycology". Edward Arnold, London, 6th Edn., 1969.
4. M. E. Hale, Jr., "The Biology of the Lichens". Edward Arnold, London, 1967.
5. J. W. Foster, "Chemical Activities of Fungi". Academic Press, New York and London, 1949.
6. K. Mosbach and I. Ljungcrantz, *Physiol. Plantarum*, 1965, **18**, 1.
7. A. Borrow, E. G. Jefferys, R. H. J. Kessell, E. C. Lloyd, P. B. Lloyd and I. S. Nixon, *Can. J. Microbiol.*, 1961, **7**, 227.
8. J. D. Bu'Lock, "Essays in Biosynthesis and Microbial Development", John Wiley, London, 1967.
9. J. D. Bu'Lock, D. Shepherd and D. J. Winstanley, *Can. J. Microbiol.*, 1967, **15**, 279.

10. B. D. Davies, *Advances in Enzymology*, 1955, **16,** 247.
11. J. D. Bu'Lock, "The Biosynthesis of Natural Products". McGraw-Hill, London, 1965 (see especially Chapter 3 and p. 82).
12. P. W. Brian, *Soc. Gen. Microbiol.* 7th Symposium, Microbial Ecology, 1957 p. 168.
13. M. R. Pollock, *Brit. Med. J.*, 1967, **4,** 71.
14. J. D. Bu'Lock, *Advances in Applied Microbiology*, 1961, **3,** 293.

Secondary Metabolites Derived without the Intervention of Acetate

IN CHAPTER 1 we stressed the importance of acetyl coenzyme A as an intermediate in the biosynthesis of fungal secondary metabolites, and compounds derived from acetate will form the subject of Chapters 4, 5, 6 and 7. We also

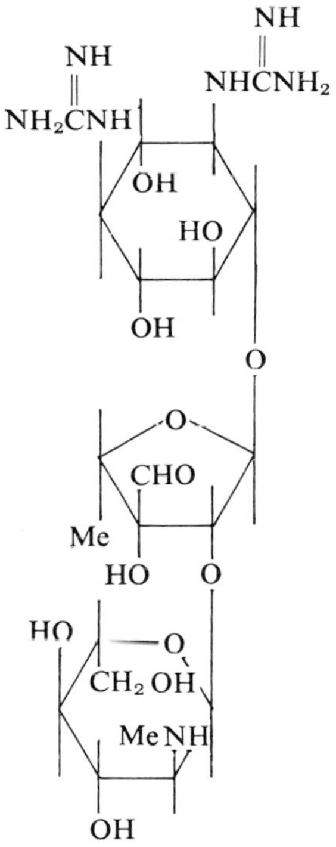

1

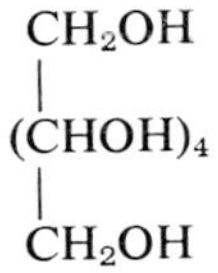

$$CH_2OH$$
$$|$$
$$(CHOH)_4$$
$$|$$
$$CH_2OH$$

2

$$CO_2H$$
$$|$$
$$(CHOH)_4$$
$$|$$
$$CH_2OH$$

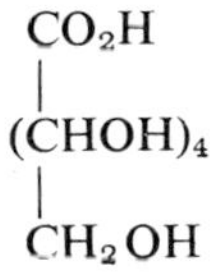

3

saw that three classes of primary metabolites—(1) those derived directly from glucose, (2) the aromatic amino-acids derived via shikimic acid and (3) the purine and pyrimidine bases—are derived from intermediates between carbohydrate and acetate. There are secondary metabolites related to each of these classes and these are discussed in this chapter.† We shall also outline the processes by which the intermediates become available, emphasizing the correlation of the carbon atoms of the intermediates with those of glucose.

A. SECONDARY METABOLITES DERIVED DIRECTLY FROM GLUCOSE

Although carbon from glucose is incorporated indirectly into all fungal metabolites, few secondary metabolites of fungi incorporate the intact

4

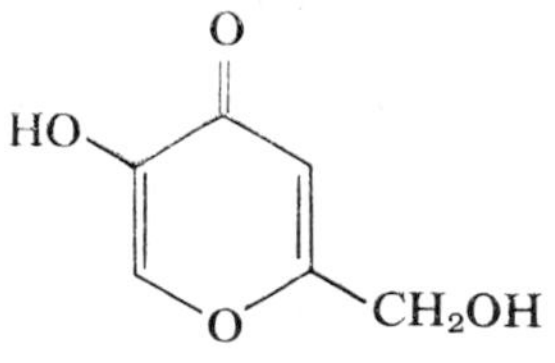

kojic acid L[8-10]
Aspergillus spp.[2]

5

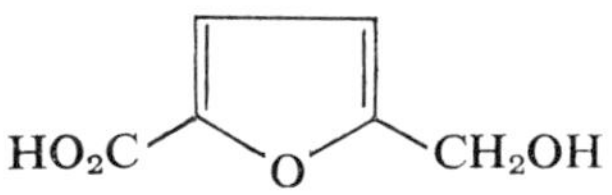

2-hydroxymethylfuran-5-carboxylic acid
Aspergillus spp.[3], *Gibberella fujikuroi*[4]

6

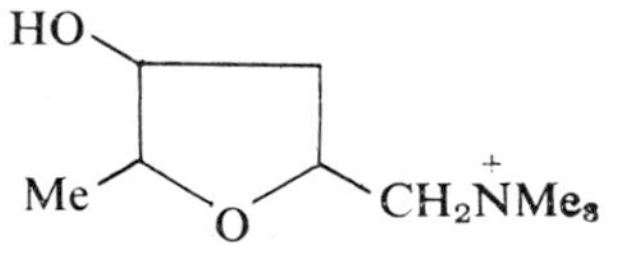

muscarine
Amanita muscaria, Inocybe spp.,
Clitocybe sp.[5,6]

7 MeCH(OH)CH(OH)CH$_2$CH$_2$CH$_2$NMe$_3^+$

muscaridine
A. muscaria[7]

carbon skeleton of glucose. This contrasts with the products of higher plants and bacteria (especially Actinomycetes) among which glycosides are of frequent occurrence while several antibiotics, for example streptomycin (**1**), derive their full carbon skeletons directly from glucose.

† Some aliphatic amino-acids are also derived without the intervention of acetate; secondary metabolites derived from these are discussed in Chapter 8.

Fungi sometimes accumulate simple derivatives of glucose, such as mannitol (2) or gluconic acid (3), in large quantity.[1] Simple disaccharides have been obtained from some fungi and lichens,[1] and a few glycosides have

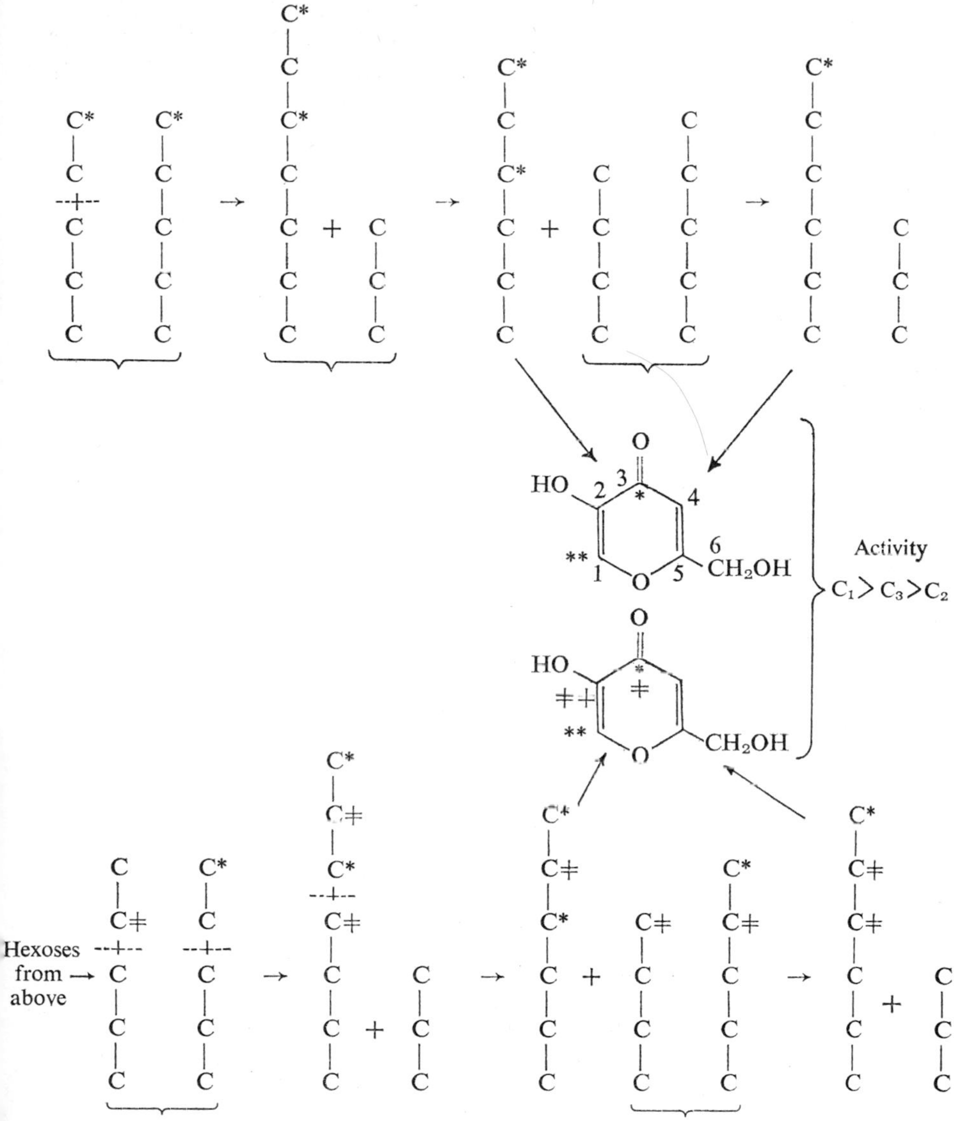

Activity
$C_1 > C_3 > C_2$

Hexoses from above →

Scheme 1: Formation of kojic acid from pentoses and C_3 compounds.[9,10] Original radio-activity of pentose is denoted by *; radioactivity in pentoses from hexoses is denoted by ⧧

been isolated from fungi—see the fusicoccins, the virescenols, the plant toxin from *Rhizoctonia*, the ustilagic acids and nemotinic acid.

In addition to these compounds, there are four fungal metabolites—(4) to (7) —whose carbon skeletons are, or appear to be, derived directly from glucose.

Kojic acid is often produced in high yield, 65% conversion of glucose having been obtained. For many years there was doubt as to whether kojic acid arises directly from glucose or from smaller molecules, notably C_3

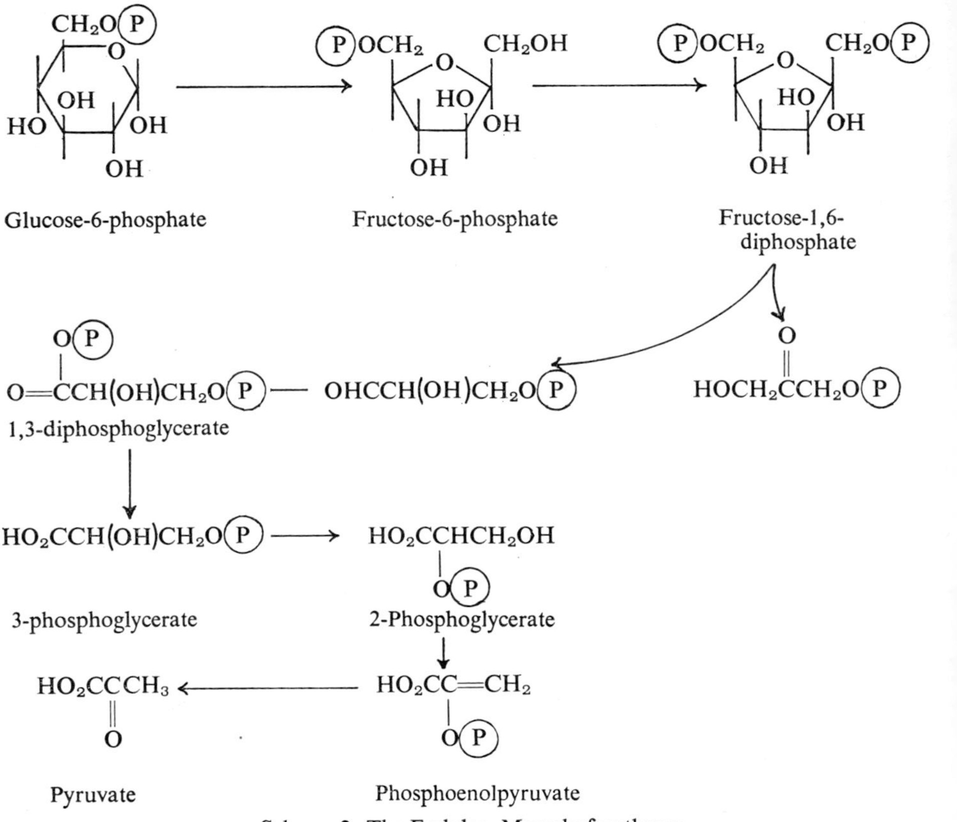

Glucose-6-phosphate   Fructose-6-phosphate   Fructose-1,6-diphosphate

Scheme 2: The Embden–Meyerhof pathway

compounds, but Arnstein and Bentley[8] used [1-^{14}C]glucose and [3,4-^{14}C]-glucose to show that the major pathway to kojic acid is by direct conversion of glucose. Further experiments with C_2 and C_3 precursors[9] and especially with pentoses[10] suggest that these precursors are incorporated into glucose via hexoses formed by the pentose phosphate pathway (see p. 31). This route plays a minor part when glucose is the substrate but becomes important with other carbon sources and explains some of the anomalous results obtained in earlier experiments.

Although no labelling experiments have been reported for the furan (**5**), the formation of the corresponding aldehyde on treatment of glucose with mineral acid makes its biosynthesis from glucose likely.

Muscarine (**6**) and the related muscaridine (**7**), too, are allocated to this section without biosynthetic support, though their biosynthesis from a desoxy-amino-sugar seems likely. Muscarine has now been produced by cultures of *Clitocybe rivulosa*[6] so that it is now amenable to biosynthetic studies. Muscarine produces characteristic effects on the parasympathetic nervous system, and has played an important part in modern pharmacology.

B. GLUCOSE CATABOLISM

In the fungi, glucose is converted to pyruvate by two major pathways—the Embden–Meyerhof (Scheme 2) and the pentose phosphate (Scheme 3)—and one minor pathway—the Entner–Douderoff (Scheme 4). The way in which pyruvate is derived from glucose by the three pathways is summarized in Scheme 5 which shows (1) that if the Embden–Meyerhof pathway is dominant then pyruvate (and hence acetate) will be labelled in the methyl group in the presence of either [1-^{14}C]- or [6-^{14}C]glucose, (2) that C-6 of glucose becomes C-3 of pyruvate in all three pathways, (3) that the relative importance of the Embden–Meyerhof and pentose phosphate pathways can be established (at least qualitatively) by experiments with [1-^{14}C]glucose, and (4) that only the Entner–Douderoff pathway yields pyruvate whose methyl group is derived from C-3 of glucose.

Since two of the steps of the pentose phosphate pathway result in the formation of glucose via fructose, the process forms a cycle (Scheme 6), the overall result of which is the oxidation of glucose to pyruvate and three molecules of carbon dioxide. A consequence of the recycling of the carbon atoms is that randomization of C-2 and C-3 of glucose occurs (Scheme 7), a point to which we shall return in connection with shikimic acid biosynthesis.

The pentose phosphate cycle not only provides an alternative route for the oxidation of glucose to pyruvate, with liberation of energy, but also yields ribose for nucleic acid synthesis and erythrose for shikimic acid synthesis.

C. THE SHIKIMIC ACID PATHWAY

The primary role of the shikimic acid pathway is to provide the essential aromatic amino-acids phenylalanine, tyrosine and tryptophan. The intermediates of the pathway can also serve as precursors of secondary metabolites which are discussed in the next section. The starting-point is the condensation

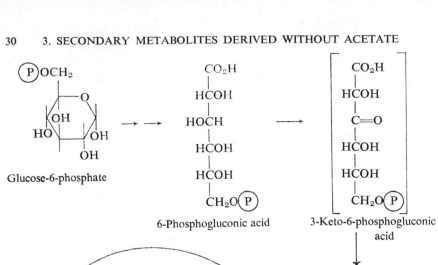

Glucose-6-phosphate

6-Phosphogluconic acid

3-Keto-6-phosphogluconic
acid

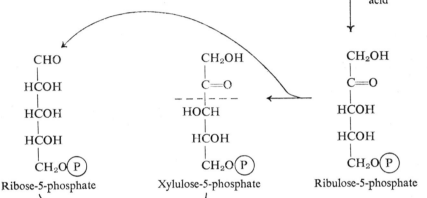

Ribose-5-phosphate

Xylulose-5-phosphate

Ribulose-5-phosphate

transketolase

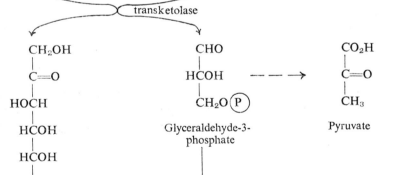

Sedoheptulose-7-phosphate

Glyceraldehyde-3-
phosphate

Pyruvate

transaldolase

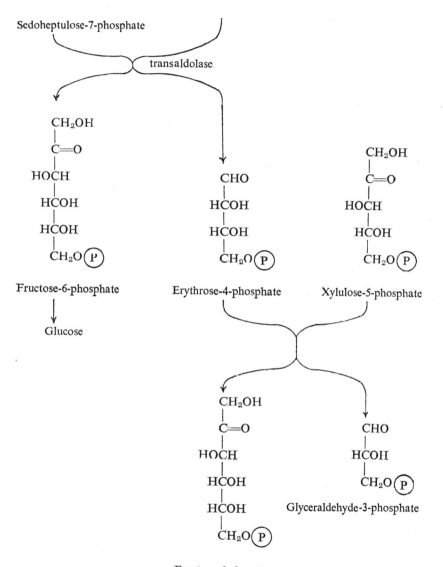

Scheme 3: The pentose phosphate pathway

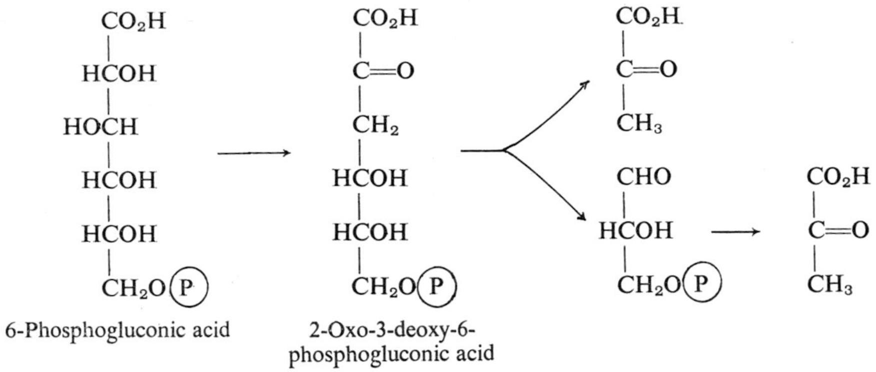

6-Phosphogluconic acid	2-Oxo-3-deoxy-6-phosphogluconic acid

Scheme 4: The Entner–Douderoff pathway

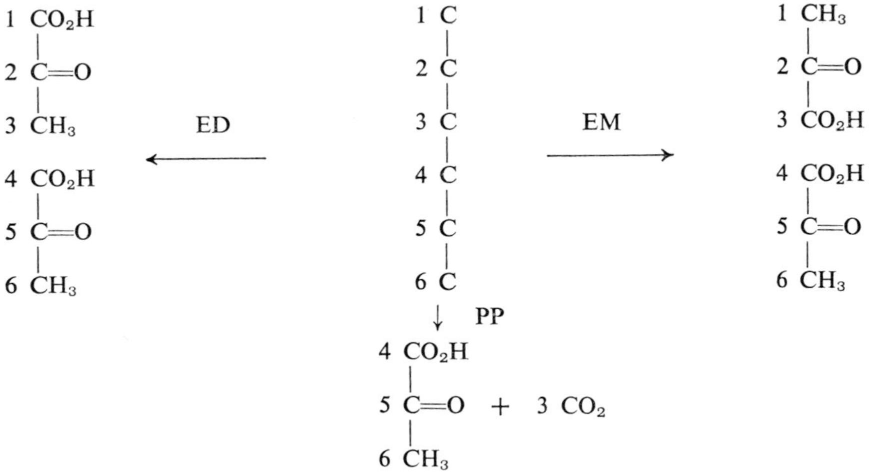

Scheme 5: The fate of the carbon atoms of glucose in the Embden–Meyerhof (EM), pentose phosphate (PP), and Entner–Douderoff (ED) pathways

of erythrose-4-phosphate, derived from the pentose phosphate pathway, with phosphoenolpyruvate, derived from the Embden–Meyerhof pathway, to give 3-deoxy-7-phospho-D-arabinoheptulosonate which then undergoes the sequence of reactions shown in Scheme 8.

By an extension of Scheme 7 the carbon atoms of glucose will be distributed in the heptulose, and hence in shikimate and phenylpyruvate, as shown in

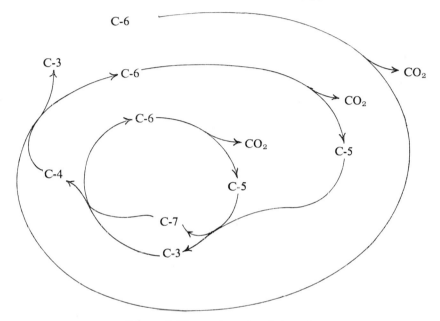

Scheme 6: The pentose phosphate cycle

Scheme 9. In fact this represents an oversimplification and carbon atoms 1,2 and 3 of glucose are incorporated into positions 2,3 and 4 of shikimate as a result of triose derived by glycolysis contributing to the biosynthesis of the tetrose. The distribution of carbon from glucose in phenylpyruvate is thus as in (8) (for a detailed discussion of the origin of the carbon atoms of shikimate see Sprinson[11]).

D. METABOLITES DERIVED FROM INTERMEDIATES OF THE SHIKIMIC ACID PATHWAY

As well as leading to the aromatic amino-acids† the shikimic acid pathway also provides intermediates for the biosynthesis of aromatic compounds.

† Metabolites which are derived from the aromatic amino-acids with retention of the amino-nitrogen atom are discussed in Chapter 8.

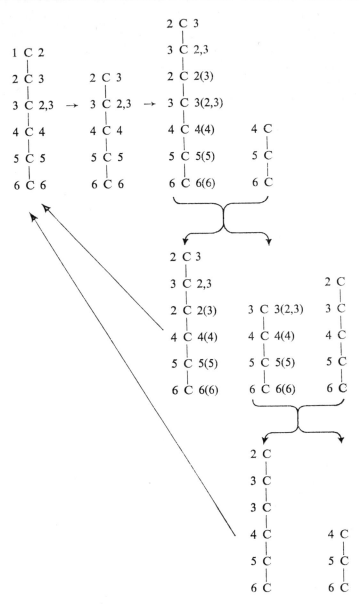

Scheme 7: The randomization of C-2 and C-3 of glucose in the pentose phosphate cycle. The figures to the left of the carbon chains refer to the first turn of the cycle and those to the right to the second turn of the cycle; the figures in parentheses are those resulting from the alternatively labelled compounds in the second cycle. One and a half turns of the cycle are illustrated

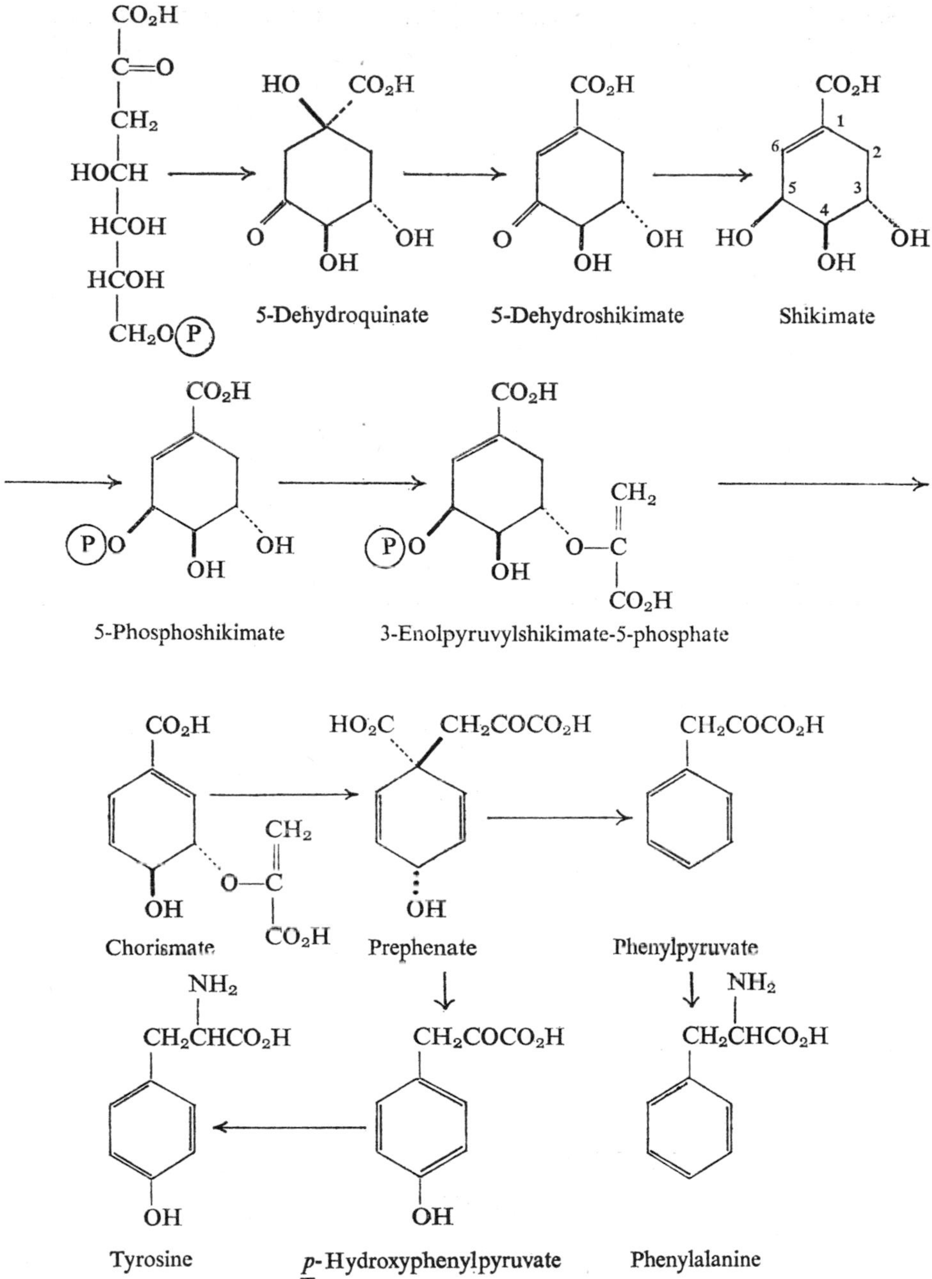

CO₂H
C=O
CH₂
HOCH
HCOH
HCOH
CH₂O(P)
HO
CO₂H
O
OH
OH
5-Dehydroquinate
5-Dehydroshikimate
HO
Shikimate
5-Phosphoshikimate
CH₂
O—C
3-Enolpyruvylshikimate-5-phosphate
HO₂C
CH₂COCO₂H
Chorismate
Prephenate
Phenylpyruvate
NH₂
CH₂CHCO₂H
Tyrosine
p-Hydroxyphenylpyruvate
Phenylalanine

Scheme 8: The biosynthesis of phenylalanine and tyrosine via shikimic acid

We shall divide the compounds into (1) simple benzene derivatives, many of which are widely distributed in nature, especially in higher plants, (2) the diphenylbenzoquinones and related compounds, which are peculiar to fungi, and (3) a few miscellaneous compounds, some of which are of types more common in higher plants while others are peculiar to fungi.

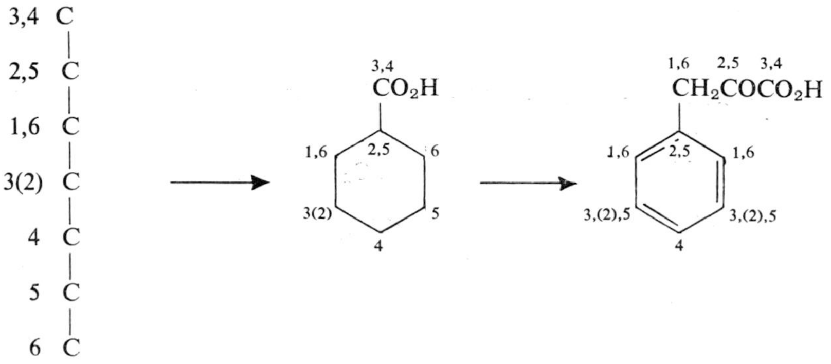

Scheme 9: Theoretical distribution of the carbon atoms of glucose in intermediates of the shikimic pathway

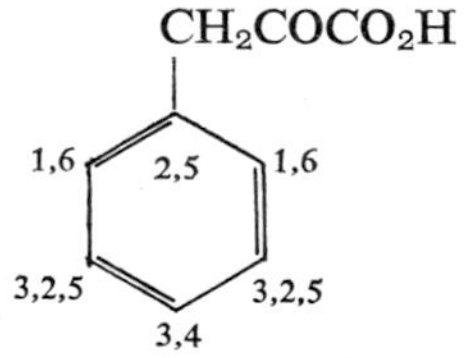

8

D.1. SIMPLE BENZENE DERIVATIVES

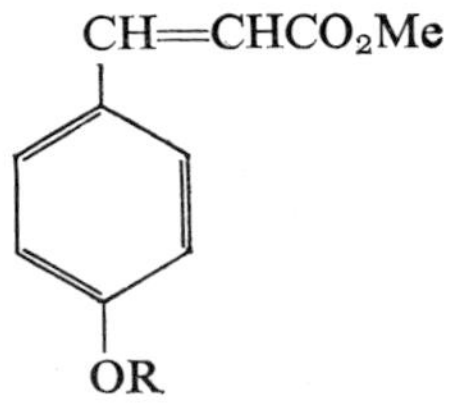

a: R=H, methyl *p*-coumarate
       Lentinus lepideus[12]

b: R=Me, methyl *p*-methoxycinnamate   *L*[29]
       Lentinus lepideus[13]

10

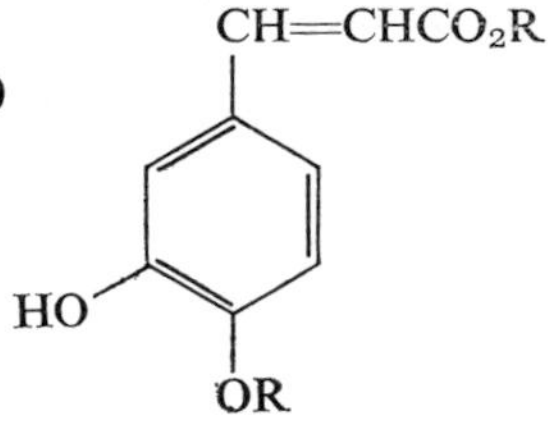

CH=CHCO₂R

HO

OR

a: R=H, caffeic acid (3,4-dihydroxycinnamic acid) L^{30}
   Boletus scaber[14]
b: R=Me, methyl isoferulate
   L. lepideus[15]

11

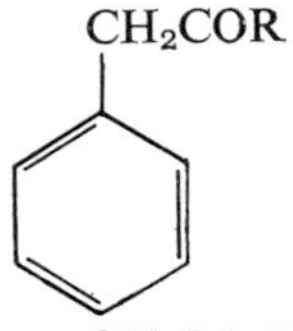

CH₂COR

a: R=OH, phenylacetic acid
   Streptomyces rimosus[16]
b: R=NH₂, phenylacetamide
   Streptomyces rimosus[16]

12

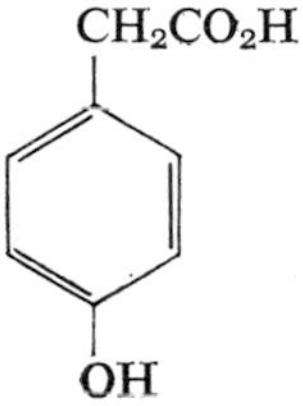

CH₂CO₂H

OH

p-hydroxyphenylacetic acid L^{34}
Pellicularia filamentosa,[17]
   Polyporus tumulosus[18]

13

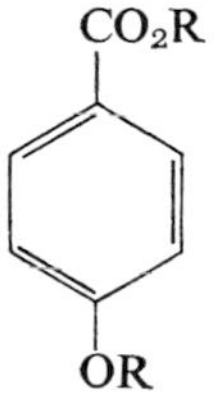

CO₂R

OR

a: R=H, *p*-hydroxybenzoic acid L^{34}
   Penicillium patulum,[19]
      Polyporus tumulosus[18]
b: R=Me, methyl anisate
   L. lepideus,[13] *Trametes suaveolens*[20]

14

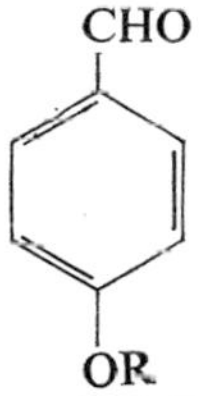

CHO

OR

a: R=Me, anisaldehyde
   Several Basidomycetes[21]
b: R=H, *p*-hydroxybenzaldehyde
   Streptomyces rimosus[16]

15

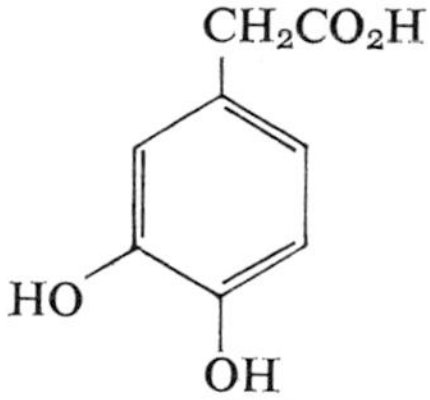

CH₂CO₂H

HO

OH

homoprotocatechuic acid L^{34}
Polyporus tumulosus[22]

16

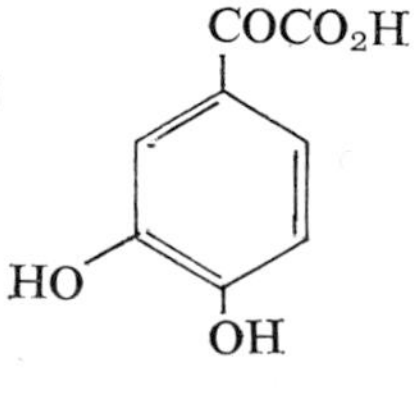

COCO₂H

HO

OH

3,4-dihydroxyphenylglyoxylic acid L^{34}
Polyporus tumulosus[18]

17

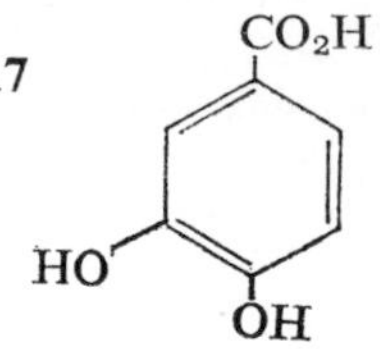

protocatechuic acid
Phycomyces blakesleeanus[23]

18

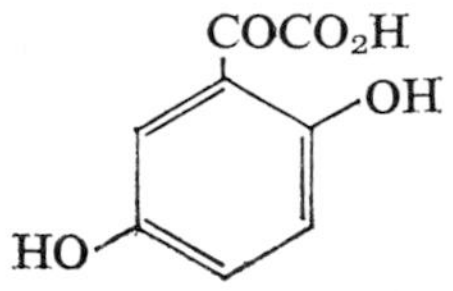

2,5-dihydroxyphenylglyoxylic acid L[34]
Polyporus tumulosus[26]

19

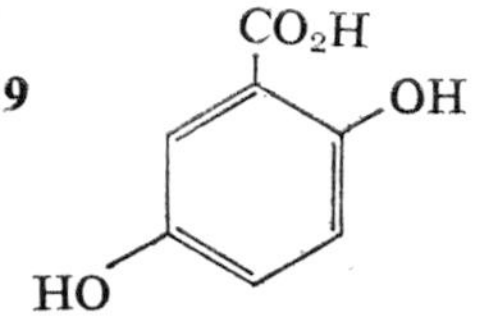

gentisic acid L[34]
Polyporus tumulosus[18]

20

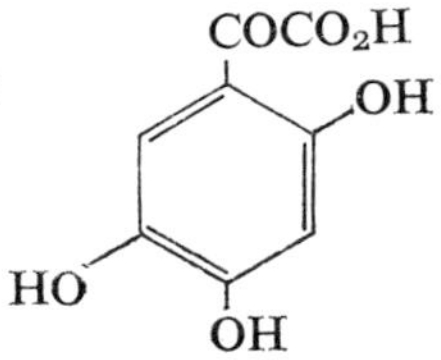

2,4,5-trihydroxyphenylglyoxylic acid
Polyporus tumulosus[22]

21

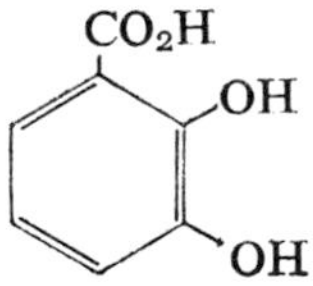

2,3-dihydroxybenzoic acid
Claviceps paspali,[27] *Streptomyces rimosus*[16]

22

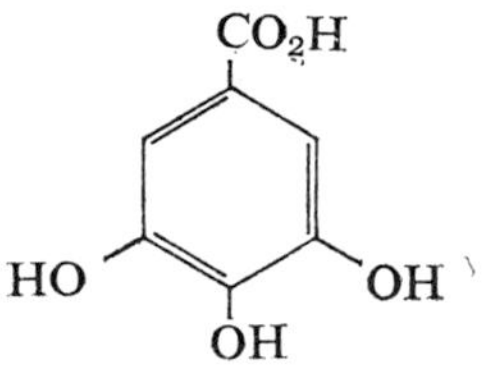

gallic acid L[23,24]
Phycomyces blakesleeanus[23]

23

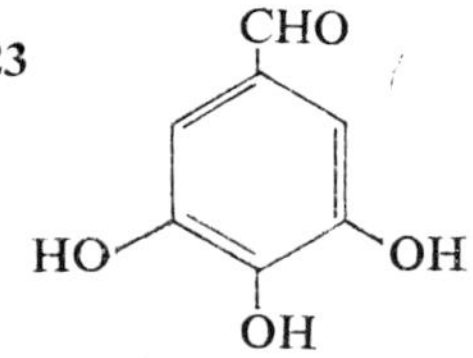

3,4,5-trihydroxybenzaldehyde
Boletus scaber[14]

24

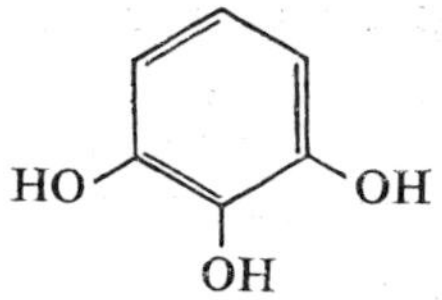

pyrogallol
Penicillium patulum[19]

The list should, perhaps, include the quinone (**25**), produced by *Lentinus degener*, since there is evidence that it is derived from tyrosine[28]. There is, however, some controversy, discussed on p. 97, concerning its origin.

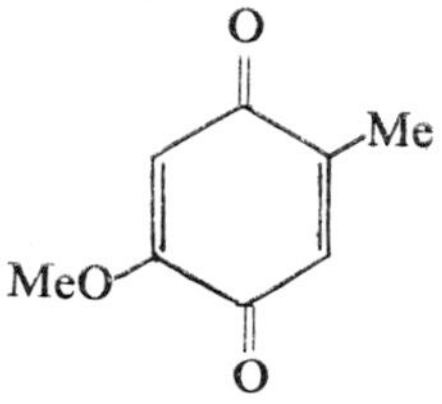

25

Many of the simple phenolic compounds also occur in higher plants, bacteria, or mammals. They are usually classified as "C_6–C_3", "C_6–C_2", or "C_6–C_1" compounds according to the length of the side-chain. The C_6–C_3 compounds are particularly widespread in higher plants where cinnamic acid and its hydroxylated derivatives are intermediates (Scheme 10) in the biosynthesis of the coumarins and lignins and can serve as "starter" acids in place of acetate

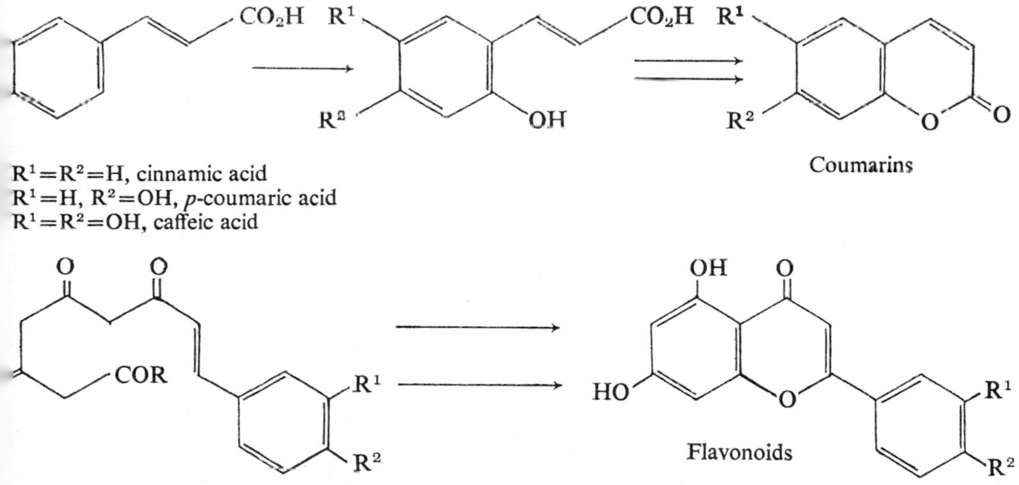

Coumarins

$R^1 = R^2 = H$, cinnamic acid
$R^1 = H$, $R^2 = OH$, *p*-coumaric acid
$R^1 = R^2 = OH$, caffeic acid

Flavonoids

Scheme 10: The formation of coumarins and flavonoids from C_6–C_3 compounds in polyketide-type biosynthesis to give, for example, the flavonoids (C_6–C_3–C_6 compounds). The flavonoids and the C_6–C_1 compound gallic acid are involved in tannin biosynthesis.

The shikimic acid pathway plays a smaller part in the secondary metabolism of fungi and in particular is rarely used by the Fungi Imperfecti, which produce a wide range of aromatic compounds by the polyketide route (see Chapter 5). Only two C_6–C_3–C_6 compounds have been isolated from fungi (see Section D.3 below).

Aromatic compounds are derived from intermediates of the shikimic acid pathway by various routes (Scheme 11) most of which have now been

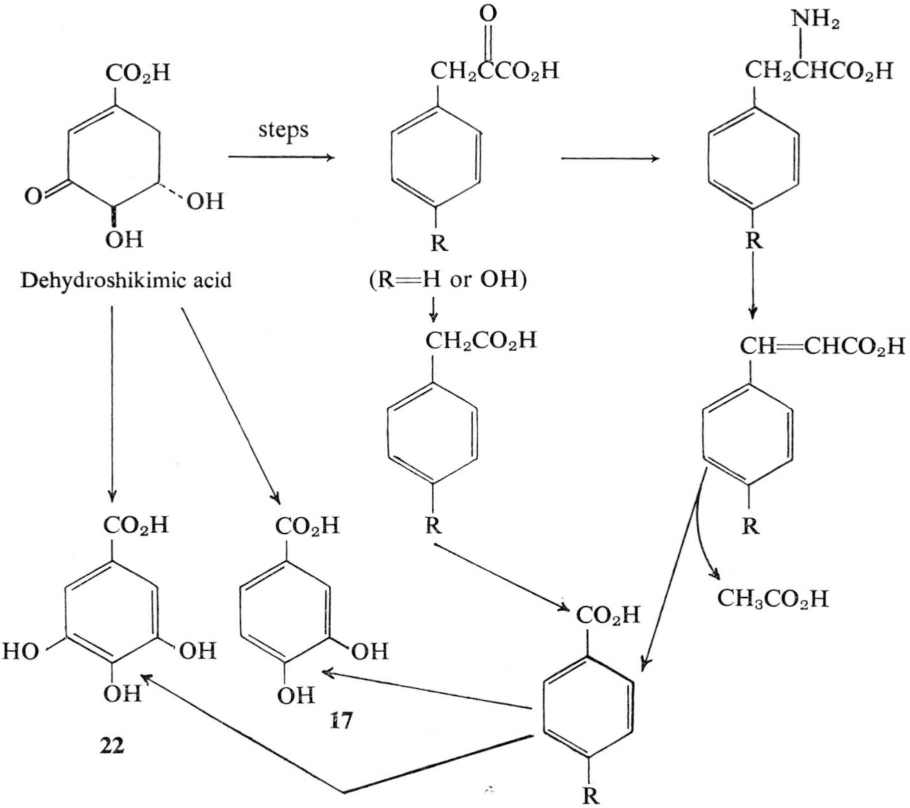

Scheme 11: The biosynthesis of aromatic compounds via the shikimic acid pathway

demonstrated in fungi. The biosynthesis of methyl p-methoxycinnamate (9b) in *Lentinus lepideus* has been studied [29] as a model for the biosynthesis of p-hydroxycinnamyl alcohol, one of the "building stones" in lignin biosynthesis; the results showed that methyl p-methoxycinnamate is derived from glucose, probably via shikimic acid. More recently it has been shown[30] that methyl isoferulate (10b), another cinnamic acid derivative produced by *L. lepideus*, is derived from glucose via phenylalanine, and that *L. lepideus* and other Basidiomycetes possess enzymes (ammonia-lyases) for the conversion of phenylalanine and tyrosine to cinnamic acids, a process which had previously been observed in higher plants. Subsequently ammonia-lyase activity was observed in cell-free extracts of *Sporobolomyces roseus*[31] and a purified ammonia-lyase was obtained from *Ustilago hordei*[32].

In higher plants the conversion of cinnamic acids to C_6–C_1 compounds and acetic acid by β-oxidation is well known. Evidence for the existence of a similar pathway in fungi, at least in Basidiomycetes, is now accumulating. Washed cells of *Sporobolomyces roseus* convert cinnamic, *p*-coumaric and caffeic acids to protocatechuic acid, probably via benzoic and *p*-hydroxybenzoic acids,[31] and *Schizophyllum commune* converts cinnamic acid, but not phenylacetic acid, to benzoic acid.[33] Moreover, during the study of the biosynthesis of 5-methoxybenzofuran (see p. 56) evidence was obtained for the formation of acetic acid from phenylalanine.

C_6–C_1 compounds can arise in other ways (Scheme 11). Protocatechuic acid (17) is formed from dehydroshikimic acid, and the results of Haslam *et al.*[23] suggest that in *Phycomyces blakesleeanus* gallic acid (22), too, is formed from dehydroshikimic acid. On the other hand, Hashem and Brücker[24] conclude from similar experiments that gallic acid is not formed directly from dehydroshikimic acid but via an aromatic amino-acid. Subsequent experiments in higher plants[25] suggest that gallic acid can arise by either pathway. C_6–C_1 compounds might also be formed by stepwise degradation of C_6–C_3 compounds via C_6–C_2 compounds; this appears to be so in *Polyporus tumulosus* (see below) but not in *S. commune* (see above).

The most detailed study of the metabolism of shikimate-derived aromatic compounds in a fungus is that of Crowden[34] who examined the incorporation of labelled substrates into seven phenols—(12), (13a), (15), (16), (18), (19) and (20)—in *Polyporus tumulosus*. He found, as expected, that the most efficient incorporation of radioactivity into the phenols was from glucose and shikimate, and that radioactivity from acetate was incorporated mainly into a non-phenolic fraction. (In view of its formation from acetate in Fungi Imperfecti, see p. 92, it is interesting that gentisic acid was particularly heavily labelled from glucose and shikimate in *Polyporus tumulosus*; gentisic acid is also formed by the shikimic acid pathway in higher plants.[35]) As a result of a study of the sequence of formation of the compounds, supplemented by experiments in which most of the logical precursors were supplied to the organism in replacement cultures, Crowden proposed Scheme 12 which summarizes both the observed conversions and their relationship to biochemical pathways already described in the literature. He concludes that

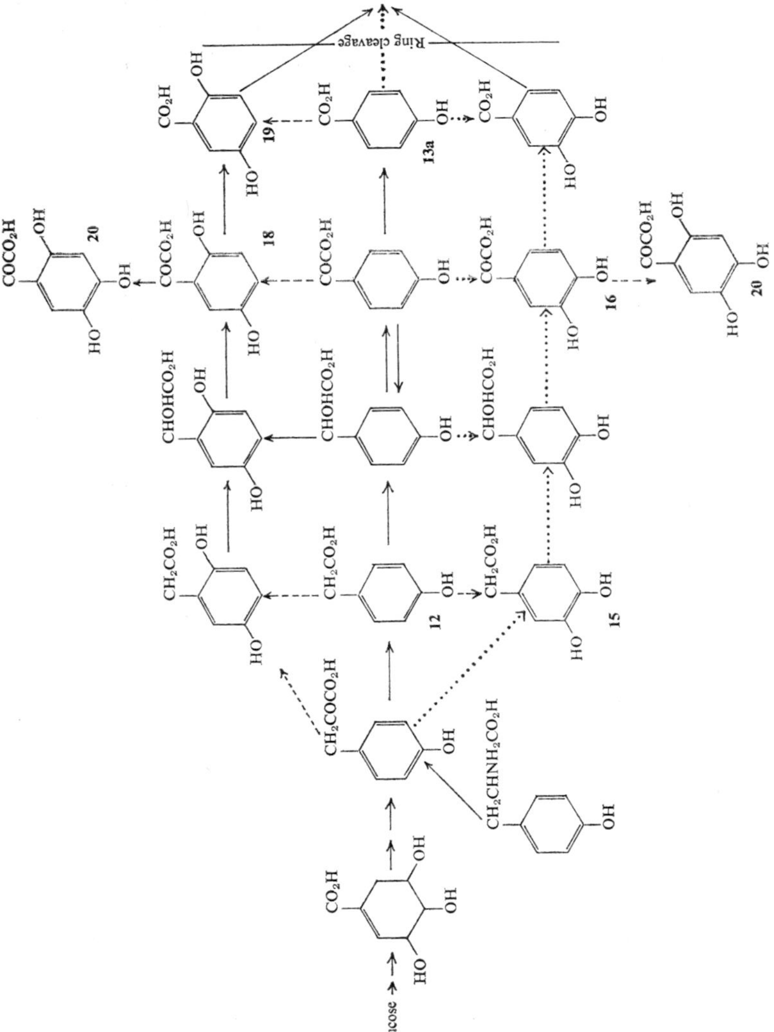

Scheme 12: Interrelationship of *Polyporus tumulosus* metabolites (numbered formulae). ⟶ Demonstrated reactions; – – ⟶ probable reactions (circumstantial evidence); ⋯⟶ possible reactions (no direct evidence)

the three alternative pathways for the metabolism of C_6–C_2 to C_6–C_1 compounds operate simultaneously, the relative importance of the pathways depending in part upon prevailing conditions, and points out that the simultaneous operation of several alternative biochemical reactions is characteristic of detoxification processes. Scheme 12 provides a good example of the factors, discussed on p. 21, which can complicate precursor experiments.

The formation of 2,5-dihydroxyphenylacetic acid from p-hydroxyphenyl-acetic acid (cf. Scheme 12) in mammalian systems is known to involve

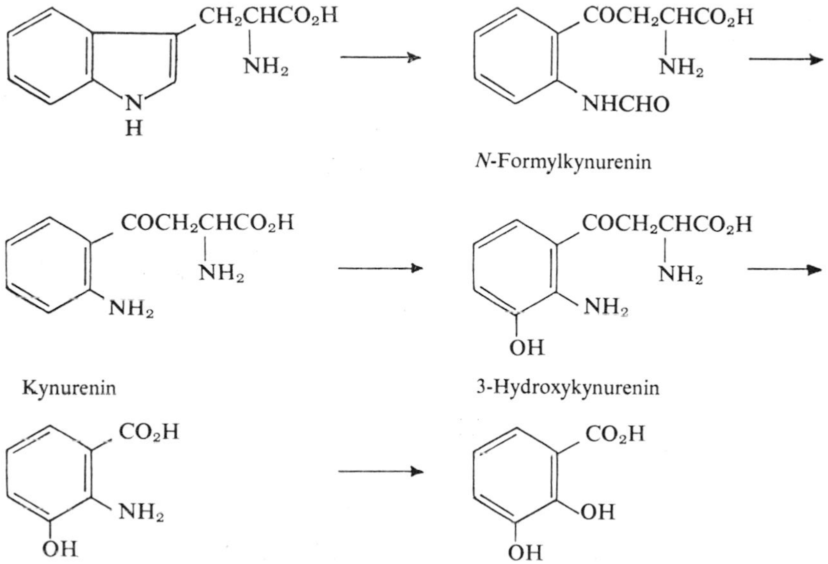

N-Formylkynurenin

Kynurenin

3-Hydroxykynurenin

Scheme 13: The formation of 2,3-dihydroxybenzoic acid from tryptophan in *Claviceps paspali*

hydroxylation at position 1 with migration of the side-chain to the *ortho*-position, and a similar process has recently been demonstrated for the formation of 2-hydroxyphenylacetic acid in higher plants.[36]

In a study of the phenolic products of mutants of *Streptomyces rimosus* no longer able to synthesize tetracyclines, Hassall and his co-workers[16] isolated four shikimate-derived metabolites [compounds (11a), (11b), (14) and (21)], a result which they take to suggest that shikimate-derived intermediates may be involved in tetracycline biosynthesis. However, the bulk of the available evidence (p. 183) points to a purely polyketide derivation of the tetracyclines.

2,3-Dihydroxybenzoic acid (21) is formed from tryptophan in *Claviceps paspali*,[37] possibly by the route of Scheme 13 which is the normal route of tryptophan breakdown; only the last step has not been demonstrated in *C. paspali*.

D.2. DIPHENYLBENZOQUINONES AND RELATED COMPOUNDS

26

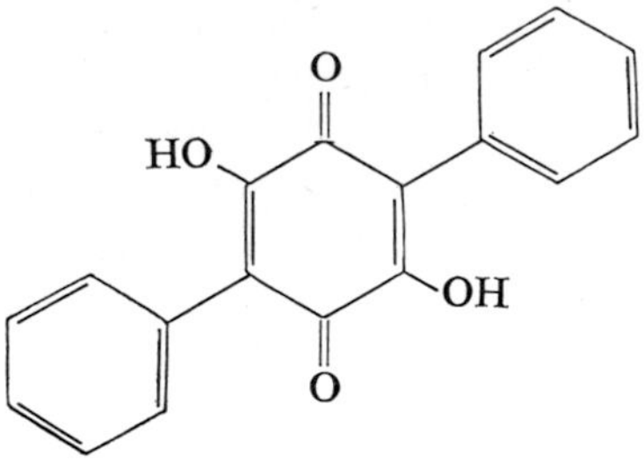

polyporic acid
Polyporus spp., *Sticta* spp.[38]

27

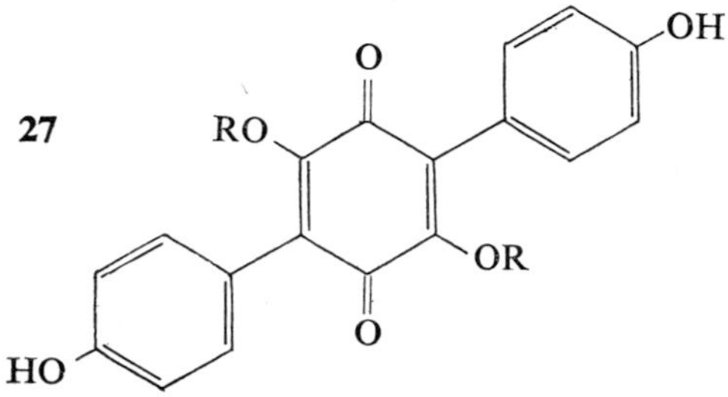

a : R = H, atromentin
        Paxillus atromentosus[38]
        Clitocybe subilludens[39]

b : R = benzoyl, aurantiacin
        Hydnum aurantiacum[40]

28

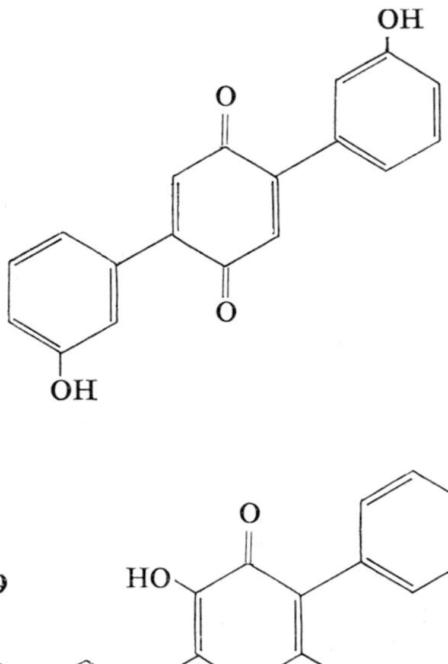

volucrisporin   L[57,60]
Volucrispora aurantiaca[41]

29

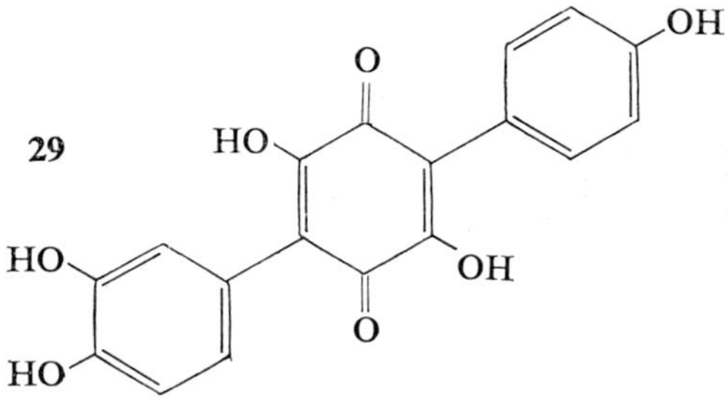

leucomelone
Polyporus leucomelas[42]

30

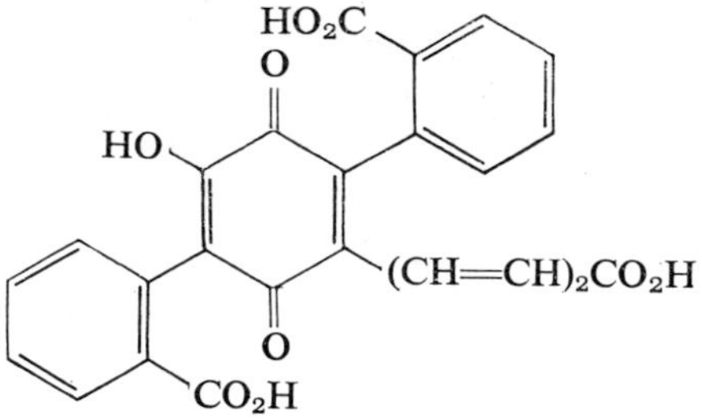

muscarufin
Amanita muscaria[38]

31

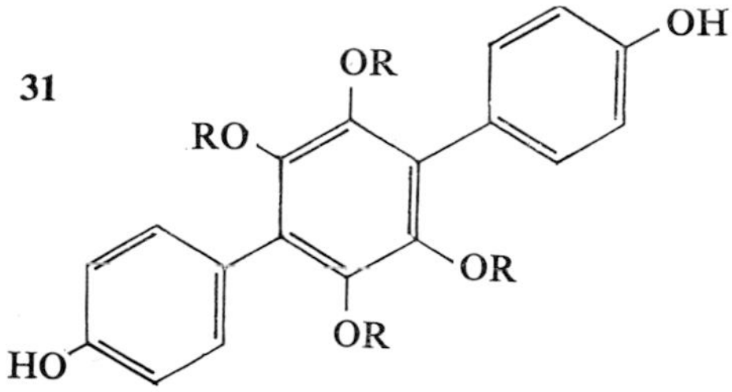

R = benzoyl,   dibenzoylleuco-
aurantiacin
Hydnum aurantiacum[43]

32

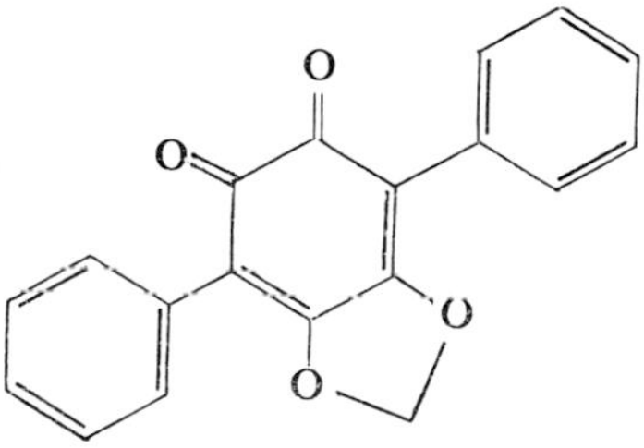

phlebiarubrone   *L*[61]
Phlebia strigosozonata

33

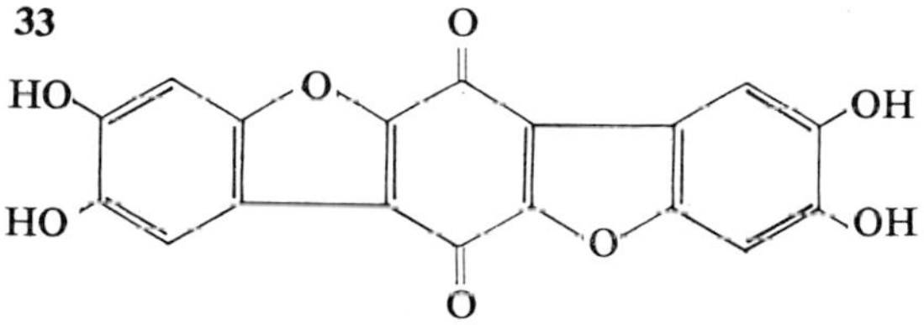

thelephoric acid
Basidiomycetes, lichens[38,45]

34

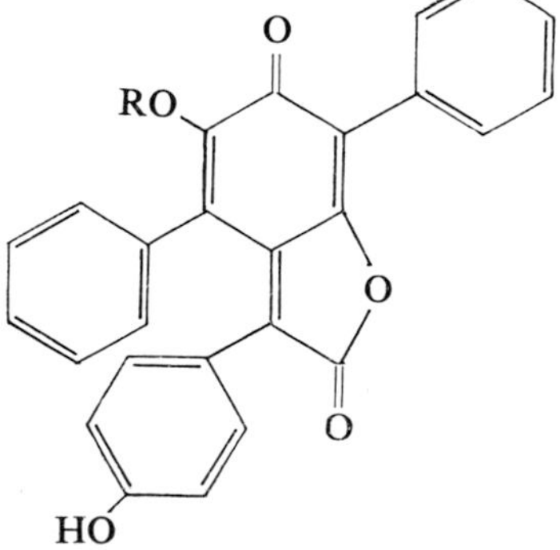

a: R = H, xylerythrin
b: R = Me, 5-*O*-methylxylerythrin
Peniophora sanguinea[46]

35

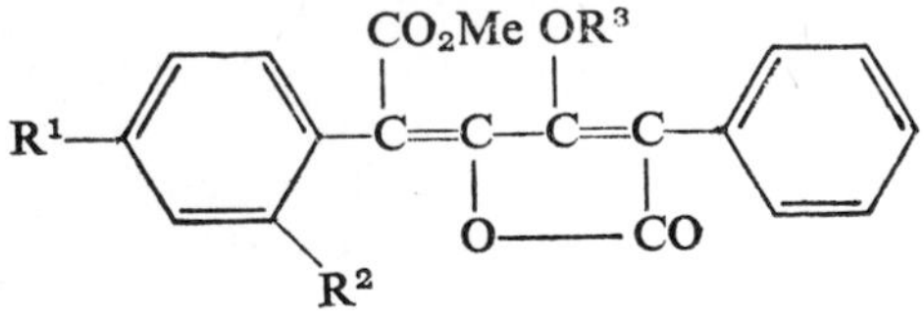

$$R^1 \text{—} \underset{R^2}{\diagdown} \text{—} \underset{\underset{O\text{——}CO}{|}}{\overset{\overset{CO_2Me \quad OR^3}{|}}{C\text{=}C\text{—}C\text{=}C}} \text{—} \bigcirc$$

	R¹	R²	R³		
a	H	H	H	vulpinic acid Lichens[47]	L^{62-64}
b	H	OMe	H	leprapinic acid *Lepraria citrina*[48]	
c	H	OMe	Me	leprapinic acid methyl ether *L. chlorina*[49]	
d	OMe	H	H	pinastric acid Lichens[47]	

36[†]

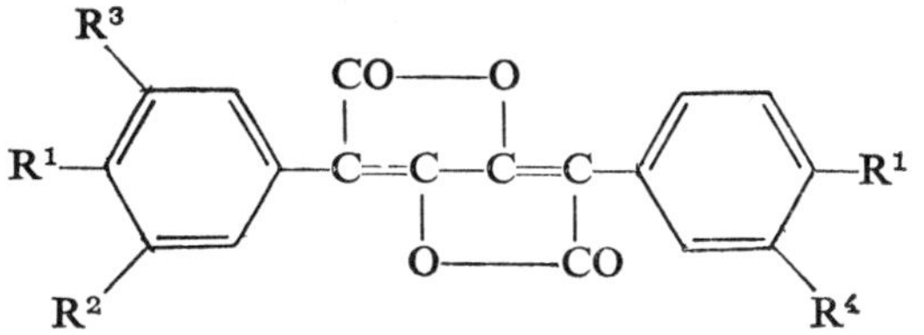

$$R^3 \text{—} \underset{R^2}{\diagdown} \text{—} \underset{\underset{O\text{——}CO}{|}}{\overset{\overset{CO\text{——}O}{|}}{C\text{=}C\text{—}C\text{=}C}} \text{—} \underset{R^4}{\diagup} \text{—} R^1$$

	R¹	R²	R³	R⁴		
a	H	H	H	H	pulvinic acid lactone (pulvic anhydride) *Sticta aurata*[50]	L^{64}
b	OH	OH	H	OH	variegatic acid lactone *Suillus (Boletus) variegatus*[51]	
c	OH	OH	H	H	xerocomic acid lactone *Xerocomus chrysenteron*,[52] *Gomphidus glutinosus*[53]	
d	OH	OH	OH	H	gomphidic acid lactone *Gomphidus glutinosus*[53]	

37

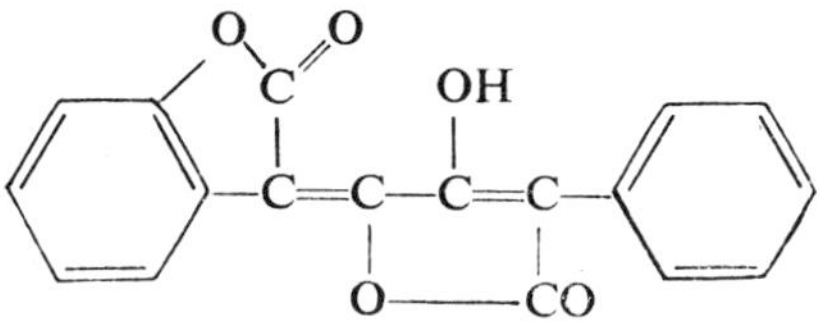

$$\underset{}{\diagdown}\text{—}\underset{\underset{O\text{——}CO}{|}}{\overset{\overset{\overset{O}{\diagdown}\overset{\diagup}{C}\quad\quad OH}{|\qquad\quad|}}{C\text{=}C\text{—}C\text{=}C}}\text{—}\bigcirc$$

calycin
Lichens[47] L^{64}

† These compounds also occur as the mono-lactones (pulvinic acid, etc.). In the case of unsymmetrical compounds (**36c**) and (**36d**) it is not known which lactone is open.

38

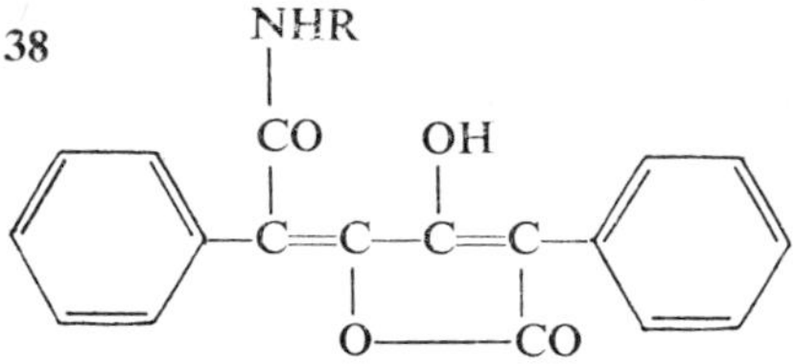

a: R = MeO$_2$CCHCH$_2$CHMe$_2$, epanorin
                          Lecanora epanora[54]
b: R = MeO$_2$CCHCH$_2$Ph, rhizocarpic acid
                          Lichens[54]

39

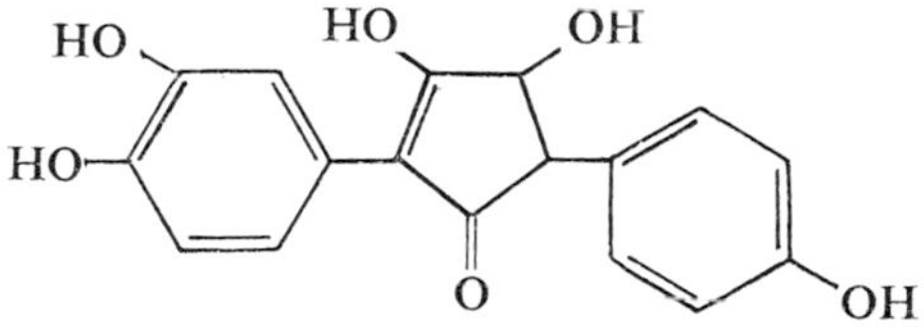

or

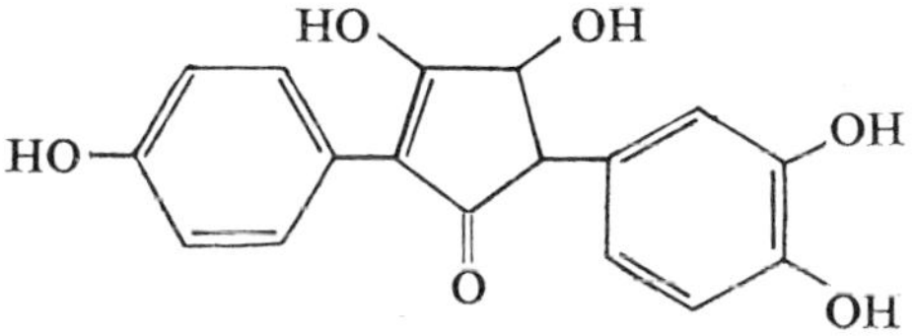

involutin
Paxillus involutus[55]

With one exception—volucrisporin (**28**), isolated from a member of the Hyphomycetes (Fungi Imperfecti)—the compounds have all been isolated from the fruiting bodies of the Basidomycetes or from lichens. The compounds can be interrelated chemically[56] (and presumably in nature) by oxidation of the quinones (probably in the *ortho*-form) to give the lactones (Scheme 14), which in turn can give rise to cyclopentane derivatives [cf. involutin (**39**)]. Until the recent isolation of variegatic acid (**36b**) and involutin (**39**) from Basidomycetes, and Mosbach's demonstration[64] that a lichen fungus grown in culture can produce pulvinic acids, the isolation of the lactones only from lichens had suggested that the quinones might be synthesized by the mycobiont of the lichens and oxidized by the phycobiont.

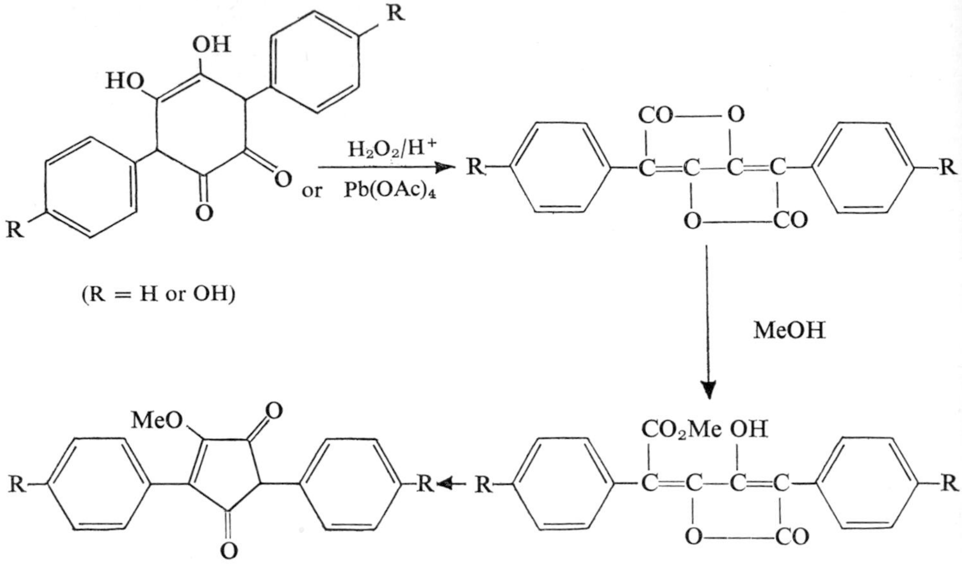

Scheme 14: The chemical conversion of the benzoquinones to lactones and cyclopentanones

Xylerythrin (**34**), which was isolated from wood infected with *Peniophora sanguinea*, obviously results from condensation of polyporic acid with *p*-hydroxyphenylacetic acid. Structure (**30**) for muscarufin is doubtful[58] and subsequent workers[59] were unable to isolate the compound.

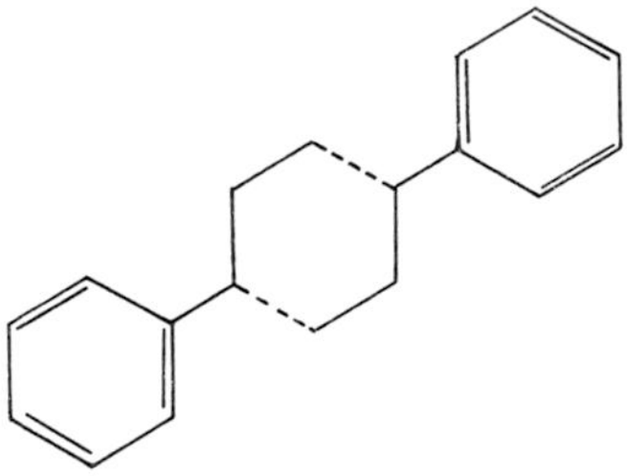

40

It has been shown[57] that shikimic acid, phenylalanine, phenyllactic acid and *m*-tyrosine will serve as precursors of volucrisporin but that cinnamic acid, *m*-hydroxycinnamic acid, caffeic acid, L-tyrosine and 3,4-dihydroxy-phenylalanine will not. These results suggest that volucrisporin might result from condensation of two phenylpyruvate moieties (**40**), and that the actual precursor might be *m*-hydroxyphenylpyruvate formed by direct hydroxylation of phenylalanine or phenylpyruvate. Further experiments,[60] in which the incorporation of labelled precursors into both volucrisporin and the mycelial aromatic amino-acids was measured, suggest that the shikimate-derived aromatic compounds of *V. aurantiaca* are interrelated as in Scheme 15.

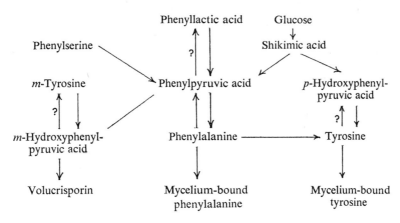

Scheme 15: Interrelationship of the shikimate-derived aromatic metabolites of *V. aurantiaca*

The incorporation of ^{13}C-labelled substrates into phlebiarubrone (32)[61] suggests that its biosynthesis is similar to that of volucrisporin.

Mosbach[62] has studied the biosynthesis of vulpinic acid in *Evernia vulpina* by suspending freshly collected lichen thallus in Czapek–Dox medium containing [*carboxy*-^{14}C]phenylalanine. Degradation of the resulting labelled

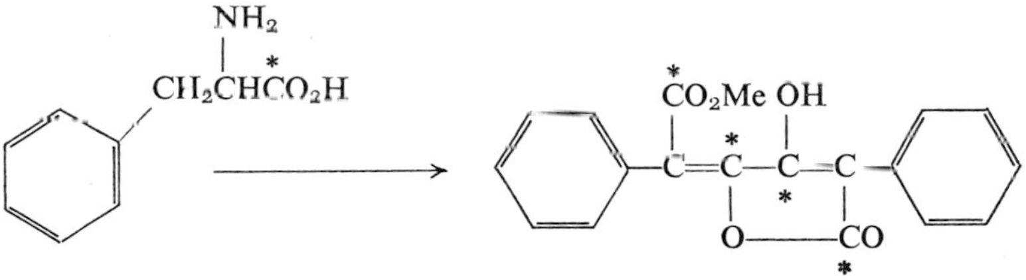

Scheme 16: Incorporation of phenylalanine into vulpinic acid by the lichen, *Evernia vulpina*

vulpinic acid showed that it possessed the labelling pattern shown in Scheme 16, in accord with formation of vulpinic acid via a diphenylbenzoquinone (since a symmetrical intermediate is involved each labelled carbon atom possesses 25% of the total radioactivity).

Similar results were obtained by Maass and his co-workers[63] using *Pseudocyphellaria crocata*. They showed, further, that polyporic acid is a precursor of pulvinic acid lactone (36a) and of calycin (37) and obtained evidence that in the formation of calycin, hydroxylation occurs after condensation of the C_6–C_3 units, probably at the pulvinic acid stage. Cinnamic acid was not converted into pulvinic acid derivatives, in accord with the finding[57] that it is not incorporated into volucrisporin.

Recently Mosbach[64] has demonstrated the production of pulvinic acid lactone, calycin, and vulpinic acid by the isolated fungus (grown on a solid malt-extract medium) of the lichen *Candellariella vitellina* and has shown that labelled phenylalanine is incorporated into all three compounds. The production of vulpinic acid was unexpected since it does not appear to be formed by the intact lichen.

D.3. MISCELLANEOUS COMPOUNDS

Hispidin (41), an orange pigment isolated from the fruiting body of *Polyporus hispidus*, probably results from condensation of caffeic acid with two C_2-units. It is the only compound of this type to have been obtained from fungi though several related compounds, including the trimethyl ether of hispidin, occur in higher plants. It has been suggested[66] that hispidin polymerizes to form a fungal "lignin" responsible for the toughening of fruit bodies of *P. hispidus*. Vegetative mycelium is non-pigmented on most culture media but contains hispidin when grown on blocks of ashwood, suggesting that the C_6–C_3 unit may be derived from the wood. (On the other hand, hispidin is produced by *P. schweinitzii* grown on "normal" media.) A fuller account of the possible role of hispidin in the fruiting bodies of *P. hispidus* has been given by Bu'Lock.[78]

Chlorflavonin (42) and dihydrochalcone (43) are the only C_6–C_3–C_6 compounds to have been isolated from fungi.

The diphenylbutadiene chromophore and symmetrical nature of xanthocillin-X (45a) suggest a derivation similar to that of the pulvinic acids. Support for this has been provided by Achenbach and Grisebach[79] who showed that radioactivity from [2-^{14}C]tyrosine, but not from [1-^{14}C]tyrosine, is incorporated into xanthocillin. The origin of the isonitrile groups, an unusual feature for a natural product, remains obscure; experiments with doubly-labelled tyrosine showed that the nitrogen atom is not derived directly from the amino-group of tyrosine, while [^{14}C]formate and [1-^{14}C]- and [2-^{14}C]acetate did not give rise to labelling of the carbon atom. Tuberin (46) can be formally derived by hydration of the isonitrile group of a monomeric precursor of xanthocillin (or xanthocillin might be derived by dehydration of a dimer of tuberin).

Of several ^{14}C-labelled compounds examined[80] the most efficient precursor for caldariomycin (47) was [U-^{14}C]shikimic acid, and the results of experiments with [1-^{14}C]-, [2-^{14}C]-, and [6-^{14}C],glucose were consistent with the

41

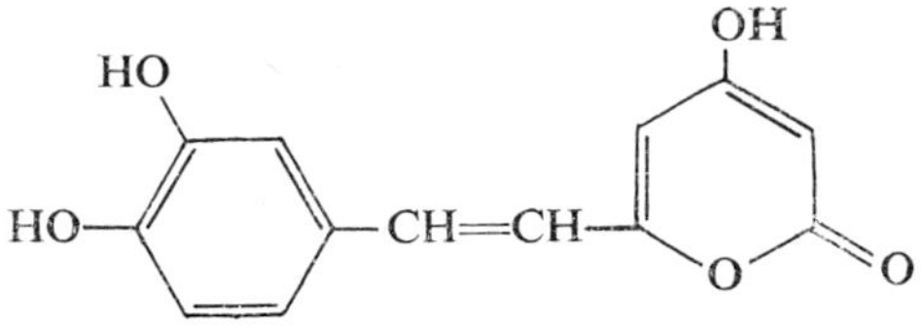

hispidin
Polyporus hispidus,[65,66]
P. Schweinitzii[67]

42

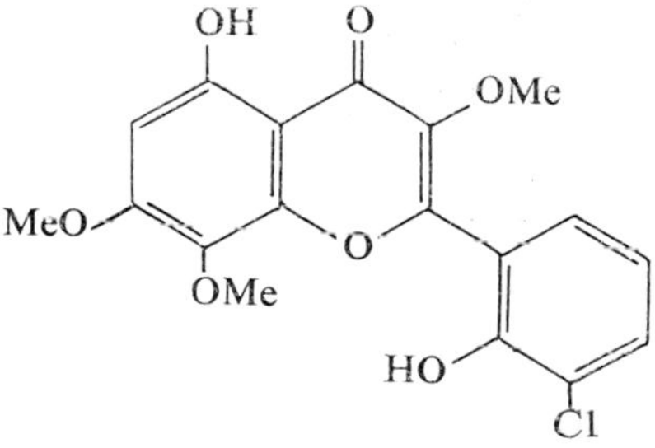

chlorflavonin
Aspergillus candidus[78]

43

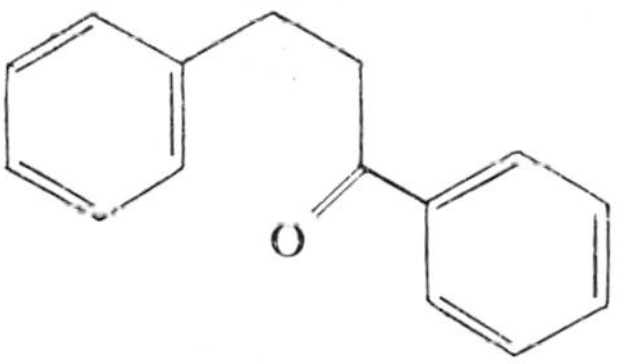

dihydrochalcone
Phallus impudicus[69]

44

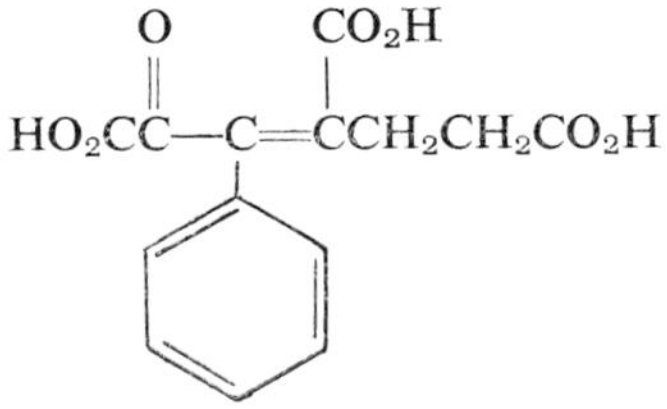

$HO_2CC-C=CCH_2CH_2CO_2H$

4-carboxy-2-oxo-3-phenylhept-
3-enedioic acid
Chaetomium indicum[70]

45 R^1O—〈benzene〉—CH=C—C=CH—〈benzene, R^3〉—OR^2
                              | |
                             CN   NC

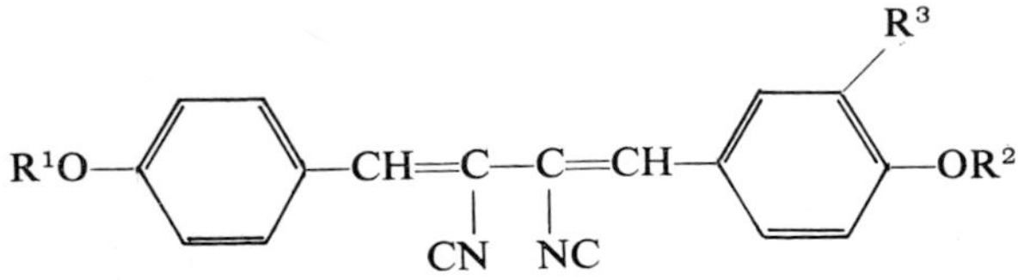

	R^1	R^2	R^3
a	H	H	H	xanthocillin-X L^{79} *Penicillium notatum*[71]
b	Me	H	H	xanthocillin-X methyl ether *Dichotomyces albus*,[72] *Aspergillus* spp.[73]
c	Me	Me	H	xanthocillin-X dimethyl ether *Aspergillus* spp.[73]
d	Me	Me	OMe	methoxyxanthocillin-X dimethyl ether *Aspergillus* spp.[73]

46 MeO—〈benzene〉—CH=CHNHCHO   tuberin
 Streptomyces amakusaensis[74]

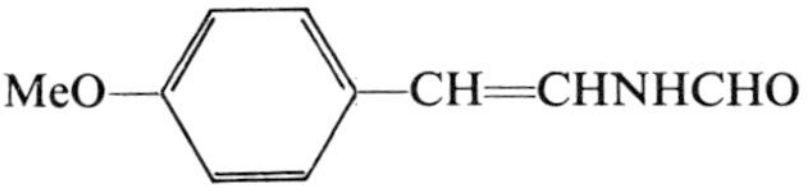

47 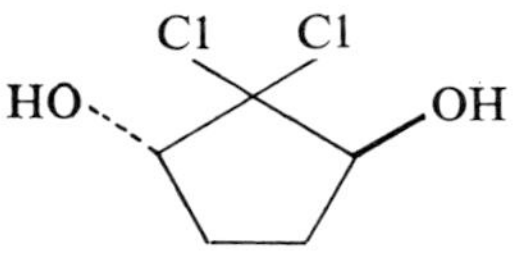   caldariomycin L^{80-83}
 Caldariomyces fumago[75]

48

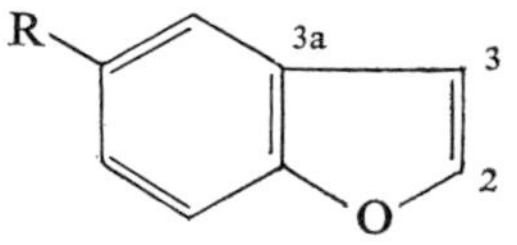

a: R = OMe, 5-methoxybenzofuran L^{77}
    Stereum subpileatum[76]
b: R = CHO, 5-formylbenzofuran
c: R = CH$_2$OH, 5-hydroxymethylbenzofuran
d: R = CH(OH)CH(OH)Me,5-(1,2-dihydroxypropyl)benzofuran
e: R = CCH(OH)Me,5-(2-hydroxypropionyl)benzofuran
       ‖
       O
f: R = CHCHMe, 5-(epoxypropyl)benzofuran
       \ /
        O

        Stereum subpileatum[77]

49

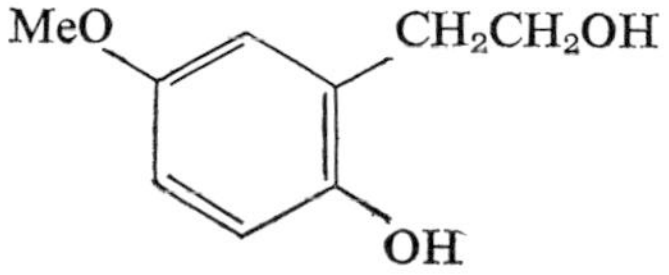

2-(2-hydroxy-5-methoxyphenyl)ethanol
S. subpileatum[77]

biosynthetic pathway of Scheme 17. The timing of the introduction of the chlorine atoms has not been established. [1,4-^{14}C]-δ-Chlorolaevulinic acid is not incorporated into caldariomycin[80] (in contrast with earlier results with [^{36}Cl]-δ-chlorolaevulinic acid[81]). Extracts of *C. fumago* catalyse the chlorination of cyclopentane-1,3-dione in two stages, one at pH 5.6 and one at pH 3, and the resulting 2,2-dichlorocyclopentane-1,3-dione is converted to caldariomycin by growing cultures of *C. fumago*[82]. However, since the chloroperoxidase of *C. fumago* (see p. 9) is not specific and will catalyse the chlorination of a variety of β-dicarbonyl compounds,[83] this result does not necessarily mean that cyclopentane-1,3-dione is a natural intermediate for caldariomycin.

5-Methoxybenzofuran (**48a**) was obtained from the Basidiomycete *Stereum subpileatum* (found growing in beer-barrels, which is as good a place as any for a "foray") and is responsible for the characteristic and penetrating odour of the fungus. The other benzofurans and the phenylethanol derivative (**49**) were isolated during a study of the biosynthesis of 5-methoxybenzofuran.[77]

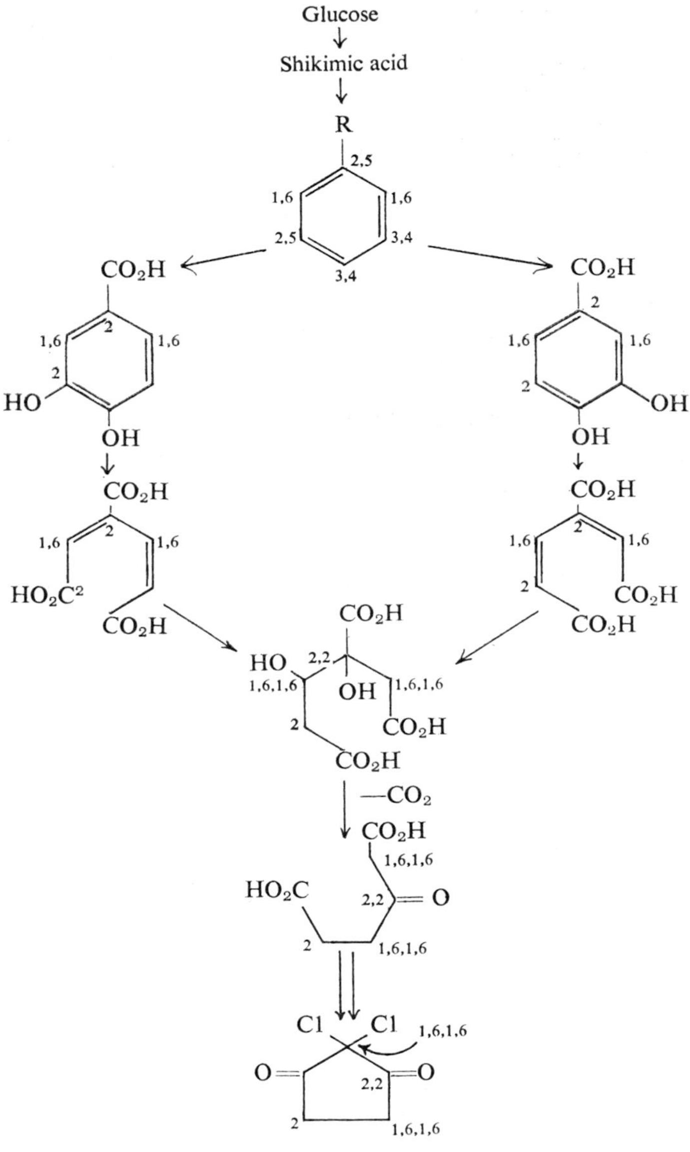

Scheme 17: The incorporation of [^{14}C]glucose into caldariomycin. The numbers refer to the carbon atoms of glucose

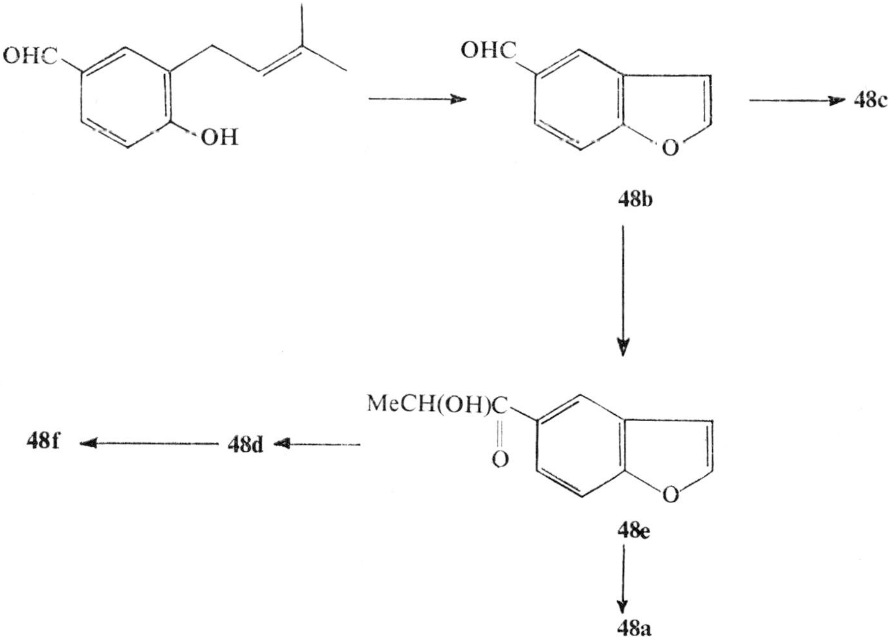

(R = H or OH)

48b

48e

48f ◄———— 48d ◄----

48a

Scheme 18: Proposed[77] biosynthesis of the benzofurans of *Stereum subpileatum*

Although the carbon skeleton and oxygenation pattern of 5-methoxy-benzofuran (**48a**) are those of homogentisic acid (2,5-dihydroxyphenylacetic acid) only 30% of the radioactivity of 5-methoxybenzofuran produced in the presence of [2-^{14}C]phenylalanine was at C-2 and 80% of radioactivity from [1-^{14}C]phenylalanine was at C-3.[77] A similar distribution of radioactivity was obtained with [2-^{14}C]- and [1-^{14}C]acetate. Thus phenylalanine is degraded to give acetate from C-1 and C-2 (see p. 41) and the aromatic ring of 5-methoxybenzofuran is not derived from acetate. The last conclusion is supported by the fact that [6-^{14}C]glucose was incorporated specifically into *part* of the benzene ring (C-3a, C-3, and C-2 were inactive). After incorporation of [U-^{14}C]phenylalanine into 5-(1,2-dihydroxypropyl) benzofuran (**48d**), C-2 and C-3 were unlabelled while C-1 contained 14.7% (= 1.03/7) of the radioactivity of (**48d**), i.e. phenylalanine contributes a C_6–C_1 unit. A sequence accounting for these observations is given in Scheme 18. The involvement of an isopentenyl intermediate, rather than a compound with a two-carbon side-chain derived directly from acetate, is postulated by analogy with the known pathway to benzofurans in higher plants.

50

a : $R^1 = NH_2$, $R^2 = OH$,   adenosine
b : $R^1 = NH_2$, $R^2 = H$,   3′-deoxyadenosine (cordycepin)
    Cordyceps militaris,[84] *Aspergillus nidulans*,[85] *Helminthosporium* sp.[86]
c : $R^1 = R^2 = NH_2$,   3′-amino-3′-deoxyadenosine
    Helminthosporium sp.[86,87]
d : $R^1 = NH_2$, $R^2 = NHAc$,   3′-acetamido-3′-deoxyadenosine
    Helminthosporium sp.[86]
e : $R^1 = H$, $R^2 = OH$,   nebularine
    Clitocybe nebularis,[88] *Streptomyces* sp.[8]

51

lentysine, lentinacin
Lentinus edodes[90,91]

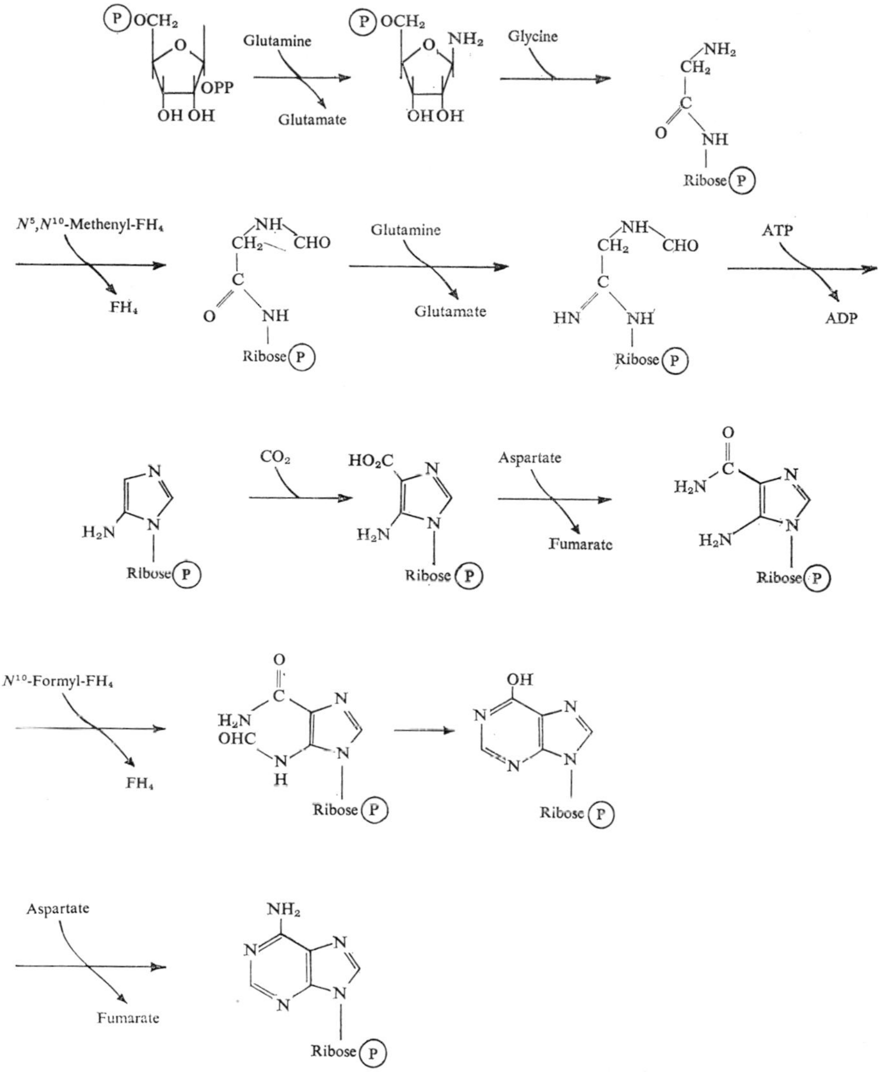

Scheme 19: The biosynthesis of adenosine

E. SECONDARY METABOLITES RELATED TO ADENOSINE

Five metabolites related to the essential purine base adenosine have been isolated from fungi and about a dozen others have been obtained from *Streptomyces* spp., nebularine **(50e)** having been obtained from both sources. 3'-Deoxyadenosine **(50b)**,[92] 3'-amino-3'-deoxyadenosine **(50c)**,[86] and 3'-acetamido-3'-deoxyadenosine **(50d)**[86] are formed from adenosine without cleavage of the C–N bond between the ribose and purine moieties. 3'-Deoxyadenosine is not a precursor of the 3'-amino and 3'-acetamido compounds.[86]

The biosynthesis of adenosine is outlined in Scheme 19.

REFERENCES

1. M. W. Miller, "The Pfizer Handbook of Microbial Metabolites", Ch. 2. McGraw-Hill, London, 1961.
2. T. Korzybski, Z. Kowszyk-Gindifer and W. Kurylowicz, "Antibiotics. Origin, Nature, and Properties", Vol. II, p. 1220. Pergamon, London, 1967.
3. Y. Sumiki, *Bull. Agr. Chem. Soc. Japan*, 1929, **5**, 10; *J. Agr. Chem. Soc. Japan*, 1931, **7**, 819.
4. A. Kawarada, N. Takahashi, H. Kitamura, Y. Seta, M. Takai and S. Tamura, *Bull. Agr. Chem. Soc. Japan*, 1955, **19**, 84.
5. S. Wilkinson, *Quart. Rev.* 1961, **15**, 153.
6. M.-L.L. Swenberg, W. J. Kelleher and A. E. Schwarting, *Science*, 1967, **155**, 1259.
7. F. Kögl, C. A. Salemink and P. L. Schuller, *Rec. Trav. Chim.*, 1960, **79**, 278; C. A. Salemink and P. K. Schuller, *Rec. Trav. Chim.*, 1960, **79**, 485.
8. H. R. V. Arnstein and R. Bentley, *Biochem. J.*, 1953, **54**, 493.
9. H. R. V. Arnstein and R. Bentley, *Biochem. J.*, 1953, **54**, 517.
10. H. R. V. Arnstein and R. Bentley, *Biochem. J.*, 1956, **62**, 403.
11. D. B. Sprinson, *Advan. Carbohydrate Chem.*, 1960, **15**, 235.
12. H. Shimazono and F. F. Nord, *Arch. Biochem. Biophys.*, 1958, **78**, 263.
13. J. H. Birkinshaw and W. P. K. Findlay, *Biochem. J.*, 1940, **34**, 82.
14. R. L. Edwards and G. C. Elsworthy, *J. Chem. Soc. C*, 1967, 410.
15. H. Shimazono, *Arch. Biochem. Biophys.*, 1959, **83**, 206.
16. E. R. Catlin, C. H. Hassall and B. C. Pratt, *Biochim. Biophys. Acta*, 1968, **156**, 109.
17. Y. S. Chen, *Bull. Agr. Chem. Soc. Japan*, 1958, **22**, 136.
18. R. K. Crowden and B. J. Ralph, *Aust. J. Chem.*, 1961, **14**, 475.
19. E. W. Bassett and S. W. Tanenbaum, *Biochim. Biophys. Acta*, 1958, **28**, 21, 247.

20. J. H. Birkinshaw, A. Bracken and W. P. K. Findlay, *Biochem. J.*, 1944, **38**, 131.
21. See J. H. Birkinshaw and P. Chaplen, *Biochem. J.*, 1955, **60**, 255.
22. B. J. Ralph and A. Robertson, *J. Chem. Soc.*, 1950, 3380.
23. E. Haslam, R. D. Haworth and P. F. Knowles, *J. Chem. Soc.*, 1961, 1854, and references cited.
24. M. Hashem and W. Brücker, *Flora*, 1962, **152**, 57.
25. For references see D. Cornthwaite and E. Haslam, *J. Chem. Soc.*, 1965, 3008; P. M. Dewick and E. Haslam, *Chem. Commun.*, 1968, 673.
26. G. F. J. Moir and B. J. Ralph, *Chem. and Ind.*, 1954, 1143.
27. F. Arcamone, E. B. Chain, A. Ferretti and P. Penella, *Nature*, 1961, **192**, 552.
28. N. M. Packter, *Biochem. J.*, 1969, **114**, 369.
29. F. F. Nord and W. J. Schubert. *In* "Biogenesis of Natural Compounds" (P. Bernfeld, ed.), 2nd edition, pp. 909–916. Pergamon, Oxford, 1967.
30. D. M. Power, G. H. N. Towers and A. C. Neish, *Can. J. Biochem.*, 1965, **43**, 1397.
31. K. Moore, P. V. Subba Rao and G. H. N. Towers, *Biochem. J.*, 1968, **106**, 507.
32. K. Moore, P. V. Subba Rao and G. H. N. Towers, *Can. J. Biochem.*, 1967, **45**, 1863,
33. K. Moore and G. H. N. Towers, *Can. J. Biochem.*, 1967, **45**, 1659.
34. R. K. Crowden, *Can. J. Microbiol.*, 1967, **13**, 181.
35. G. Billek and F. P. Schmook, *Monatsh. Chem.*, 1967, **98**, 1651.
36. H. Kindl, *Eur. J. Biochem.*, 1969, **7**, 340.
37. V. E. Tyler, K. Mothes, D. Gröger and H.-G. Floss, *Tetrahedron Lett.*, 1964, 593.
38. For an account of the history of these compounds see R. H. Thomson, "The Naturally Occurring Quinones", p. 26 ff. Butterworths, London, 1957.
39. G. Sullivan and W. L. Guess, *Lloydia*, 1969, **32**, 72.
40. J. Gripenberg, *Acta Chem. Scand.*, 1956, **10**, 1111.
41. P. V. Divekar, G. Read, L. C. Vining and R. H. Haskins, *Can. J. Chem.*, 1959, **37**, 1970.
42. M. Akagi, *J. Pharm. Soc. Japan*, 1942, **62**, 129
43. J. Gripenberg, *Acta Chem. Scand.*, 1958, **12**, 1411.
44. T. C. McMorris and M. Anchel, *Tetrahedron*, 1967, **23**, 3985.
45. J. Gripenberg, *Tetrahedron*, 1960, **10**, 135.
46. J. Gripenberg, *Acta Chem. Scand.*, 1965, **19**, 2242; J. Gripenberg and J. Martikkala, *Acta Chem. Scand.*, 1969, **23**, 2583.
47. Y. Asahina and S. Shibata, "Chemistry of Lichen Substances", p. 43. Japan Society for the Advancement of Science, Tokyo, 1954.
48. O. P. Mittal and T. R. Seshadri, *J. Chem. Soc.*, 1955, 3053; S. C. Agarwal and T. R. Seshadri, *Tetrahedron*, 1965, **21**, 3205.
49. P. K. Grover and T. R. Seshadri, *J. Sci. Ind. Res. (India)*, 1959, **18B**, 238.
50. O. Hesse, *J. Prakt. Chem.*, 1900, **170**, 334.
51. R. L. Edwards and G. C. Elsworthy, *Chem. Commun.*, 1967, 373; P. C. Beaumont, R. L. Edwards and G. C. Elsworthy, *J. Chem. Soc. C*, 1968, 2968.
52. W. Steglich, W. Furtner and A. Prox, *Z. Naturforsch.*, 1968, **23B**, 1044.
53. W. Steglich, W. Furtner and A. Prox, *Z. Naturforsch.*, 1969, **24B**, 941.
54. R. L. Frank, S. M. Cohen and J. N. Coker, *J. Amer. Chem. Soc.*, 1950, **72**, 4454.

55. R. L. Edwards, G. C. Elsworthy and N. Kale, *J. Chem. Soc. C*, 1967, 405.
56. F. Kögl, H. Becker, A. Detzel and G. de Voss, *Ann. Chem.*, 1928, **465**, 211; R. L. Frank, G. R. Clark and J. N. Coker, *J. Amer. Chem. Soc.*, 1950, **72**, 1824; O. P. Mittal and T. R. Seshadri, *Current Sci. (India)*, 1957, **26**, 4.
57. G. Read, L. C. Vining and R. H. Haskins, *Can. J. Chem.*, 1962, **40**, 2357.
58. M. Nilsson and T. Norin, *Acta Chem. Scand.*, 1960, **14**, 2243.
59. G. Talbot and L. C. Vining, *Can. J. Bot.*, 1963, **41**, 639.
60. P. Chandra, G. Read and L. C. Vining, *Can. J. Biochem.*, 1966, **44**, 403.
61. A. K. Bose, K. S. Kanchandani, P. T. Funke and M. Anchel, *Chem. Commun.*, 1969, 1347.
62. K. Mosbach, *Biochem. Biophys. Res. Commun.*, 1964, **17**, 363.
63. W. S. G. Maass, G. H. N. Towers and A. C. Neish, *Ber. Deut. Botan. Ges.*, 1964, **77** (Sondernummer, 1. Generalversammlungshaft), 157; W. S. G. Maass and A. C. Neish, *Can. J. Bot.*, 1967, **45**, 59.
64. K. Mosbach, *Acta Chem. Scand.*, 1967, **21**, 2331.
65. R. L. Edwards, D. G. Lewis and D. V. Wilson, *J. Chem. Soc.*, 1961, 4995.
66. J. D. Bu'Lock and H. G. Smith, *Experientia*, 1961, **17**, 553; J. D. Bu'Lock, P. R. Leeming and H. G. Smith, *J. Chem. Soc.*, 1962, 2085.
67. A. Ueno, S. Fukushima, Y. Saiki and T. Harada, *Chem. Pharm. Bull. (Tokyo)*, 1964, **12**, 376.
68. M. Richards, A. E. Bird and J. E. Munden, *J. Antibiotics (Tokyo)*, 1969, **22**, 388; A. E. Bird and A. C. Marshall, *J. Chem. Soc. C*, 1969, 2418.
69. P. H. List and B. Freund, *Planta Med.*, 1968 (Suppl.), 123.
70. D. H. Johnson, A. Robertson and W. B. Whalley, *J. Chem. Soc.*, 1953, 2429.
71. W. Rothe, *Deut. Med. Wochschr.*, 1954, **79**, 1080; I. Hagedorn and H. Tönjes, *Pharmazie*, 1957, **12**, 567; I. Hagedorn, U. Eholzer and A. Lüttringhaus, *Chem. Ber.*, 1960, **93**, 1584.
72. K. Ando, S. Suzuki, A. Takatsuki, K. Arima and G. Tamura, *J. Antibiotics (Tokyo)*, 1968, **21**, 582; K. Ando, G. Tamura and K. Arima, *J. Antibiotics (Tokyo)*, 1968, **21**, 587.
73. A. Takatsuki, S. Suzuki, K. Ando, G. Tamura and K. Arima, *J. Antibiotics (Tokyo)*, 1968, **21**, 671.
74. K. Ohkuma, K. Anzai and S. Suzuki, *J. Antibiotics (Tokyo)*, 1962, **15A**, 115; K. Anzai, *J. Antibiotics (Tokyo)*, 1962, **15A**, 117, 123.
75. P. W. Clutterbuck, S. L. Mukhopadhyay, A. E. Oxford and H. Raistrick, *Biochem. J.*, 1940, **34**, 664; S. M. Johnson, I. C. Paul, K. L. Rinehart and R. Srinivasan, *J. Amer. Chem. Soc.*, 1968, **90**, 136.
76. J. H. Birkinshaw, P. Chaplen and W. P. K. Findlay, *Biochem. J.*, 1957, **66**, 188.
77. J. D. Bu'Lock, A. T. Hudson and B. Kaye, *Chem. Commun.*, 1967, 814.
78. J. D. Bu'Lock, "Essays in Biosynthesis and Microbial Development", pp. 2–7. Wiley, London, 1967.
79. H. Achenbach and H. Grisebach, *Z. Naturforsch.*, 1965, **20B**, 137.
80. J. R. Beckwith, R. Clark and L. P. Hager, *J. Biol. Chem.*, 1963, **238**, 3086.
81. P. D. Shaw, J. R. Beckwith and L. P. Hager, *J. Biol. Chem.*, 1959, **234**, 2560.
82. J. R. Beckwith and L. P. Hager, *J. Biol. Chem.*, 1963, **238**, 3091.
83. D. R. Morris and L. P. Hager, *J. Biol. Chem.*, 1966, **241**, 1763; L. P. Hager, D. R. Morris, F. S. Brown and H. Eberwein, *J. Biol. Chem.*, 1966, **241**, 1769.
84. K. G. Cunningham, S. A. Hutchinson, W. Manson and F. S. Spring, *J. Chem. Soc.*, 1951, 2299.

85. E. A. Kaczka, E. L. Dulaney, C. O. Gitterman, H. B. Woodruff and K. Folkers, *Biochem. Biophys. Res. Commun.*, 1964, **14**, 452; E. A. Kaczka, N. R. Trenner, B. Arison, R. W. Walker and K. Folkers, *Biochem. Biophys. Res. Commun.*, 1964, **14**, 456.
86. B. M. Chassy and R. J. Suhadolnik, *Biochim. Biophys. Acta*, 1969, **182**, 316.
87. N. N. Gerber and H. A. Lechevalier, *J. Org. Chem.*, 1962, **27**, 1731.
88. N. Löfgren, B. Lüning, and H. Hedstrom, *Acta Chem. Scand.*, 1954, **8**, 670.
89. K. Isono and S. Suzuki, *J. Antibiotics (Tokyo)*, 1960, **13**, 270.
90. T. Kamiya, Y. Saito, M. Hashimoto and H. Seki, *Tetrahedron Lett.*, 1969, 4729.
91. I. Chibata, K. Okumara, S. Takeyama and K. Kotera, *Experientia*, 1969, **25**, 1237.
92. R. J. Suhadolnik, G. Weinbaum and H. P. Meloche, *J. Amer. Chem. Soc.*, 1964, **86**, 948.

CHAPTER 4

Secondary Metabolites Derived from Fatty Acids

IN CHAPTER 1 we emphasized the key position of acetate as a biosynthetic intermediate, and here and in the next three chapters we deal with compounds derived from acetate. A comprehensive list of the fatty acids of fungi has been compiled by Shaw.[1] The most abundant are palmitic, stearic, oleic and linoleic, but small amounts of the other common saturated and unsaturated straight-chain acids of chain lengths from fourteen to twenty-four carbon atoms also occur.

The fungi also produce a few species-specific derivatives of the fatty acids (see Section B.1), and two groups of compounds—the polyacetylenes (Section B.2) and the cyclopentanes (Section B.3)—which are formed from the fatty acids by desaturation and chain-shortening or cyclization processes. The role of fatty acids in the biosynthesis of the polyacetylenes and cyclopentanes has only recently been recognized and it may well be that other compounds, especially some of those classed as polyketides, belong in this chapter (this point is further discussed in Section B.4). A further group of compounds whose biosynthesis probably involves fatty acid intermediates, is discussed in Chapter 7.

A. FATTY ACID BIOSYNTHESIS

A.1. THE BIOSYNTHESIS OF THE SATURATED FATTY ACIDS

For many years it was thought that fatty acids were formed from acetate by the reverse of their breakdown by β-oxidation (Scheme 1). In 1958 it was realized that malonate is involved in fatty acid biosynthesis, and since then

$$RCH_2CH_2CO_2H \longrightarrow RCH_2CH_2COSCoA \longrightarrow RCH{=}CHCOSCoA$$
$$\longrightarrow RCH(OH)CH_2COSCoA \longrightarrow RCOCH_2COSCoA$$
$$\longrightarrow RCOSCoA + CH_3COSCoA$$

Scheme 1: The breakdown of fatty acids by β-oxidation

many of the details of the process have been elucidated using cell-free systems and purified enzymes.[2] The mechanism of fatty acid biosynthesis will be considered in some detail because of its implications for the biosynthesis of polyketides, the largest class of fungal secondary metabolites (Chapter 5).

The steps involved are shown in Scheme 2. The carboxylation of acetyl coenzyme A (step 1) is mediated by biotin, a cofactor in other carboxylation reactions, e.g. the formation of oxaloacetate from pyruvate. An acetyl residue is then transferred from coenzyme A first to acyl carrier protein (ACP) and then to an enzyme (steps 2 and 3). Malonate is similarly transferred from coenzyme A to ACP (step 4) and condensation occurs to give acetoacetyl ACP and carbon dioxide (step 5). The simultaneous decarboxylation makes the condensation process energetically favourable; in the corresponding

1. $CH_3COSCoA + CO_2 \longrightarrow HO_2CCH_2COSCoA$

2. $CH_3COSCoA + HSACP \longrightarrow CH_3COSACP + HSCoA$

3. $CH_3COSACP + HSE \longrightarrow CH_3COSE + HSCoA$

4. $HO_2CCH_2CO_2SCoA + HSACP \longrightarrow HO_2CCH_2COSACP + HSCoA$

5. $CH_3COSE + CH_2(CO_2H)COSACP \longrightarrow$
$\quad\quad\quad CH_3COCH_2COSACP + CO_2 + HSE$

6. $CH_3COCH_2COSACP \longrightarrow CH_3CH(OH)CH_2COSACP$

7. $CH_3CH(OH)CH_2COSACP \longrightarrow CH_3CH{=}CHCOSACP$

8. $CH_3CH{=}CHCOSACP \longrightarrow CH_3CH_2CH_2COSACP$

9. $CH_3CH_2CH_2COSE + CH_2(CO_2H)COSACP \longrightarrow$
$\quad\quad\quad CH_3CH_2CH_2COCH_2COSACP$

Scheme 2: Fatty acid biosynthesis. ACP = acyl carrier protein and E = enzyme

reaction in the β-oxidation pathway (Scheme 1) the equilibrium lies in favour of the degradative process. The protein-bound acetoacetate now undergoes, successively, reduction (step 6), dehydration (step 7), and reduction (Step 8) to give butyryl ACP. The butyryl residue is then transferred to the enzyme and can then condense with further malonyl ACP (step 9) to give a C_6 intermediate bound to ACP. The process ends with the transfer of a C_{16} or C_{18} fatty acid from ACP to CoA (in yeast systems) or with the liberation of the free fatty acid (in bacterial and mammalian systems).

It will be seen from Scheme 2 that there are no free intermediates between acetate (and malonate) and the final long-chain acids, and that the first two-carbon unit of the growing chain is derived directly from acetate whereas the remainder are derived from malonate. These features of fatty acid biosynthesis are also characteristic of the biosynthesis of secondary metabolites by the polyketide route.

Fatty acid synthetase is thus an enzyme complex able to carry out a sequence of reactions on protein-bound intermediates. The complex obtained from yeast cannot be broken down into its active components whereas the complex obtained from bacteria can be readily fractionated by the normal techniques of protein purification. In both cases the linkage of the acyl residue to ACP has been shown to be through 4′-phosphopantotheine to serine in the protein (**1**) [the resemblance between this prosthetic group and acyl coenzyme A (**2**) is striking]. In the bacterial system both the acetyl and the malonyl residue are bound in this way during the condensation step (5 in

$$\begin{array}{c} \text{NH} \quad\quad\quad \text{O} \\ | \quad\quad\quad\quad\quad || \\ \text{CHCH}_2\text{O---P---OCH}_2\text{C(Me)}_2\text{CH(OH)CONHCH}_2\text{CH}_2\text{CONHCH}_2\text{CH}_2\text{SCOR} \\ | \quad\quad\quad\quad\quad | \\ \text{CO} \quad\quad\quad \text{OH} \end{array}$$

1

Scheme 2) but in the yeast system Lynen has found that acetyl is transferred to a thiol group of a cysteinyl residue in one of the enzymes before condensation. Lynen refers to the two thiol groups as "central" (pantotheine),

$$\begin{array}{c} \text{CH}_2\text{O---P---O---P---OCH}_2\text{C(Me)}_2\text{CH(OH)CONHCH}_2\text{CH}_2\text{CONHCH}_2\text{CH}_2\text{S} \\ || \quad\quad || \\ \text{O} \quad\quad \text{O} \\ | \quad\quad\quad | \\ \text{OH} \quad\quad \text{OH} \end{array}$$

2

attached to ACP, and "peripheral" (cysteine), attached to the condensing enzyme, and visualizes fatty acid biosynthesis as follows.

The process starts with the transfer of acetate from coenzyme A first to the central and then to the peripheral thiol group. Malonate is then transferred to the central thiol group and condensation occurs to give acetoacetate attached to the central thiol group. The acetoacetate is reduced to butyrate while remaining on the central thiol group, and the butyrate is then transferred to the peripheral thiol group, making way for attachment of a fresh molecule of malonate to the central thiol group and repetition of the whole process. Lynen suggests that the function of the pantotheine residue is to provide a flexible linkage which permits the bound intermediates to come into proximity with the active sites of the enzymes of the complex. He also has evidence for a second, non-sulphydryl, point of attachment of malonate to ACP which prevents its transfer to the peripheral thiol group.

A.2. The Biosynthesis of the Unsaturated Fatty Acids[3]

The unsaturated fatty acids can arise either by dehydrogenation of the saturated fatty acids of appropriate chain-length or from a branch in the pathway to the saturated acids (Scheme 3). The latter route involves dehydration of an intermediate enzyme-bound β-hydroxy-acid to give a $\beta\gamma$-unsaturated intermediate instead of the $\alpha\beta$-unsaturated intermediate involved in the biosynthesis of the saturated fatty acids. The $\beta\gamma$-unsaturated intermediate is

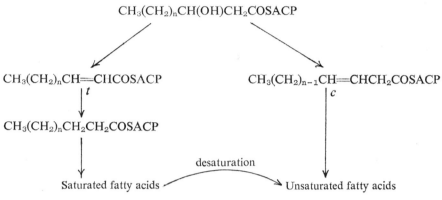

Scheme 3: Alternative routes to the unsaturated fatty acids

then converted to the unsaturated fatty acids by the normal chain-elongation processes.

The desaturation process, which is the more important process for the biosynthesis of secondary metabolites from fatty acids, is catalysed by a mixed-function oxidase (p. 5), but no oxygenated intermediate has yet been implicated in the process.

B. SECONDARY METABOLITES

B.1. Simple Derivatives of Fatty Acids

3   $Me(CH_2)_{11}CO(CH_2)_4CO_2H$ lactarinic acid
                                       Lactarius rufus[4]

4   $Me(CH_2)_9CH{=}CH(CH_2)_9COSH$ fagicladosporic acid
                                       Cladosporium fagi[5]

5   $Me(CH_2)_7CH{=}CH(CH_2)_{13}COSH$ epicladosporic acid
                                       C. fagi[5]

6   $HOCH_2CH(CH_2)_{12}CH(OH)CO_2H$ a: R = H, ustilic acid A
    | b: R = OH, ustilic acid B
    R   *Ustilago zeae*[6]

The ustilic acids (**6**) were isolated as components of the ustilagic acids (**7**), which are β-glycosidic esters of the ustilic acids with acylated cellobiose. In the major components the cellobiose is esterified, at unknown positions, with one acetyl and one β-hydroxyhexanonyl or β-hydroxyoctanoyl residue.

(R = ustilic acid A or B)

7

B.2. THE POLYACETYLENES

The term "polyacetylenes" is used rather loosely to denote a large group of natural products which possess straight carbon chains containing conjugated acetylenic or acetylenic–ethylenic systems, or compounds which are formally, and probably naturally, derived from such systems. The compounds are readily detected by virtue of their characteristic and usually intense ultraviolet spectra and this has formed the basis of systematic searches among higher plants by Sörensen and by Bohlmann, and among fungi by E. R. H. Jones. As a result, over 400 natural acetylenic compounds are known, about one-fifth having been obtained from fungi. With one or two exceptions, the fungal acetylenes have been obtained from Basidiomycetes, both from mycelial cultures (where the earliest examples were detected as a result of their antibiotic activity) and from fruiting bodies.[7a]

A complete list of the fungal acetylenes will not be given here, though the various types will be exemplified. On the other hand the biosynthesis of the acetylenes will be discussed in detail since it is not only of interest *per se* but has implications for the biosynthesis of other fungal metabolites. The subject has been extensively reviewed.[7]

B.2.a. *Structural Types*

The chain-lengths of the natural acetylenic compounds vary from C_6, in hexatriyne (**8**) isolated from *Fomes annosus*,[8] to C_{18}. The most common chain-lengths are C_9 and C_{10} for fungal acetylenes and C_{13} for plant acetylenes. The compounds often occur in closely related groups which are exemplified in the Table. These three groups of compounds also illustrate most of the commonly encountered variations—(1) an oxygen function is present at one end of the chain while the other end may be methyl, hydroxymethyl, aldehyde or carboxylic acid, (2) the acids may be decarboxylated to

Families of acetylenic compounds produced by fungi

12	$CH_3CH=CH[C=C]_2CH=CHCO_2H^{a,b}$	18	$HOCH_2[C=C]_3CH=CHCH_2OH$	24 $CH_3[C=C]_3CH=CHCO_2H$
13	$CH_3CH=CH[C=C]_2CH=CHCH_2OH^c$	19	$OHC[C=C]_3CH=CHCH_2OH$	25 $CH_3[C=C]_3CH=CHCHO$
14	$HOCH_2CH=CH[C=C]_2CH=CHCO_2H$	20	$H[C=C]_3CH=CHCHO$	26 $CH_3[C=C]_3CH=CHCH_2OH$
15	$HO_2CCH=CH[C=C]_2CH=CHCO_2H^a$	21	$H[C=C]_3CH=CHCH_2OH$	27 $HOCH_2[C=C]_3CH=CHCO_2H$
16	$HOCH_2CH_2CH_2[C=C]_2CH=CHCO_2H^a$	22	$H[C=C]_3CH—CHCH_2OH$	*Pleurotus ulmarius*[11]

For compound 22:

$$H[C=C]_3CH \overset{\displaystyle O}{\diagdown\diagup} CHCH_2OH$$

17 $HO_2CCH_2CH_2[C=C]_2CH=CHCO_2H$	23 $H[C=C]_3CH(OH)CH(OH)CH(OH)CH_2OH$
Polyporus anthracophilus[9]	*Coprinus quadrifidus*[10]

[a] These compounds also occur as their methyl esters.
[b] The *cis*-isomer occurs in higher plants.
[c] Also occurs in higher plants.

give chains containing odd numbers of carbon atoms and terminating in an ethinyl group, (3) in some series, pairs of carbon atoms may be linked by a single, double, or triple bond, (4) double bonds may be epoxidized, and

$$H[C{\equiv}C]_3H \qquad\qquad HC{\equiv}CC{\equiv}CCH{=}C{=}CHCH(OH)CH_2CH_2CO_2H$$

8   **9**

—$[C{\equiv}C]_2CH{=}CHMe$

10   **11**

triple bonds may isomerize to allenes e.g. in nemotinic acid (**9**)[12a], which occurs mainly as a xyloside.[12b] Oxygenation is less frequent in the acetylenic compounds from higher plants, but further transformations to give aromatic compounds, e.g. (**10**), or heterocyclic compounds, e.g. (**11**), are more common.

B.2.b. *Biosynthesis*

The incorporation of acetate into nemotinic acid (**9**)[13] and *trans-trans*-matricaria ester (**12**, Table),[14] and of malonate into dehydromatricarianol (**26**, Table),[15] has been demonstrated; in the last experiment a clear "starter" effect was observed. The labelled ester (**28**) undergoes the transformations shown in Scheme 4 when supplied to mycelium in replacement cultures,[16] and the decarboxylation of the hydroxy-acid (**29**) to the alcohol (**30**) has been carried out using a cell-free system from *Coprinus quadrifidus*.[17] Similar

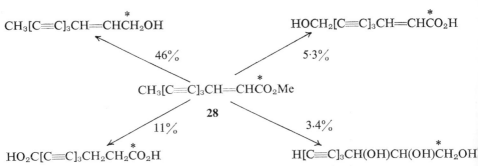

Scheme 4: Transformations of methyl [1-14C]dec-2-ene-4,6,8-triynoate (**28**) by fungi[16]

transformations have been observed in higher plants.[7c] Thus the origin of the

$$HO_2C[C\equiv C]_3CH=CHCH_2OH \longrightarrow H[C\equiv C]_3CH=CHCH_2OH$$

$$\textbf{29} \qquad\qquad\qquad\qquad\qquad\qquad \textbf{30}$$

carbon atoms and the inter-relationships of the acetylenic compounds are those which were predictable from inspection of their structures. The more interesting question is that of the origin of the triple bond and the relationship of the compounds to the fatty acids and polyketides, which are also derived from acetate plus malonate.

The isolation of crepenynic acid (**31**, Scheme 5) (first from the seed-oils of *Crepis* spp. but now known to be widespread) and of other C_{18} and C_{17} acetylenic compounds led Bu'Lock to suggest that the acetylenic compounds might be derived from the fatty acids by a series of dehydrogenations and prototropic rearrangements on one side of the double bond of oleic acid, with a series of β-oxidations to give the acetylenic compounds with shorter chains (Scheme 5). The key steps of this sequence have been confirmed by Bu'Lock and Smith,[18] who have shown that [10-^{14}C]oleic acid is converted

$$CH_3(CH_2)_{16}CO_2H$$
$$\downarrow -2H$$
$$CH_3(CH_2)_7CH=CH(CH_2)_7CO_2H$$
$$\downarrow -2H$$
$$CH_3(CH_2)_4CH=CHCH_2CH=CH(CH_2)_7CO_2H$$
$$\downarrow -2H$$

31 $\qquad CH_3(CH_2)_4C\equiv CCH_2CH=CH(CH_2)_7CO_2H$
$$\downarrow -2H$$

32 $\qquad CH_3(CH_2)_2CH=CHC\equiv CCH_2CH=CH(CH_2)_7CO_2H$
$$\downarrow -6H$$

$[CH_3(C\equiv C)_3CH_2CH=CH(CH_2)_7CO_2H]$
$\qquad\downarrow$ 2β-oxidation cycles
$[CH_3(C\equiv C)_3CH_2CH=CH(CH_2)_3CO_2H] \longrightarrow \begin{matrix} C_{14} \\ C_{13} \end{matrix}$ acetylenes
$\qquad\downarrow \beta$-oxidation
? $\qquad [CH_3(C\equiv C)_3CH_2CH=CHCH_2CO_2H] \longrightarrow \begin{matrix} C_{12} \\ C_{11} \end{matrix}$ acetylenes
$\qquad\downarrow$
$\qquad CH_3(C\equiv C)_3CH=CHCO_2H \longrightarrow \begin{matrix} C_{10} \\ C_9 \end{matrix}$ acetylenes

Scheme 5: Possible sequence of steps for the derivation of acetylenic compounds from fatty acids[18] by *Tricholoma grammopodium* into [10-^{14}C]crepenynic acid and [2-^{14}C] dehydromatricarianol (26, Table); they also showed that the proposed intermediate dehydrocrepenynic acid (32) is present in polyacetylene-producing fungi.

B.3. CYCLOPENTANES DERIVED FROM FATTY ACIDS

33

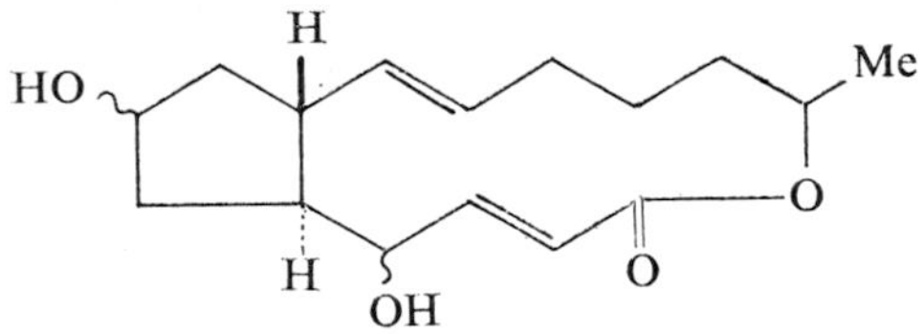

brefeldin A (cyanein, decumbin) L^{28-}
Penicillium brefeldianum,[19]
  P. cyaneum,[20] *P. decumbens*,[21]
  P. simplicissimum,[22] *Nectria
  radicicola*,[23] *Curvularia lunata*,[24]
  C. subulata[24]

34

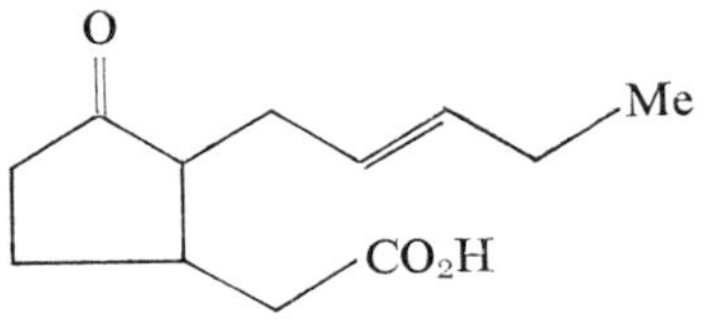

jasmonic acid
Lasiodiplodia theobromae[25]

Jasmonic acid (34) also occurs as its methyl ester in the essential oil of jasmine (*Jasmium grandiflora*)[26] along with the related jasmone (35) which is probably derived from jasmonic acid in nature. The alcohol moieties of the

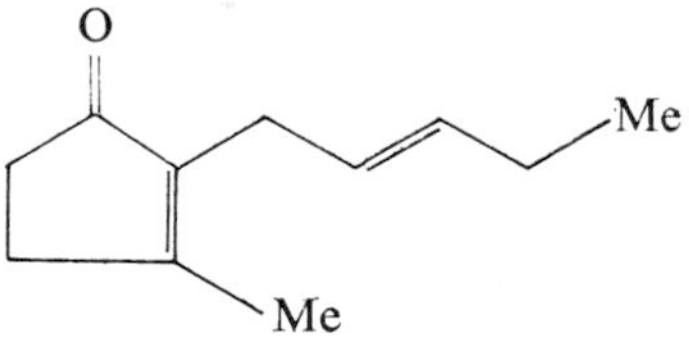

35

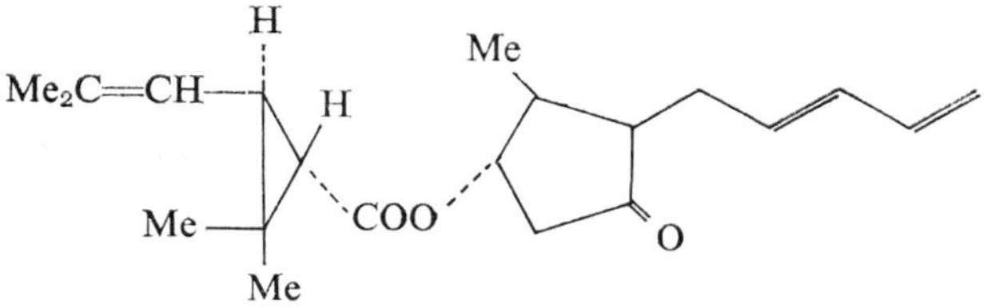

36

Chrysanthemum insecticides, e.g. pyrethrin I (**36**), are also related to jasmonic acid. Brefeldin A (**33**) is structurally similar to the prostaglandins, e.g. prostaglandin E_2, which are physiologically active compounds present in mammals and derived by oxidative cyclizations of unsaturated fatty acids, e.g. arachidonic acid (Scheme 6).[27]

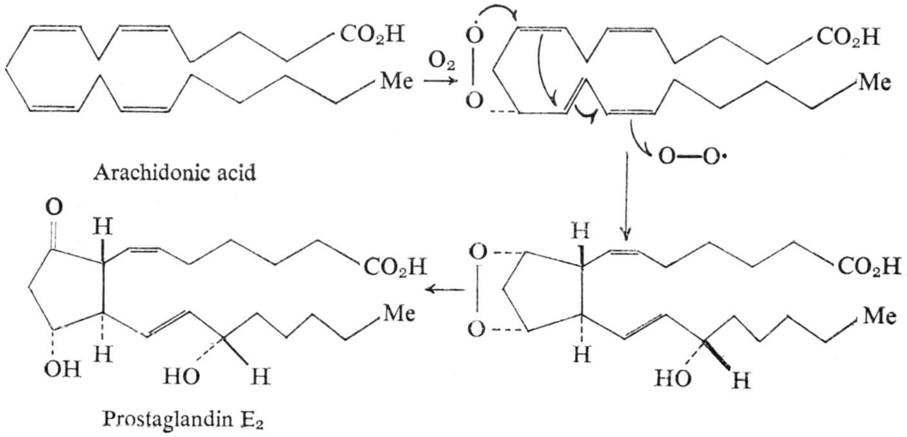

Scheme 6: The biosynthesis of prostaglandin E_2 from arachidonic acid

Labelled acetate is incorporated into brefeldin A[28,29] and detailed degradation of the radioactive brefeldin shows the labelling pattern to be as in Scheme 7.[29] More interesting is the fact that [9-^{14}C]palmitate is specifically

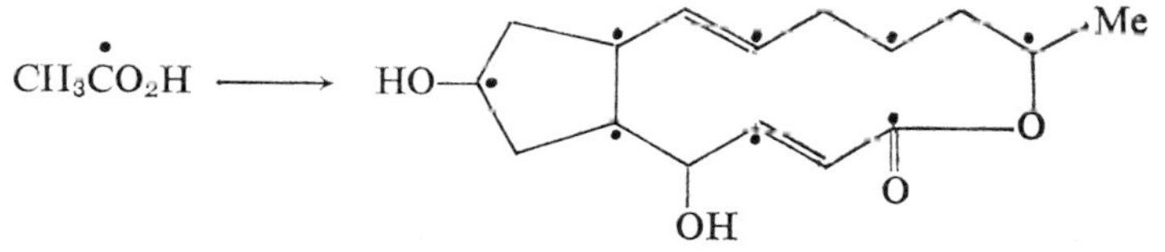

Scheme 7: The incorporation of [1-^{14}C]acetate into brefeldin A

incorporated into brefeldin,[30] bringing the biosynthesis of brefeldin into line with that of the prostaglandins (for a possible mechanism, see ref. 30). Jasmonic acid presumably arises by a pathway similar to that leading to brefeldin, but whether by cyclization of a C_{12} intermediate or of a (more common) C_{16} or C_{18} intermediate followed by β-oxidations remains to be seen.

B.4. OTHER COMPOUNDS WHICH MIGHT BE DERIVED FROM FATTY ACIDS

The demonstration that fatty acids are intermediates in the biosynthesis of the acetylenes and of brefeldin raises the question[30] of whether they might be involved in the biosynthesis of other types of secondary metabolite, especially some which are classed at present as polyketides. This applies particularly to the more reduced compounds such as palitantin (37), which co-occurs

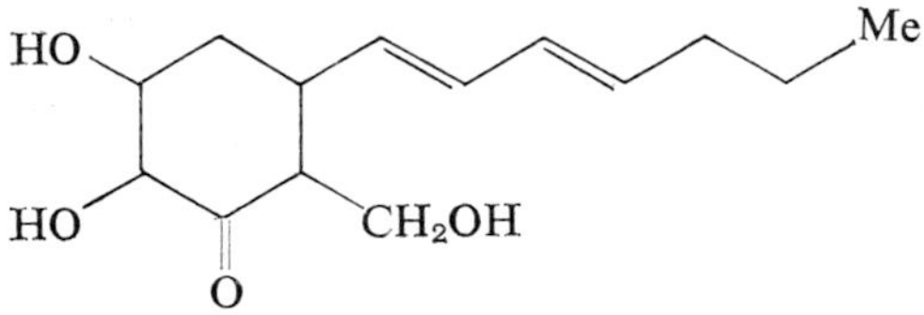

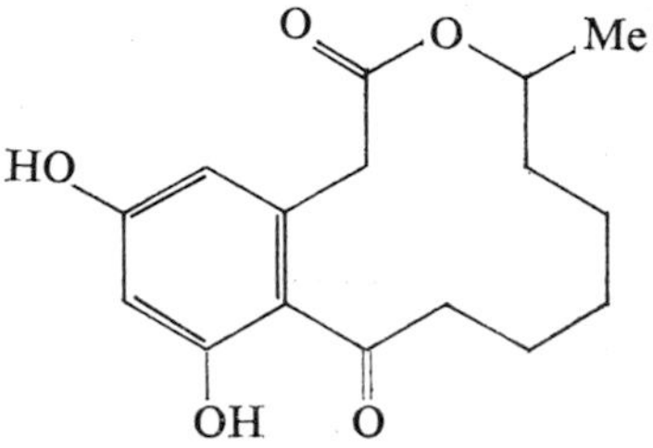

with brefeldin in *P. brefeldianum*, or curvularin (38) which, like brefeldin, is a metabolite of *C. lunata*, though the compounds have not been obtained from the same fermentation. The cyclization of acetylenic precursors to give phenyl rings, e.g. in (10), has already been noted, and cyclization of acetylenic precursors in higher plants to give compounds, such as (39),[31]

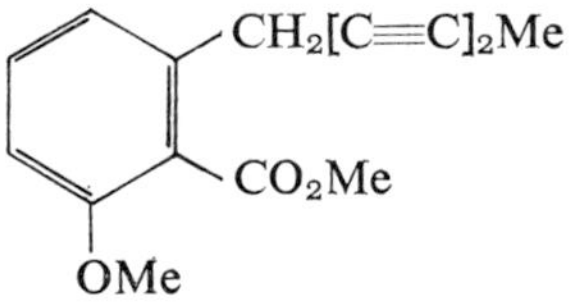

which have typically polyketide oxygenation patterns, lends credence to the possibility. Clearly, the incorporation of specifically-labelled fatty acids into palitantin and curvularin must be studied in order to settle the question.

REFERENCES

1. R. Shaw, *Advan. Lipid Res.*, 1966, **4**, 116.
2. F. Lynen, *Pure Appl. Chem.*, 1967, **14**, 137; P. W. Majerus and P. R. Vagelos, *Advan. Lipid Res.*, 1967, **5**, 1.
3. K. Bloch, *Accounts Chem. Res.*, 1967, **2**, 193.
4. J. Zellner, *Monatsh.*, 1928, **50**, 211.
5. L. E. Olifson, *Chem. Abs.*, 1962, **56**, 5195.
6. R. V. Lemieux, *Can. J. Chem.*, 1953, **31**, 396.
7. Recent reviews are (a) J. D. Bu'Lock. *In* "Biosynthesis of Antibiotics" (J. F. Snell, ed.), Vol. I, p. 141; Academic Press, London and New York, 1966. (b) M. Anchel. *In*. "Antibiotics. Vol. II. Biosynthesis" (D. Gottlieb and P. D. Shaw, eds.), p. 189; Springer-Verlag, Berlin, 1967. (c) F. Bohlmann, *Prog. Chem. Org. Nat. Prod.*, 1967, **25**, 1.
8. A. T. Glen, S. A. Hutchinson and N. J. McCorkindale, *Tetrahedron Lett.*, 1966, 4223.
9. J. D. Bu'Lock, E. R. H. Jones and W. B. Turner, *J. Chem. Soc.*, 1957, 1607.
10. E. R. H. Jones and J. S. Stephenson, *J. Chem. Soc.*, 1959, 2197; E. R. H. Jones, J. S. Stephenson, W. B. Turner and M. C. Whiting, *J. Chem. Soc.*, 1963, 2048.
11. J. N. Gardner, E. R. H. Jones, P. R. Leeming and J. S. Stephenson, *J. Chem. Soc.*, 1960, 691.
12. (a) J. D. Bu'Lock, E. R. H. Jones and P. R. Leeming, *J. Chem. Soc.*, 1955, 4270; (b) J. D. Bu'Lock and H. Gregory, *Experientia*, 1959, **15**, 420.
13. J. D. Bu'Lock and H. Gregory, *Biochem. J.*, 1959, **72**, 322.
14. J. D. Bu'Lock, D. C. Allport and W. B. Turner, *J. Chem. Soc.*, 1961, 1654.
15. J. D. Bu'Lock and H. M. Smalley, *J. Chem. Soc.*, 1962, 4662.
16. P. Hodge, E. R. H. Jones and G. Lowe, *J. Chem. Soc.*, 1966, 1216.
17. J. N. Gardner, G. Lowe and G. Read, *J. Chem., Soc.*, 1961, 1532.
18. J. D. Bu'Lock and G. N. Smith, *J. Chem. Soc.*, 1967, 332.
19. E. Härrie, W. Loeffler, H. P. Sigg, H. Stähelin and Ch. Tamm, *Helv. Chim. Acta*, 1963, **46**, 1235; H. P. Sigg, *Helv. Chim. Acta*, 1964, **47**, 1401.
20. V. Betina, P. Nemec, J. Dobias and Z. Baráth, *Folia Microbiol.*, 1962, **7**, 353.
21. V. L. Singleton, N. Bohonos and A. J. Ullstrup, *Nature*, 1958, **181**, 1072.
22. V. Betina, J. Fuska, A. Kjaer, M. Kutkova, P. Nemec and R. H. Shapiro, *J. Antibiotics (Tokyo)*, 1966, **19**, 115.
23. R. N. Mirrington, E. Ritchie, C. W. Shoppee, S. Sternhell and W. C. Taylor, *Aust. J. Chem.*, 1966, **19**, 1265.
24. R. G. Coombe, J. J. Jacobs and T. R. Watson, *Aust. J. Chem.*, 1968, **21**, 783.
25. D. C. Aldridge, D. Giles, S. Galt and W. B. Turner, *J. Chem. Soc. C*, in the press.
26. E. Demole, E. Lederer and D. Mercier, *Helv. Chim. Acta*, 1962, **45**, 675; E. Demole and M. Stoll, *Helv. Chim. Acta*, 1962, **45**, 692.
27. For a review of the chemistry and biochemistry of the prostaglandins, see S. Bergstrom, *Science*, 1967, **157**, 382.
28. R. G. Coombe, P. S. Foss and T. R. Watson, *Chem. Commun.*, 1967, 1229; R. G. Coombe, P. S. Foss, J. J. Jacobs and T. R. Watson, *Aust. J. Chem.*, 1969, **22**, 1943.
29. U. Handschin, H. P. Sigg and Ch. Tamm, *Helv. Chim. Acta*, 1968, **51**, 1943.
30. J. D. Bu'Lock and P. T. Clay, *Chem. Commun.*, 1969, 237.
31. F. Bohlmann, R. Jente, W. Lucas, J. Laser and H. Schulz, *Chem. Ber.*, 1967, **100**, 3183.

CHAPTER 5

Polyketides

WHILE MOST of the biosynthetic processes discussed in this book lead to primary as well as secondary metabolites and are therefore used by most living systems, the polyketide route leads almost exclusively to secondary metabolites and is used mainly by fungi and to a lesser extent by bacteria and higher plants. Even among the fungi the ability to produce polyketides is not evenly distributed, for polyketides are the characteristic secondary metabolites of the Fungi Imperfecti and the Ascomycetes (which are often equivalent, see Chapter 2), and occur relatively rarely in the Basidiomycetes.

A. POLYKETIDE BIOSYNTHESIS

Because of the limited distribution and secondary nature of polyketides, the concepts of polyketide biosynthesis were developed by organic chemists, and the problem has received little attention from biochemists. For this reason, although the basic tenets of the hypothesis have been firmly established by labelling experiments, we know less about the detailed mechanisms of polyketide biosynthesis than of any other biosynthetic pathway. We shall start, therefore, by presenting the polyketide hypothesis in the chemist's terms of starting materials and products and will then discuss possible mechanisms for the processes involved. The biosynthesis of individual compounds is discussed in the appropriate part of the chapter and here we shall refer only to those which illustrate general principles. Many reviews of polyketide biosynthesis have appeared[1] and A. J. Birch, who was primarily responsible for its modern development, has presented an interesting account of the history of the concept.[2]

A.1. THE POLYKETIDE HYPOTHESIS

Polyketides are formed by condensation of an acetyl unit, or other acyl unit (see below), with malonyl units, with concomitant decarboxylation as in fatty acid biosynthesis but without obligatory reduction of the intermediate β-dicarbonyl system. The resulting poly-β-ketomethylene (polyketide) system, e.g. (1), possesses activated methylene groups which can take part in internal aldol-type condensations to give aromatic compounds, and which are susceptible to electrophilic substitution. The cyclization process is illustrated (Scheme 1) by the formation of orsellinic acid or acetyphloroglucinol by alternative reactions of the intermediate (**1**); clearly with longer chains a greater variety of cyclizations become possible, but we shall see (p. 198) that not all of the theoretically possible cyclizations have been observed.

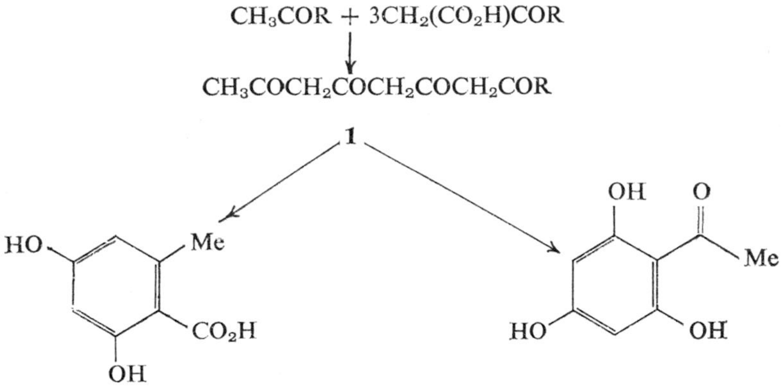
$CH_3COR + 3CH_2(CO_2H)COR$
$CH_3COCH_2COCH_2COCH_2COR$
1
HO
Me
CO_2H
OH
OH
O
Me
HO
OH

Orsellinic acid Acetylphloroglucinol

Scheme 1: The formation of orsellinic acid and acetylphloroglucinol from a polyketide intermediate (**1**)

The fact that only the first two carbon atoms of a polyketide chain are derived directly from acetate means that acetate should be more efficiently, and malonate less efficiently, incorporated into this position than into the rest of the chain. This "starter effect" has been clearly demonstrated in experiments with purified enzymes (p. 80) and in some cases with whole

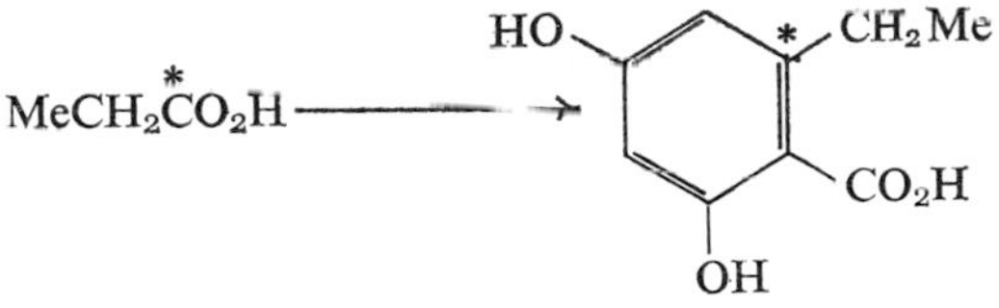
$Me\overset{*}{C}H_2CO_2H$
HO
*
CH_2Me
CO_2H
OH

2

cells, but often the equilibration of acetate and malonate in whole cells is too rapid for a clear effect to be observed.

We noted above that an acyl group other than acetyl might initiate the polyketide chain-building process. In higher plants this is often an aromatic acid derived from shikimic acid (as in the flavonoids, p. 39), and in bacteria it might be propionate (as in the anthracyclines, p. 190) or malonamoate (as in the tetracyclines, p. 183). In fungi, however, the intervention of an acyl group other than acetyl is rare (but for a reservation to this statement, see p. 83). An interesting example is provided by *homo*-orsellinic acid (2) which is produced, along with the normal metabolite orsellinic acid, by *Penicillium baarnense* grown in the presence of propionate.[3]

In bacteria, not only can propionate initiate chain-formation, but methylmalonate can replace malonate in the chain-extending process. Striking, and important, examples are provided by the macrolide antibiotics[4] whose aglycones, e.g. erythronolide (3), are derived mainly or wholly from "propionate units". This process has not been observed in fungi.

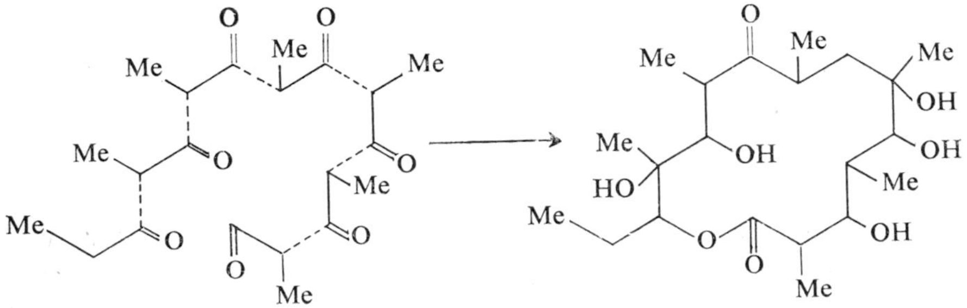

3

Further structural variety can be introduced into polyketide-derived molecules by the following reactions:

    i. Loss of one or more oxygen atoms from the polyketide chain, possibly by reduction of the ketone and dehydration of the resulting alcohol as in fatty acid biosynthesis. 6-Methylsalicylic acid (4) can arise from the

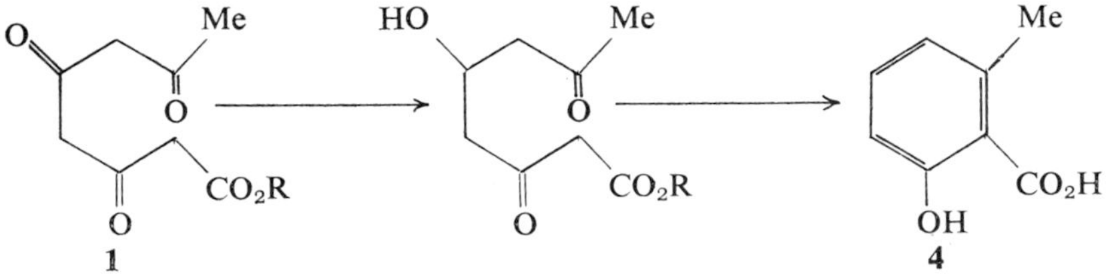

orsellinic acid precursor (1) by such a process. In other metabolites the dehydration may be followed by reduction of the double bond (again as in fatty acid biosynthesis) to give saturated systems, e.g. the side-chain of flavoglaucin (5). Flavoglaucin also illustrates other features, discussed below, of polyketide metabolites.

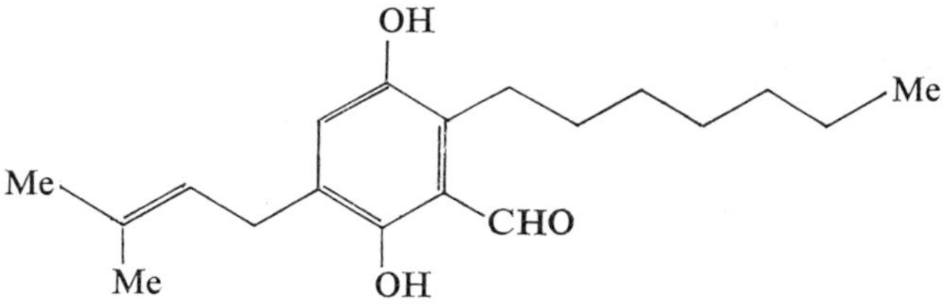

OH

Me

Me

Me   CHO

Me   OH

5

ii. Electrophilic introduction of alkyl groups, notably methyl from methionine (see p. 6) and isoprenoid units from their pyrophosphates (see p. 215), and of halogen (see p. 9). For example, 4,6-dihydroxy-2,3-dimethylbenzoic acid (6) can be formed by methylation of the orsellinic acid precursor, while the dimethylallyl group of flavoglaucin (5) is derived from dimethylallylpyrophosphate. C-Alkylation always occurs at positions

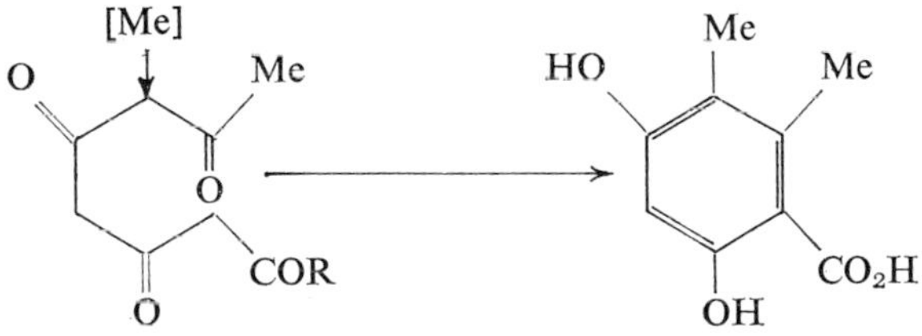

[Me]   Me   Me

O   HO   Me

O

COR CO_2H

O   OH

6

corresponding to the methylene groups of the polyketide chain, a fact which is of great assistance in determining the structure of a polyketide metabolite.

iii. Alkyl groups may be oxidized and carboxyl groups reduced. For example, during the biosynthesis of flavipin (7) an extra methyl group

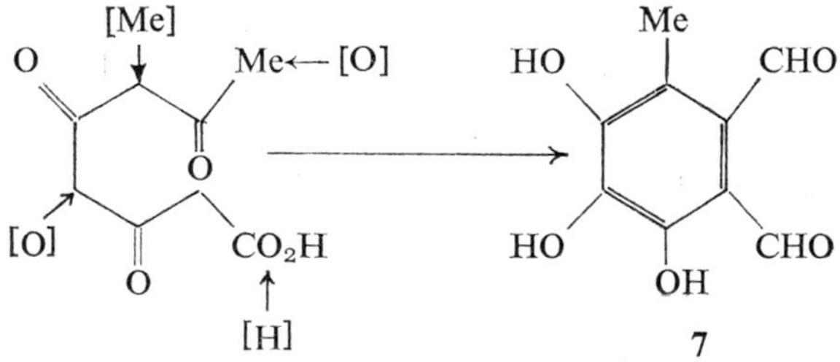

[Me]   Me←[O]   Me

O   HO   CHO

O

[O] CO_2H HO   CHO

O   OH

[H]   **7**

is introduced, a methyl group is oxidized to an aldehyde, and a carboxyl group is reduced to an aldehyde; in flavoglaucin (5), too, a carboxyl group has been reduced to an aldehyde.

iv. Flavipin also illustrates a further common reaction of polyketide biosynthesis—the introduction of "extra" oxygen atoms. The process is probably catalysed by a mixed-function oxygenase (p. 5).

v. Decarboxylation, as in the formation of orcinol (8) from orsellinic acid, commonly occurs. Coupled to the oxidation of methyl to carboxyl this can result in the loss of a methyl group.

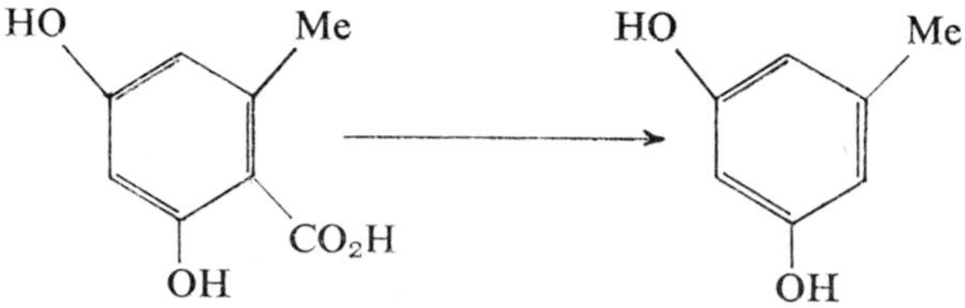

vi. Inter- or intra-molecular oxidative coupling can occur with formation of carbon–carbon or carbon–oxygen bonds, e.g. in the formation of griseofulvin (9), or of dimers such as oosporein (10). Oxidative coupling has recently been reviewed.[5]

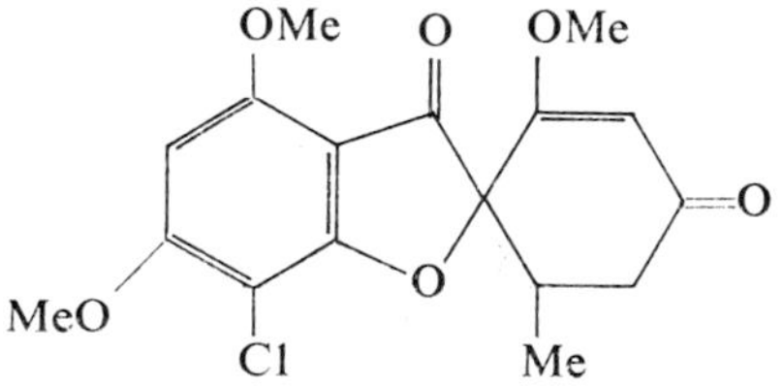

9 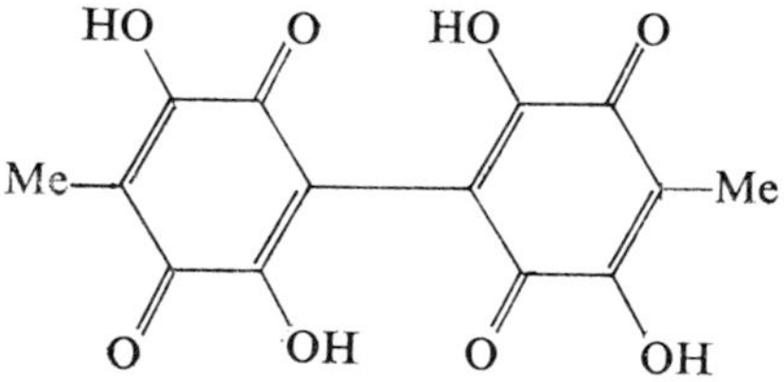   10

vii. More drastic modifications of polyketide-derived skeletons can occur, as in the formation of patulin (11) from 6-methylsalicylic acid.

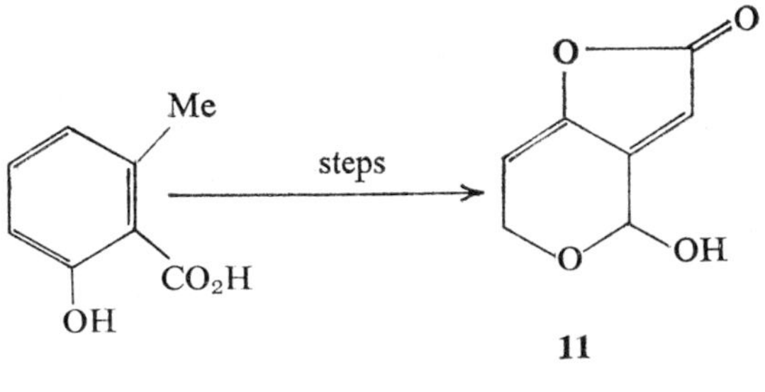

11

This formal picture of polyketide biosynthesis, developed largely by organic chemists, has been of immense value in the determination and correlation of the structures of natural products.

A.2. THE MECHANISM OF POLYKETIDE BIOSYNTHESIS

Little is known about the biochemical processes which underly the poly-ketide hypothesis, but it is worth examining some aspects of the problem if only to underline our ignorance. In this discussion we shall first examine the similarity between polyketide and fatty acid biosynthesis; from this similarity we shall draw conclusions concerning the assembly of the polyketide chains; and finally we shall consider the timing of the various processes by which the polyketide chain is modified.

A.2.a. *The Relationship of Polyketide and Fatty Acid Biosynthesis*

The closest analogy for polyketide biosynthesis is provided by fatty acid biosynthesis, and the analogy goes deeper than the fact that both polyketides and fatty acids arise by linear condensation of acetyl and malonyl units. In the first place, experiments with cell-free systems (see below) have shown that the active acetyl and malonyl units of polyketide biosynthesis, as of fatty acid biosynthesis, are the coenzyme A derivatives. Secondly, all the evidence from experiments with cell-free and whole-cell systems suggests that there are no free intermediates between acetate and fully-assembled and stabilized polyketides. Thirdly, 6-methylsalicylic acid and fatty acids co-produced in the presence of $[1-^{14}C,2-^{3}H]$acetate have distributions of radioactivity consistent with their formation by the same condensation process.[7] Finally, it has recently been shown[8] that, in the absence of NADPH, highly purified fatty acid synthetase produces triacetic acid lactone **(12)**, which is a stabilized polyketide chain.

OH

12

Whereas the details of fatty acid biosynthesis have been studied using purified enzyme systems it has not yet proved possible to isolate a pure enzyme capable of polyketide biosynthesis. Cell-free extracts capable of

5. POLYKETIDES

synthesizing orsellinic acid,[9] 6-methylsalicylic acid[10] and alternariol (**13**)[299,300] have been prepared, and partial purification of the enzymes concerned has been achieved. In each case there is evidence for a multi-enzyme complex as

13

in fatty acid biosynthesis. The 6-methylsalicyclic acid synthetase requires NADPH, presumably for the reduction of the carbonyl group which is lost during the biosynthesis of 6-methylsalicylic acid. Cell-free systems for the synthesis of the more complex metabolite patulin[11] and stipitatic acid[12] have also been obtained but these contain a mixture of enzymes and no fractionation was achieved.

A.2.b. *Chain Assembly*

It seems likely that the polyketide backbones are formed by stepwise condensations of malonate with a growing protein-bound chain as in fatty acid biosynthesis, and that the processes of reduction and/or cyclization to give stable products take place on protein-bound intermediates. Granted this, then there is only one possible sequence of events leading to a compound such as orsellinic acid which requires cyclization of the polyketide precursor with no other modification. In such a case, the only problems are the stabilization of the intermediate and the specificity of cyclization, and Bu'Lock[13] has presented an interesting discussion of both these problems in terms of metal chelation.

A.2.c. *Loss of Oxygen*

For a compound such as 6-methylsalicylic acid, whose biosynthesis involves only the loss of one oxygen atom relative to orsellinic acid, *six* possible pathways become available (Scheme 2) depending on the timing of the reduction, dehydration and cyclization (aldol) steps. Admittedly not all of these pathways are equally likely, nor are they necessarily mutually exclusive, but none can be ruled out on *a priori* grounds. If a situation as complex

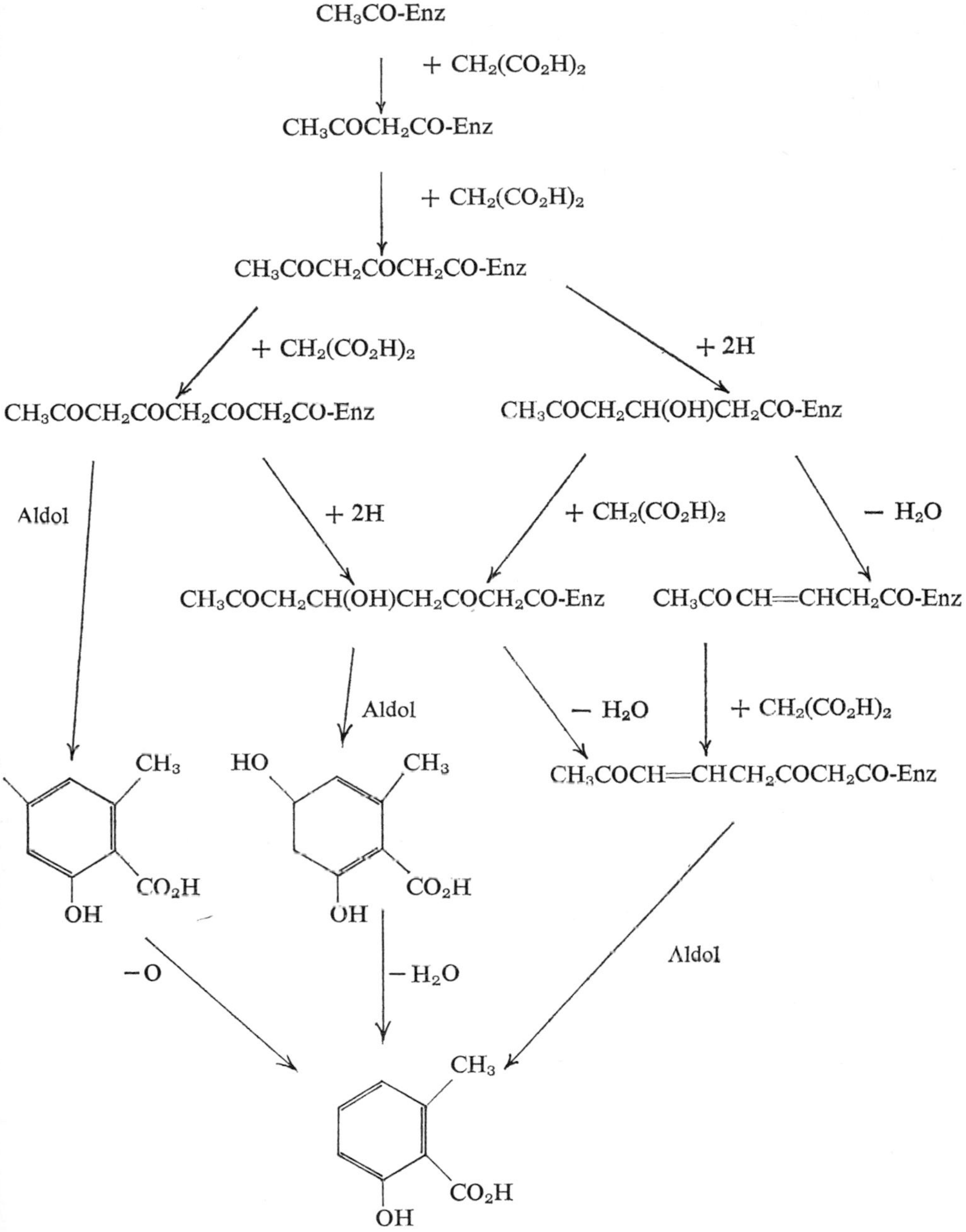
CH3CO-Enz
+ CH2(CO2H)2
CH3COCH2CO-Enz
+ CH2(CO2H)2
CH3COCH2COCH2CO-Enz
+ CH2(CO2H)2
+ 2H
CH3COCH2COCH2COCH2CO-Enz
CH3COCH2CH(OH)CH2CO-Enz
Aldol
+ 2H
+ CH2(CO2H)2
− H2O
CH3COCH2CH(OH)CH2COCH2CO-Enz
CH3CO CH=CHCH2CO-Enz
Aldol
− H2O
+ CH2(CO2H)2
CH3COCH=CHCH2COCH2CO-Enz
CH3
HO
CH3
CO2H
OH
CO2H
OH
−O
− H2O
Aldol
CH3
CO2H
OH

Scheme 2: Possible pathways to 6-methylsalicylic acid

as this exists for a compound as simple as 6-methylsalicylic acid, imagine the permutation of steps which is possible in the biosynthesis of a compound such as sclerotiorin (14).

14

A.2.d. *Timing of Methylations*

Sclerotiorin introduces a further variable—the timing of the methylations. There is now a body of evidence which suggests that the aromatic C-methyl groups of polyketides are introduced at a pre-aromatic stage, cf. 4,6-dihydroxy-2,3-dimethylbenzoic acid (p. 102), clavatol (p. 110), atranorin (p. 108), usnic acid (p. 111), the tropolones (p. 114) and the tetracyclines (p. 184). But even if we accept that methylation occurs at a pre-aromatic stage, we are left with two possibilities—the methyl groups can be introduced either onto the growing chain or after chain-assembly is complete. The isolation of methyl triacetic acid lactone as a byproduct of tropolone biosynthesis (p. 113) might be taken to suggest that methylation occurs before the introduction of the fourth (and final) malonate unit. On the other hand, a cell-free preparation from an *Aspergillus* sp.† which produces flavipin (7), catalyses the methylation of a protein-bound substrate believed to be a tetra-acetic acid,[14] so that in this case methylation seems to occur after complete formation of the polyketide chain. Similar ambiguity exists for the introduction of aliphatic C-methyl groups; for instance the methyl groups of sclerotiorin (14) could be introduced at the "active methylene" positions before reduction of the carbonyl groups or at double bonds later in the biosynthetic sequence.

A.2.e. *Timing of the Introduction of Oxygen*

The evidence concerning the introduction of the "extra" oxygen atoms which many polyketides possess leads to two conclusions—the oxygen is introduced late in the biosynthetic sequence, often after the first stabilized products have been formed, and the enzymes catalysing the process are

---

† Gatenbeck *et al.*[14] refer to the organism as *A. flaviceps* but this is probably a typographical error since they cite earlier work where the organism is *A. flavipes*.

relatively non-specific. The latter point has been discussed already (pp. 5 and 43).

A.2.f. *Summary*

The picture of polyketide biosynthesis which is emerging from the mass of indirect evidence which is available may be described as follows. The backbone is built by successive condensations of malonate onto a growing protein-bound chain. Reductions, dehydrations and methylations can occur either on the growing chain or on the fully-formed chain; little evidence is available on the timing of these processes but it seems certain that protein-bound intermediates are involved. The precursor is then stabilized either by aromatization or by reduction, and is liberated from the protein. Further oxidations (and perhaps reductions) can then occur to give series of related compounds. The picture is blurred and requires experiments with pure enzymes to bring it into focus.

A.3. "TWO-CHAIN" THEORIES

In the foregoing discussion we have assumed that the metabolites arise by suitable folding of a single polyketide chain. There are, however, a few compounds which can only arise by condensation of two chains, and some where evidence for a two-chain derivation is either not convincing (e.g. sclerin) or is contradicted by independent experiments (e.g. sulochrin). Whenever a compound can be formally derived from a single chain, and provided there is no undisputed evidence to the contrary, we have assumed in the following classification that only one chain is involved.

15   16

Ambiguity also arises for compounds which possess saturated straight-chain residues. Many of the best examples are provided by the depsides, e.g. perlatolic acid (15). Such compounds could arise either by the normal polyketide route, with complete reduction of the first carbonyl groups of the chain, or by using a preformed fatty acid (hexanoic in the case of perlatolic acid) as a "starter" unit in place of acetate. In the only example studied experimentally, hexanoate is not incorporated into pulvilloric acid (16).[244] The incorporation of a preformed fatty acid would not be regarded as an example of "two-chain" biosynthesis as discussed above but rather as an example of the use of "starters" other than acetate for polyketide bio-synthesis.

B. POLYKETIDE METABOLITES

Following the policy adopted throughout this book the polyketides are classified here on biosynthetic grounds rather than according to chemical type. This is achieved by grouping the compounds first according to the number of "C_2-units" which have contributed to the polyketide chain and then according to the type of cyclization which the (hypothetical) precursor has undergone. The terpenes have, of course, long been classified on a similar basis, and here the analogy is taken a stage further by using the terms trike-tide, tetraketide, pentaketide, etc., to denote compounds derived from three, four, five, etc., "C_2-units". The introduction of new terminology should not be undertaken lightly but the terms described above have proved so useful in this chapter, obviating the need for expressions such as "compounds derived from n C_2-units", that their use seems justified.

B.1. TRIKETIDES

17

a: R = Me, 3-methyltriacetic acid lactone
    Penicillium stipitatum[15]

b: R = H, triacetic acid lactone
    P. stipitatum[16]

No C_6 polyketide aromatic compounds have been detected. The triacetic acid lactones (17) were isolated during studies of tropolone biosynthesis and will be further discussed in that connection. Triacetic acid lactone (17b) has also been detected in *P. patulum*,[17] and is produced by fatty acid synthe-tase in the absence of NADPH. A dihydrotriacetic acid lactone residue is present in alternaric acid (162) and a (formal) triketide chain is present in colletodiol (23).

B.2. Tetraketides

B.2.a. *Open-chain Compounds*†

18 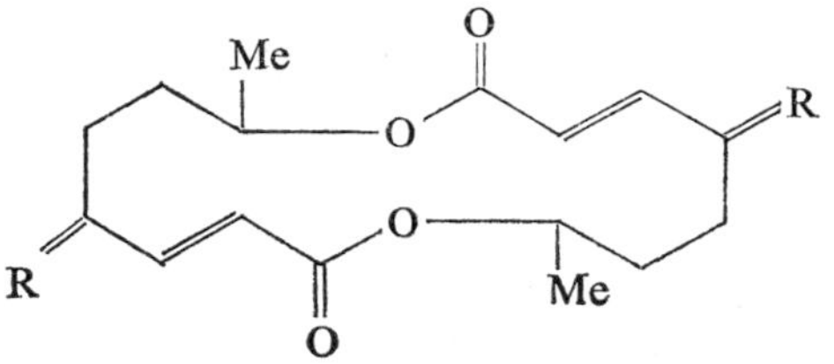

a: R = O, pyrenophorin
Pyrenophora avenae,
Stemphyllium radicinum[18]
b: R = H, OH; pyrenophorol
Byssochlamys nivea[19]

19 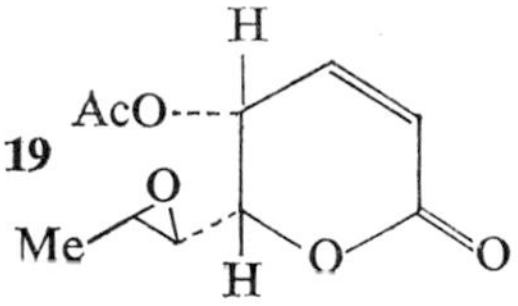

asperline (antibiotic U-13,933)
Aspergillus nidulans[20]

20 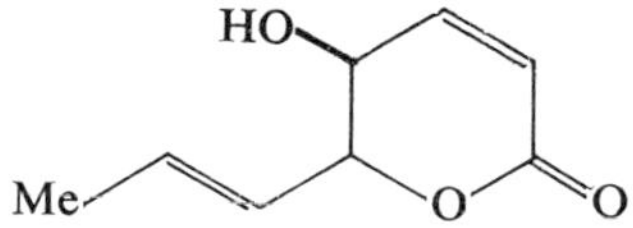

6-allyl-5,6-dihydro-5-hydroxypyran-2-one
Nigrospora sp.[21]

21 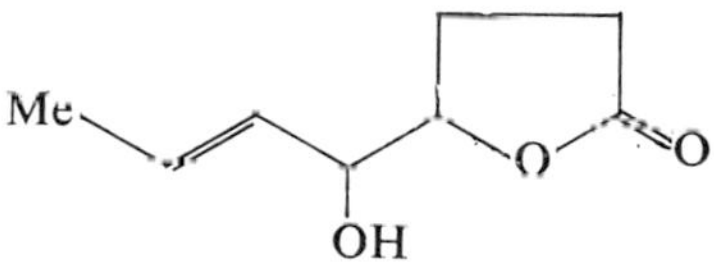

4,5-dihydroxyoct-6-enoic acid, 1,4-lactone
Nigrospora sp[21]

22 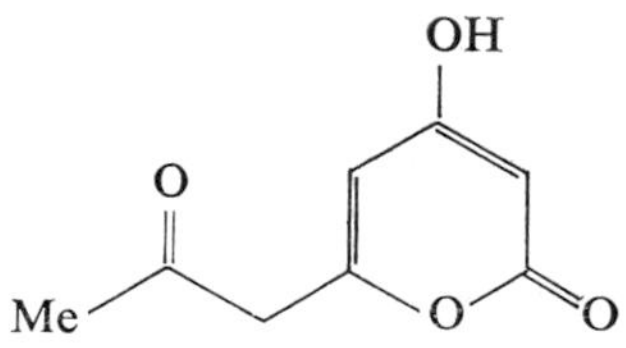

tetra-acetic acid lactone
Penicillium stipitatum[16]

† For our purpose "open-chain" means possessing no *carbon* ring.

23   colletodiol
          Colletotrichum capsici[22]

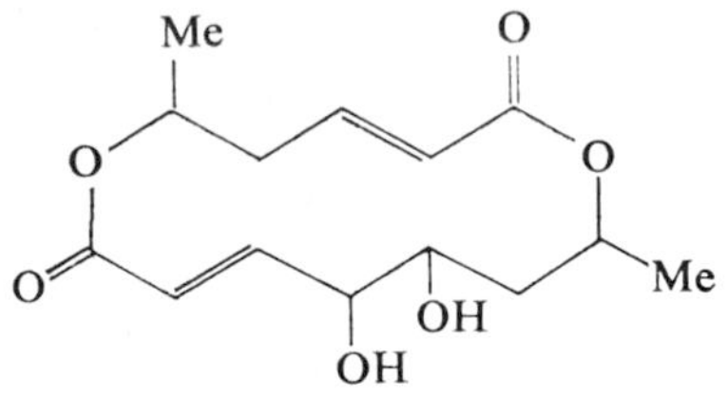

24   radicinin
          Stemphyllium radicinum[18b]

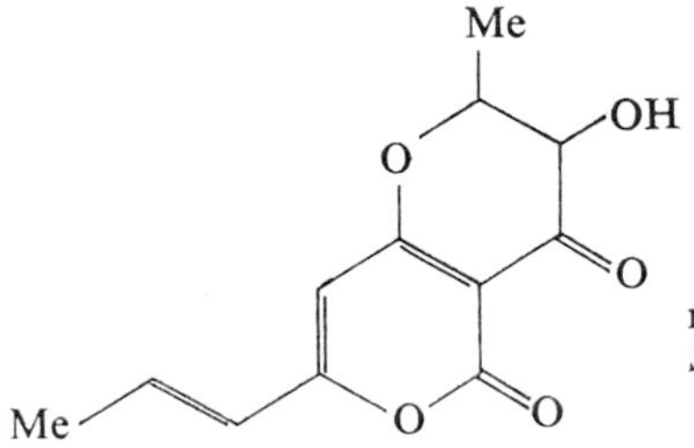

Although no labelling experiments have been reported, the derivation of these compounds from a C_8 polyketide precursor seems clear. Indeed tetra-acetic acid lactone (**22**), which was obtained with triacetic acid lactone when tropolone biosynthesis in *P. stipitatum* was blocked by ethionine, has an unchanged polyketomethylene chain. In colletodiol (**23**) a C_8 chain is joined through ester linkages to a C_6 chain presumably derived from three "C_2-units", and in radicinin (**24**), a co-metabolite of pyrenophorin in *S. radicinum*, a C_8 chain has condensed with a C_4 chain.

B.2.b. *Compounds of Type*

Most tetraketides result from this mode of cyclization, often with subsequent modification of the initial aromatic product. Indeed this is the largest class of polyketide-derived compounds and has probably received the greatest attention from the point of view of the biosynthesis and interrelationship of its members.† The class is conveniently subdivided according to the number of carbon atoms lost from, or introduced into,‡ the C_8 precursor (before or after cyclization).

† The large size of the class is, of course, partly a result of these biosynthetic studies (see p. 197).
‡ That is, with the formation of a carbon–carbon bond by an intermediate from the C_1-pool; ether formation is neglected for the purpose of this classification.

B.2.b.i. *Without Loss or Gain of Carbon*

25

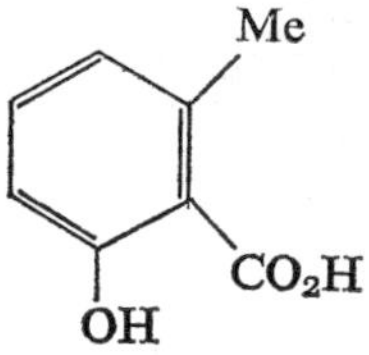

6-methylsalicylic acid
Widespread

$L^{23,24}$

26

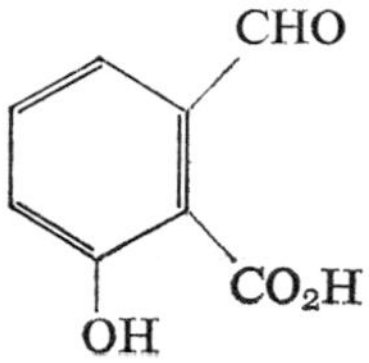

6-formylsalicylic acid
P. patulum[25]

27

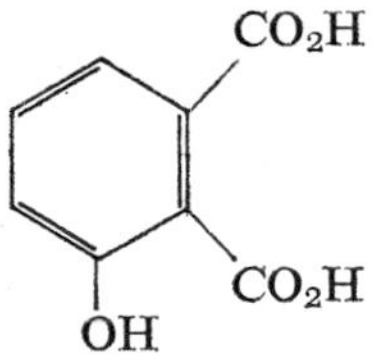

3-hydroxyphthalic acid
P. patulum,[25] *P. islandicum*[26]

L^{27}

28

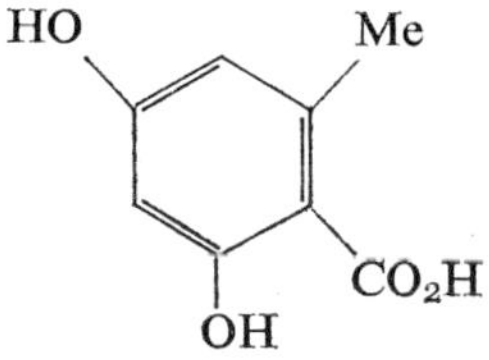

orsellinic acid
Widespread[28]

$L^{29,30\ 31}$

29

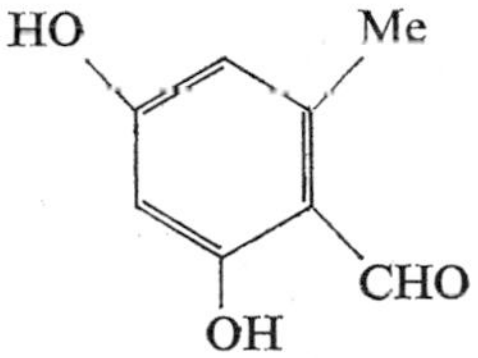

orsellinaldehyde
A. rugulosus[32]

30

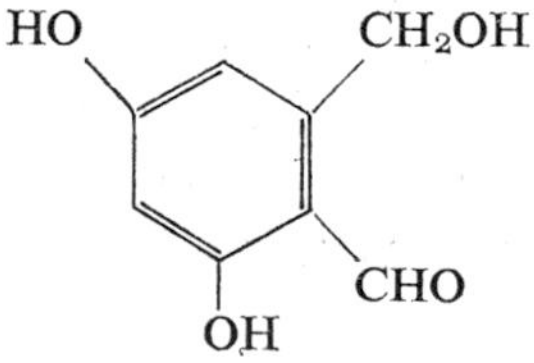

2,4-dihydroxy-6-(hydroxymethyl)-
   benzaldehyde
A. rugulosus[32]

4

31

a: R = H, asperugin B
b: R = Me, asperugin
            A. rugulosus[33,34]

Orsellinic acid (**28**) is the parent compound derived by this cyclization since it retains all the carbon and oxygen atoms (except that lost by dehydration during aromatization) of the polyketide precursor. In spite of the simplicity of its derivation, orsellinic acid was not, in fact, isolated until 1958,[35] though orsellinic acid residues had long been known to be present in lichen products (see p. 108) and a dimethyl derivative, sparassol, had also been isolated from lichens. Orsellinic acid itself has since been obtained from an isolated lichen fungus[36] and has been detected in a number of Fungi Imperfecti. It is rarely produced in high yield, a fact which may be a result of the further transformations which it can undergo (see below). The derivation of orsellinic acid from acetate has been confirmed[29,30,31] by experiments with [¹⁴C]acetate and also[28] with [¹⁸O]acetate, when it was found that the phenolic oxygens and one of the carboxyl oxygens were labelled as expected (one of the carboxyl oxygens would be lost by formation and hydrolysis of the coenzyme A derivative). That malonate is involved in the usual way may be inferred from the study by Bentley and Keil[37] of the biosynthesis of penicillic acid which is derived from orsellinic acid (see p. 106).

6-Methylsalicyclic acid (**25**), formed with loss of one of the oxygen atoms of the tetraketide precursor, is of special interest as the compound with which the "acetate",[23] and later the "acetate–malonate",[24] hypothesis were first tested experimentally. It has also been the subject of experiments with cell-free systems (p. 80) and of studies of the secondary metabolism of fungi (p. 19).

Both 6-methylsalicylic acid and orsellinic acid are important as precursors of other metabolites, the simplest of which are 6-formylsalicylic acid (**26**), 3-hydroxyphthalic acid (**27**), and the aldehydes (**29**) and (**30**). The acetate origin of 3-hydroxyphthalic acid (but not the intermediacy of 6-methylsalicylic acid) has been confirmed;[27] it could be derived by oxidation of a compound with a longer side-chain rather than by oxidation of the methyl group of 6-methylsalicylic acid (see also 3,5-dihydroxyphthalic acid, p. 123). In this case ambiguity exists because the experiments necessary to resolve it have not been carried out. But we shall see that for fumigatin, the *Lentinus degener* quinones, aurantiogliocladin and flavipin, experiments by different groups have given different results, so that there is considerable confusion as to whether 6-methylsalicylic acid or orsellinic acid is the precursor of these compounds.

B.2.b.ii. *With Loss of One Carbon*

32

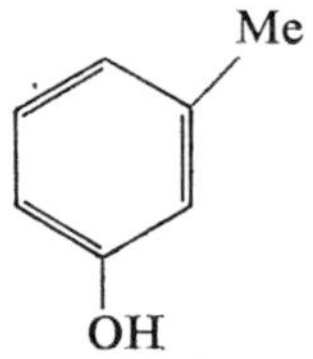

m-cresol L^{38}
A. fumigatus[38]

33

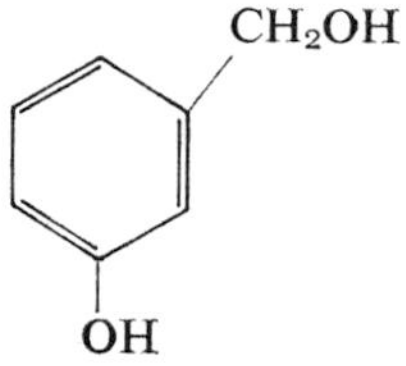

m-hydroxybenzyl alcohol
P. patulum[39]

34

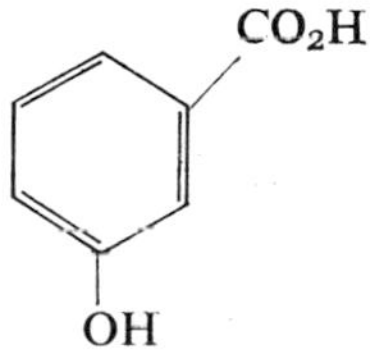

m-hydroxybenzoic acid
P. griseofulvum[40]

35

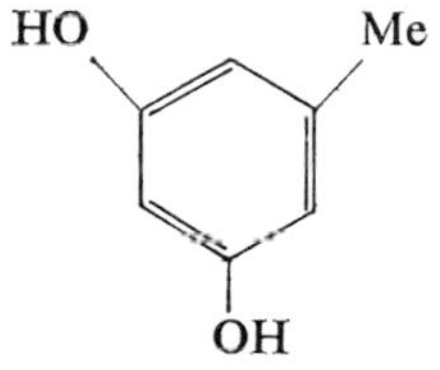

orcinol L^{41}
A. fumigatus[41]

36

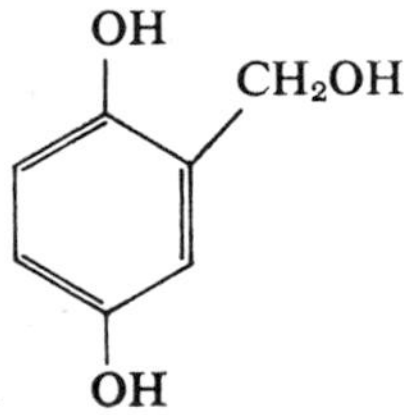

gentisyl alcohol
Penicillium spp.[426] *Phoma* spp.[43]

37

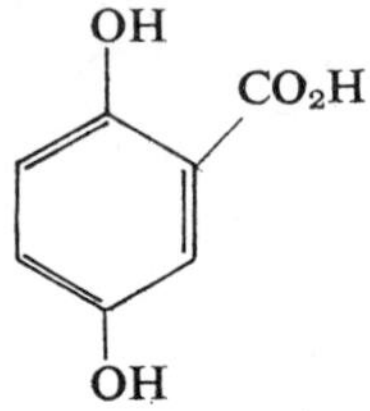

gentisic acid
Penicillium spp.[42]

L^{44}

38

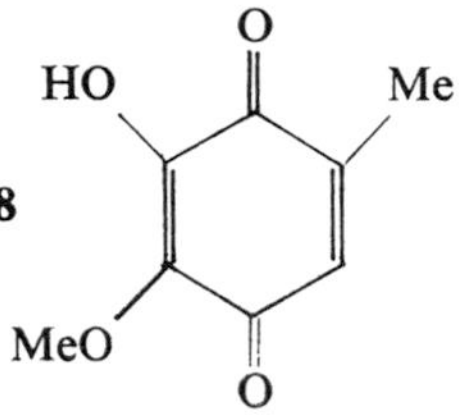

fumigatin
A. fumigatus[45]

L^{46-52}

39

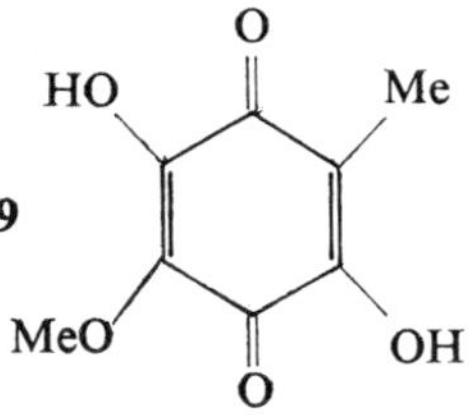

spinulosin
P. spinulosum,[45] *A. fumigatus*[54]

$L^{51,56}$

40

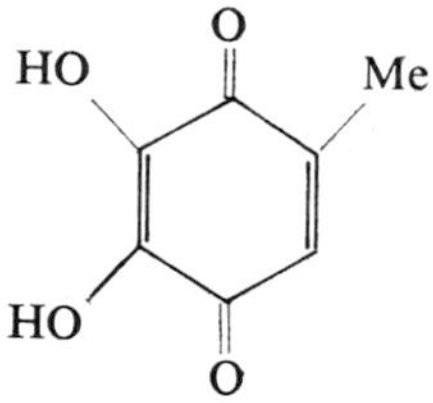

3,4-dihydroxytoluquinone
A. fumigatus[48,50]

$L^{48,50,51,53}$

41

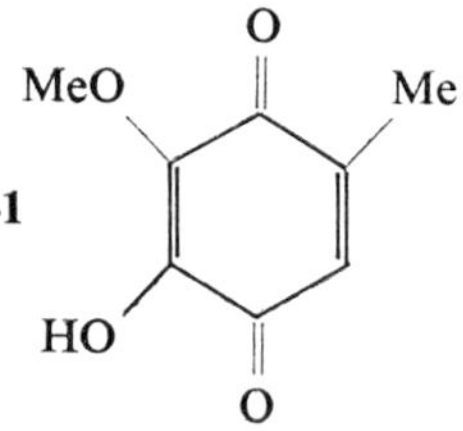

4-hydroxy-3-methoxytoluquinone
A. fumigatus[50]

$L^{50\ 51\ 53}$

42 3-hydroxytoluquinone $L^{50,51,53}$
A. fumigatus[50]

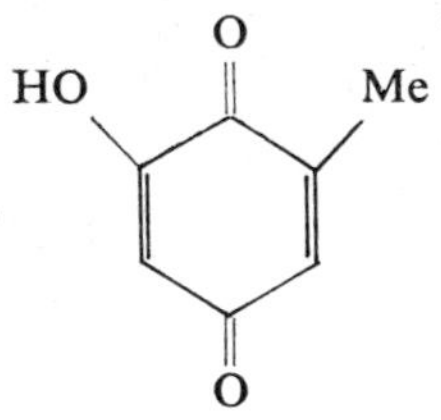

43 3,6-dihydroxytoluquinone L^{50-53}
A. fumigatus[50]

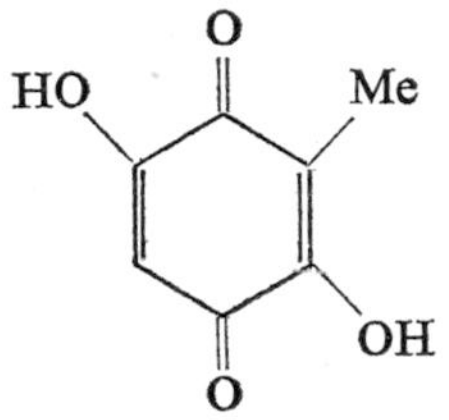

44 4-methoxytoluquinone L^{58-61}
Lentinus degener, Coprinus similis[57]

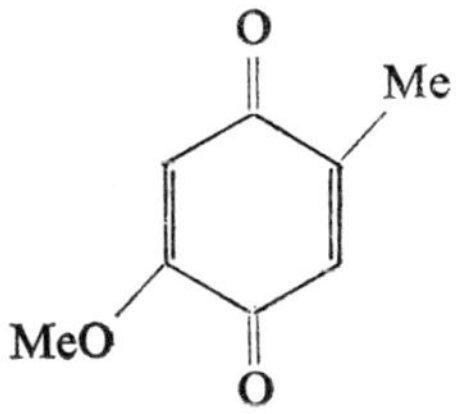

45 6-hydroxy-4-methoxytoluquinone $L^{59,60}$
L. degener[59]

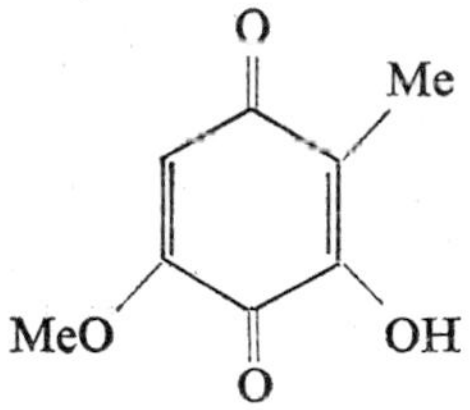

46 terreic acid L^{63}
A. terreus[62]

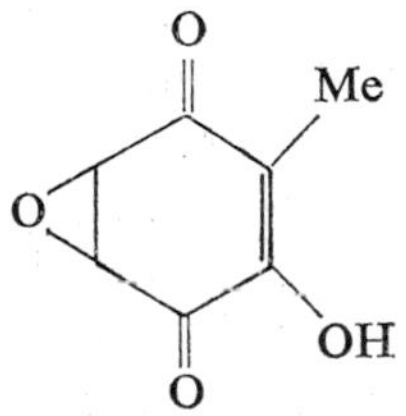

47

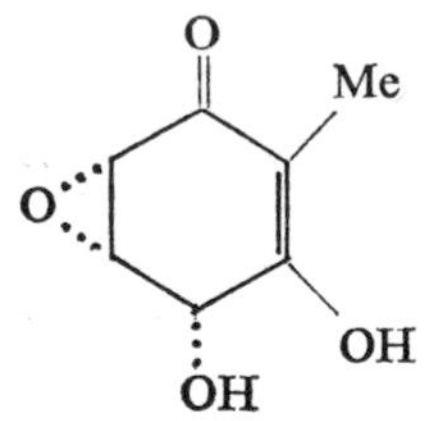

terremutin
A. terreus[64]

48

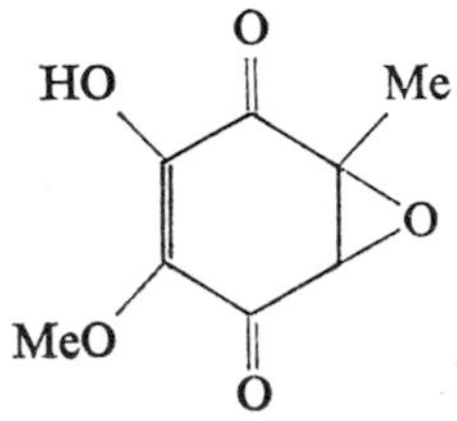

3-hydroxy-4-methoxytoluquinone-
1,6-epoxide
A. fumigatus[65]

49

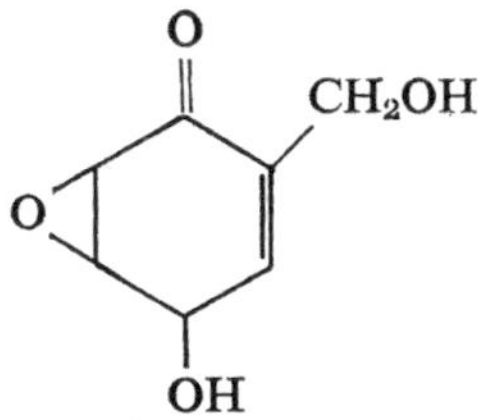

epoxydon, sphaeropsidin
Phoma spp.[43]

The carbon skeletons of these compounds can be derived by decarboxylation of 6-methylsalicylic acid or orsellinic acid, the first products being *m*-cresol (32) and orcinol (35). The process has been demonstrated *in vivo*[47,66,67] and orsellinic acid decarboxylases have been isolated both from a fungus[68] and from a lichen.[69] Nuclear hydroxylation and oxidation of the methyl group of *m*-cresol can give gentisyl alcohol (36) and gentisic acid (37). The acetate origin of gentisic acid in *Penicillium urticae* has been confirmed,[44] but in other fungi it is derived from shikimate (see p. 41).

Further oxidation and *O*-methylation, either before or, more likely, after decarboxylation (see below), gives the well-known quinones fumigatin (38) and spinulosin (39), whose origins have been the subject of detailed study by several groups. As we shall see, although the overall picture seems well established, there are some points of ambiguity and disagreement.

Although the compounds were first isolated in the quinone form (spinulosin was the first fungal product to be recognized as a quinone), it has been suggested[48,70] that the true metabolites are the quinols which undergo non-enzymatic autoxidation in the medium. In support of this Pettersson[71] has shown that the mycelium of *A. fumigatus* actually inhibits the oxidation of fumigatol to fumigatin and has isolated a cell-free system carrying out the reduction of fumigatin by NADH. He suggests that the biological function of the enzyme might be to keep the compounds in the reduced state during the production phase, either as a necessary step in the biosynthesis of fumigatin

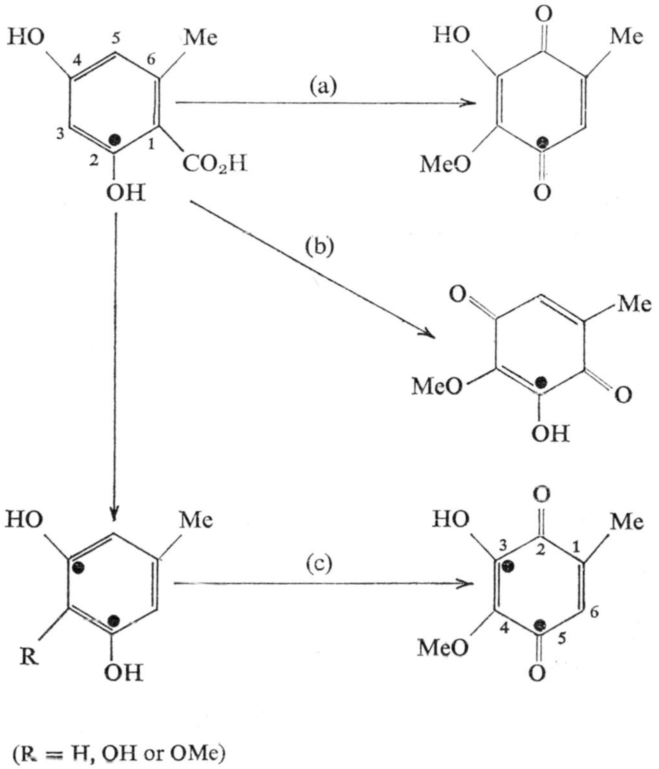

(R = H, OH or OMe)

Scheme 3: Alternative routes for the formation of fumigatin from orsellinic acid; labelling experiments[49] show that route (a) is used

or perhaps to prevent accumulation of toxic quantities of quinones within the cell.† The same situation obtains for the related quinones **(40)** to **(43)** (we have written the compounds in the quinone form for historical reasons).

Pettersson has shown that acetate,[49] methionine,[49] malonate[51] and orsellinic acid[49] are incorporated into fumigatin, but that 6-methylsalicylic

† On the other hand, the presence of this reducing system in the mycelium makes it possible that the quinones are, after all, true metabolites and are reduced before transfer to the medium.

acid and orcinol are not.[52] In the presence of [2-^{14}C]orsellinic acid, fumigatin is labelled exclusively at C-5 (for numbering, see Scheme 3), so that orsellinic acid is converted to fumigatin via pathway (a) of Scheme 3; this experiment also confirms that orsellinic acid is incorporated intact. As a result of incorporation experiments[50,51,53] with fumigatin and the quinones (40) to (43), Pettersson[53] proposed the sequence of Scheme 4 for the biosynthesis of

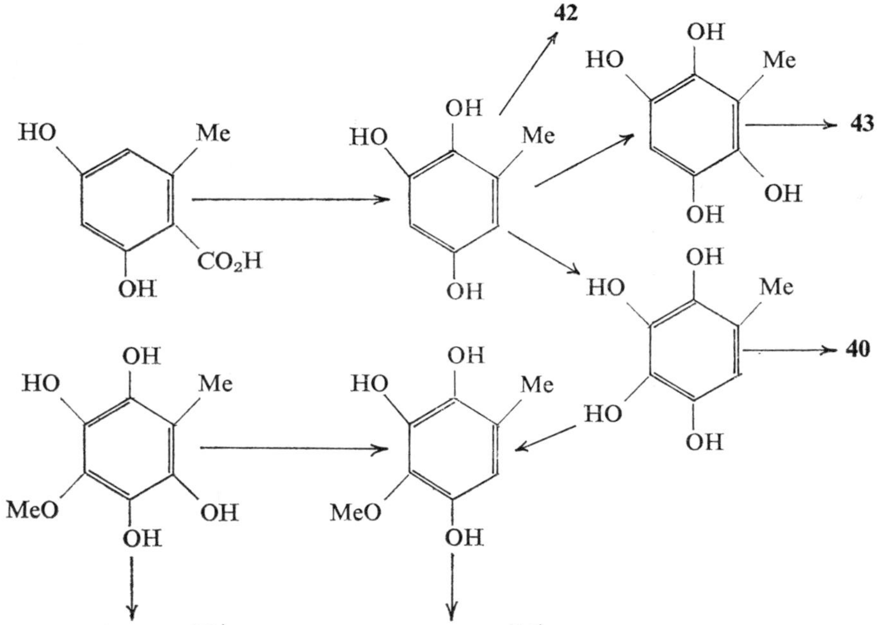

Scheme 4: The biosynthetic relationship of the *A. fumigatus* quinones[53]

the compounds, though he points out that while his results provide strong evidence that two hydroxylation steps precede the *O*-methylation process, they do not exclude the possibility that 2,3,4,5-tetrahydroxy-6-methylbenzoic acid serves as a natural substrate for the *O*-methyltransferase system, i.e. that two hydroxylations precede decarboxylation. The biosynthesis of spinulosin in *P. spinulosum* follows a similar pathway to that in *A. fumigatus*.[56]

Many of the above results have been confirmed by Packter.[46-48] However, in contrast with Pettersson,[52] Packter[47] finds that 6-methylsalicylic acid *is* incorporated into fumigatin. Since incorporation occurs only after a delay and is inhibited by orsellinic acid and orcinol, Packter suggests that 6-methylsalicylic acid is first hydroxylated to orsellinic acid. This being so, it seems unlikely that 6-methylsalicylic acid is a natural precursor of fumigatin, since orsellinic acid is more readily formed directly from acetate and malonate; this may be an example of a "microbiological transformation" (see p. 21). We shall see that there is doubt as to the intermediacy of 6-methylsalicylic acid in the biosynthesis of other members of this class of compound.

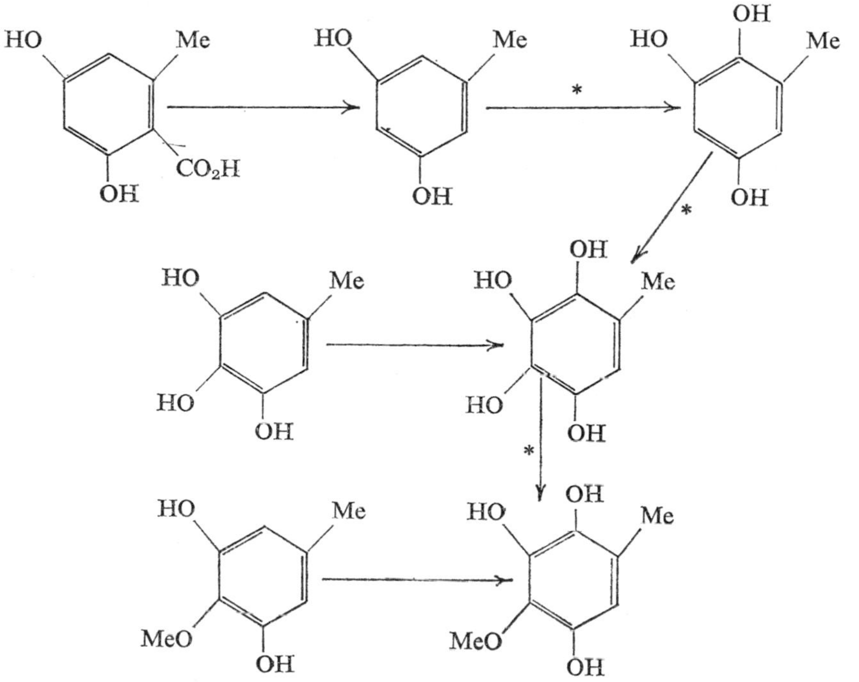

Scheme 5: Transformations observed in *A. fumigatus*.[72] Those marked with an asterisk form the most likely natural sequence

The biosynthesis of fumigatin has also been studied by Simonart and Verachtert[72] by an examination both of the order of appearance of metabolites of *A. fumigatus*, and of transformation of possible intermediates by replacement cultures. Except for the proposed involvement of orcinol, their results, summarized in Scheme 5, are in agreement with those of Pettersson. Though they could not rigidly define the hydroxylation sequence, the accumulation of 6-hydroxyorcinol at the expense of 4,6-dihydroxyorcinol in the presence of isonicotinic acid hydrazide, suggests that 6-hydroxyorcinol is a natural precursor. The other transformations shown in Scheme 5 illustrate the ambiguity of such experiments.

The quinones (44) and (45) appear to be two of the relatively few polyketides to have been isolated from Basidiomycetes. Birch has reported[58] without experimental details, that 6-methylsalicylic acid is a precursor of (44) in *L. degener* and Pettersson[60] has confirmed the incorporation of

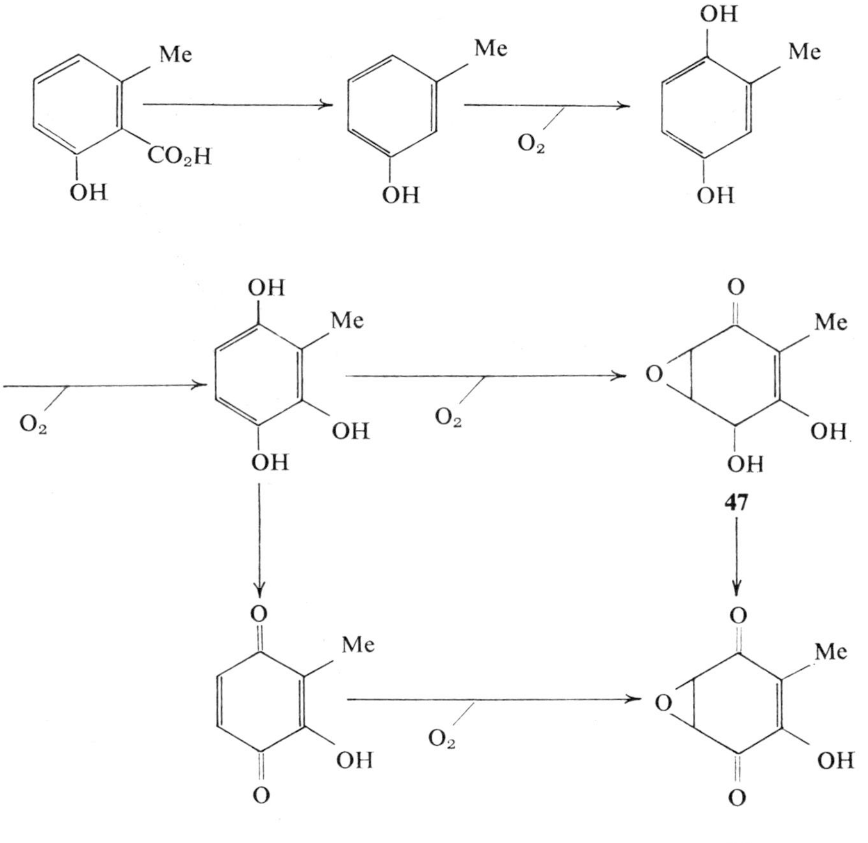

Scheme 6: Biosynthesis of terreic acid (46) from 6-methylsalicylic acid;[63] the epoxidation could occur either at the hydroquinone level to give terremutin (47) as an intermediate, or at the quinone level

generally labelled 6-methylsalicylic acid into (44) and (45) and of specifically labelled 6-methylsalicyclic acid (produced by *P. urticae* in the presence of [2-¹⁴C]malonate) into (45), the major quinone of his strain of *L. degener*. Pettersson also found[59] that a non-pigmented strain of *L. degener* produced 6-methylsalicylic acid and, as with the *A. fumigatus* compounds, that the quinones are formed from the initially produced quinols.

In striking contrast with these results, Packter[61,67] has found that neither acetate nor 6-methylsalicylic acid is incorporated into the quinone (44), but that the aromatic ring and C-methyl group are derived from tyrosine. This is interesting because on chemotaxonomic grounds one would expect a Basidiomycete to use the shikimate rather than the polyketide route to aromatic compounds. A possible explanation for the discrepancy is that the groups have used different organisms (Packter points out that the organism used by Pettersson grew much quicker than is normal for a Basidiomycete). If this is so then the situation would correspond to that observed for gentisic acid which is derived from acetate in *Penicillium urticae* but from shikimate in Basidiomycetes.

Because of the discrepancies between his own and other (notably Packter's) results, discrepancies which will appear again when we discuss aurantio-gliocladin (62) (Section B.2.b.iv), Pettersson has re-examined[60] the incorporation of 6-methylsalicylic acid into quinones by *A. fumigatus, P. spinulosum* and *Gliocladium roseum* (aurantiogliocladin). He found only poor incorporation, possibly after breakdown to acetate. Pettersson points out, very reasonably, that one might expect 6-methylsalicylic acid to be a precursor of those quinones [(44) and (45)] which lack the 3-hydroxyl group, and orsellinic acid to be a precursor of the rest. On balance, this seems to represent a fair summary of the present position, especially if the suggested explanation for the discrepancy observed for (44) proves correct.

The epoxides (46) to (49) are formally related to the toluquinones, and on the basis of time sequence studies with *A. fumigatus* Yamamoto *et al.*[74] have suggested that (48) might be an intermediate between orsellinic acid and fumigatin and that spinulosin is an artefact derived from it. In the light of the experiments discussed above, this seems unlikely. Acetate and 6-methylsalicylate, but not orsellinate, are incorporated into terreic acid (46), and experiments with $^{18}O_2$ suggest the sequence shown in Scheme 6[63], in accord with Pettersson's suggestion that 6-methylsalicylic acid, rather than orsellinic acid, is involved in the formation of toluquinones which lack the 3-hydroxyl group.

B.2.b.iii. *With gain of one or more carbons*

50

4,6-dihydroxy-2,3-dimethyl-benzoic acid   L[66]
A. terreus,[75] *Gliocladium roseum*[66]

5. POLYKETIDES

51

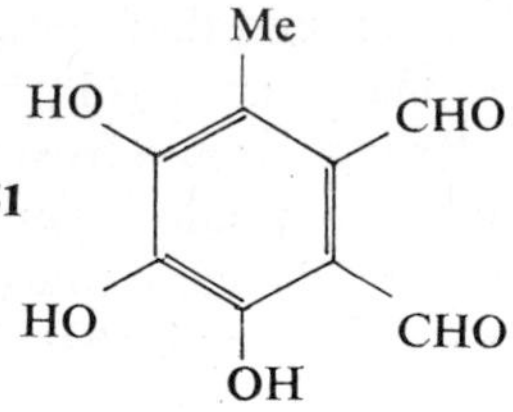

flavipin
A. flavipes[76]

L[77]

52

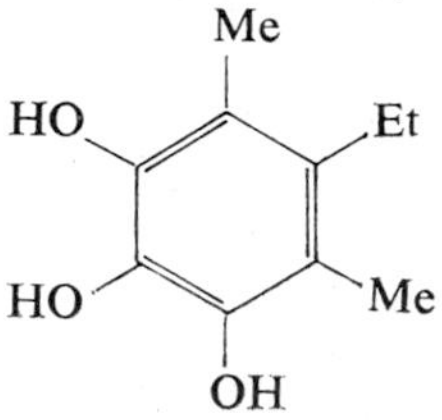

barnol
P. baarnense[78]

L[79]

53

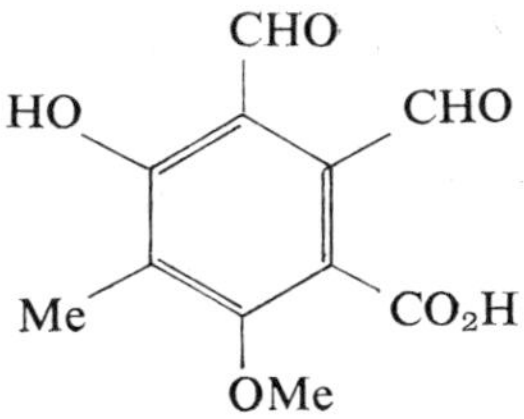

cyclopaldic acid
P. cyclopium[80]

L[81]

54

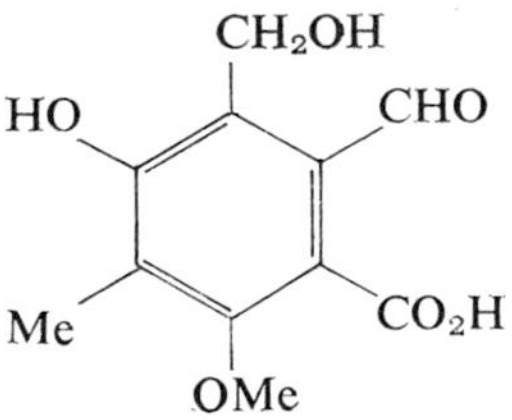

cyclopolic acid
P. cyclopium[81]

55

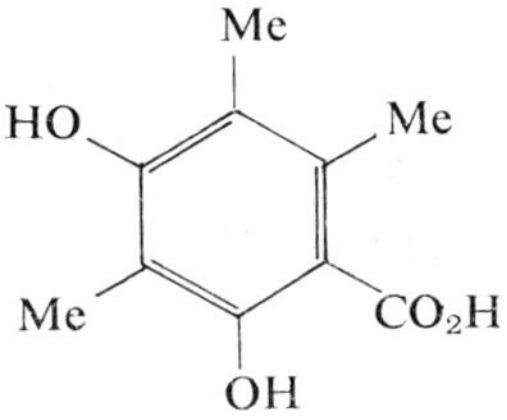

2,3,5-trimethyl-4,6-dihydroxybenzoic acid
Mortierella ramanniana[82]

56

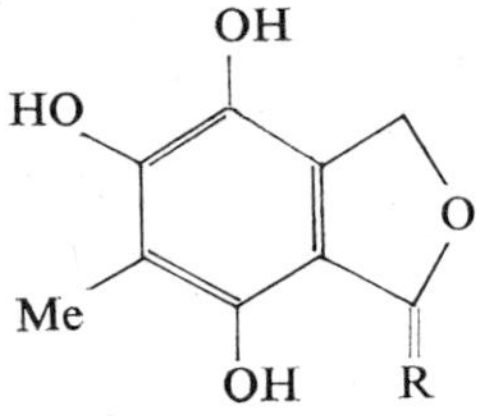

a: R = H$_2$, 1,3-dihydro-5-methyl-4,6,7-
   trihydroxyisobenzofuran
b: R = O, 6-methyl-4,5,7-trihydroxyphthalide
   A. terreus[83]

57

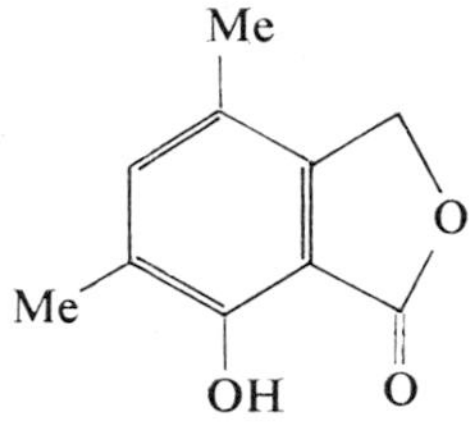

3,5-dimethyl-6-hydroxyphthalide   *L*[85]
P. gladioli[84]

58

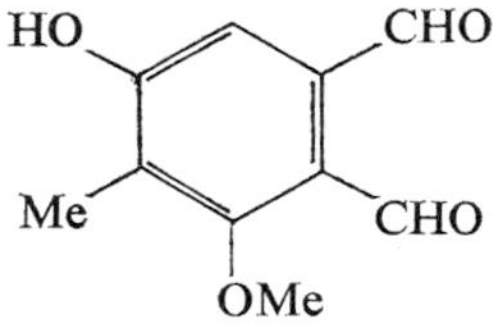

quadrilineatin
A. quadrilineatus[86]

58a

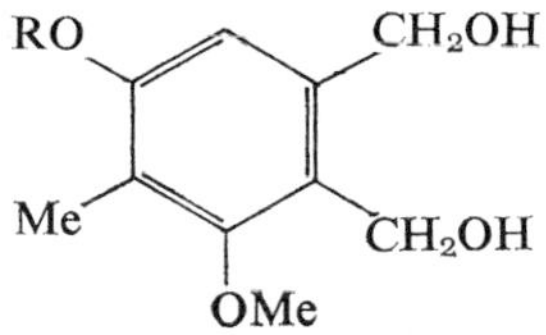

R = Me$_2$C=CHCH$_2$, zinniol
                    Alternaria zinniae[87]

59

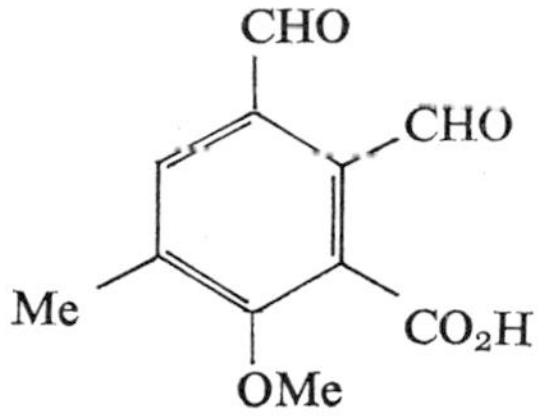

gladiolic acid
P. gladioli[88]

60

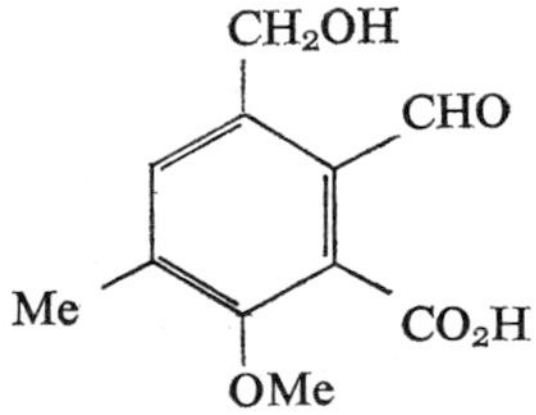

dihydrogladiolic acid
P. gladioli[84]

The simplest member of this group, formally derived from orsellinic acid by introduction of a *C*-methyl group, is 4,6-dihydroxy-2,3-dimethyl-benzoic acid (50) which occurs in small amounts in *A. terreus* mutants isolated in connection with biosynthetic studies on geodoxin, and in *G. roseum* strains which produce aurantiogliocladin. Its origin from acetate/malonate and methionine has been confirmed by Pettersson[66] and it will be discussed further in connection with aurantiogliocladin. The derivation of cyclopaldic acid (53) and of 3,5-dimethyl-6-hydroxyphthalide (57) from acetate, and of flavipin (51) from acetate and methionine, has been confirmed.

Pettersson found that orsellinic acid and 4,6-dihydroxy-2,3-dimethyl-benzoic acid (50) were present in small quantity in cultures of *A. flavipes* one to two days before flavipin production commenced, and that the [*carboxyl*-^{14}C]compounds were incorporated into flavipin. He could not however, obtain direct evidence for the conversion of orsellinic acid to the homologue (50), though the efficient incorporation of orsellinic acid into flavipin suggested that it undergoes direct *C*-methylation (this point is further discussed below). The corresponding aldehydes were incorporated into flavipin more rapidly than the acids, but this may be a permeability effect. Pettersson was unable to conclude at which stage hydroxylation occurs. Recent experiments with a cell-free system from *A. flavipes* suggest that *C*-methylation of orsellinic acid does *not* occur (see p. 8).

Barnol (52) presents two features of interest from a biosynthetic point of view—the complete reduction of a carboxyl group to methyl and the introduction of a methyl group onto what was the methyl group, rather than the more usual methylene group, of a polyketide precursor. The origin of the carbon atoms, as in Scheme 7, was proved by the labelling experiments of Mosbach and Ljungcrantz,[79] who suggest that the methylation might occur on an intermediate quinone-methide. The reduction of a carboxyl group to methyl

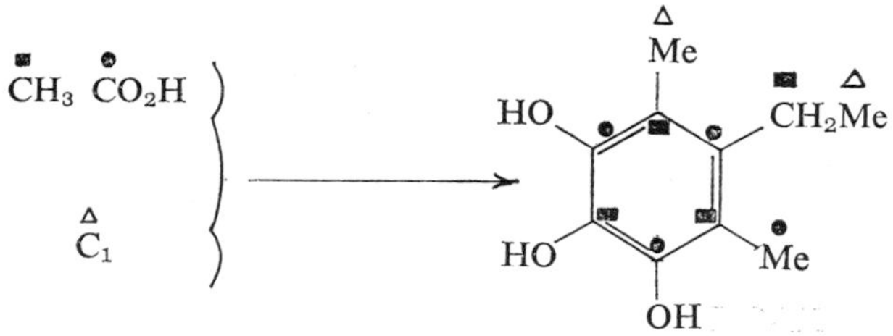

Scheme 7: The incorporation of radioactivity from acetate and methionine into barnol[79]

has been demonstrated in one other compound, javanicin (187), and appears also to have occurred during formation of the tetrahydrofuran (160). The effect of the medium on the production of barnol is discussed on p. 16.

B.2.b.iv. *With Loss of One Carbon and Gain of One Carbon*

61

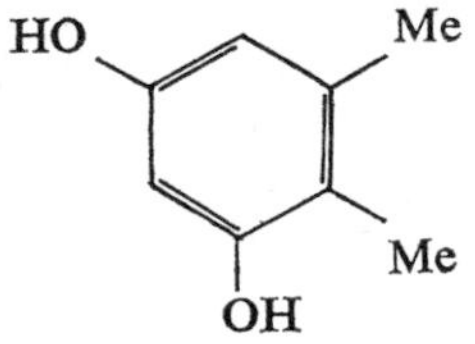

3,5-dihydroxy-1,2-dimethylbenzene L^{66}
G. roseum,[66] *A. terreus*[89]

62

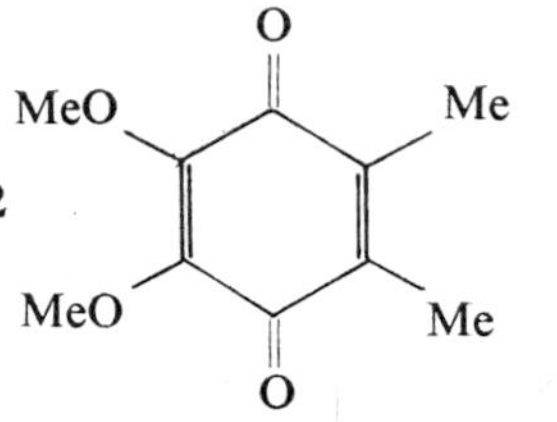

aurantiogliocladin $L^{67,91-93}$
G. roseum[90]

63

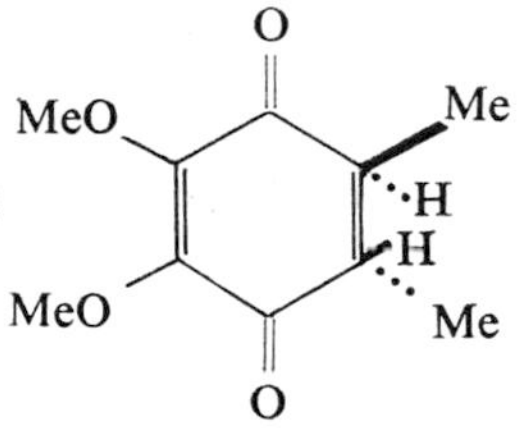

gliorosein $L^{67,91}$
G. roseum[90]

64

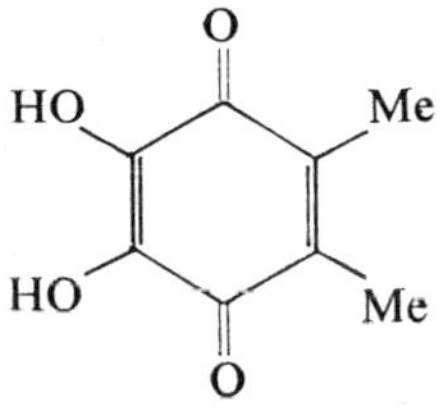

2,3-dihydroxy-5,6-dimethyl- L^{93}
    benzoquinone
G. roseum[94]

65

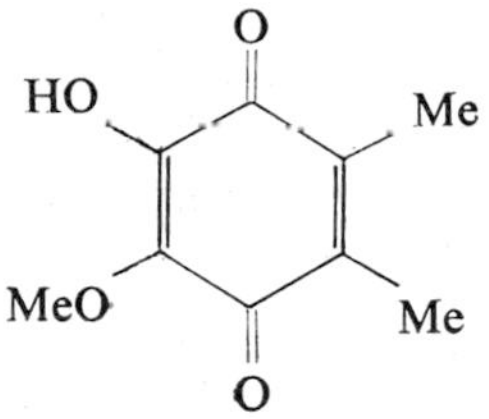

2-hydroxy-3-methoxy-5,6- L^{93}
    dimethylbenzoquinone
G. roseum[94]

66

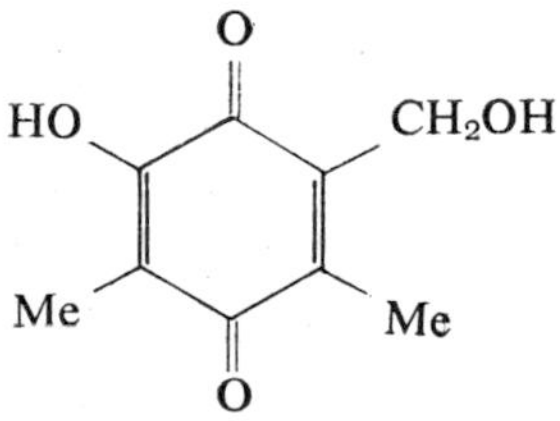

shanorellin
Shanorella spirotricha[95]

The total reduction of a carboxyl group to a methyl group (see preceding paragraph) *appears* to have occurred during the biosynthesis of this group of compounds. In fact it has been shown that the second methyl group in the compounds arises from the C_1-pool, the original carboxyl group having been lost. The biosynthesis of shanorellin (66) appears to involve two methylations and oxidation of the acetate-derived methyl group to hydroxy-methylene.

The derivation of aurantiogliocladin (62) from acetate and the C_1-pool was established by Birch and his co-workers[91] and confirmed by Bentley and Lavate[92] who also showed that malonate is involved in the usual way. Pettersson[93] isolated orcinol and its homologue (61) from *Gliocladium*

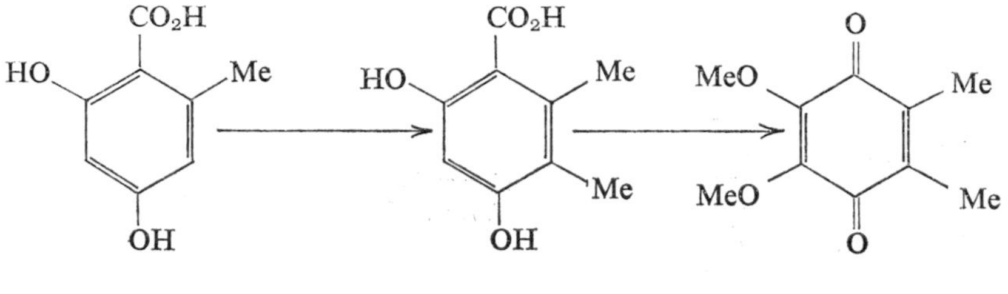

50

Scheme 8: The biosynthesis of aurantiogliocladin from orsellinic acid according to Pettersson

roseum and showed that the organism rapidly decarboxylates orsellinic acid and its homologue (50). He also found, as in the case of fumigatin and the *L. degener* compounds (see above), that aurantiogliocladin and the minor products (64) and (65) are formed in the medium by autoxidation of the corresponding quinols. Orsellinic acid and its homologue (50) were incorporated (0·07 and 0·9% respectively) into aurantiogliocladin without prior breakdown, but the corresponding decarboxy-compounds were not. Nor were the quinones (39), (40), (41), (64) and (65) or the corresponding quinols. As expected, none of the metabolites were labelled in the presence of [*carboxyl*-14C]orsellinic acid or its [*carboxyl*-14C]homologue. Pettersson suggests that the biosynthesis of aurantiogliocladin follows the route of Scheme 8.

Packter and Steward[67] have studied the biosynthesis of gliorosein (63), and since they find that gliorosein is formed from aurantiogliocladin (glioro-sein is optically active so that the process is enzymatic) their observations apply equally to aurantiogliocladin. Unlike Pettersson they find that quinone (65) and the corresponding quinol *are* incorporated into gliorosein. They suggest that 4,6-dihydroxy-2,3-dimethylbenzoic acid (50) is the first *aromatic* compound to be formed (i.e. *C*-methylation occurs at a pre-aromatic stage) and that the biosynthesis of gliorosein follows the route shown in Scheme 9.

Steward and Packter[97] subsequently found, again in contrast with Pettersson,[93] that 4,6-dihydroxy-2,3-dimethylbenzoic acid (50) was *not* incorporated into gliorosein whereas the corresponding aldehyde was incorporated to the extent of 36%; the acid (50) was rapidly decarboxylated (as Pettersson had already found). They also found that the quinone (64) was incorporated to the extent of 20% (whereas Pettersson observed no incorporation). They again suggest that 4,6-dihydroxy-2,3-dimethylbenzoic acid (50) is the first aromatic compound to be formed, that it is present only in a bound form, and that as soon as it is released from the enzyme it is decarboxylated. A

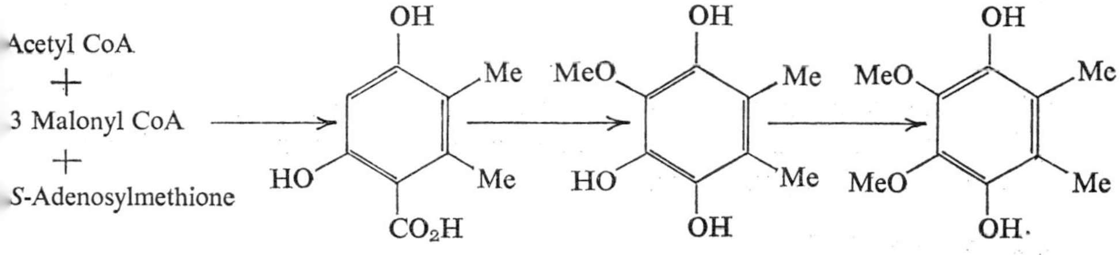

50

Scheme 9: The biosynthesis of aurantiogliocladin according to Packter and Steward; orsellinic acid is not thought to be an intermediate

similar hypothesis of enzyme-bound substrates has been advanced to account for the non-incorporation of some of the likely intermediates in the biosynthesis of griseofulvin (see p. 144). Steward and Packter point out that the percentage incorporation of orsellinic acid and its homologue (50) into aurantiogliocladin in Pettersson's experiments is very low and might be accounted for by an impurity in the product.

The present position concerning the steps involved in the biosynthesis of aurantiogliocladin and gliorosein is thus most unsatisfactory and is further confused by Birch's claim,[58] without supporting evidence, that 6-methylsalicylic acid, but not orsellinic acid, is incorporated into aurantiogliocladin.

Steward and Packter[97] have also made a detailed study of the *C*- and *O*-methylations involved in the biosynthesis of gliorosein. They found that the labelling of the *C*-methyl and *O*-methyl groups was unequal (in agreement with Birch *et al.*[91]). More important, using [*methyl*-³H₃,¹⁴C]methionine they found that whereas the *O*-methyl groups retained three tritium atoms as expected, the *C*-methyl group retained only two. This result is surprising, and prompted a re-investigation of the problem[98] by mass spectroscopic analysis of aurantiogliocladin produced in the presence of [*methyl*-²H₃]methionine;

in this case the results are consistent with retention of three deuterium atoms during both *O*- and *C*-methylation.

Aurantiogliocladin is structurally related to coenzyme **Q** (**67**), in which one of the methyl groups of aurantiogliocladin is replaced by a polyisoprene chain, and Birch[91] has suggested that the aromatic ring of both compounds

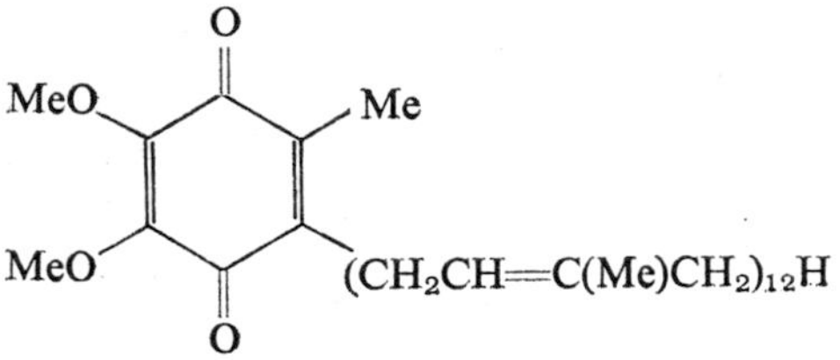

67

might arise by the same pathway. However, Bentley[92] has shown that the quinone ring of coenzyme Q, isolated from *G. roseum*, is not derived from acetate, but probably via shikimate.

B.2.b.v. *Dimeric Compounds*

68   oosporein   *L*[99,10]
              Widespread[99]

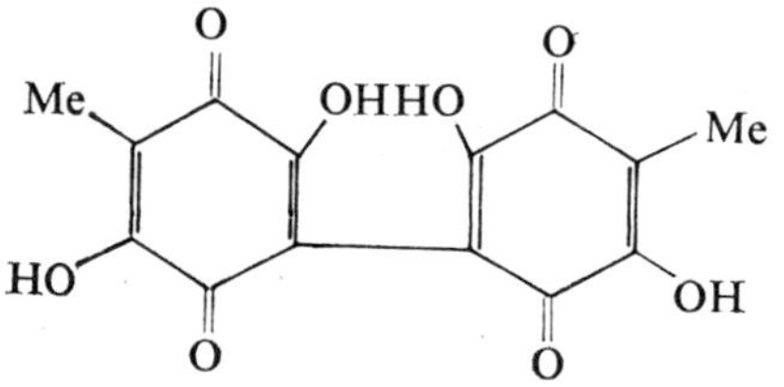

69   phoenicin   *L*[106]
              Penicillium phoëniceum,
                  P. rubrum[101]

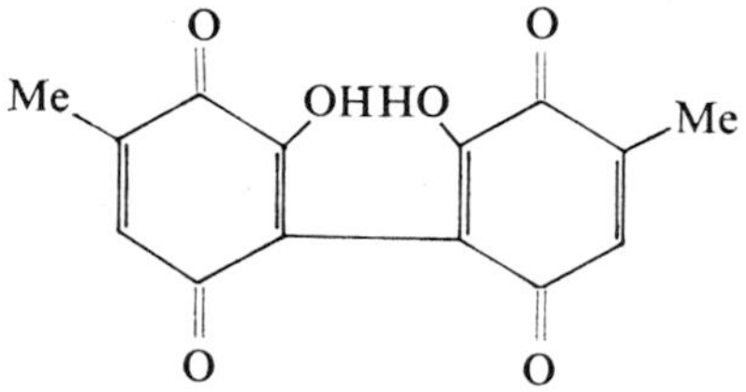

70   CO₂H a: R = H, pannaric acid
                    Crocynea membranacea[102]

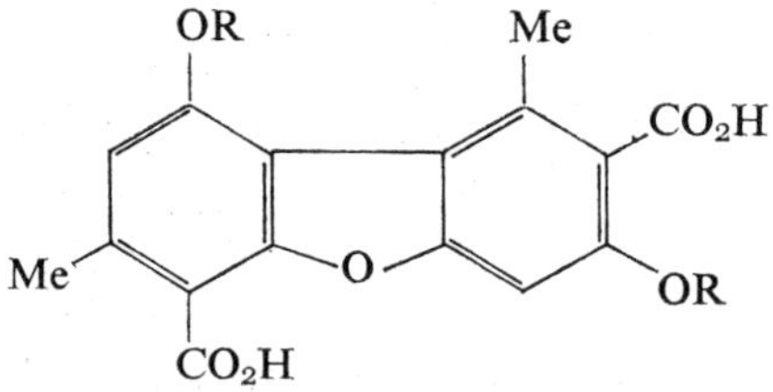

              b: R = Me, schizopeltic acid
                    Shizopelta californica[103]

71 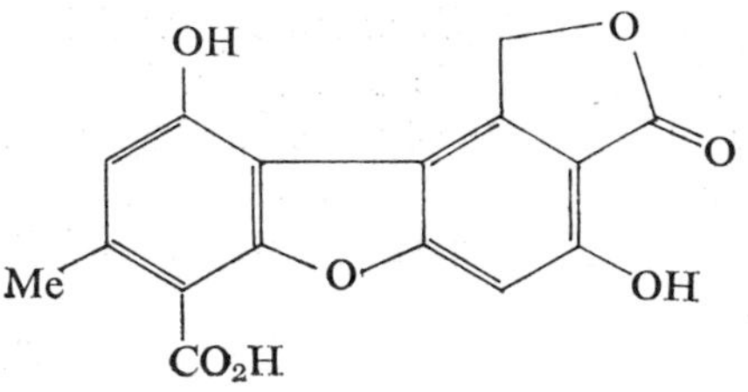   porphyrilic acid
      Haematomma porphyrium[104]

72 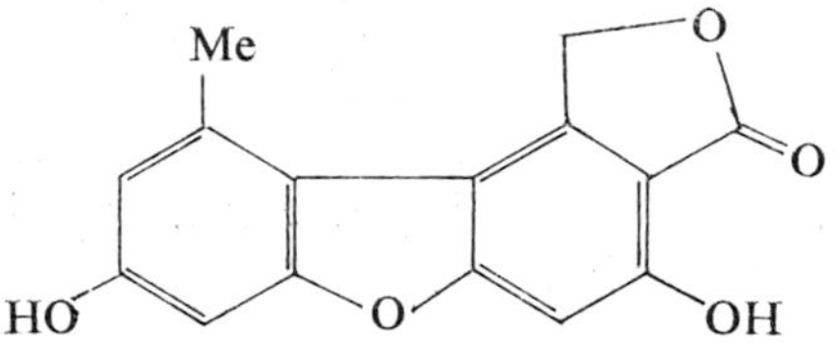   strepsilin
      Cladonia strepsilis[105]

Oosporein (**68**) has been obtained from several Fungi Imperfecti and from one genus—*Phlebia*—of the Basidiomycetes (see ref. 99). The labelling experiments discussed below show that the two dibenzoquinones are formed by coupling of precursors of the 6-methylsalicylic acid/orsellinic acid type. The dibenzofurans are presumably formed by coupling followed by ether formation (cf. usnic acid, p. 111). Although dibenzofurans have been isolated only from lichens their production has been observed in isolated mycobionts.

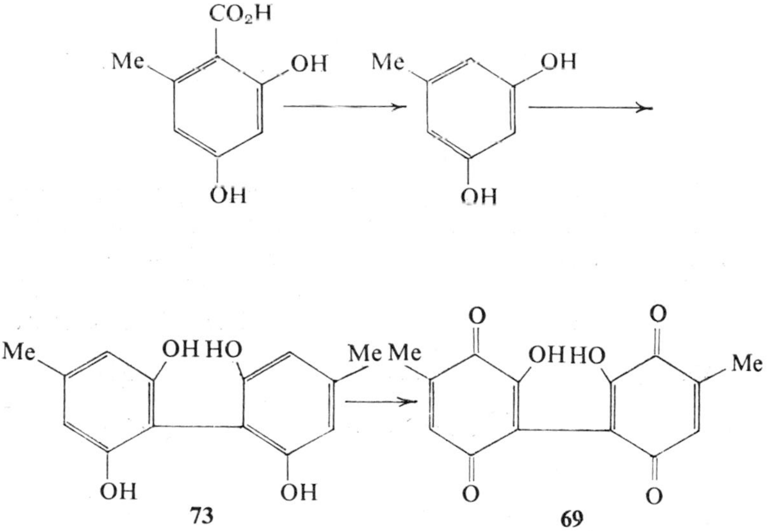

Scheme 10: The biosynthesis of phoenicin (**69**)

Charollais *et al.*[106] have shown that in *P. phoeniceum* phoenicin (**69**) is formed from acetate, and that [1-^{14}C]orsellinic and (to a lesser extent) [6-^{14}C]orcinol are specifically incorporated into phoenicin. They suggest the sequence of Scheme 10 which is supported by the observation that addition of the ditolyl (**73**) leads to increased yields of phoenicin, whereas addition of 2,3,5-trihydroxytoluene does not.

Essentially similar results were obtained by El Basyouni and Vining[99] from a study of the biosynthesis of oosporein (**68**) in *Beauvaria bassiana*. The incorporation of orsellinic acid into oosporein was very efficient (66%) but the incorporation of orcinol (8%) was again lower. The lower incorporation of orcinol in each case may be the result of permeability effects, but it is possible that coupling of orsellinic acid or of orcinol represent two alternative pathways to dibenzoquinones, i.e. that a metabolic grid exists. Experiments with cell-free systems might resolve this point.

B.2.b.vi. *Compounds Formed by Ring-opening*

74

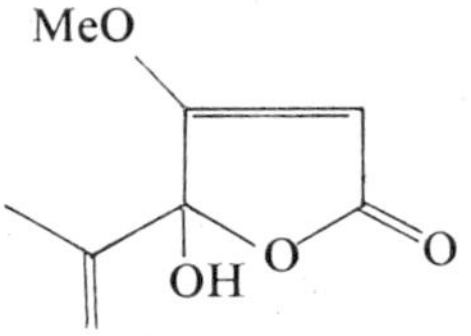

penicillic acid L[26,37,109,110]
P. puberulum,[107] and others

75

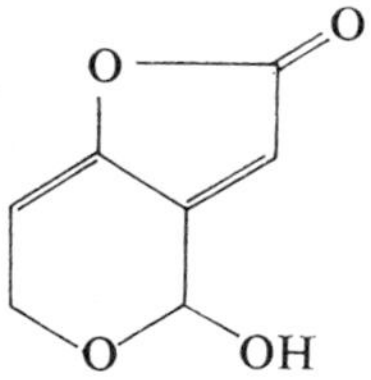

patulin L[113,116]
P. patulum, and others

Penicillic acid was isolated in 1913 from *Penicillium puberulum*[107] and its structure was established by Raistrick and his co-workers[108] who obtained it from *P. cyclopium*. The acetate origin of penicillic acid was established by Birch *et al.*[109] who suggested that it might be derived from orsellinic acid (Scheme 11). Mosbach[110] confirmed that orsellinic acid is converted to

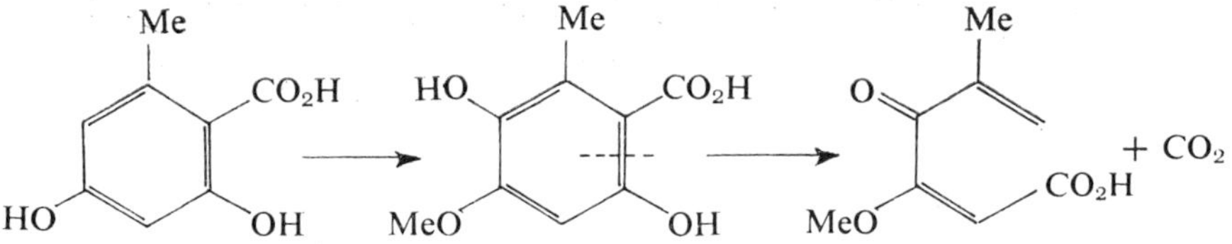

Scheme 11: Suggested derivation of penicillic acid from orsellinic acid[109]

penicillic acid by *P. baarnense*, but found that cleavage of the aromatic ring is as shown in Scheme 12 and not as suggested by Birch *et al*. Similar results (but no evidence concerning the ring-opening) were obtained using orsellinic acid formed in the presence of labelled acetate,[26] and Bentley and Keil[37]

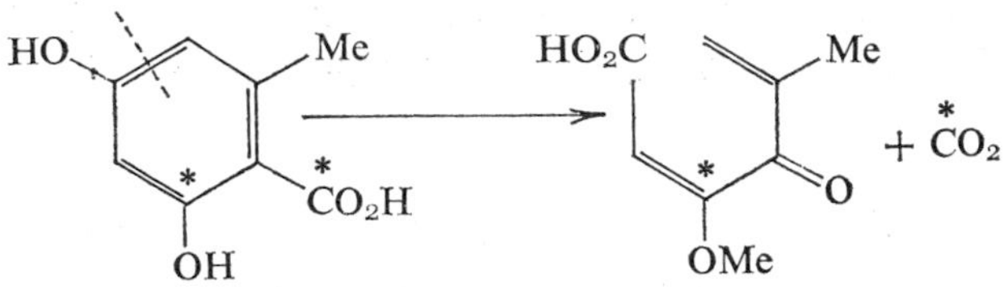

Scheme 12: The incorporation of labelled orsellinic acid into penicillic acid[110]

have shown the involvement of malonate (this was the first demonstration that malonate is involved in polyketide biosynthesis).

Patulin (**75**) has been isolated (under a variety of names) from many members of the Fungi Imperfecti as a result of its pronounced antibacterial activity *in vitro*, and its structure was finally elucidated by Woodward and

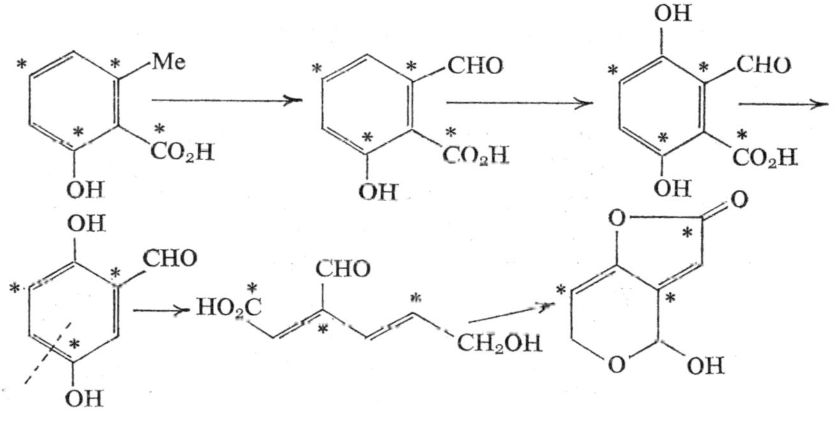

Scheme 13: The formation of patulin from 6-methylsalicylic acid[113]

Singh.[111] The first suggestion that patulin might be derived from an aromatic precursor was made by Birkinshaw[112] and this was confirmed by Bu'Lock and Ryan[113] who showed that 6-methylsalicylic acid is converted to patulin by *P. patulum* and suggested the sequence shown in Scheme 13. A similar sequence was proposed independently by Bassett and Tanenbaum[114] on the basis of the sequence of formation of the compounds by *P. patulum*, but later work with a cell-free system[115] led them to suggest that 6-methyl-salicylic acid is not a precursor of patulin.

More recently, Scott and Yalpani[116] have used deuterium-labelled precursors and mass spectroscopic analysis of the products to study the biosynthesis of patulin. They found that *m*-cresol is converted to patulin and that deuterium from *m*-cresol is also incorporated into toluquinol, gentisyl alcohol and gentisaldehyde, suggesting the sequence of Scheme 14.

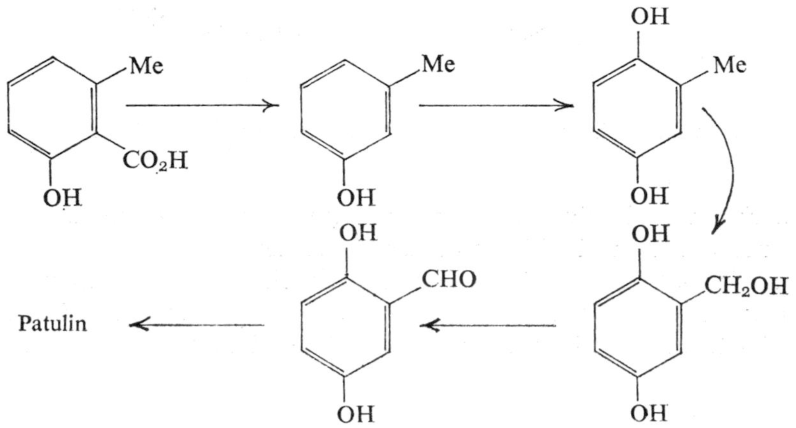

Scheme 14: Suggested sequence between 6-methylsalicylic acid and patulin[116]

B.2.b.vii. *Depsides and Depsidones*

The depsides are esters of polyketide aromatic acids with polyketide phenols, e.g. lecanoric acid (76) in which two molecules of orsellinic acid are linked through an ester group. The depsidones, e.g. variolaric acid (77), are compounds in which ether formation in a depside has produced a seven-membered ring. In some cases, e.g. gyrophoric acid (79), three aromatic units are involved. With the exception of nidulin the depsides and depsidones are produced only by lichens. About sixty are now known and a full list is not given in this book; instead a few examples are given in the appropriate places in this Chapter to illustrate the various cyclization patterns—here, for example, we are concerned with orsellinic acid types. The most recent complete list is that of Huneck.[117]

The biosynthesis of lecanoric acid (76) from acetate,[118] of atranorin (78) from acetate and formate,[118] and of gyrophoric acid (79) from malonate[119] has been confirmed. Since tritiated orsellinic acid was incorporated into lecanoric acid but not into atranorin while tritiated 2,4-dihydroxy-3,6-dimethylbenzoic acid was incorporated into atranorin but not into lecanoric acid,[120] it seems that the C_1-units of atranorin are introduced at a pre-aromatic stage.

It has often been suggested that the aromatic units of the depsides are elaborated by the mycobiont and the ester linkage formed by the phycobiont. However the recent production of a depsidone, salazinic acid, by the isolated

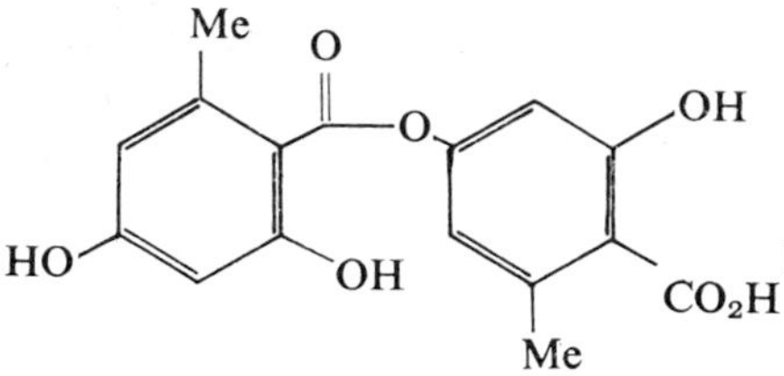

76 Lecanoric acid 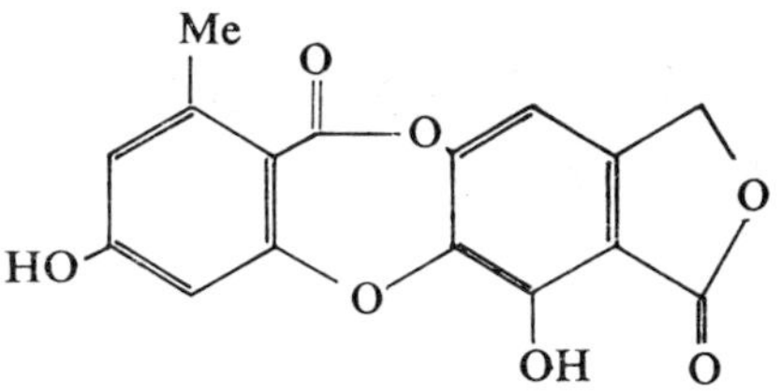   **77** Variolaric acid

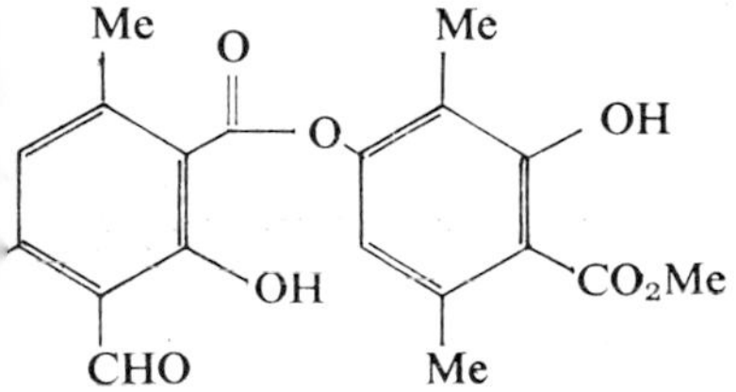

78 Atranorin 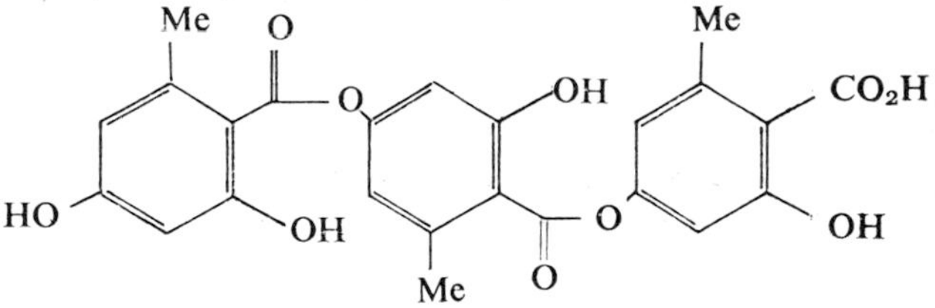   **79** Gyrophoric acid

mycobiont of *Ramalina crassa*[121] shows that, in this case at least, the mycobiont can carry out both processes.

B.2.c. *Compounds of Type*

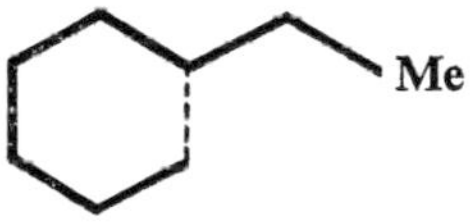

80 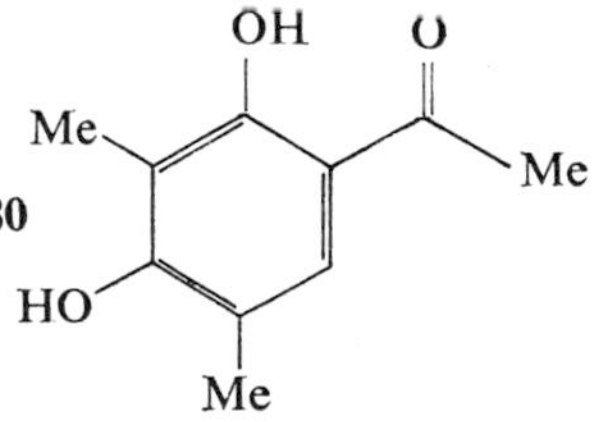   clavatol   L[126]
                                                 Aspergillus clavatus[122]

81 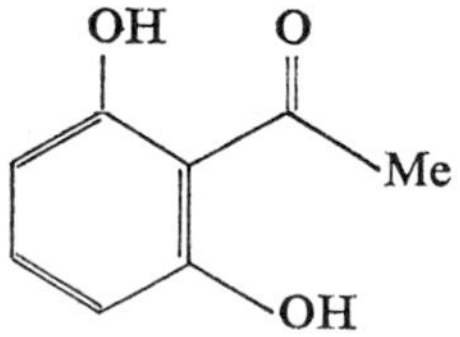   2,6-dihydroxyacetophenone
                                                 Daldinia concentrica[123]

82   usnic acid
Widespread in lichens[124]

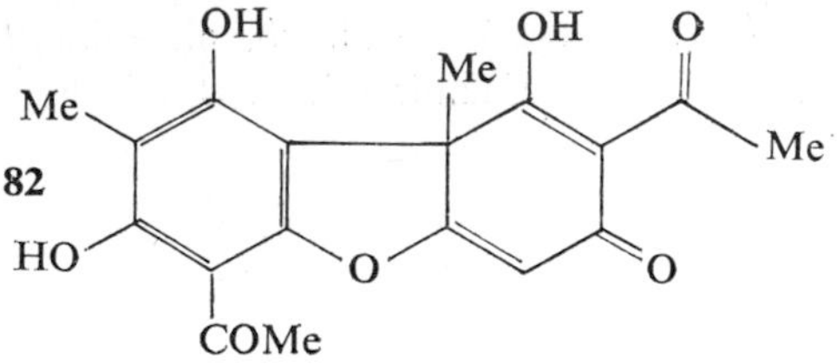

83   isousnic acid
Cladonia spp.[125]

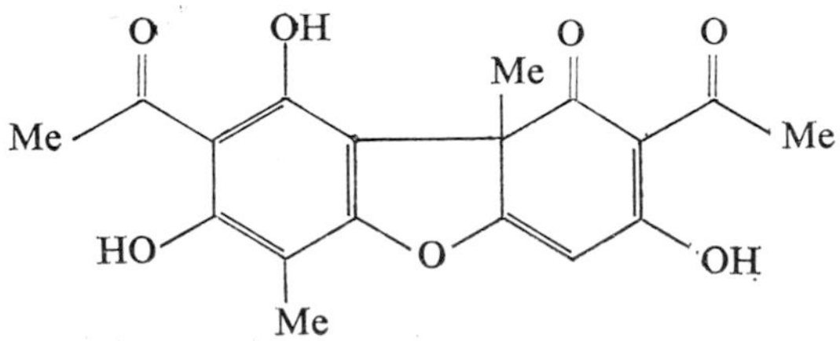

2,6-Dihydroxyacetophenone might be derived from the β-dicarbonyl derivative (**139**) with which it co-occurs in *D. concentrica*, rather than from a tetraketide precursor. Radioactivity from [2-^{14}C]malonate and [*methyl*-^{14}C]-methionine is incorporated into clavatol (**80**) in the expected manner. [1′-^{14}C]2,4-Dihydroxyacetophenone (**84**) and [1′-^{14}C]2,4-dihydroxy-3-methyl-acetophenone (**85**), though metabolized by *A. fumigatus*, are not incorporated

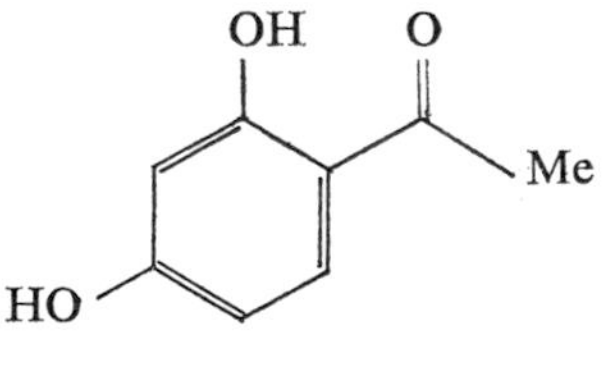

84 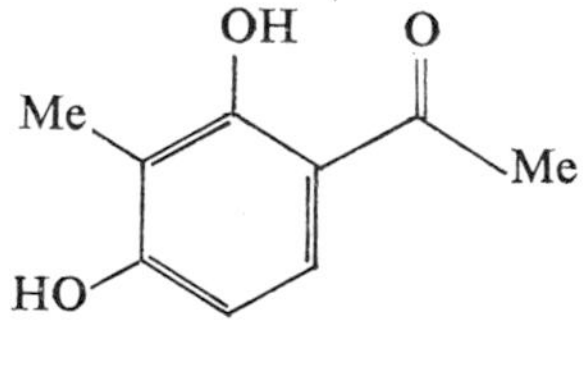   **85**

into clavatol, suggesting that in the biosynthesis of clavatol methylation occurs before cyclization and inhibits the more usual cyclization to a compound of the orsellinic acid type. On the other hand, replacement cultures of *Streptomyces rimosus* (a tetracycline-producing organism) convert compounds (**84**) and (**85**) to clavatol, so that this organism is capable of aromatic methylation (though in the biosynthesis of the tetracyclines, methylation seems to occur at a pre-aromatic stage).

Both (+)- and (−)-isomers of usnic acid (**82**) are of common occurrence among lichens. (+)-Isousnic acid (**83**) co-occurs with (+)-usnic acid in

Cladonia spp., but not in *Usnea, Parmelia* or *Evernia* spp., and (−)-isousnic acid co-occurs with (−)-usnic acid in *C. pleurota*. Usnic and isousnic acids can result from alternative cyclizations of the dimeric intermediate (**86**) (Scheme 15). The polyketide derivation of usnic acid has been confirmed[127]

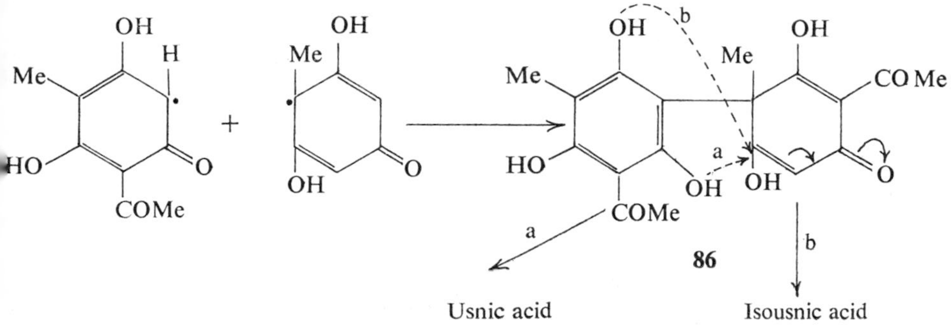

Scheme 15: The biosynthesis of usnic acid by oxidative coupling.

using labelled precursors applied either in Czapek–Dox medium shaken with fresh lichen or in aqueous ethanol dropped onto lichen thallus. The presence in the lichen of the monomer, methylphloroacetophenone, was detected by isotope dilution. [1′-¹⁴C]Methylphloroacetophenone was incorporated into usnic acid but [1′-¹⁴C]phloroacetophenone was not, suggesting that methylation occurs at a pre-aromatic stage as in clavatol biosynthesis.

B.2.d. *The Tropolones*

87

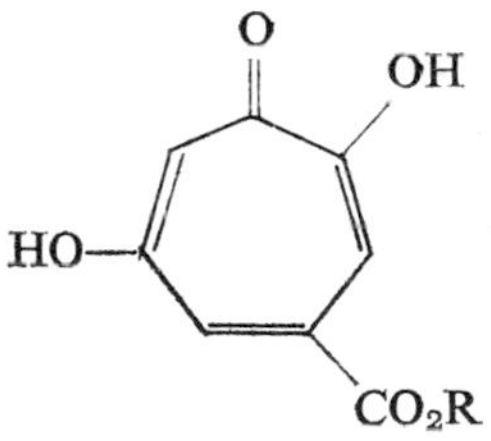

a: R = H, stipitatic acid
        P. stipitatum[128]

b: R = Et, ethyl stipitatate
        P. stipitatum[137]

L[129,130,139]

88

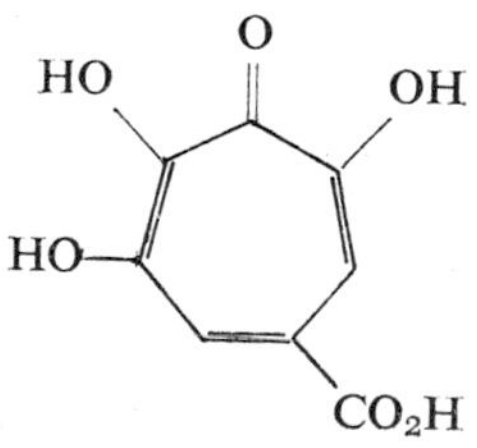

puberulic acid
Penicillium spp.[131]

L[132]

89 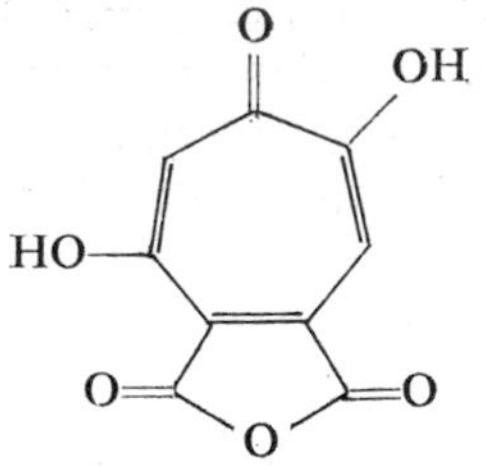   stipitatonic acid $L^{129,130}$
P. stipitatum[131]

90 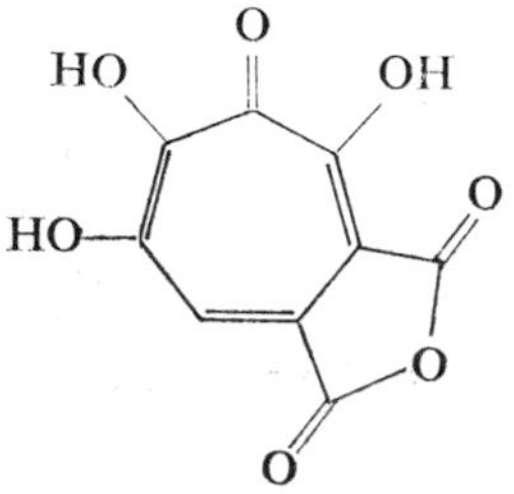   puberulonic acid L^{132}
Penicillium spp.[131]

Stipitatic acid (**87a**) was isolated from *P. stipitatum* by Raistrick and his collaborators, who carried out extensive chemical degradations without arriving at a structure for the compound. They did, however, note its resemblance to puberulic acid (**88**) which had been isolated, along with puberulonic acid (**90**), from *P. puberulum* some years earlier by Birkinshaw and Raistrick.[131] The correct structure of stipitatic acid was proposed in 1945 by Dewar,[134] who showed that the published properties of the compound could be explained in terms of the hitherto unrecognized pseudoaromatic system for which he coined the name "tropolone". The structure of puberulic acid was deduced by Todd and his co-workers[135] and the correct structure for puberulonic acid was first proposed by Aulin-Erdtmann.[136] More recently, Segal[133] isolated stipitatonic acid from *P. stipitatum* and Tanenbaum and his co-workers isolated ethyl stipitatate from the same organism.[137] Sepedonin, a pentaketide tropolone, is discussed on p. 134.

As well as the chemical interest of their structures, the fungal tropolones present a fascinating biosynthetic problem, aspects of which remain unsolved. The biosynthesis of stipitatic and stipitatonic acids has received the greatest attention and two facts seem firmly established[129,130,133,138,139] (Scheme 16)—that the compounds are formed from acetate, malonate and methionine, and that stipitatonic acid is a precursor of stipitatic acid.

Robinson[140] suggested that the tropolone nucleus might arise by introduction of a C_1-unit onto a preformed aromatic ring followed by ring-expansion, and this proposal was further discussed by Seshadri[141] who suggested that the aromatic precursor might be 3,5-dihydroxyphthalic acid (97), a metabolite of *P. brevi-compactum*. Following its isolation from *Penicillium* spp., orsellinic acid seemed a more likely precursor (see e.g.[129]). The possibility was strengthened by the observation of Bentley *et al.*[142] that addition of ethionine,

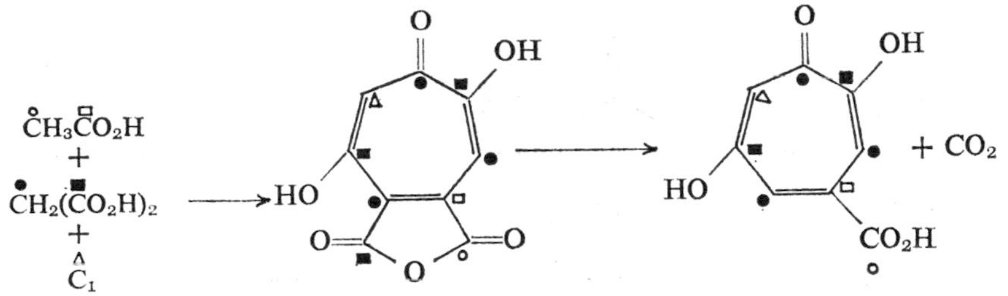

Scheme 16: The incorporation of labelled precursors into stipitatic acid (87a)

a known inhibitor of biological methylation, to cultures of *P. stipitatum* inhibited the production of tropolones and led to the formation of orsellinic acid. On the other hand,[129] labelled orsellinic acid was not incorporated into stipitatic acid, and a further examination[16] of the ethionine-inhibited cultures revealed the presence of triacetic acid lactone (17b) and tetra-acetic acid lactone (22). Since the lactone (22) was readily converted *in vitro* to orsellinic

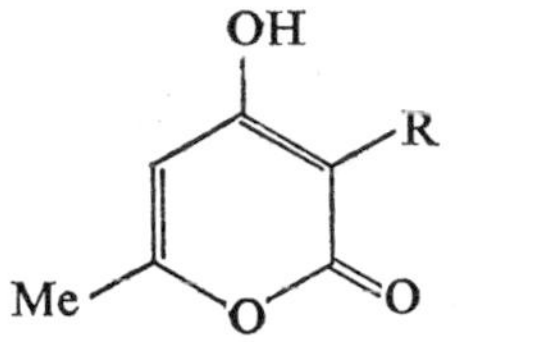

17   a: R = Me
     b: R = H

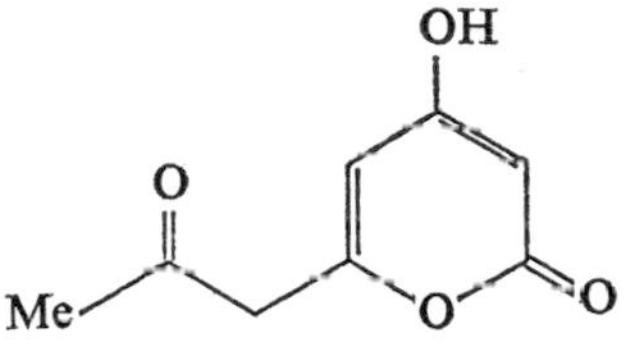

22

acid, it seemed possible that it is the true precursor of the tropolones and that orsellinic acid is a reversion product. However, a study[143] of the incorporation of [14C]-labelled lactones (17b) and (22) into stipitatic acid and into ergosterol showed that incorporation occurred only after breakdown to acetate. It would appear[143] that under normal conditions the polyketo-methylene intermediates remain bound to an enzyme complex during methylation and rearrangement (e.g. Scheme 17), and are only stabilized and released as lactones (17b) and (22) when the methylation process is blocked.

Similarly, the lactone (17a), which appears in fermentations of *P. stipitatum* before the tropolones,[15] might represent release of a triketide precursor before the final malonate unit has been incorporated. Evidence for the

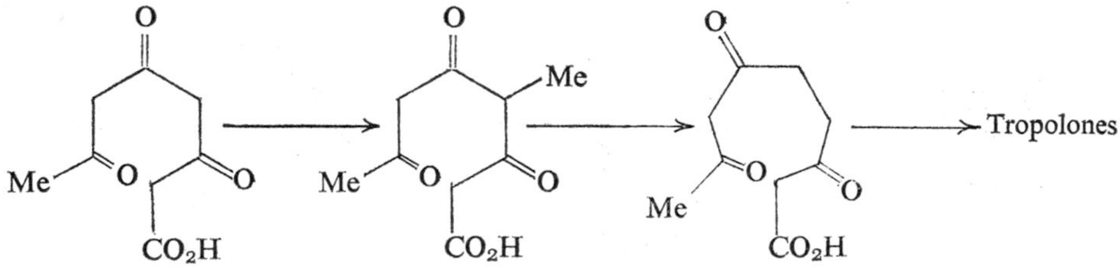

Scheme 17: Possible formation of the tropolone ring-system from an open-chain precursor

common origin of the methyl group of lactone (17a) and the C_1-derived carbon of the tropolone ring has been obtained by Marx and Tanenbaum,[144] who showed that the compounds had similar specific molar radioactivities when produced in the presence of [^{14}C]formate and that the individual C_1-derived carbon atoms, obtained by degradation, had almost identical radioactivities.

As in the case of stipitatic acid, the tropolone ring of puberulic (88) and puberulonic acid (90) is derived from acetate and formate.[132] However, the

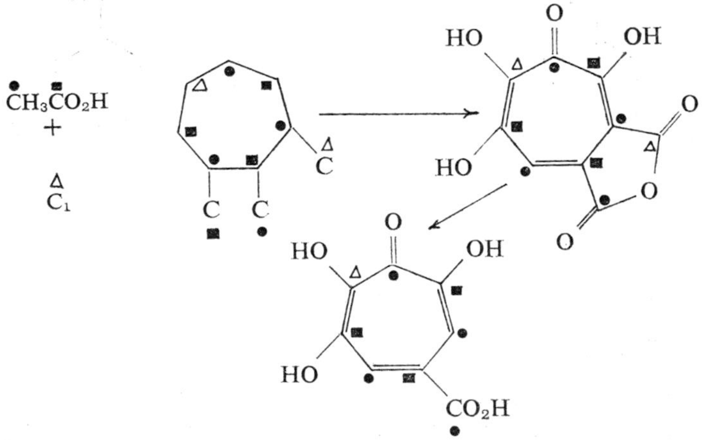

Scheme 18: The incorporation of labelled precursors into puberulic and puberulonic acids

labile carboxyl group of puberulonic acid (i.e. that which is lost during formation of puberulic acid), unlike that of stipitatonic acid, is not derived from acetate; nor was formate incorporated into this carbon atom though C-1 of glucose was. Subsequent discussions[129,145,146] of this problem accept that the labile carboxyl group is, in fact, derived from a C_1-pool and the currently accepted distribution of labelling in puberulic and puberulonic acids is shown in Scheme 18. It is interesting that the carbon atoms of molecules as closely related as stipitatonic and puberulonic acids should be derived from different sources.

B.2.e. *Mycophenolic Acid and Related Compounds*

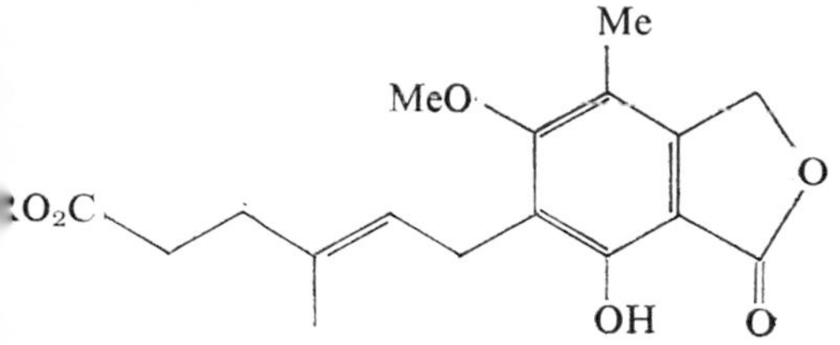

a: R = H, mycophenolic acid   *L*[148,149,152]
   P. brevi-compactum, P. stoloniferum[147]

b: R = Et, ethyl mycophenolate
   P. brevi-compactum[151]

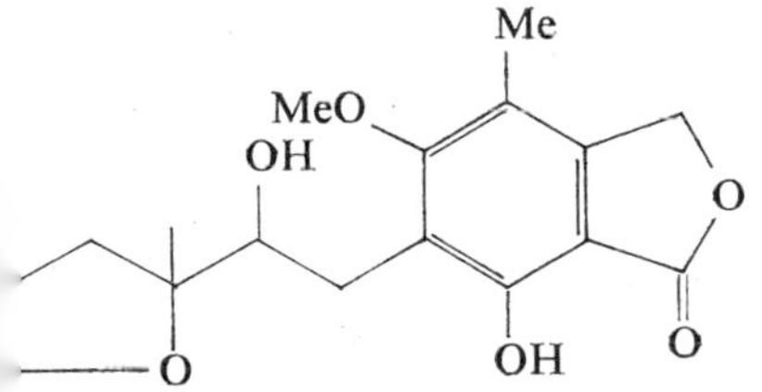

mycophenolic acid diol lactone
P. brevi-compactum[151]

93

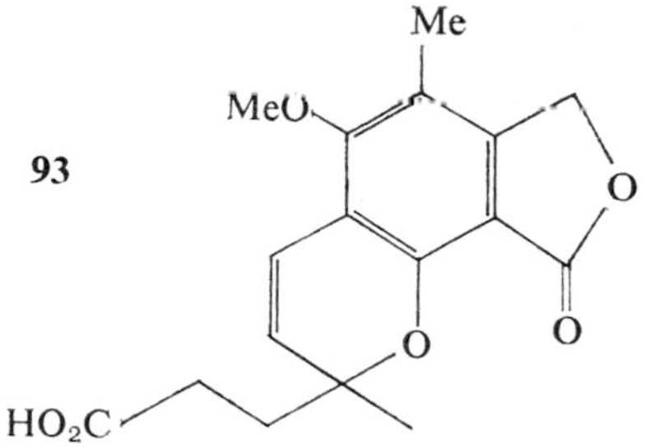

mycochromenic acid
P. brevi-compactum[151]

The aromatic ring of mycophenolic acid (**91**) is derived from acetate[148] and the side-chain from mevalonate.[149] Since radioactive acetone was obtained from the fermentations carried out in the presence of [2-^{14}C]-mevalonate, the side-chain is probably derived from geraniol, though more extensive oxidation of a longer chain cannot be excluded. Since mycophenolic acid co-occurs with compounds (94) to (97) it is conceivable that a C_{10} rather than a C_8 unit is the precursor of the phthalide system, though Birch[149,150] has claimed without experimental evidence that (a) orsellinic acid is specifically incorporated into mycophenolic acid and (b) that a "starter" effect is detected in experiments with [^{14}C]malonate. [*methyl*-^{2}H$_3$]Methionine provides the *O*- and *C*-methyl groups of mycophenolic acid without loss of deuterium.[152] Compounds (92) and (93) can be derived from mycophenolic acid.

B.3 PENTAKETIDES

B.3.a. *Compounds of Type*

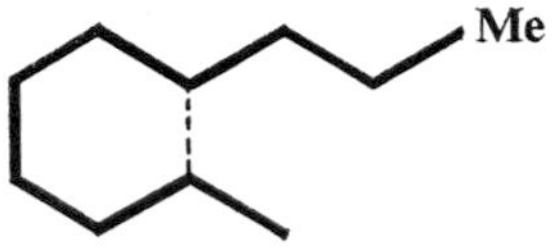

As with the tetraketides, this mode of cyclization is the most common, though fewer biosynthetic studies have been reported. The basic skeletons, exemplified by (94), which is an acetylorsellinic acid, and (104), which is an acetyl-6-methylsalicylic acid, undergo many of the changes discussed for orsellinic and 6-methylsalicylic acids, including cleavage of the aromatic ring. All the compounds in the series retain a free hydroxyl group *ortho* to the original carboxyl group. With the exceptions of reticulol (109) and

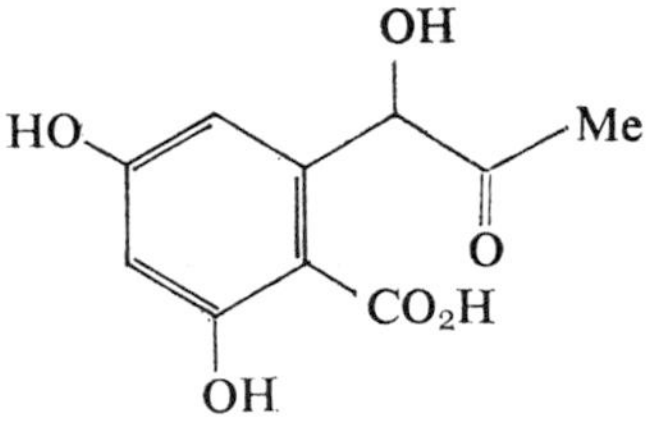

2-carboxy-3,5-dihydroxyphenyl
acetyl carbinol
P. brevi-compactum[153]

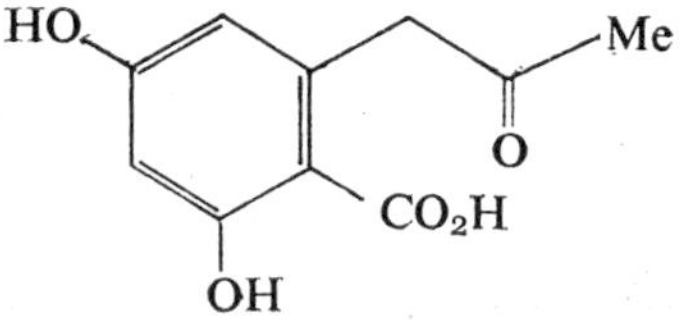

2-carboxy-3,5-dihydroxybenzyl
methyl ketone
P. brevi-compactum[153]

96

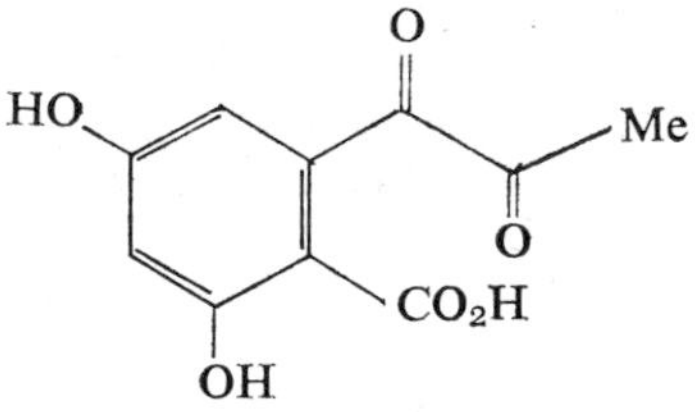

2,4-dihydroxy-6-pyruvylbenzoic acid
P. brevi-compactum[153]

97

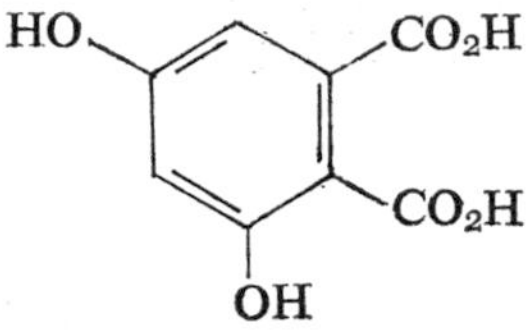

3,5-dihydroxyphthalic acid L[155]
P. brevi-compactum[154]

98

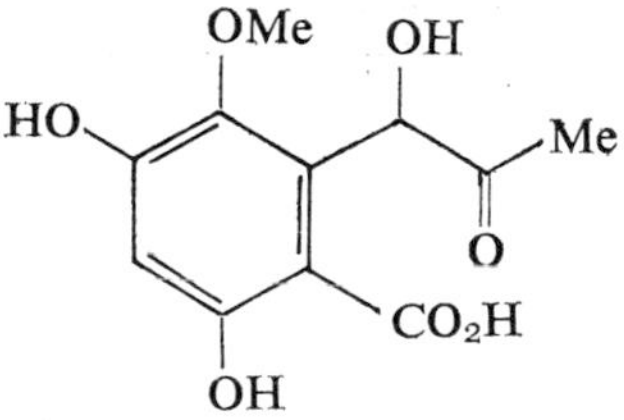

ustic acid
A. ustus,[156] *Paecilomyces victoriae*[157]

99

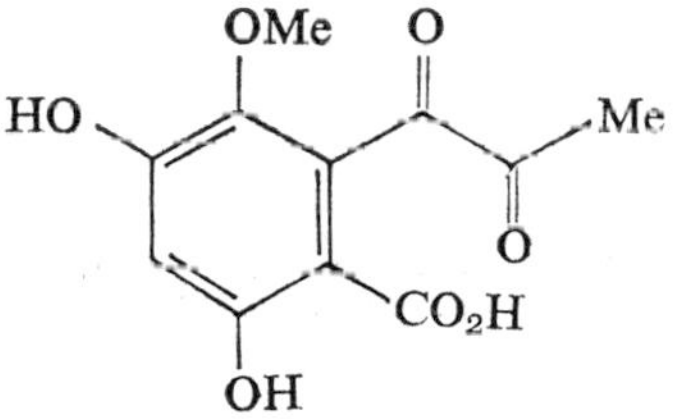

dehydroustic acid
Paecilomyces victoriae[157]

100

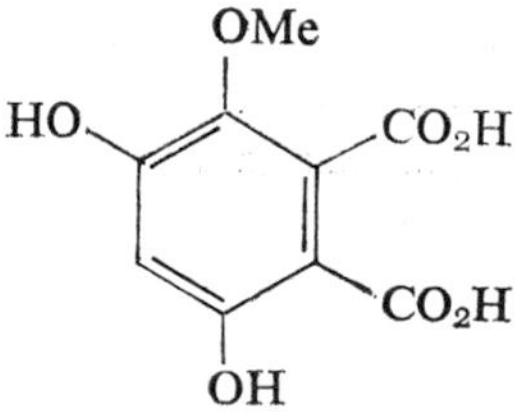

4,6-dihydroxy-3-methoxyphthalic acid
Paecilomyces victoriae[157]

101

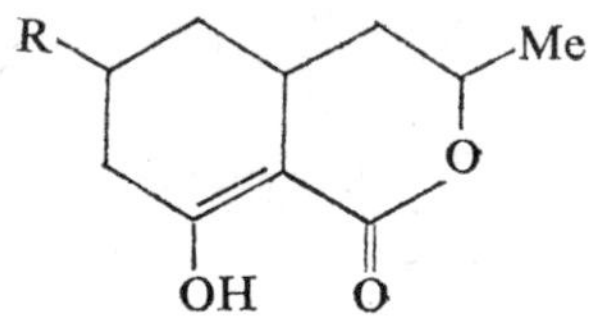

a: R = H, ramulosin
  Pestalotia ramulosa[158]
b: R = OH, hydroxyramulosin
  P. ramulosa[159]

102

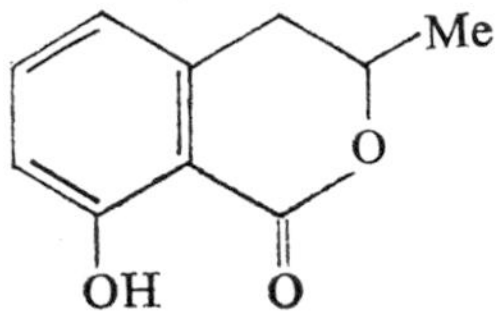

mellein
A. melleus[160] and other Imperfecti
(+)-mellein,
Unidentified fungus[161]

103

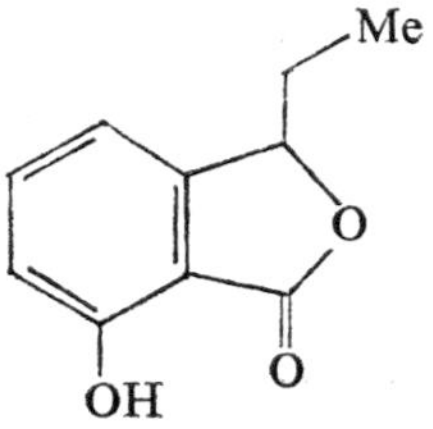

isoochracein
Hypoxylon coccineum[162]

104

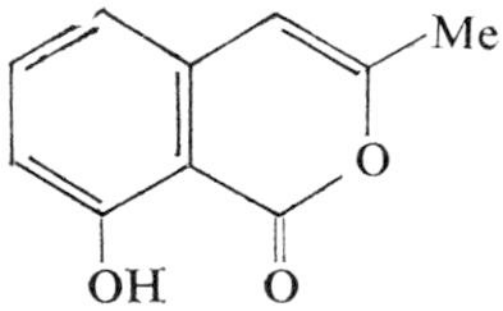

8-hydroxy-3-methylisocoumarin
Marasmius ramealis[163]

105

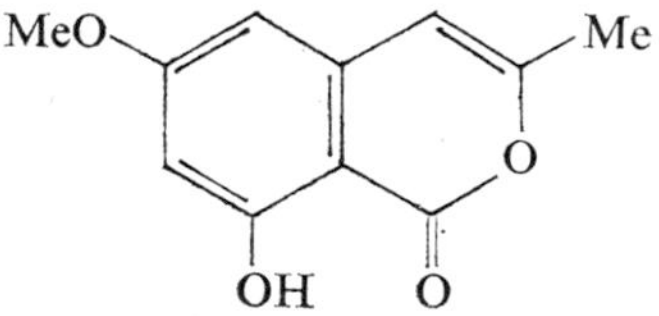

8-hydroxy-6-methoxy-3-methylisocoumarin
Ceratocystis fimbriata[164]

106

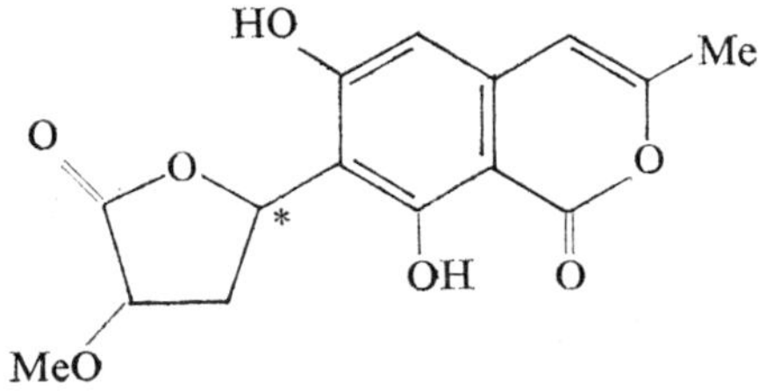

canescin
Penicillium canescens,
  Aspergillus malignus[165]

a: R = H, (—)-3,4-dihydro-6,8-dihydroxy
-3-methyl-isocoumarin
A. terreus[75]

b: R = Me, 6-methoxymellein
Sporormia bipartis,[166] *S. affinis*[171]

4-hydroxymellein
cis-isomer, *Lasiodiplodia theobromae*[167]
trans-isomer, *Apiospora camptospora*[168]

reticulol
Streptomyces rubrireticulae[169]

a: R^1 = Cl, R^2 = H, 5-chloro-8-hydroxy-6-methoxy-
3-methyl-3,4-dihydroisocoumarin
Periconia macrospinosa[170]

b: R^1 = H, R^2 = Cl, 7-chloro-8-hydroxy-6-methoxy-
3-methyl-3,4-dihydroisocoumarin.
Sporormia affinis[171]

c: R^1 = R^2 = Cl, 5,7-dichloro-8-hydroxy-6-methoxy-
3,4-dihydroisocoumarin
Sporormia affinis[171]

actinobolin
Streptomyces griseoviridus var. *atrofaciens*[172]

112

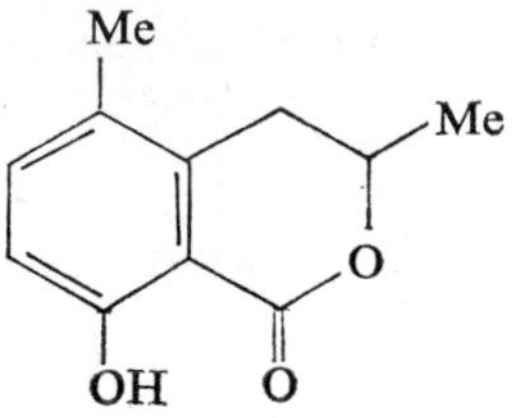

5-methylmellein
Fusicoccum amygdali[173]

113

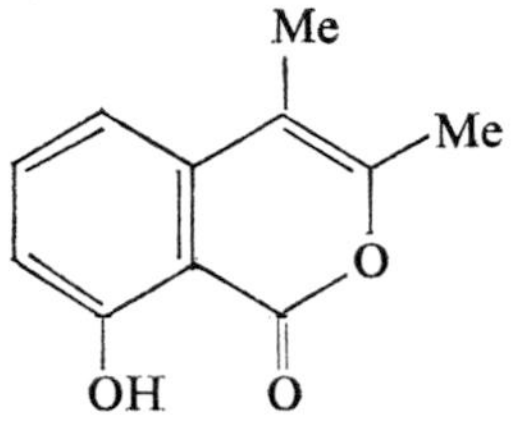

oospolactone
Oospora astringens[174]

114

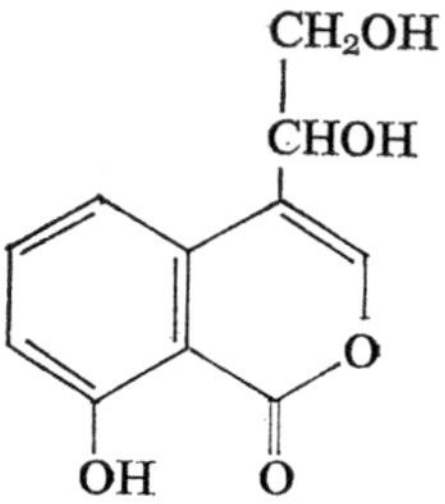

oospoglycol
Oospora astringens[176]

115

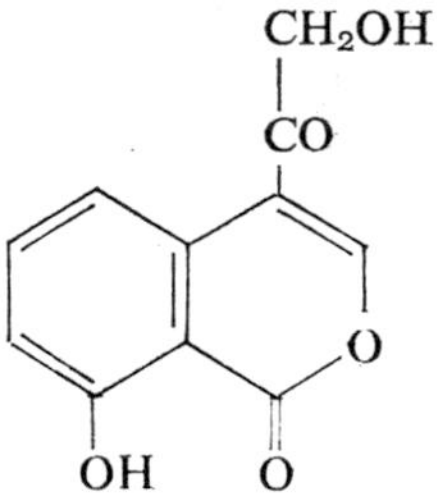

oosponol
Oospora astringens[177]

116

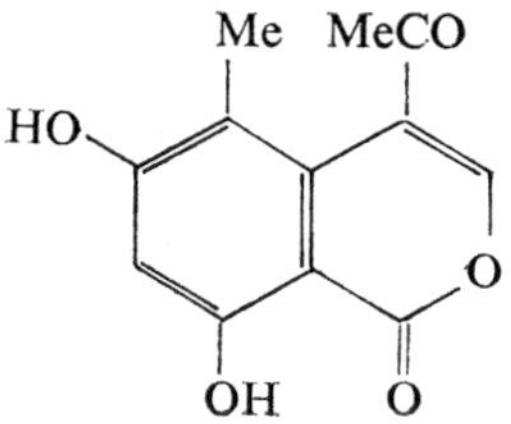

4-acetyl-6,8-dihydroxy-5-methyliosocoumarin
A. viridinutans[178]

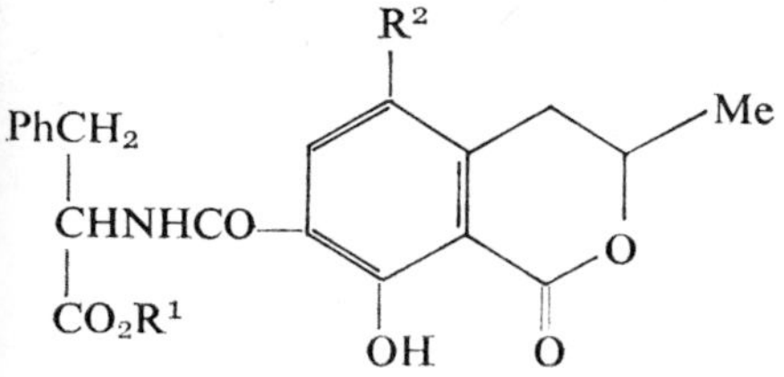

PhCH$_2$
|
CHNHCO
|
CO$_2$R^1

R^2 — (ring) — Me

OH   O

a: R^1 = H, R^2 = Cl, ochratoxin A
    A. ochraceus,[179] *P. viridicatum*[180]
b: R^1 = Et, R^2 = Cl, ochratoxin C
    A. ochraceus[179]
c: R^1 = R^2 = H, ochratoxin B
    A. ochraceus[179]

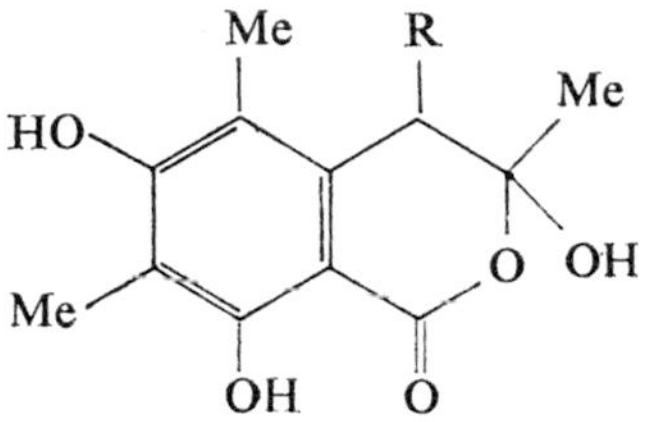

Me   R   Me
HO
Me
OH   O   OH   O

a: R = H, sclerotinin B
    Sclerotinia sclerotiorum[181]
b: R = Me, 3,6,8-trihydroxy-3,4,5,7-
    tetramethyl-3,4-dihydroisocoumarin
    (sclerotinin A)
    P. citrinum,[182] *Sclerotinia sclerotiorum*[181]

Me   Me   Me
HO
HO$_2$C
OH   O

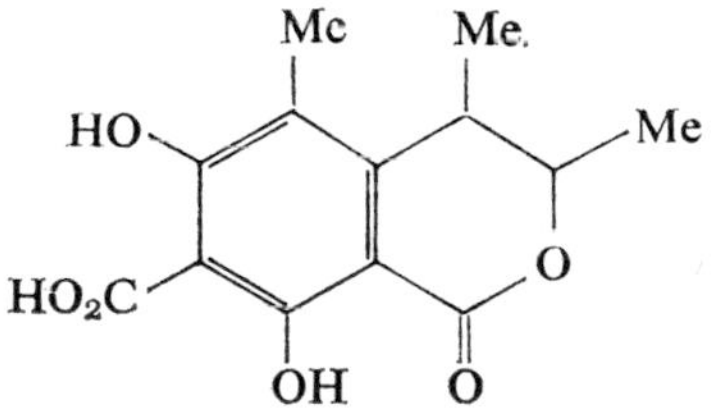

dihydrocitrinone
A. terreus,[183] *P. citrinum*[182]

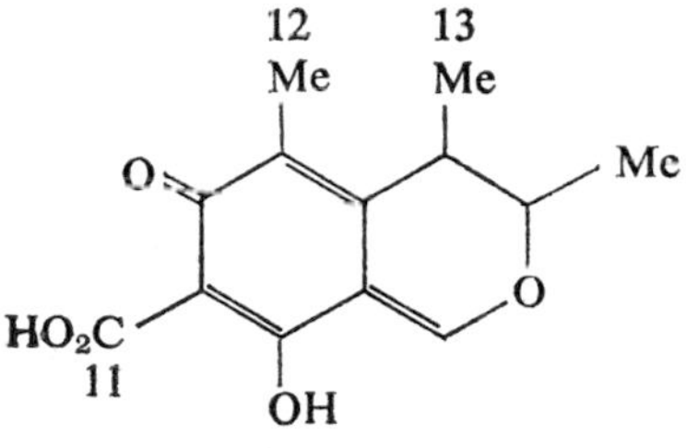

12   13
Me   Me   Me
O
HO$_2$C
11   OH   O

citrinin
P. citrinum[184] and other Imperfecti,
Crotalaria crispata[105]

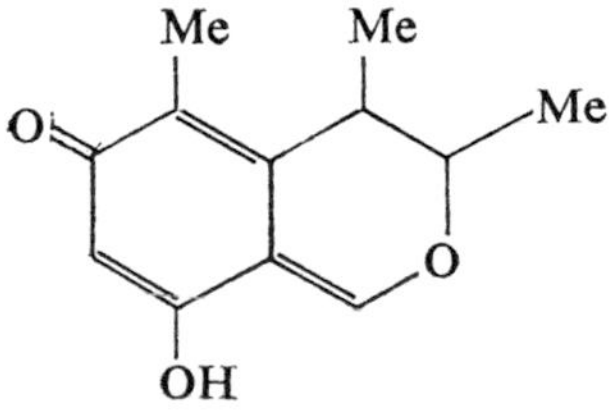

L[186-188]

Me   Me   Me
O
OH   O

decarboxycitrinin
P. citrinum[182]

122

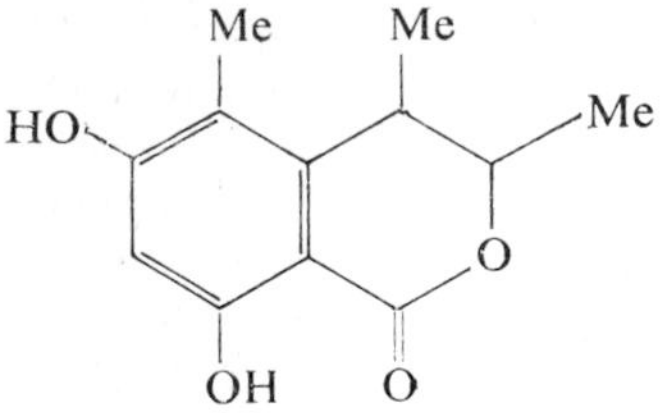

decarboxydihydrocitrinone
P. citrinum[182]

123

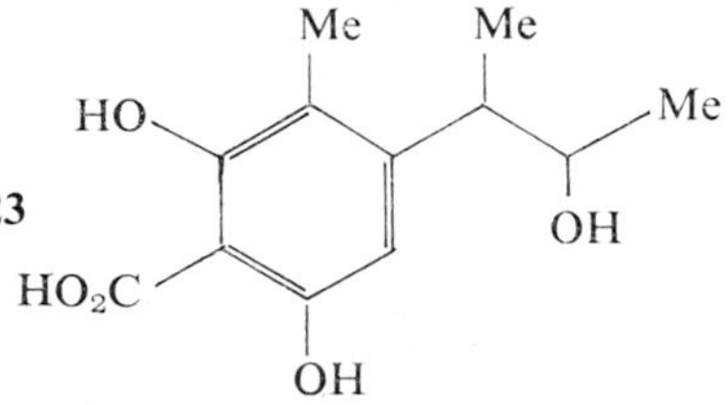

(−)-2,6-dihydroxy-4-(3-hydroxy-2-butyl)-
m-toluic acid
P. citrinum[182]

124

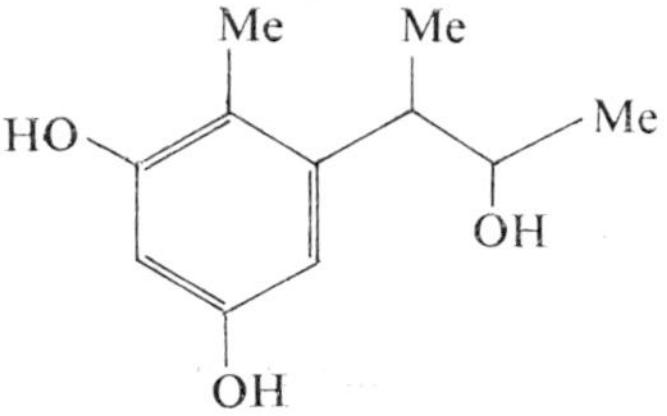

"phenol A"
P. citrinum[182]

125

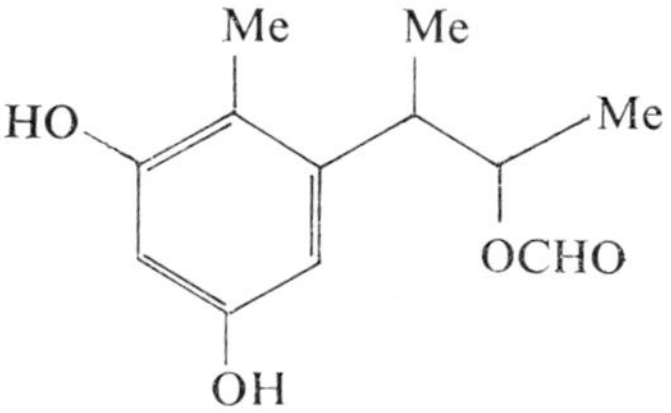

5-(2-formyloxy-1-methylpropyl)-4-methyl-
resorcinol
P. citrinum[182]

actinobolin (**111**) (from Actinomycetes) and of 8-hydroxy-3-methylisocoumarin (**104**) (from a Basidiomycete) all the compounds are produced by Fungi Imperfecti or Ascomycetes. Examples of this type of cyclization also occur among the depsides (see p. 108), e.g. in divaricatic acid (**126**).

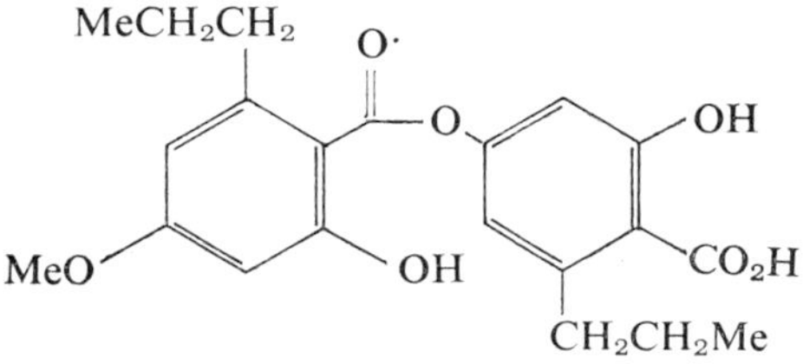

126

Oxidation of the side-chain is illustrated by compounds (94) to (96) (the so-called "Raistrick phenols") and compounds (98) to (100). The phthalic acids (97) and (100) could be derived by oxidation of the methyl group of orsellinic acid or of methoxyorsellinic acid respectively, but their co-occurrence with the C_{10} compounds and the unusual substitution pattern of (100) suggest that they are formed by oxidative cleavage of the corresponding diketones. Experiments with [^{14}C]malonate might clarify this point which has already been raised in connection with 3-hydroxyphthalic acid and mycophenolic acid.

Many of the compounds are isocoumarins or 3,4-dihydroisocoumarins. Mellein (102) occurs in both (+) and (−) forms, and 4-hydroxymellein (108) has been obtained in the *cis* form from *Lasiodiplodia theobromae*, which also produces (−)-mellein, and in the *trans* form from *Apiospora camptospora*, which also produces (+)-mellein. Presumably the hydroxyl group has the same absolute configuration in the two isomers of (108). Isoochracein (103) was first obtained during chemical degradation of mellein, but since the compound isolated from *H. coccineum* is optically active it is not an artefact.

Citrinin (120), a common mould metabolite with marked antibiotic activity *in vitro*, has also been isolated from the higher plant *Crotalaria crispata*.[185] The quinone-methide system of citrinin will be encountered several times among the hexaketides and octaketides, and its chemistry has been reviewed.[189] The acetate-C_1 origin of citrinin has been established by Birch *et al.*[186] and by Schwenk *et al.*[187] The former group found that C-12 and C-13 (measured together) and C-11 were not equally labelled from [^{14}C]formate, suggesting sequential introduction of the C_1 groups, but the latter group, using [*methyl*-^{14}C]methionine, did not detect any difference in the radioactivity of the introduced carbon atoms. The problem has been re-examined by Rodig *et al.*[188] using [1-^{14}C]- and [6-^{14}C]glucose. Their results confirm the acetate origin of citrinin, and the relative activities of the C_1-groups suggest that they are introduced in the order C-11, C-12, C-13; no evidence was obtained concerning the timing of the oxidation of C-11 or of the cyclizations. In an attempt to answer these questions Hassall and his colleagues[182] examined mutants of *P. citrinum*, from which they isolated compounds (118b), (119) and (121) to (125). They suggest the biosynthetic sequence of Scheme 19 but in the absence of experiments with labelled precursors this must be regarded as speculative.

Oospoglycol (114) and oosponol (115) are derived from five "C_2-units" and one "C_1-unit",[175] and oosponol is the precursor of oospoglycol.[176]

Canescin (106) occurs as a mixture of stereoisomers—canescin-A (127) and canescin-B (128)—which can be separated as their acetates. The biosynthesis of the side-chain of canescin is unusual.[190] The carbon atom

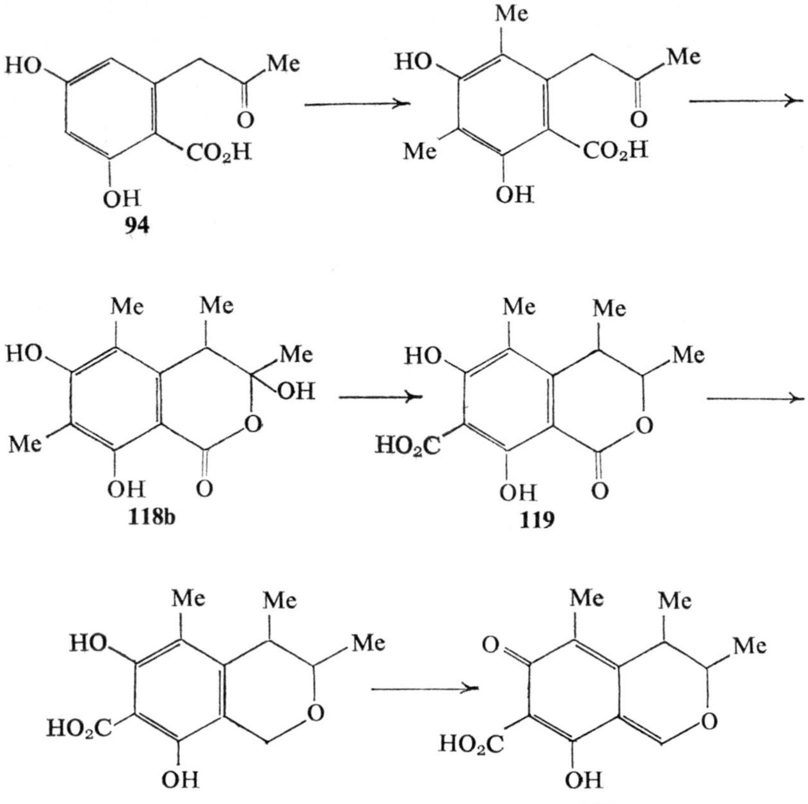

Scheme 19: A possible biosynthetic route to citrinin (**120**)[182]

(asterisk in **106**) corresponding to the carboxyl group of citrinin is derived from the C_1-pool; the remaining carbon atoms of the lactone ring can be derived from succinate or malate. Mechanisms involving either succinate or

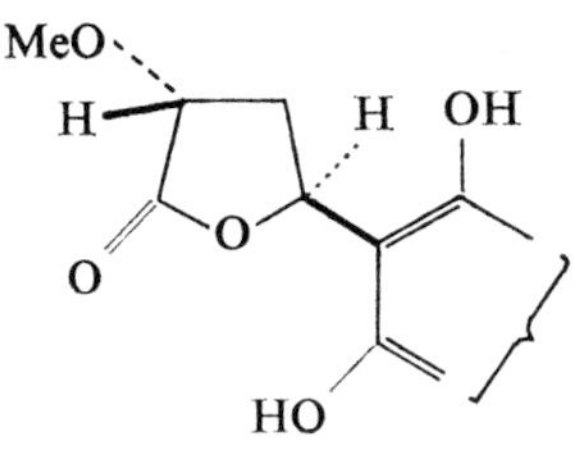

127

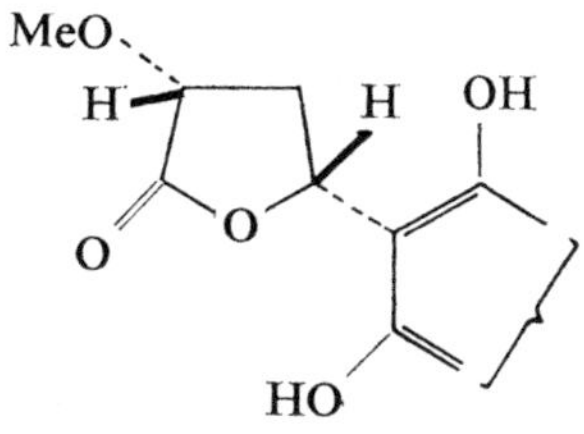

128

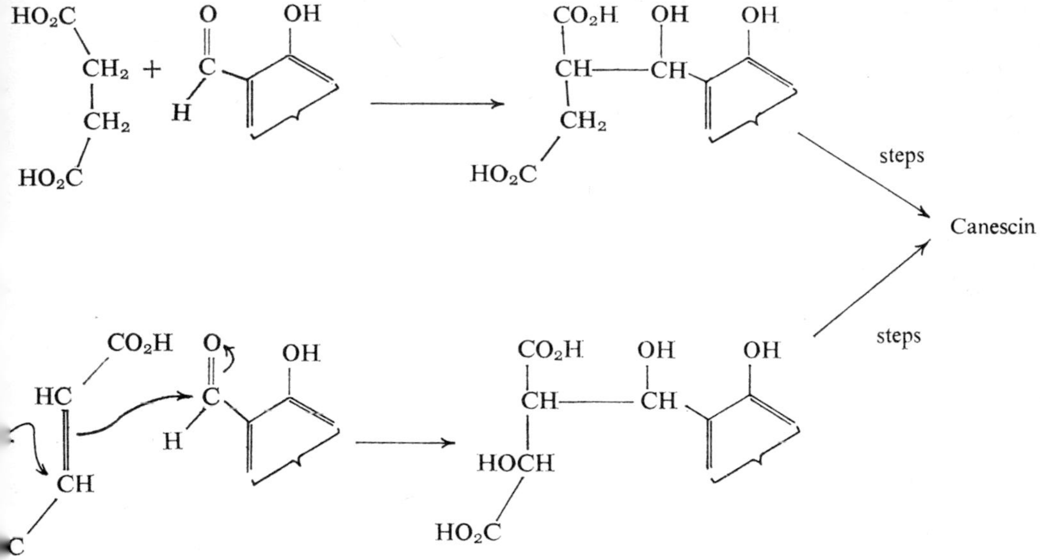

Scheme 20: Possible mechanism for the incorporation of a C_4 acid into canescin[100]

fumarate (Scheme 20) have been proposed[190] for the incorporation of a C_4 acid into canescin. In view of the biosynthetic similarity between canescin and citrinin it is interesting that a strain of *P. canescens* which originally produced canescin now produces only citrinin.[165b]

B.3.b. *Cyclopentane Derivatives Possibly Derived from Pentaketide Intermediates*

129

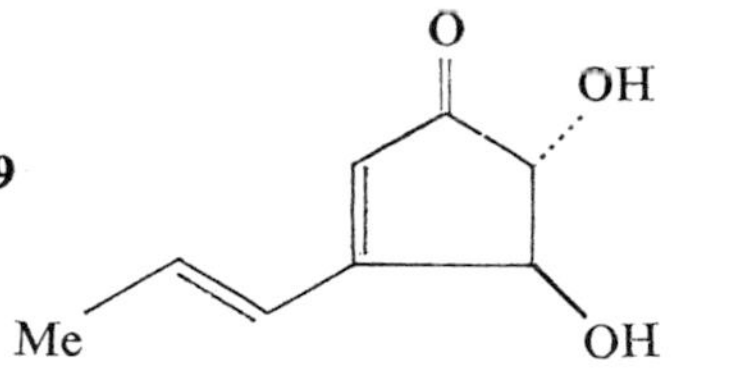

terrein L[193]
Aspergillus terreus[191] and other Fungi
    Imperfecti

130

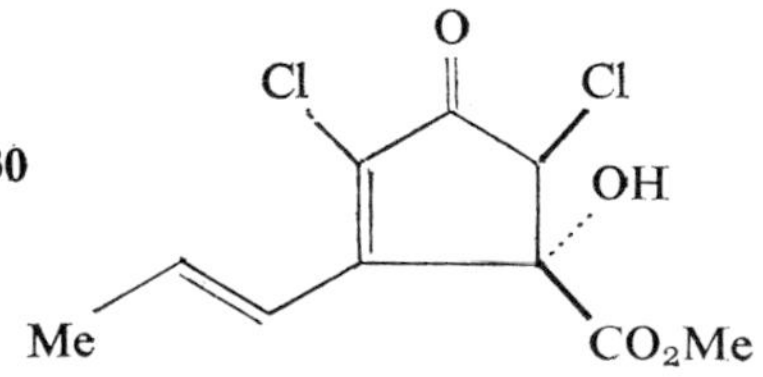

methyl 2-allyl-3,5-dichloro-1-hydroxy-
    4-oxocyclopent-2-enoate
Sporormia affinis[192]

131

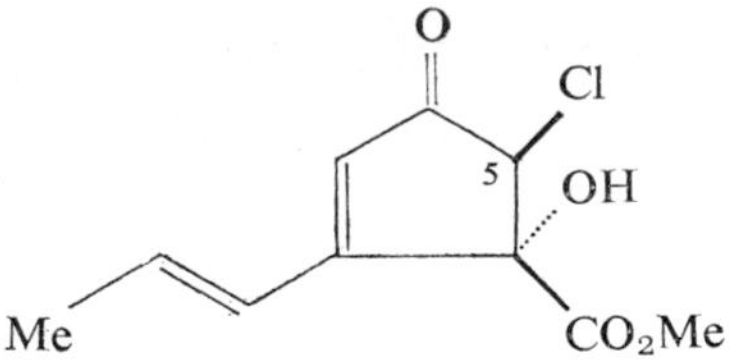

methyl 2-allyl-5-chloro-1-hydroxy-
4-oxocyclopent-2-enoate
S. affinis[192]

132

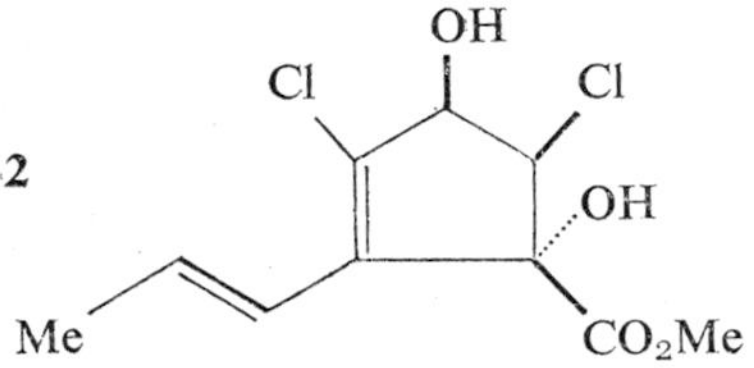

methyl 2-allyl-3,5-dichloro-1,4-dihydroxy-
cyclopent-2-enoate
Periconia macrospinosa[170]

Birch *et al.*[193] found that incorporation of labelled acetate into terrein **(129)** gives the labelling pattern of Scheme 21, with the unusual feature of adjacent carbon atoms derived from C-1 of acetate and suggested such a pattern could

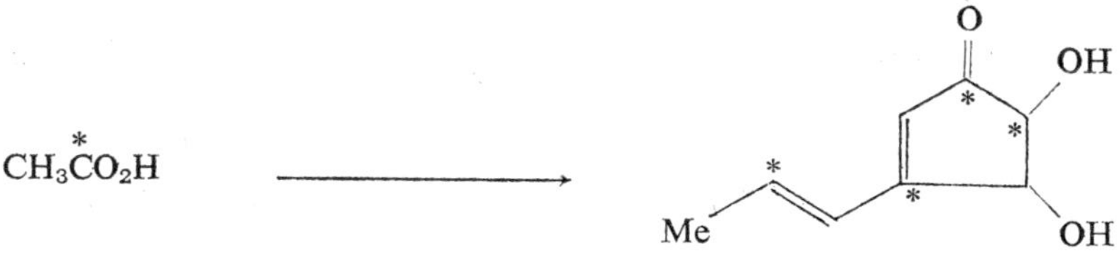

$$CH_3CO_2H \longrightarrow$$

Scheme 21: The distribution of radioactivity from acetate in terrein

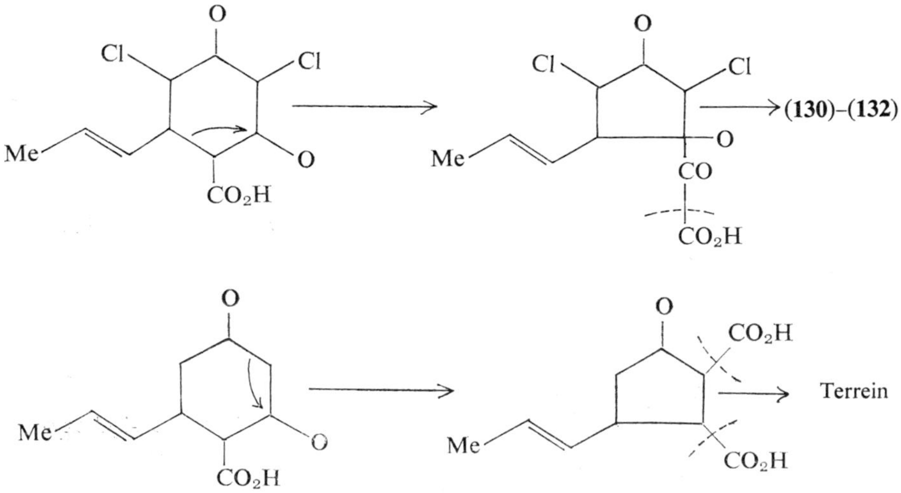

$$\longrightarrow (130)-(132)$$

Scheme 22: The formation of cyclopentanes by alternative ring-contractions arise by contraction of a six-membered ring. Similarly, the co-occurrence of the chlorinated cyclopentene compounds (130) and (131) with the chlorinated isocoumarins (110b) and (110c) in *Sporormia affinis* and of the cyclopentene (132) with the isocoumarin (110a) in *Periconia macrospinosa* suggest that the cyclopentanes might arise from the isocoumarins or from a common intermediate. The chlorine atoms serve to identify "methylene" groups of the polyketide, i.e. atoms derived from C-2 of acetate, so that if terrein and the chlorinated compounds arise from isocoumarin-type precursors they do so by different routes (Scheme 22).

Chemical analogy for the formation of the chloro-compounds is provided[194] by the reaction of phenol with chlorine in alkali to give, *inter alia*, compound (133) which bears a striking resemblance to the fungal metabolites.

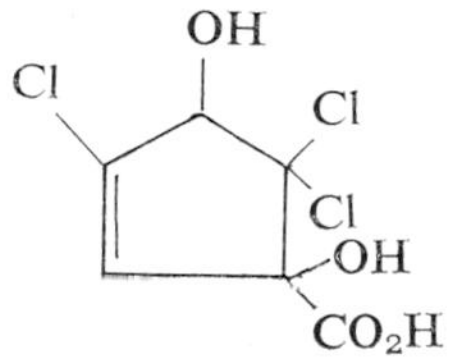

133

B.3.c. *Compounds of Type*

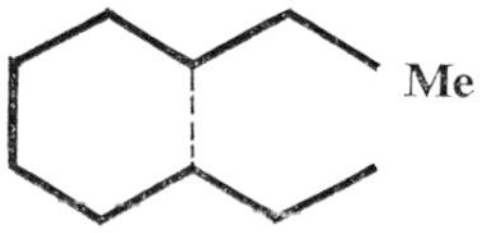

134

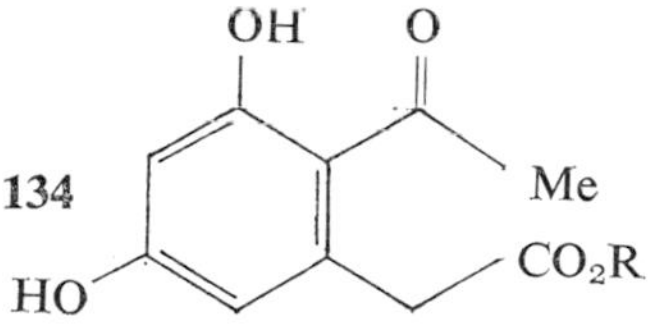

a: R = H, curvulinic acid
        Curvularia siddiqui[195–197]

b: R = Et, curvulin
        Curvularia siddiqui[195,196,198]

135

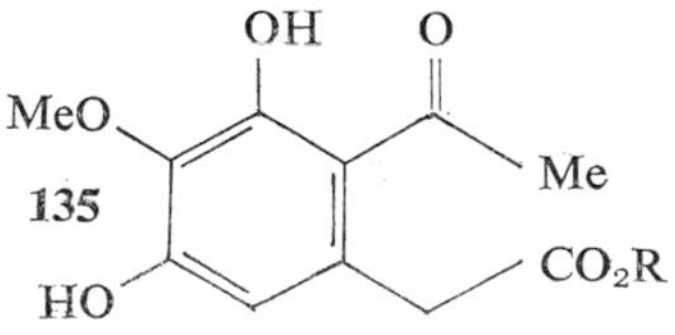

a: R = H, curvulic aicd
        Curvularia siddiqui[196,198,199]

b: R = Et, curvin
        Curvularia siddiqui[198,199]

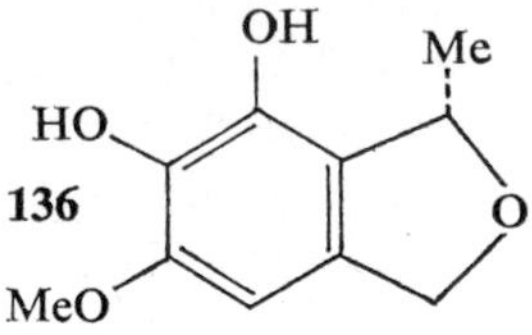

curvulol
Curvularia siddiqui[196,200]

fuscin   *L*[202]
Oidiodendron fuscum[201]

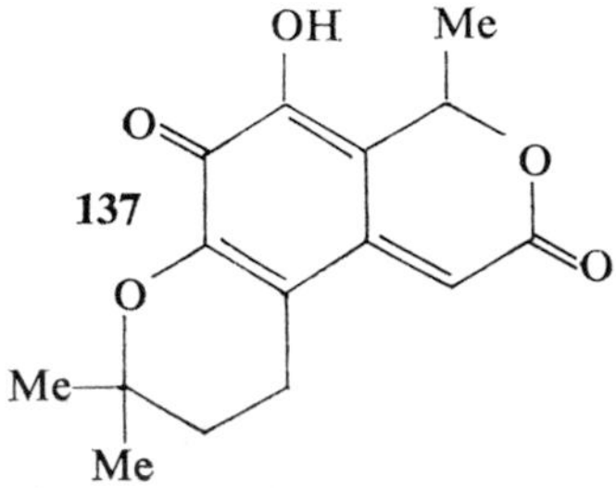

Although no labelling experiments have been reported for the *C. siddiqui* compounds, their polyketide origin seems certain, curvulol (**136**) presumably resulting by modification of an acetylphenylacetic acid. Fuscin (**137**) is formed by introduction of an isopentane unit, derived from mevalonate, onto a pentaketide nucleus.

B.3.d. *Compounds of Type*

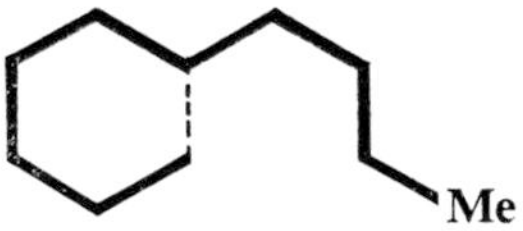

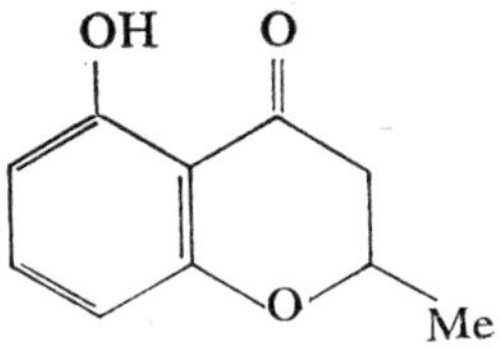

5-hydroxy-2-methylchromanone
Daldinia concentrica[203]

139

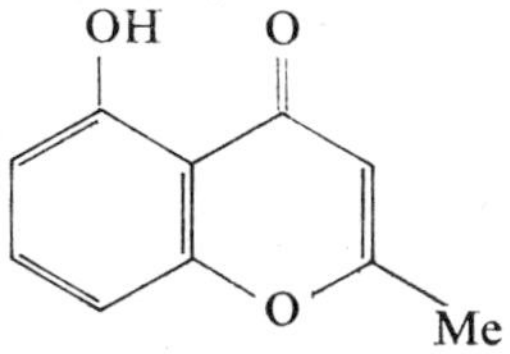

5-hydroxy-2-methylchromone
D.·concentrica[203]

140

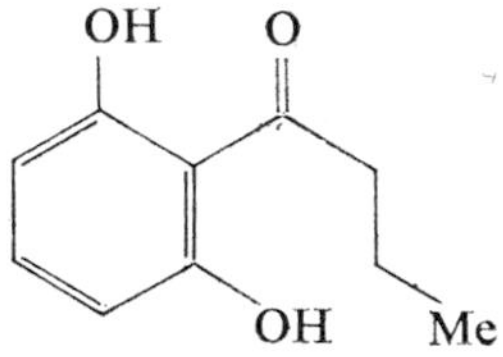

2,6-dihydroxybutyrophenone   *L*[203]
D. concentrica[203]

141

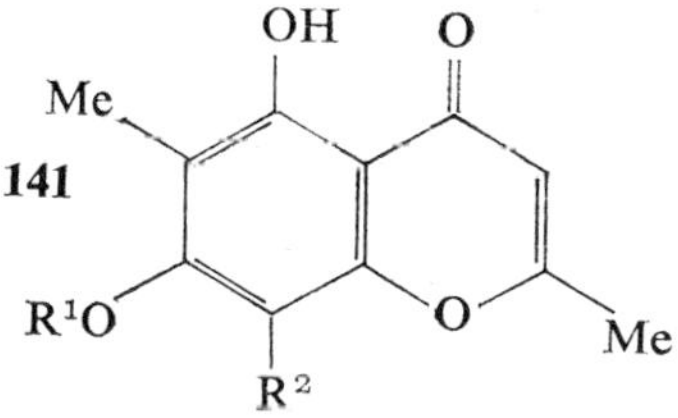

a : $R^1 = Me, R^2 = H$, eugenetin
　　　　　　　　　Lecanora rupicola[206]

b : $R^1 = H, R^2 = Cl$, sordidone (rupicolon)
　　　　　　　　L. sordida,[204] *L. rupicola*[206]

c : $R^1 = R^2 = H$, eugenitol
　　　　　　　　　L. rupicola[206]

142

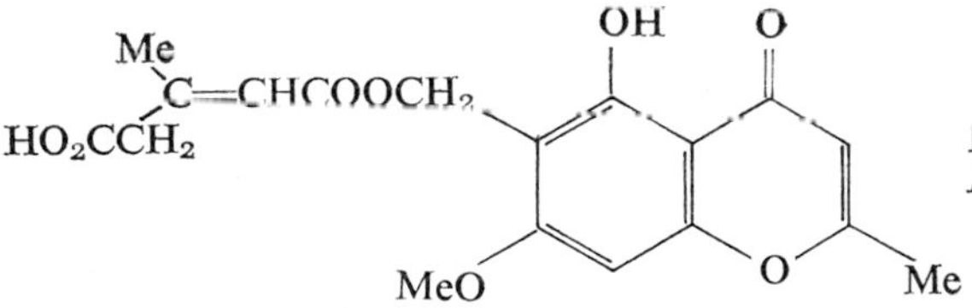

lepraric acid (fuciformic acid)
Lepraria latebrarum[205a]
　　Roccella fuciformis[205b]

The *D. concentrica* metabolites are discussed further in connection with the naphthalenes produced by the organism (Section B.3.e); the biosynthesis of the butyrophenone (**140**) from acetate has been confirmed.[203] Sordidone (**141b**) has been isolated from the mycobiont of *L. rupicola,*[206] along with eugenetin (**141a**) and eugenitol (**141c**) which have not been obtained from lichen thallus.

B.3.e. *Naphthalenes and Naphthoquinones*

143

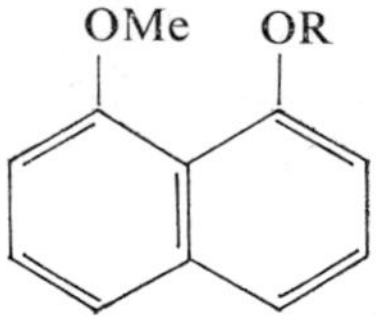

a: R = H, 8-methoxy-l-naphthol
b: R = Me, 1,8-dimethoxynaphthalene
    Daldinia concentrica[203]

144

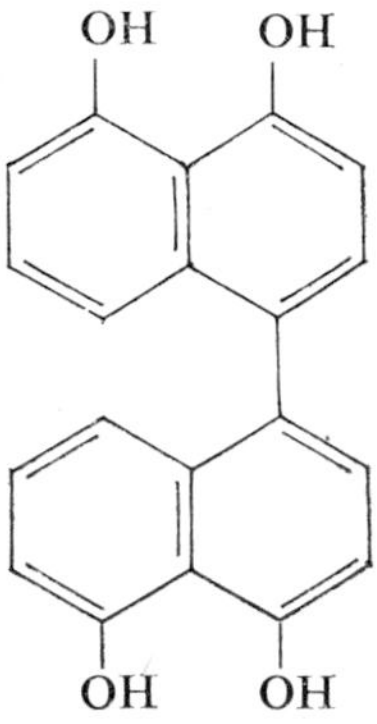

4,5,4′,5′-tetrahydroxydinaphthyl
Daldinia concentrica[203,207]

145

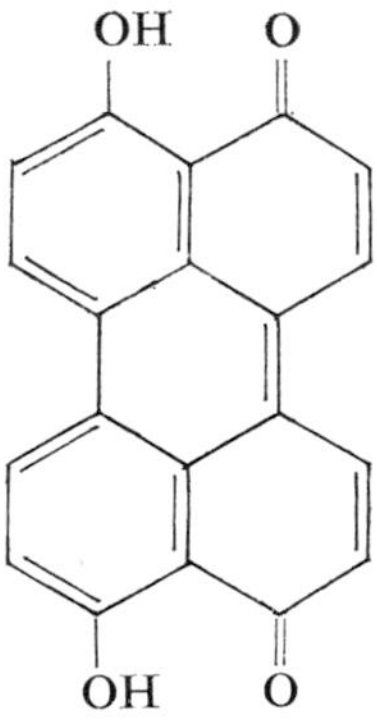

4,9-dihydroxyperylene-3,10-quinone
Daldinia concentrica[203,208]

146

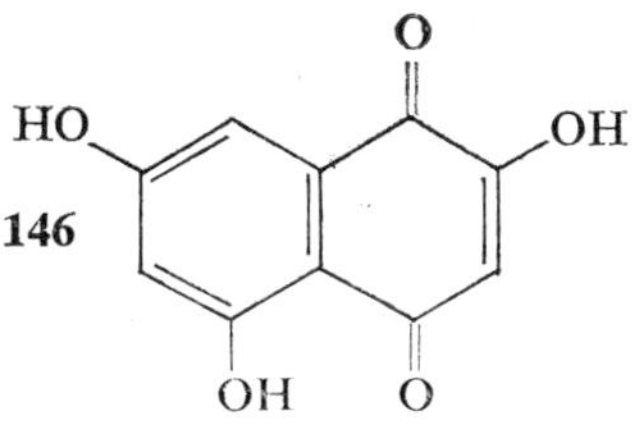

flaviolin
A. citricus[209]

mompain
Helicobasidium mompa[210] L[210b]

2,7-dimethoxy-5-hydroxy-1,4-naphtho-
    quinone
An Actinomycete[209]

6-ethyl-5-hydroxy-2,7-dimethoxy-1,4-naphthoquinone
Hendersonula torulidea[211]

5,8-dihydroxy-2,7-dimethoxy-1,4-
    naphthoquinone
An Actinomycete[212]

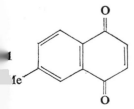

6-methyl-1,4-naphthoquinone
Marasmius gramineum[213]

152

OH   OH

sclerone
Sclerotinia sclerotiorum[214]

153

OH   O

O

mycochrysone
A discomycete[215]

O

O

OH   OH

There is no direct evidence that the C_{10} naphthalenes and related compounds which are included in this section are polyketide in origin, though there is circumstantial evidence that the *D. concentrica* compounds are so derived (see below). Nor is it possible, assuming their polyketide origin, to deduce the mode of cyclization of the open-chain precursor by inspection of their structures since several possibilities are available (Scheme 23), each

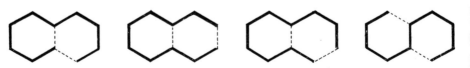

Scheme 23: Alternative cyclizations of a pentaketide to give naphthalenes

of which could give rise to the observed oxygenation patterns. Only in the case of mompain (**147**) have biosynthetic studies been reported.[210b] Mompain produced in the presence of [2-^{14}C]acetate or [2-^{14}C]malonate was converted to the dimethyl compound (**154**) which was degraded by Kuhn–Roth oxidation. For a polyketide derivation the resulting acetic acid should have contained 20% of the radioactivity of mompain, but in fact contained about 10% of the radioactivity, suggesting equal distribution of the label (the authors suggest that randomization may have occurred during the rather long incubation times involved). Structurally related compounds, e.g. juglone (155) and lawsone (156), from higher plants are derived in part from shikimate.[216]

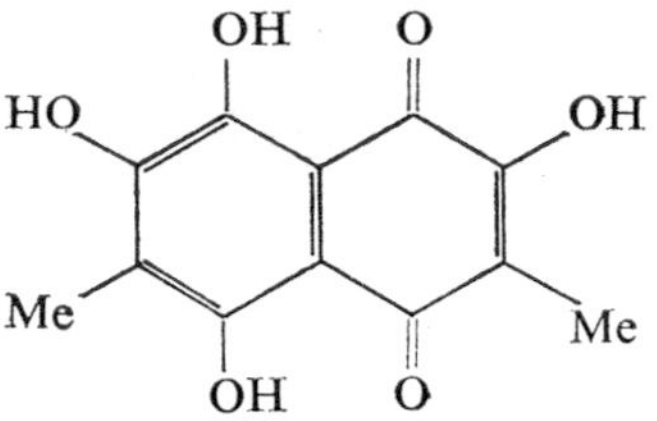

154 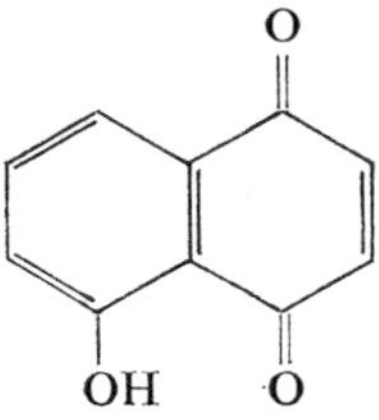   **155** 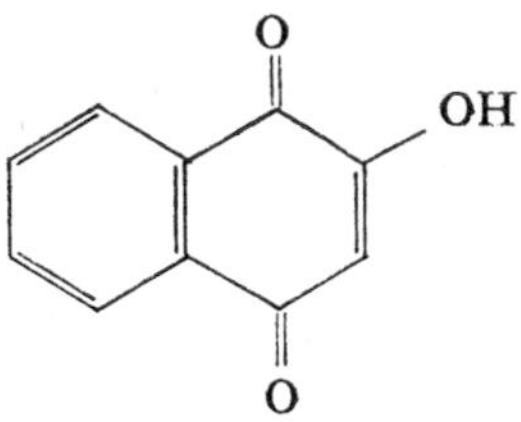   **156**

The naphthoquinone (149) is related to the pigments of the sea-urchins for example 2,3,7,-trihydroxy-6-ethyljuglone (157a). The ethyl substituent is an unusual feature, but the co-occurrence of compounds (157a) and (157b) suggests that it is formed by reduction of an acetyl group.

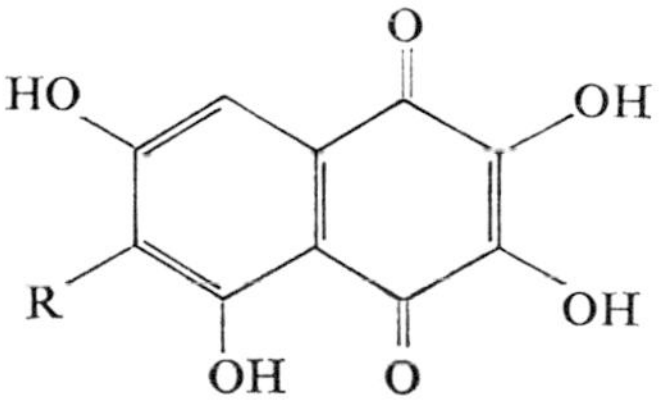

a: R = Et
b: R = Ac

157

Bu'Lock and his colleagues[203] distinguish three variants of *D. concentrica*—those which produce the butyrylresorcinol derivatives (138) to (140) and the acetylresorcinol (81), those which produce the naphthalene (143) and those,

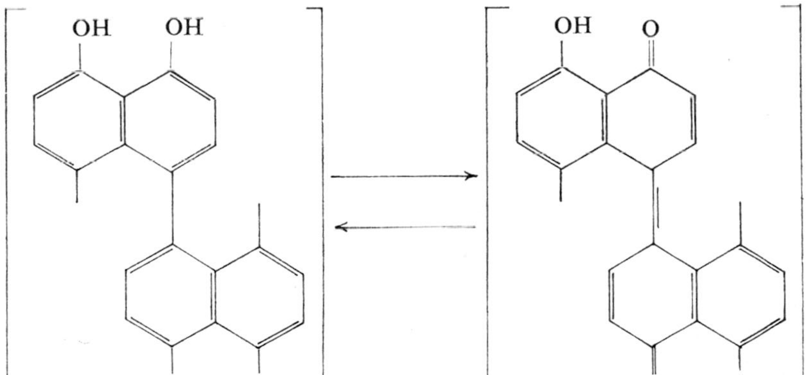

Scheme 24: Redox polymer formed by coupling of a binaphthyl

including "wild" strains, which produce the dimers (144) and (145). They suggest that the resorcinols and the naphthalenes result from alternative (genetically-controlled) cyclizations of a pentaketide precursor, but in view of the labelling experiments with mompain and with the naphthoquinones from higher plants experimental proof of the pentaketide origin of the naphthalenes seems desirable.

By further coupling, the binaphthyl (144) can form a redox polymer of the type shown in Scheme 24. Such a polymer could have a structural function in the fungus, an interesting example of a "secondary metabolite" to which some utility has been ascribed. A detailed account of this topic, and of the interrelationship of the *D. concentrica* metabolites, has been given by Bu'Lock.[217] Aspergilline, a black, high molecular weight substance from spores of *Aspergillus niger*, contains the perylenequinone (145) together with several amino-acids.[218]

B.3.f. *Sepedonin*

The biosynthesis of the tropolone sepedonin (158), isolated from *Sepedonium chrysopermum*,[219] has been studied using [1-^{13}C]- and [2-^{13}C]acetate and [^{13}C]formate.[220] Analysis of the product by n.m.r. spectroscopy gave

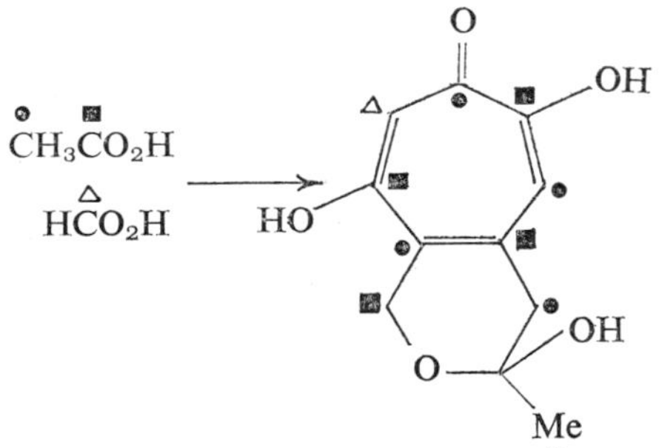

158

results consistent with the labelling pattern shown, i.e. the biosynthesis of sepedonin is analogous to that of stipitatonic acid (cf. Scheme 16).

B.4. HEXAKETIDES

There are relatively few examples of compounds derived from six "C$_2$-units". Some of the compounds included in this section are tentatively classified by inspection of their structures without the support of tracer experiments, and may belong elsewhere.

B.4.a. *Open-chain Compounds*

159   MeCH$_2$CH$_2$CH$_2$CH(OH)CH=C(Me)CH=CHCH=CHCON

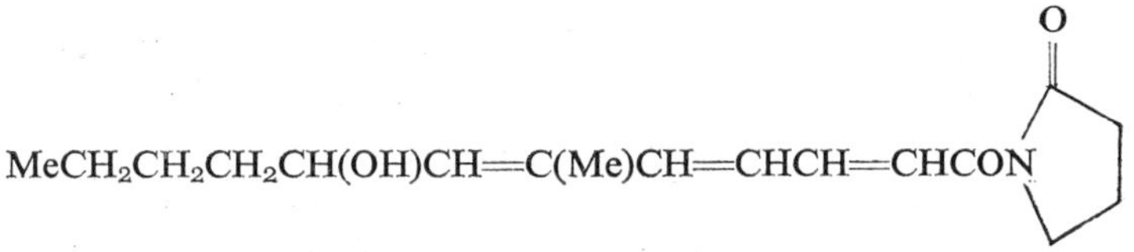

$L^{225,226}$

variotin   *Paecilomyces varioti*[221]

160   MeCH=CHCH=CH   OH

CH=CHCH=CH$_2$

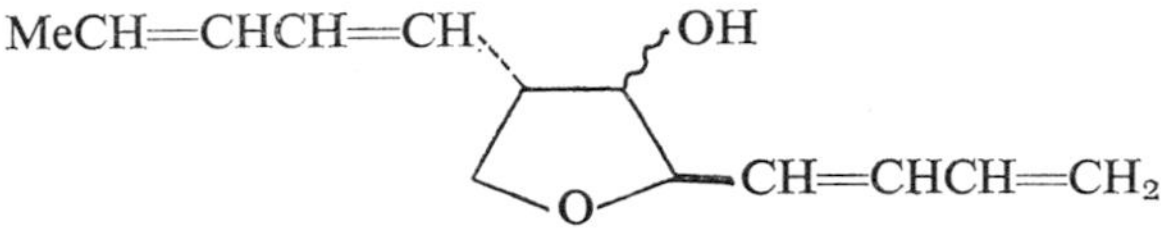

2-(buta-1,3-dienyl)-3-hydroxy-4-(penta-1,3-dienyl)tetrahydrofuran

Chaetomium coarctatum[222]

161   MeCH=CHCH$_2$CH$_2$CH=CHCH$_2$COCH-CHCONH$_2$

O

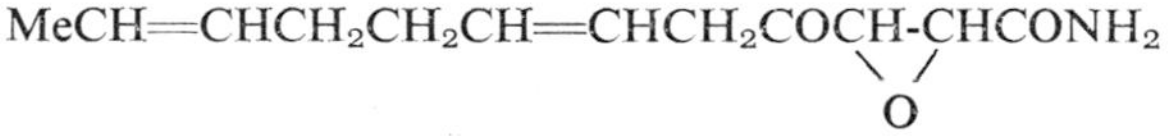

cerulenin
Cephalosporium caerulens[223]

162   MeCH$_2$CH(Me)CH(OH)C(OH)CH=CHCH$_2$CCH$_2$CH$_2$CO

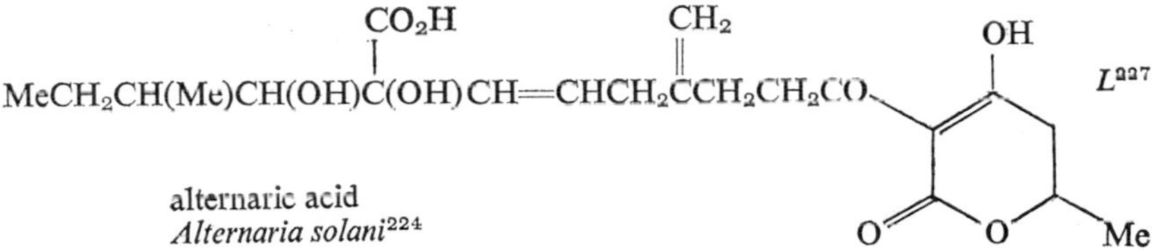

CO$_2$H   CH$_2$ OH L^{227}

alternaric acid
Alternaria solani[224]

O   O   Me

One of the compounds whose derivation from six "C$_2$-units" has been established is the antifungal compound variotin (**159**), the C$_{13}$ hydroxyacid moiety of which is derived in the expected way from acetate and methionine.[225] The γ-aminobutyric acid moiety is derived by decarboxylation of glutamic acid.[226]

The tetrahydrofuran (**160**) has the same carbon skeleton as the C$_{13}$ hydroxyacid moiety of variotin. If, as seems likely, the compound is formed by the same biosynthetic route as variotin, then the terminal methyl group results from complete reduction of a carboxyl group, a process which is known to occur in the formation of only two other compounds (barnol and javanicin).

[1-^{14}C]Acetate, [2-^{14}C]acetate and [^{14}C]formate are incorporated into alternaric acid (162), a phytotoxic metabolite of *Alternaria solani*, in a manner consonant with its formation by acylation of dihydrotriacetic acid lactone by a hexaketide-derived acid, though the alternative derivation (Scheme 25) by condensation of acetoacetate with a heptaketide-derived β-keto-acid cannot be excluded.

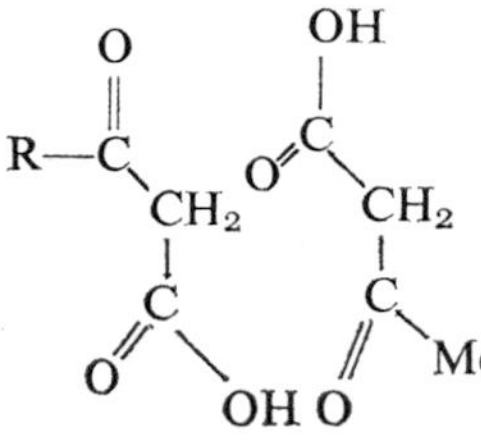

Scheme 25: An alternative biosynthesis of alternaric acid

B.4.b. *Compounds of Type*

The only example of this mode of cyclization appears to be sorbicillin (163) which was isolated [228] from an early sample of clinical sodium penicillin

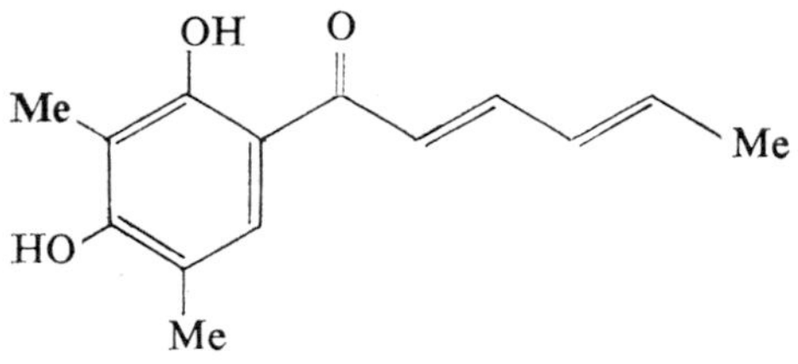

163

and is presumably a metabolite of *P. notatum* [the compound certainly looks like a mould metabolite, cf. clavatol (80)].

B.4.c. *Compounds of Type*

164

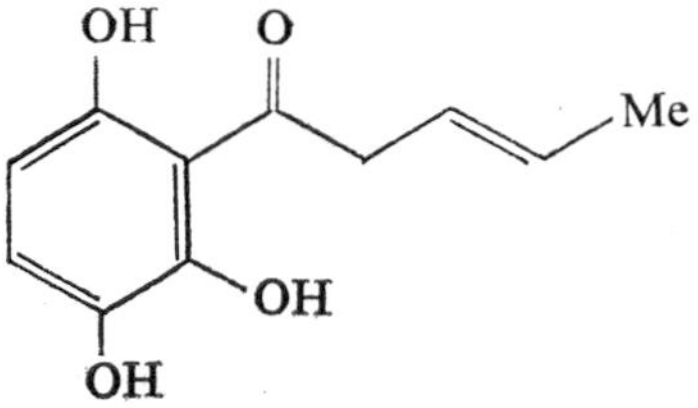

maltoryzine
A. oryzae var. *microsporus*[229]

165

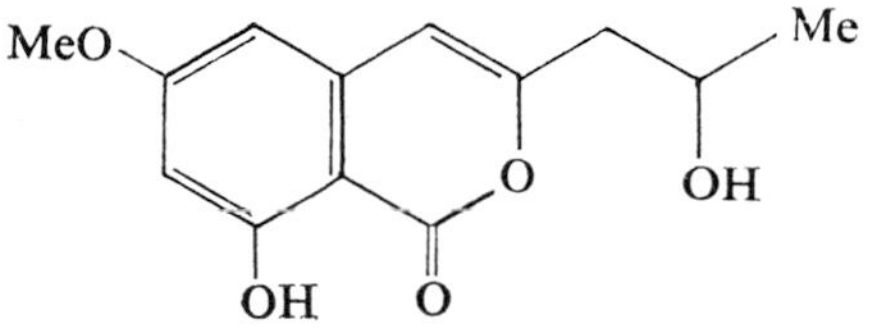

diaporthin
Endothia parasitica[230]

166

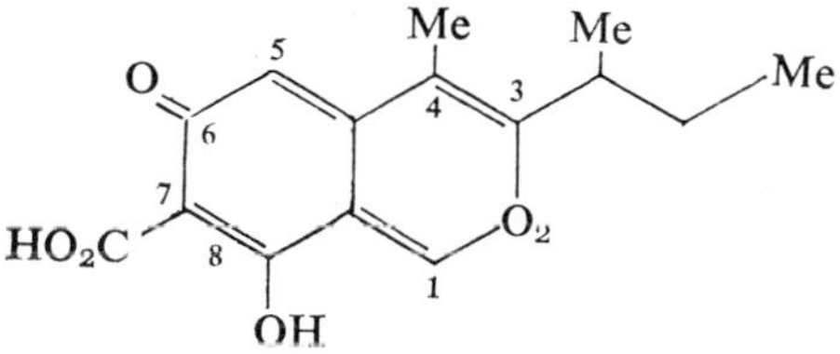

ascochitine
Ascochyta fabae[231]

167

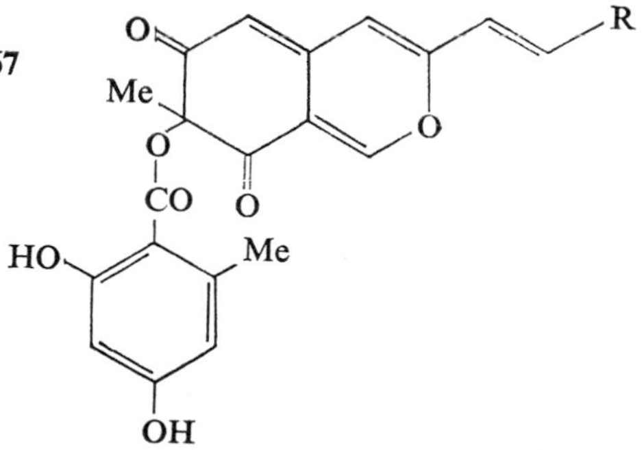

a : R = Me, mitorubrin
b : R = CH$_2$OH, mitorubrinol
        Penicillium rubrum[232]
c : R = CO$_2$H, mitorubrinic acid
        P. funiculosum[233]

168   rubropunctatin L^{238}
       Monascus rubropunctatus[234]

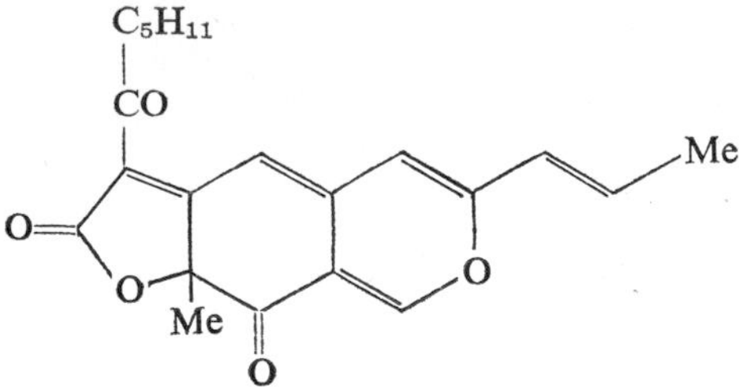

169   monascorubrin
       Monascus purpureus[235]

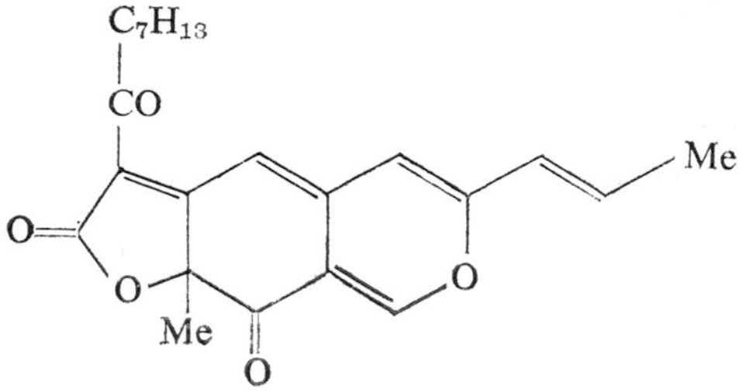

170   monascin (monascoflavin)
       Monascus spp.[236]

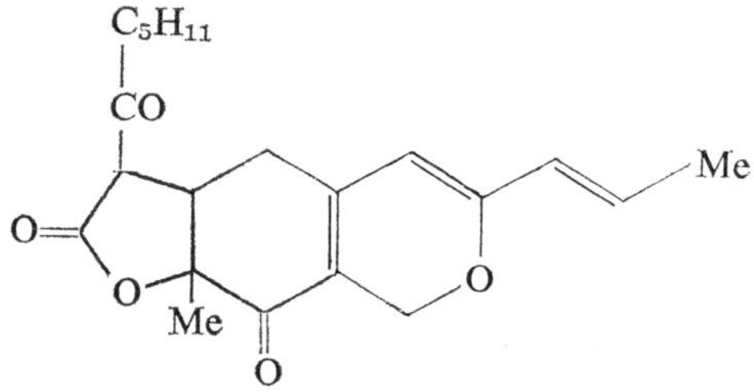

The only compound whose classification is doubtful is maltoryzine (164) which, if it does belong to this class, has lost a carbon atom during its biosynthesis; the oxygenation pattern, too, is unusual. The evidence[229] for structure (164) is not compelling; moreover the molecular formula is given as $C_{11}H_{14}O_4$ [(164) is $C_{11}H_{12}O_4$] and the trimethyl ether as $C_{14}H_{19}O_4$ (*sic*).

The remaining members represent obvious extensions by one "C_2-unit" of structural types (isocoumarins and quinonemethides) encountered previously. In compounds (167) to (170), enolization of the 8-keto-group has been blocked by the introduction of a methyl and an aroyloxy- or acyloxy-group at position 7 to give "azaphilones",† a class to which sclerotiorin and rotiorin also belong (see p. 174). In compounds (168) to (170), the *O*-acyl group has condensed with the ketone group at position 6 to form a new ring.

---

† This generic name is now thought to be too restrictive. For a discussion of this point and a review of the field, see reference.[237]

It has been pointed out[232] that *a priori* mitorubrin (**167, R = Me**) and rubropunctatin (**168**) could be formed from a common intermediate (**171**) (Scheme 26). However, the biosynthesis of rubropunctatin has been studied[238a] using [1-^{14}C]malonate (experiments with labelled acetate had

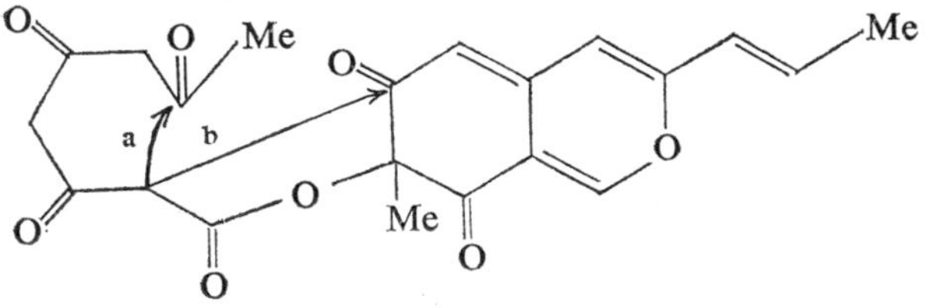

171

167a   **168**

Scheme 26: Suggested formation of mitorubrin (**167a**) and rubropunctatin (**168**) from a common precursor

been reported previously[238b]), and while the chromophore is derived as expected from one acetyl and five malonyl units, the β-keto-octanoyl residue appears to arise by condensation of hexanoic acid (derived in the normal way from one acetyl and two malonyl units) with an acetyl unit, precluding an intermediate of type (**171**). The β-ketodecanoyl moiety of monascorubrin (**169**) is similarly derived from octanoate and acetate, and an analogous situation is encountered with rotiorin (p. 174).

This type of cyclization is also involved in the formation of some depsides, e.g. perlatolic acid (**172**).

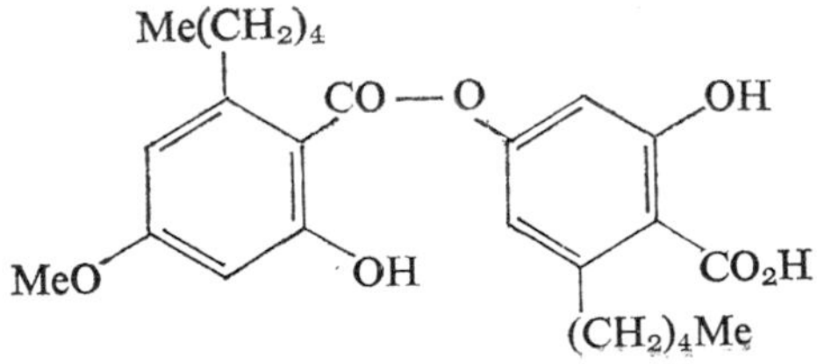

172

B.5. Heptaketides

B.5.a. *Compounds of Type*

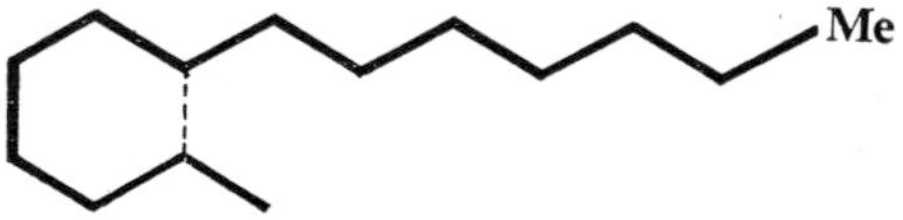

173

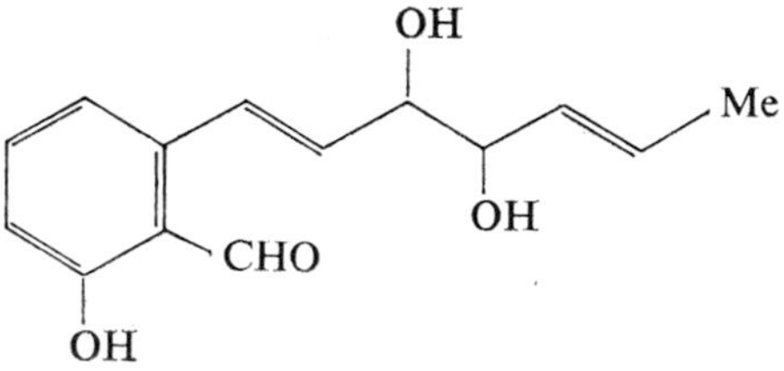

pyriculol
Pyricularia oryzae[239]

174

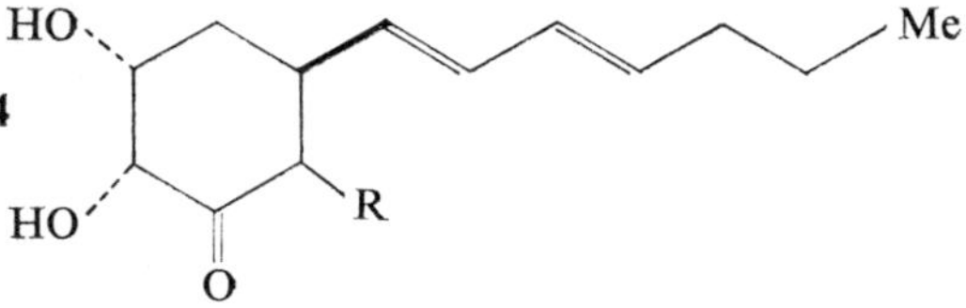

a: R = CH$_2$OH, palitantin
   P. palitans[240] and other
      Penicillia

b: R = CHO, frequentin
   P. frequentans[241]

175

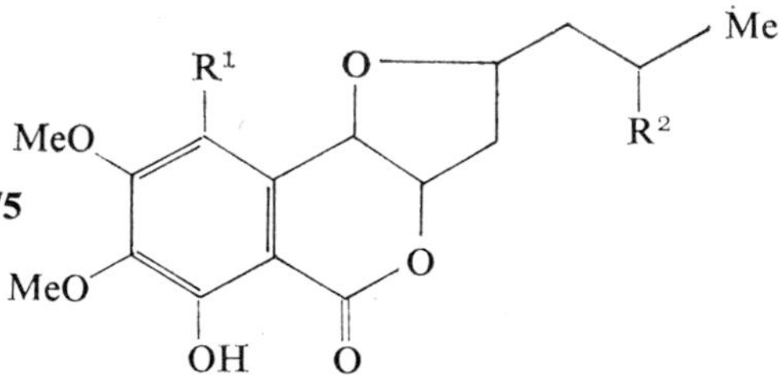

a: R^1 = H,R^2 = H$_2$, monocerin
b: R^1 = OH,R^2 = H$_2$, hydroxymonocerin
c: R^1 = H,R^2 = O, monocerone
   Helminthosporium monoc

176

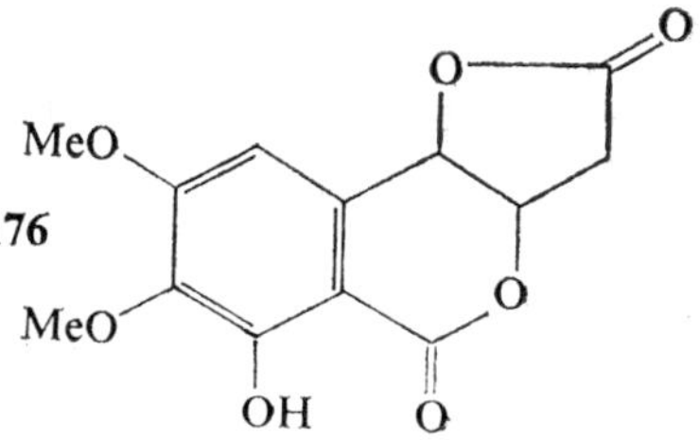

monocerolide
H. monoceras[243]

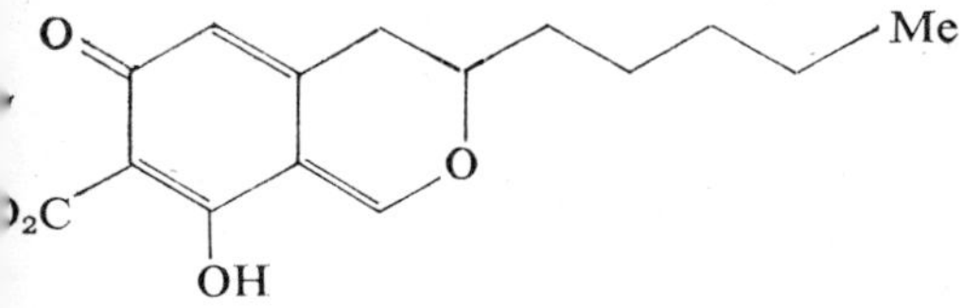

pulvilloric acid
P. pulvillorum[244] L[245]

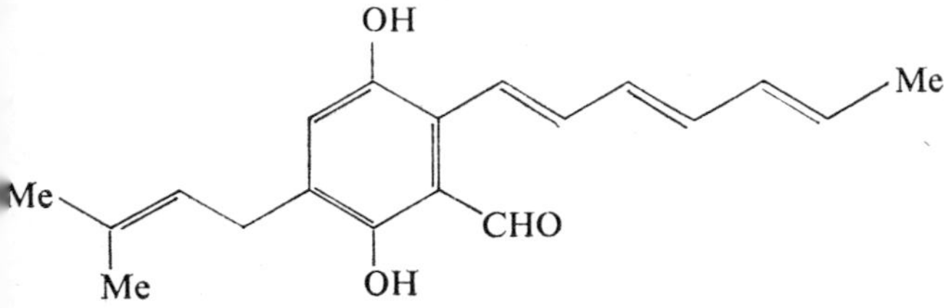

auroglaucin
A. glaucus[246] L[247]

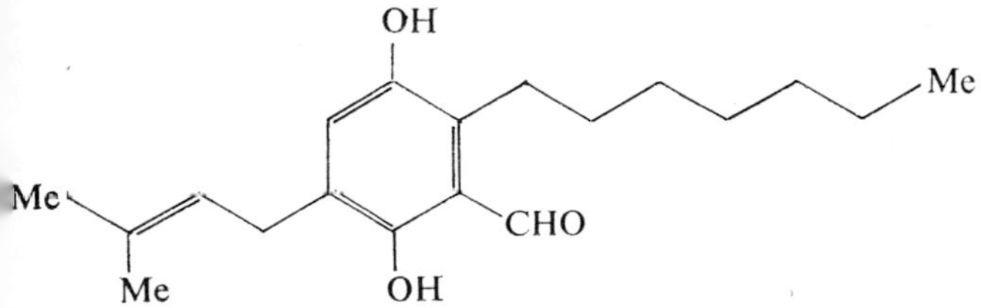

flavoglaucin
A. glaucus[246]

The biosynthesis of palitantin (**174a**) from acetate[242] and of auroglaucin (**178**) from acetate and mevalonate[247] has been confirmed. Pulvilloric acid (**177**), which is related to citrinin (**120**) and ascochitine (**166**), is derived from acetate and formate as expected[245] but hexanoate, which might have been the precursor of the side-chain, is incorporated only after breakdown to acetate.

Microphyllic acid (**180**) and sphaerophorin (**181**) are depsides which contain aromatic units derived by this type of cyclization.

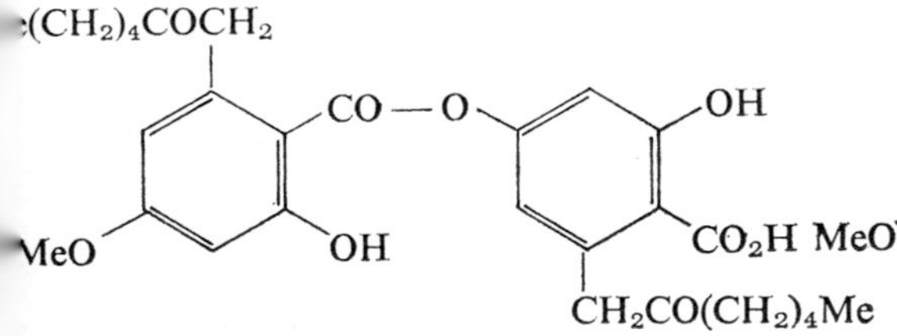

180

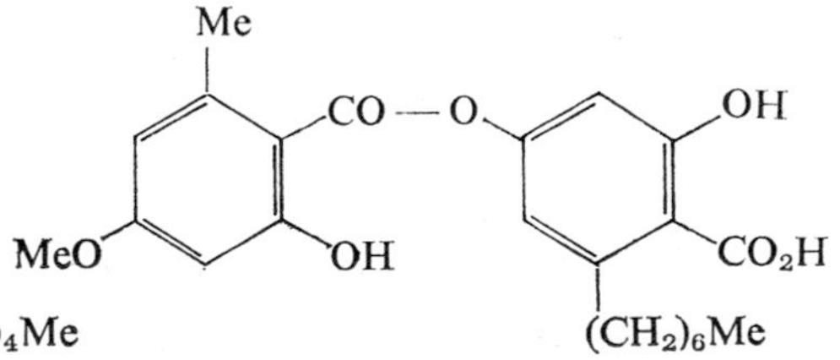

181

B.5.b. *Compounds of Type*

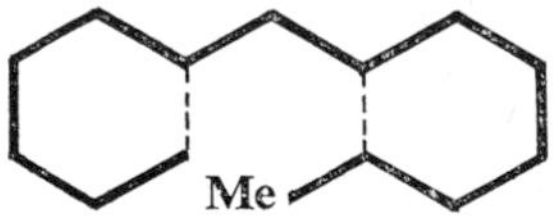

182

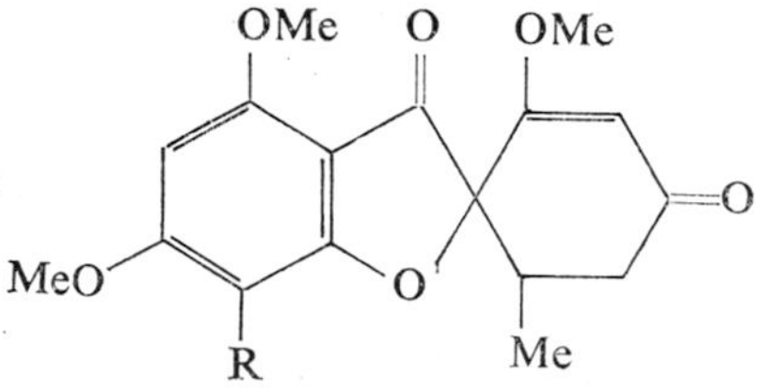

a: R = Cl, griseofulvin   *L*[249]
                 Penicillium spp.[248]

b: R = Br, bromogriseofulvin
                 Penicillium spp.[251]

c: R = H, dechlorogriseofulvin
                 Penicillium spp.[252]

183

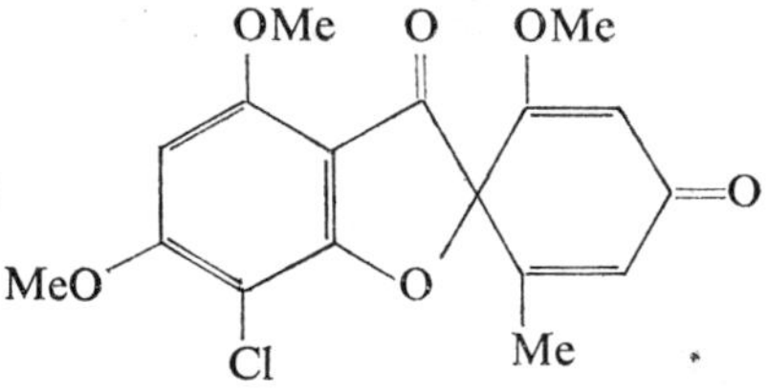

dehydrogriseofulvin
P. patulum[253]

184

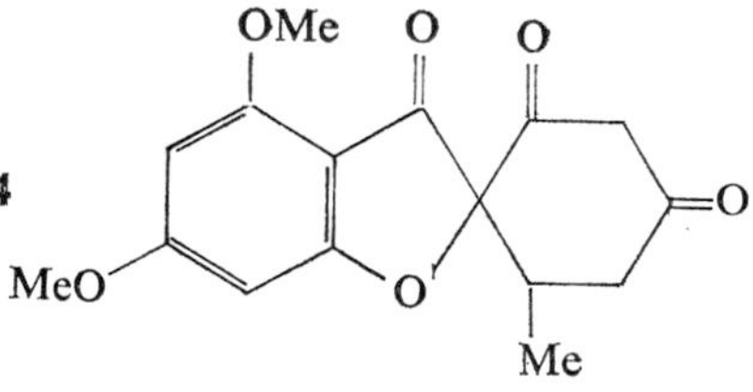

4,6-dimethoxy-2′-methylgrisan-
3,4′,6′-trione
P. patulum[253]

185

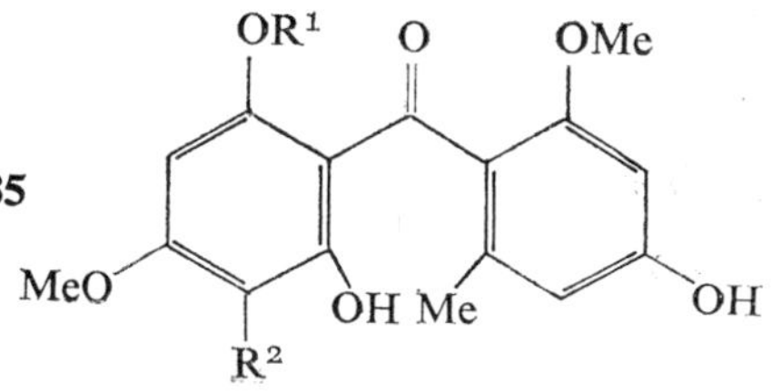

a: R¹ = Me, R² = Cl, griseophenone A
                            P. patulum[253,254]

b: R¹ = H, R² = Cl, griseophenone B
                            P. patulum[254]

c: R¹ = R² = H, griseophenone C
                       P. patulum[254]

186

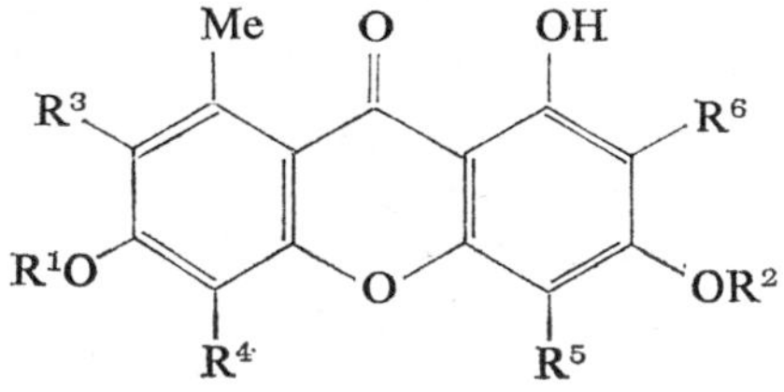

	R^1	R^2	R^3	R^4	R^5	R^6	
a	H	H	H	H	H	H	norlichexanthone *Lecanora reuteri*[255]
b	Me	H	H	H	H	H	griseoxanthone C *P. patulum*[253]
c	Me	Me	H	H	H	H	lichexanthone Several lichens[256]
d	H	H	H	H	H	Cl	2-chloronorlichexanthone *L. straminea*[257]
e	H	H	H	H	Cl	Cl	2,4-dichloronorlichexanthone *L. straminea*[257]
f	H	H	Cl	H	H	Cl	2,7-dichloronorlichexanthone *L. straminea*[258]
g	H	H	Cl	H	Cl	Cl	arthetholin *L. straminea*[259]
h	H	H	Cl	Cl	H	Cl	2,5,7-trichloronorlichexanthone Several lichens[260]
i	H	H	Cl	Cl	Cl	Cl	thiophanic acid Several lichens[259]
j	Me	H	H	H	Cl	Cl	thiophaninic acid *Pertusaria* spp.[261]
k	H	Me	H	Cl	H	Cl	3-*O*-methyl-2,5-dichloronorlichexanthone *L. contractula*[262]
l	H	Me	Cl	Cl	H	Cl	3-*O*-methyl-2,5,7-trichloronorlichexanthone Several lichens[260]
m	H	Me	Cl	H	Cl	Cl	thuringione *Lecidea carpathica*[263]
n	Me	Me	Cl	H	H	Cl	2,7-dichlorolichexanthone *Buellia glaziovana*[260]
o	Me	Me	H	Cl	H	Cl	2,5-dichlorolichexanthone *L. populicola*[262]

With the exception of griseoxanthone C, the xanthones have all been isolated from lichens, largely as a result of the screening programme of Santesson, which is based on the mass spectra and thin layer chromatography of lichen extracts.

The important antifungal antibiotic griseofulvin (**182a**) has been the subject of several reviews.[264] Griseofulvin was first isolated (as were so many of the products of *Penicillia*) by Raistrick and his colleagues from *P. griseofulvum*[248] and subsequently re-isolated as "curling factor", a name descriptive of the effect griseofulvin produces on the hyphae of *Botrytis allii*, by Brian and his co-workers[265] at the Akers Research Laboratories of I.C.I. Chemical studies at the same laboratories led to structure (**182a**), the absolute configuration being subsequently deduced[266] and confirmed by X-ray analysis[267] of bromogriseofulvin (**182b**).

Barton and Cohen[268] suggested that griseofulvin might arise by intramolecular phenol coupling (one-electron oxidation) of the heptaketide-derived benzophenone (**185a**), to give first dehydrogriseofulvin (**183**) and then griseofulvin by enzymatic reduction. The acetate origin of griseofulvin has been confirmed by Birch and his co-workers[249] using [1-^{14}C]acetate, and though the incorporation of malonate has not been studied, Birch[269] has re-interpreted his results in terms of the derivation of griseofulvin from a single chain formed from one acetyl and six malonyl units.

Whatever the initial linking of acetyl and malonyl units, the first aromatic product during the biosynthesis of griseofulvin seems to be griseophenone C (**185c**) which is converted[250] first to griseophenone B (**185b**) and then to griseofulvin; griseophenones B and C are also converted to grieseophenone A (**185a**), but [^{36}Cl]griseophenone A is not converted to griseofulvin. These results led Rhodes and his co-workers[350] to suggest that the final stage in griseofulvin biosynthesis involves the binding of griseophenone B to a multi-enzyme complex which can effect oxidation, reduction and methylation of enzyme-bound intermediates. Free griseophenone A and dehydrogriseofulvin are thus by-products of griseofulvin biosynthesis. Griseophenone A is converted to dehydrogriseofulvin by a variety of microorganisms[270] or by homogenized potato peelings.[271]

B.5.c. *Compounds of Type*

187

javanicin
Fusarium javanicum[272]

L[223]

188

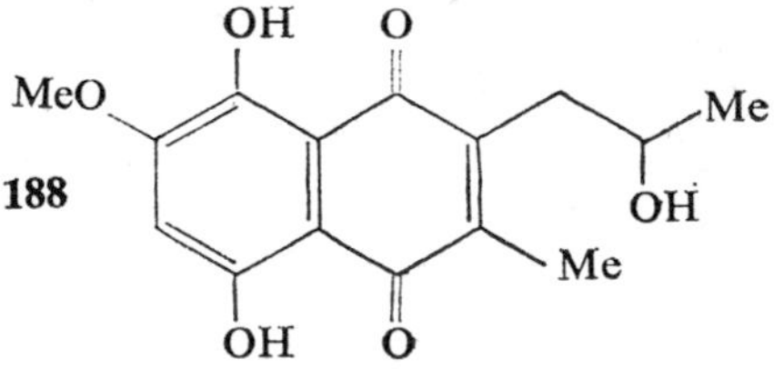

(+)-solaniol
F. solani[274]

189

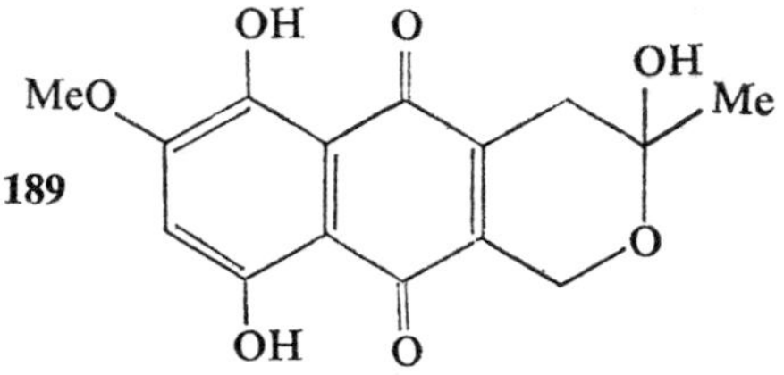

fusarubin
F. javanicum[272]

190

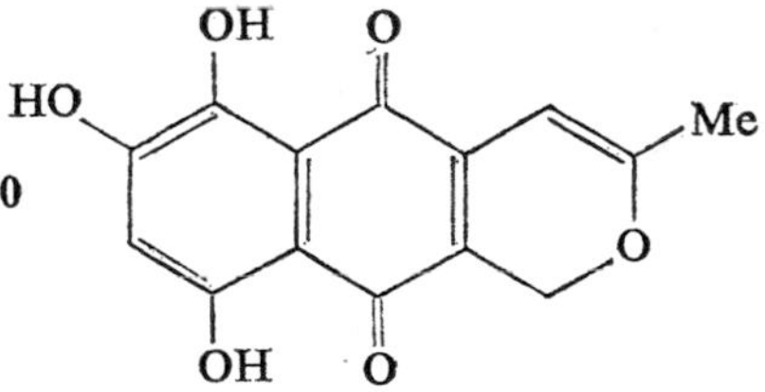

O-demethylanhydrofusarubin
Gibberella fujikuroi[275]

191

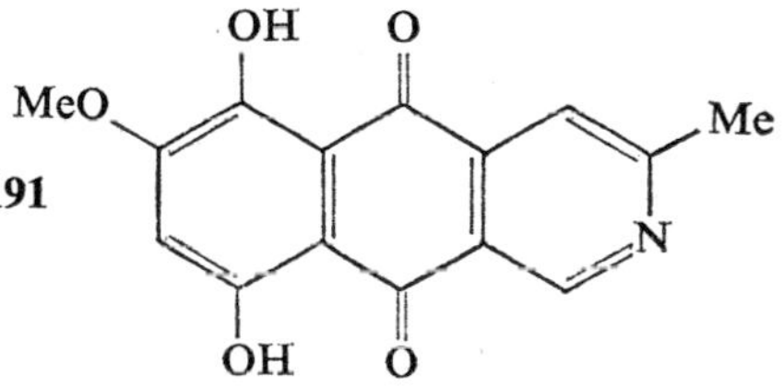

bostrycoidin
F. bostrycoides,[276] *F. solani*[277]

192

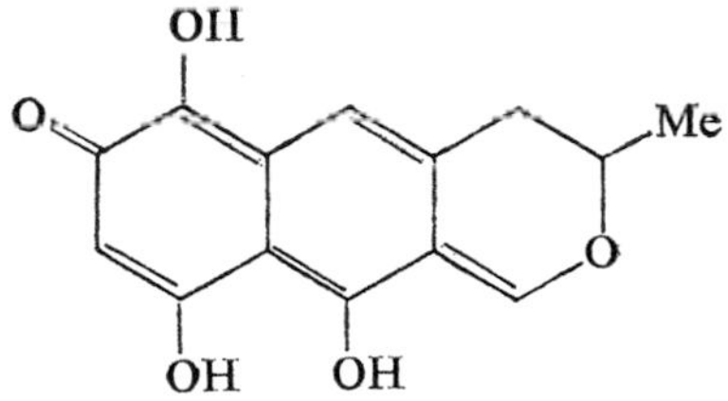

purpurogenone
P. purpurogenum[278]

193

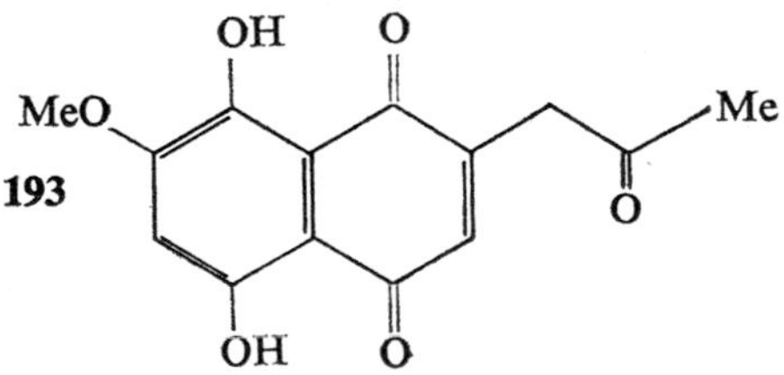

norjavanicin
Fusarium spp.[279]

194   novarubin
Fusarium spp.[279]

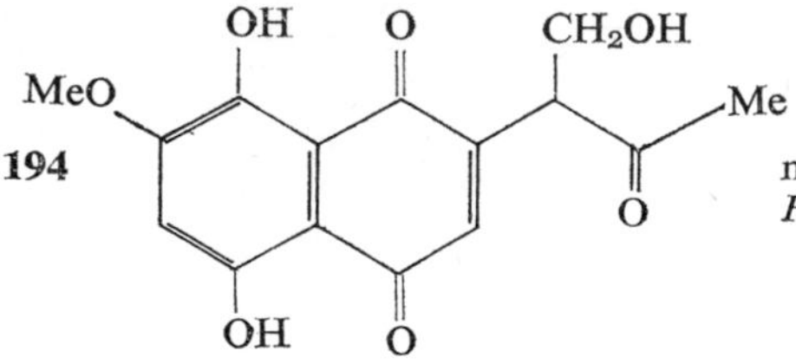

195

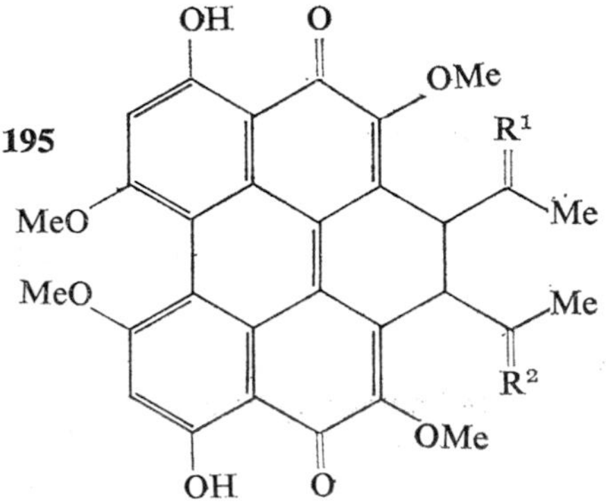

a: $R^1 = R^2 = O$, elsinochrome A L[281]
b: $R^1 = O, R^2 = H, OH$, elsinochrome B
c: $R^1 = R^2 = H, OH$, elsinochrome C
                       Elsinoe spp.[280]

196   vioxanthin
Trichophyton violaceum[282]

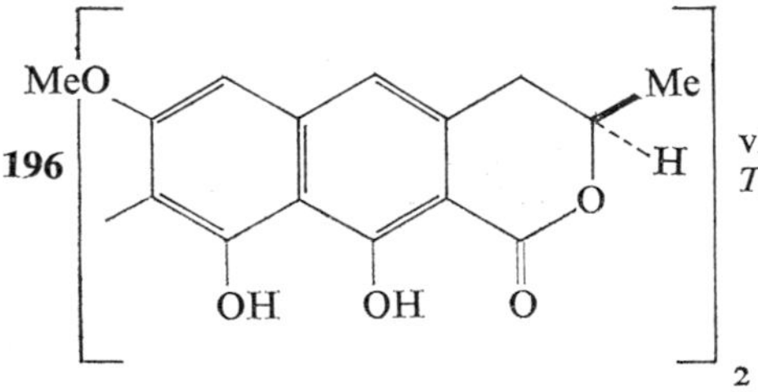

197   viopurpurin
T. violaceum[282]

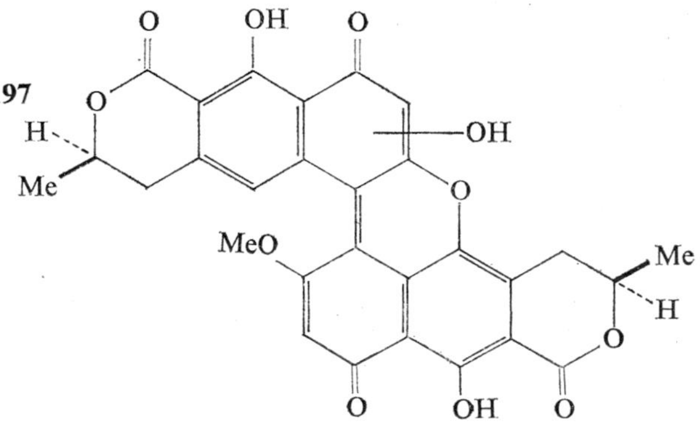

The derivation of javanicin from acetate and malonate has been confirmed by Gatenbeck and Bentley,[273] though they were unable to detect differential labelling of the "starter" acetate unit, possibly due to equilibration of acetate and malonate during the long incubation. They did, however, establish that the aromatic methyl group is derived by complete reduction of the terminal carboxyl group of the polyketide precursor [see also barnol (52) and the tetrahydrofuran (160)].

Elsinochrome A (195a) is identical with phycaron, isolated from *Phyllosticta caryae*, the conidial stage of *Elsinoe randii*. The structure of elsinochrome A suggests that it might be formed by dimerization of a javanicin-type precursor, and this is supported[281] by labelling experiments with [1-14C] acetate. The timing of the decarboxylation and dimerization steps is unknown.

Vioxanthin (196) and viopurpurin (197) co-occur with xanthomegnin (207)

B.5.d. *Compounds of Type*

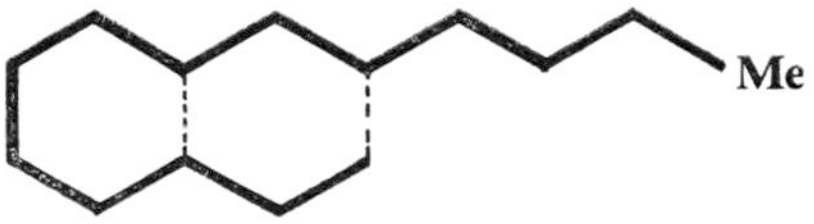

198

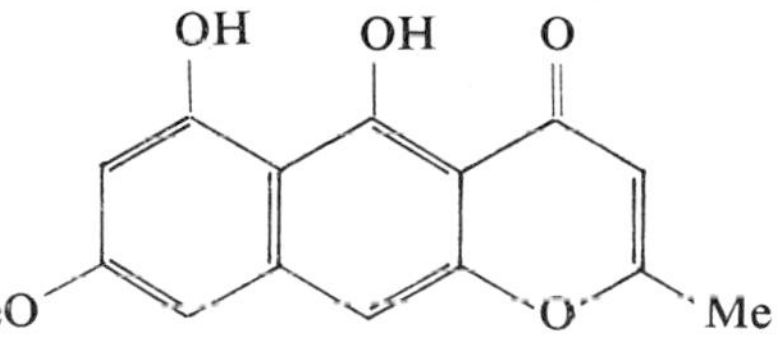

rubrofusarin
Fusarium culmorum, F. graminearum[283]

199

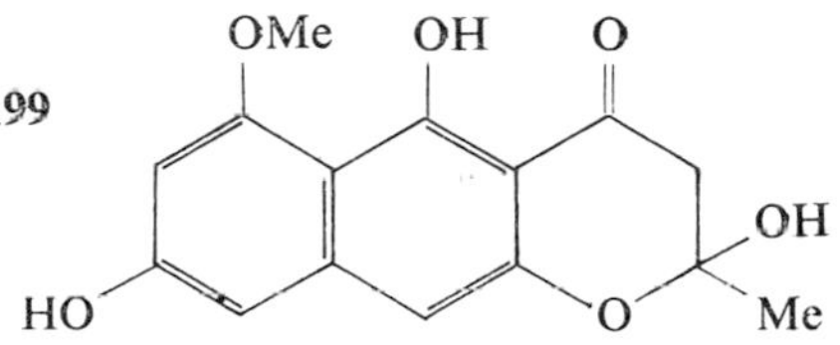

fonsecin
A. fonsecaeus[284]

200

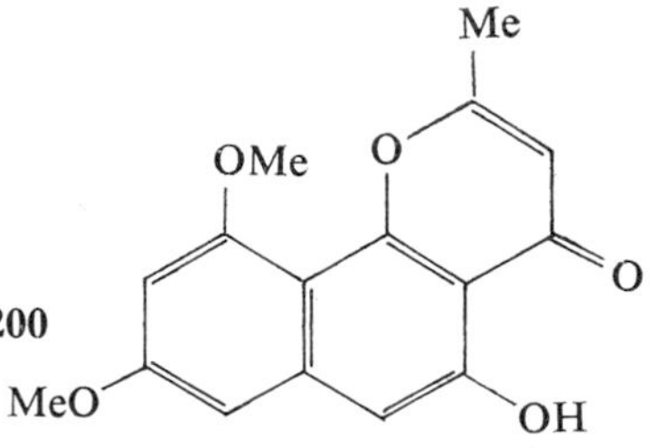

flavasperone (asperxanthone)
A. niger[285]

201

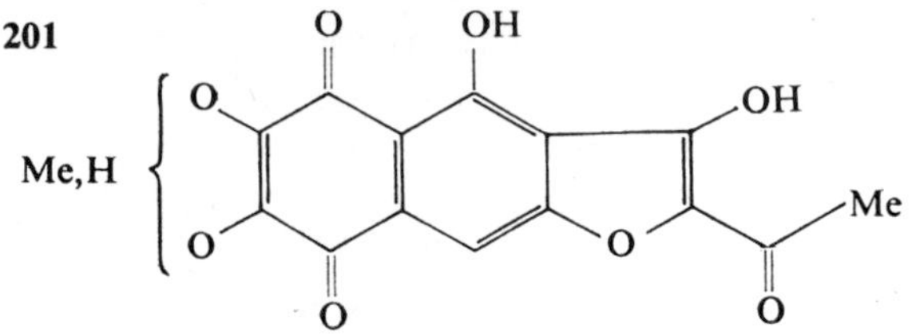

Me,H

rhodocladonic acid
Cladonia spp.[286]

202

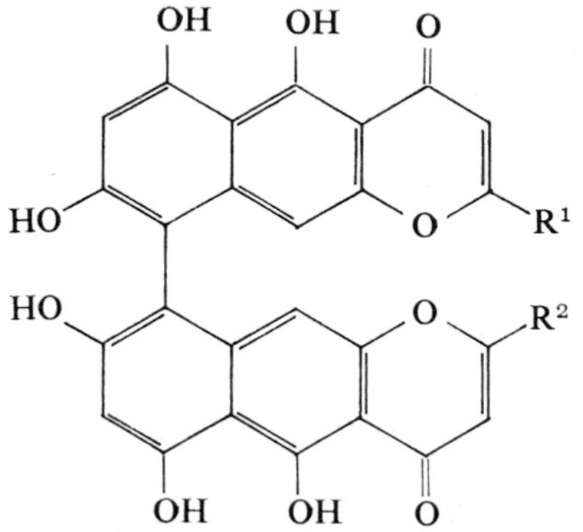

a: $R^1 = R^2 = Me$, ustilaginoidin A
b: $R^1 = Me, R^2 = CH_2OH$, ustilaginoidin B
c: $R^1 = R^2 = CH_2OH$, ustilaginoidin C
Ustilaginoida virens[287]

203

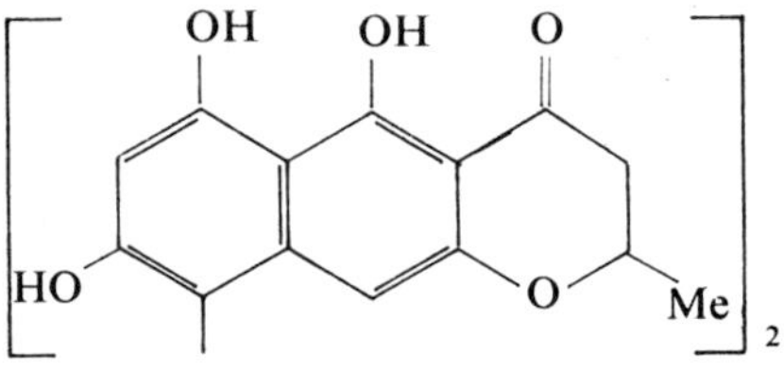

cephalochromin
Cephalosporium sp.[288]

204

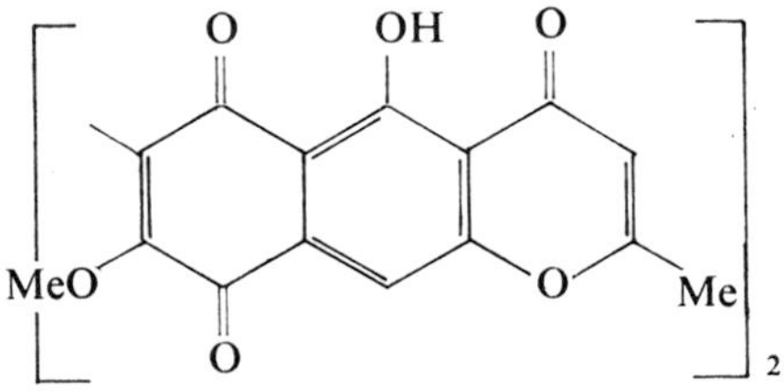

aurofusarin
Fusarium spp.[289], *Hypomyces*
   rosellus[290]

205

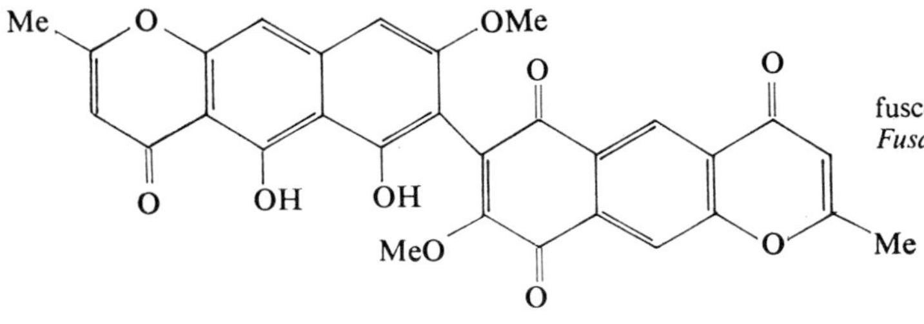

fuscofusarin
Fusarium culmorum

206

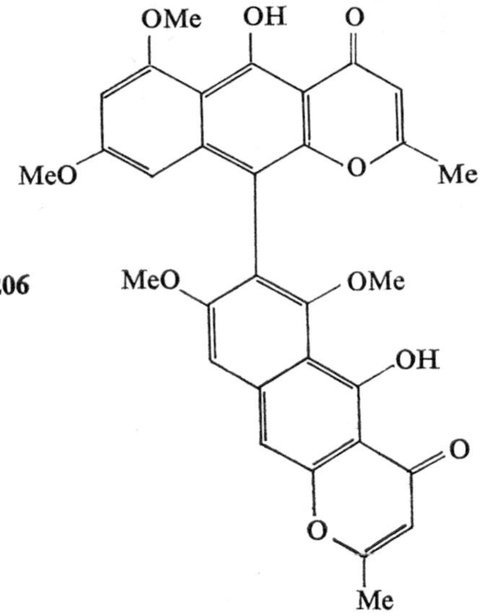

aurasperone A
A. niger, A. awamori[292]
(aurasperone B has the
pyrone double bonds hydrated
as in fonsecin)

A feature of this class is the frequency with which dimers are formed, presumably by phenol coupling (cf. the anthraquinones). The structures of the dimers have all been elucidated since 1964, aurofusarin being studied by no fewer than four independent groups,[290,293] and their chemistry has been reviewed.[294] No labelling experiments with either the monomers or the dimers have been reported.

B.5.c. *Compounds of Type*

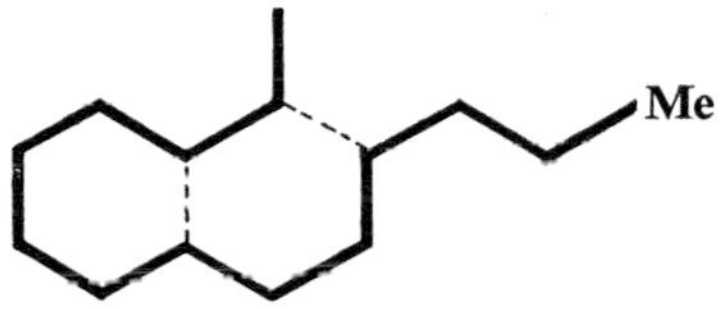

207

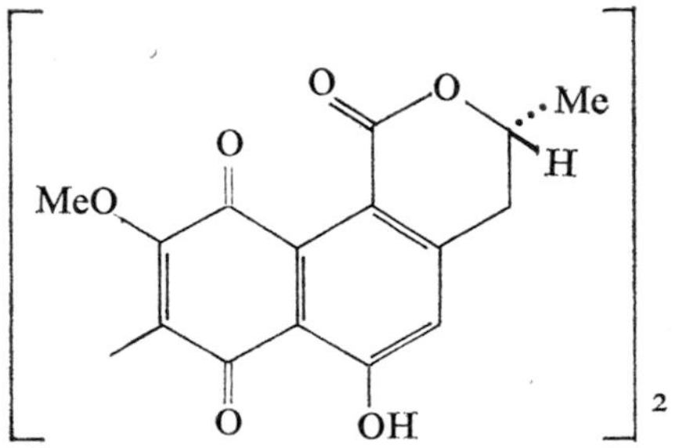

xanthomegnin
Trichophyton megnini,[295a]
T. rubrum,[295b] *T. violaceum*[282]

B.5.f. *Compounds of Type*

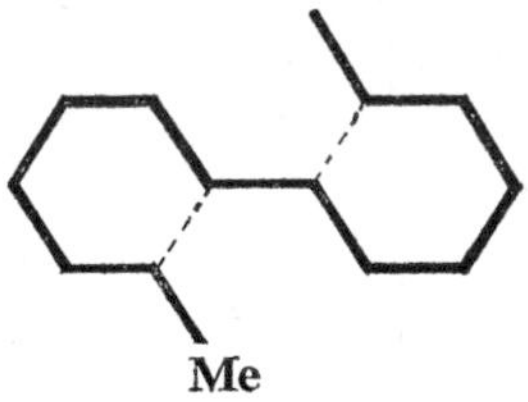

Me

208

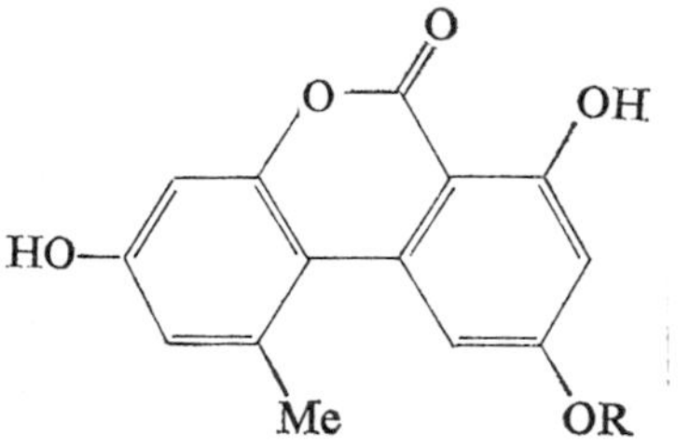

a: R = H, alternariol   L[299,300]
    Alternaria tenuis,[296] *A. dauci*[297]
b: R = Me, alternariol methyl ether

209

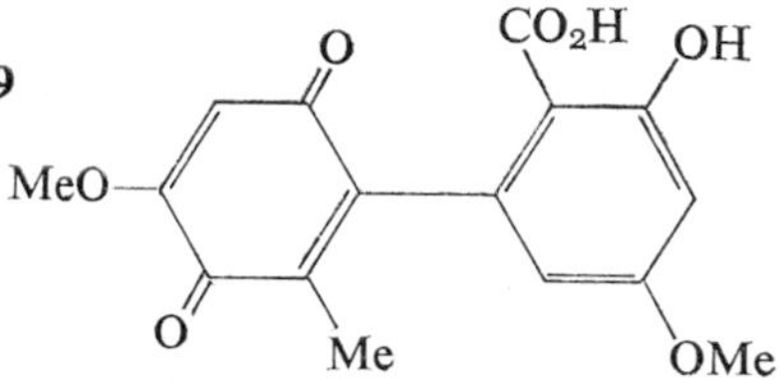

botrallin
Botrytis allii[298]

The biphenyl system, a rare structural feature among mould metabolites, can clearly be derived from seven "C$_2$-units" and this was confirmed by Thomas[299] using [1-^{14}C]acetate. Thomas found that neither orsellinic acid nor its ethyl ester (both labelled at the carboxyl group) was incorporated into alternariol, and a subsequent study[300] of the incorporation of [^{14}C]malonate into alternariol by a cell-free preparation from *A. tenuis* has confirmed that the compound is formed from one chain. The synthesis of alternariol by a purified enzyme system has also been studied.[301] Structure (**209**) for botrallin rests partly on analogy with that of alternariol.

B.5.g. *Atrovenetin and Related Compounds*

210

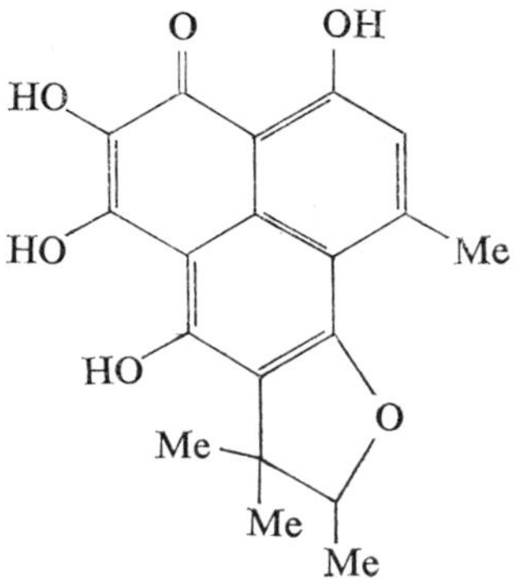

atrovenetin   L[307]
P. atrovenetum,[302] *P. herquei*[303]

211

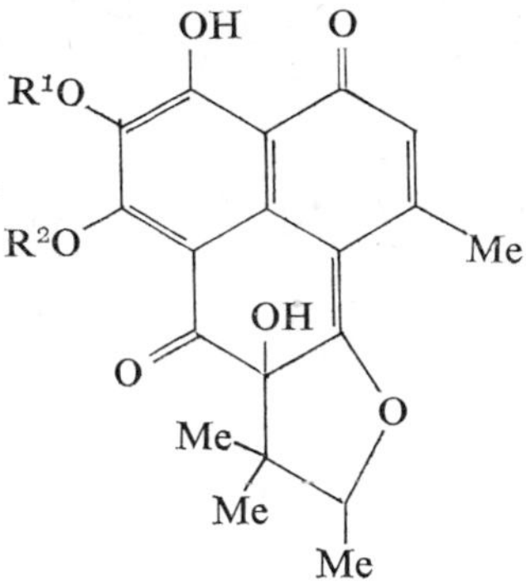

a : R^1 or R^2 = Me, herqueinone
b : R^1 = R^2 = H, norherqueinone
                    P. herquei[304,305]

212

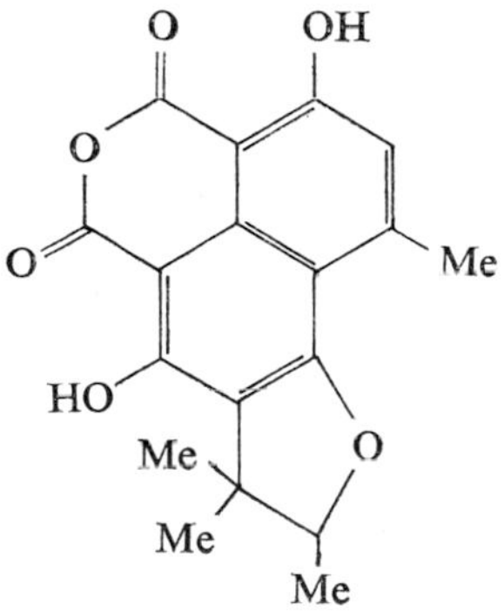

"naphthalic anhydride"
P. herquei[306]

213

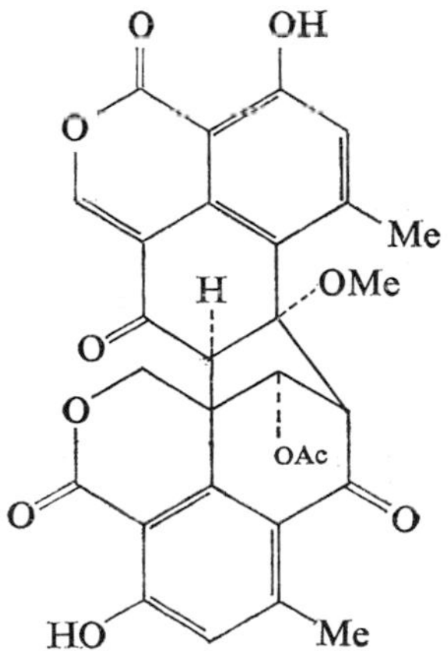

duclauxin L[309]
P. duclauxi[308]

6

O   OH

O

O

O   Me

HO

214 O

OAc

O   O

HO   Me

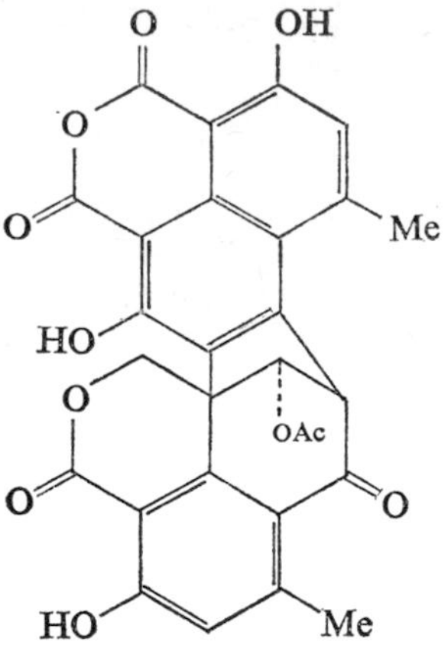

xenoclauxin
P. duclauxi[310]

O   OH

O

O

Me

OH OMe

215 O

OAc

O   O

HO   Me

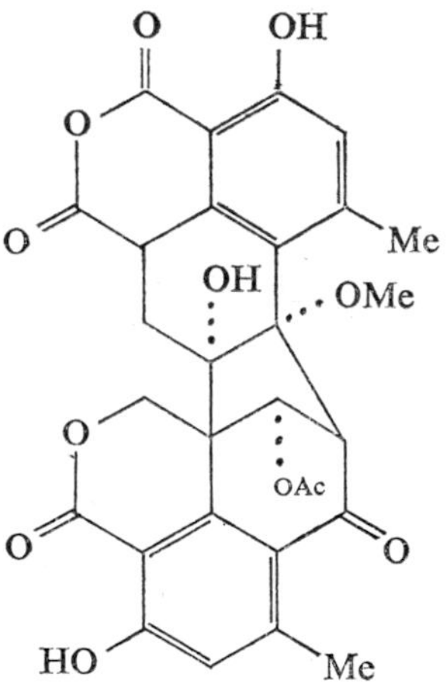

cryptoclauxin
P. duclauxi[310]

The structure of atrovenetin was finally established by X-ray analysis[311] after extensive chemical studies by Barton and his co-workers[312] had established the relationship between atrovenetin and herqueinone and had led them to propose the structure in which the fusion of the furan ring is reversed. Barton *et al.* suggested that atrovenetin might arise by introduction of a mevalonate-derived unit onto a heptaketide chain (*a* in Scheme 27),† and this has been confirmed by Thomas.[307]

The structure of duclauxin (**213**) was established by X-ray analysis.[236] Tracer experiments[309] have shown that duclauxin is biosynthesized by

† The alternative folding (*b*) of the polyketide chain seems at least as likely.

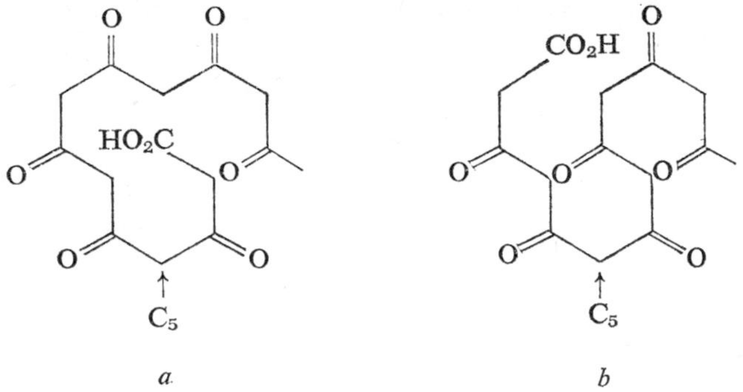

Scheme 27: Polyketide derivations of atrovenetin

dimerization of a unit formed as in Scheme 28 (though folding of the hep-takctide chain as in *b* of Scheme 27 again seems more likely). The oxidation

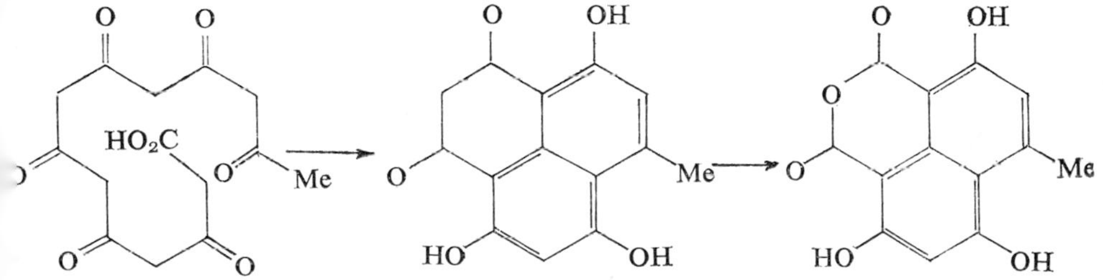

Scheme 28: The biosynthesis of the duclauxin monomer

of atrovenetin to the anhydride (212) with alkaline hydrogen peroxide highlights the close relationship between the *P. herquei* and *P. duclauxi* pigments.

B.5.h. *Citromycetin and Fulvic Acid*

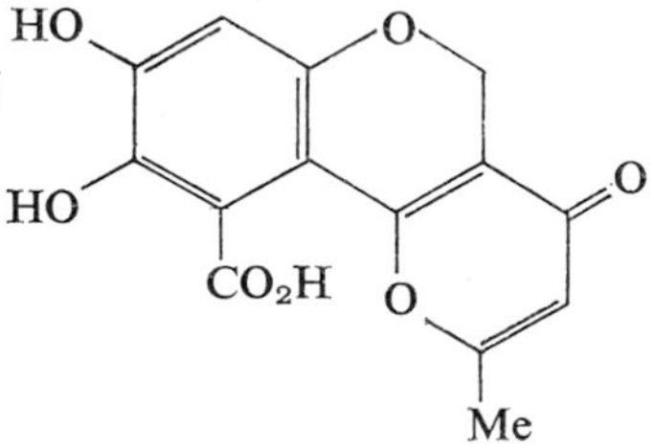

citromycetin L[186,316]
P. frequentans, and others[314]

217

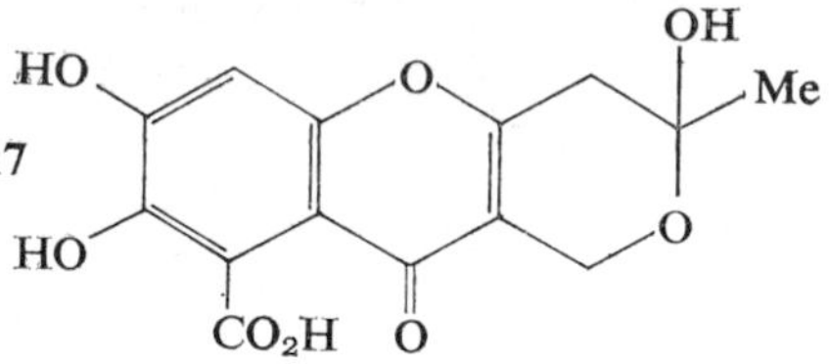

fulvic acid
P. griseofulvum, P. flexuosum,
P. brefeldianum[315]

Citromycetin (**216**) and fulvic acid (**217**) possess the same carbon skeleton. Birch and his co-workers[186] showed that the skeleton of citromycetin is derived from seven acetate units without introduction of carbon from the C_1-pool, leaving two possibilities to be considered (Scheme 29)—condensation

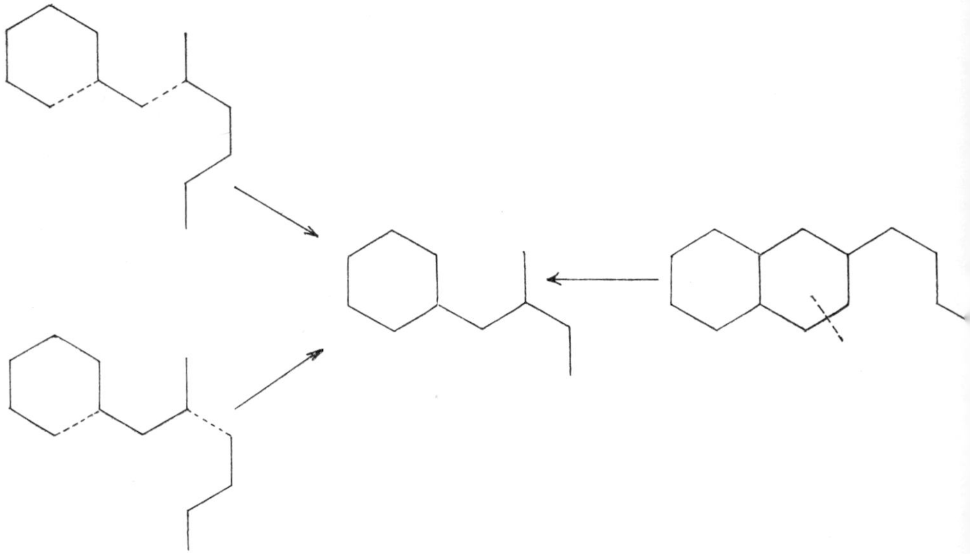

Scheme 29: Alternative biosyntheses of citromycetin and fulvic acid. A two-chain derivation has been established

of two separate acetate-derived chains or oxidative ring-opening of a rubro-fusarin-type precursor. The two-chain hypothesis is supported by experiments with [2-^{14}C]malonate[316] which reveal that *two* acetate "starter" units are involved in the biosynthesis of citromycetin.

B.5.i. *Other Theories for the Biosynthesis of Heptaketides*

With the exception of citromycetin (and, by analogy, fulvic acid) we have assumed in the foregoing discussion that the compounds derived from seven

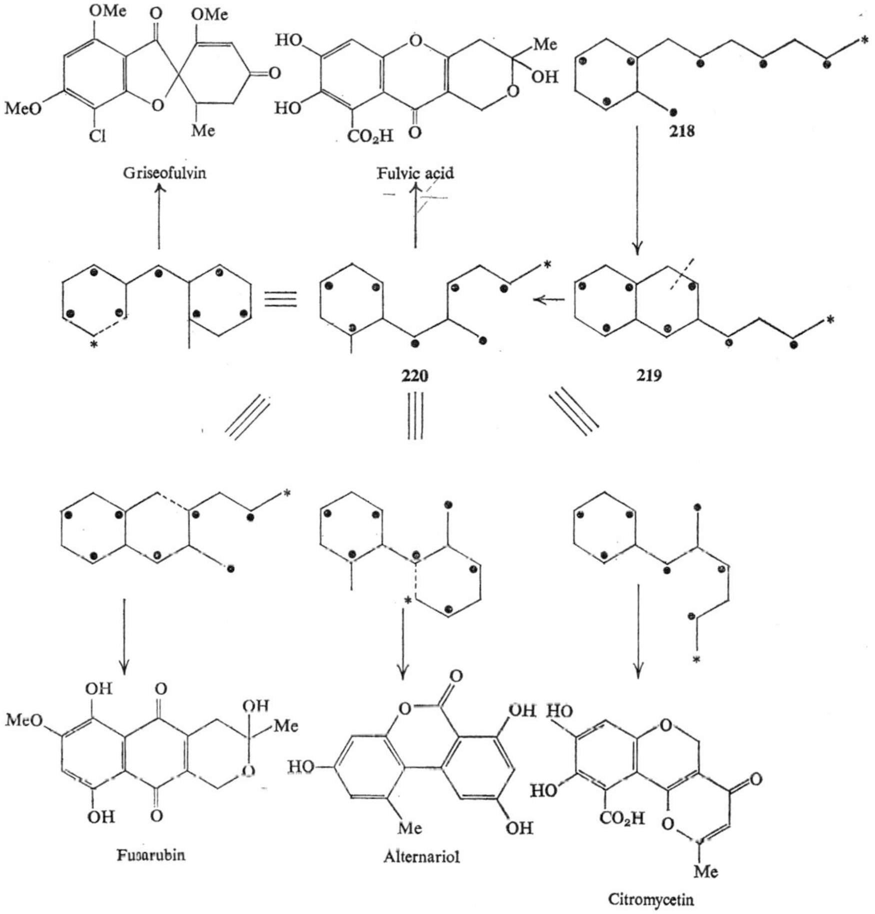

Scheme 30: An alternative scheme which relates all the heptaketides

"C_2-units" arise by appropriate folding of a straight-chain precursor, but two alternative theories have been advanced.

In the first, developed by Jones,[317] oxidative cleavage of an intermediate (219, Scheme 30) of the rubrofusarin type, itself derivable from an intermediate (218) of the palitantin type, gives a third intermediate (220) from which, by appropriate cyclizations, most of the skeletons derivable from seven "C_2-units" may be formed. The attraction of the scheme is that it provides a rationale for the production of apparently unrelated compounds, e.g. griseofulvin and fulvic acid, frequentin and citromycetin, by the same organism. It also leads to the same distribution of labelling from acetate (carbonyl groups are marked •) as would arise from a simple straight-chain precursor. On the other hand the acetate "starter" unit (marked *) occupies a different position in some of the molecules, and malonate experiments should serve to distinguish between the two alternatives. The only results of this type so far reported (for alternariol and for citromycetin) are not in accord with the theory.

According to the second theory, due to Vanek and Soucek,[318] several of the compounds formally derivable from a heptaketide chain, including griseofulvin, citromycetin and fusarubin, arise by condensation of two shorter chains. This theory, too, can be tested by malonate experiments and in the case of citromycetin, the only example examined, has been shown to be valid.

In the absence of further evidence, it seems reasonable to assume that those compounds which can be derived by cyclization of a straight-chain precursor are formed in this manner, reserving the two-chain hypothesis for compounds, such as citromycetin, which cannot be simply derived from a straight chain.

B.6. OCTAKETIDES

B.6.a. *Compounds of Type*

Almost all the octaketides are of this type or may, in principle, be derived from compounds of this type, though we shall see in connection with sulochrin that there is still debate on the latter point. The largest chemical class in this group comprises the anthraquinones and related compounds and it is these which we consider first.

B.6.a.i *Anthraquinones and Related Compounds*

221

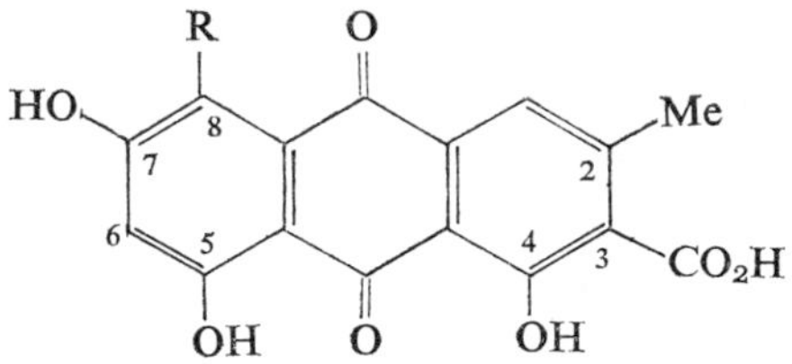

a: R = H, endocrocin
Nephromopsis endocrocea,[319]
Claviceps purpurea,[320]
Fungi Imperfecti

b: R = OH, clavorubin
Claviceps purpurea[321]

222

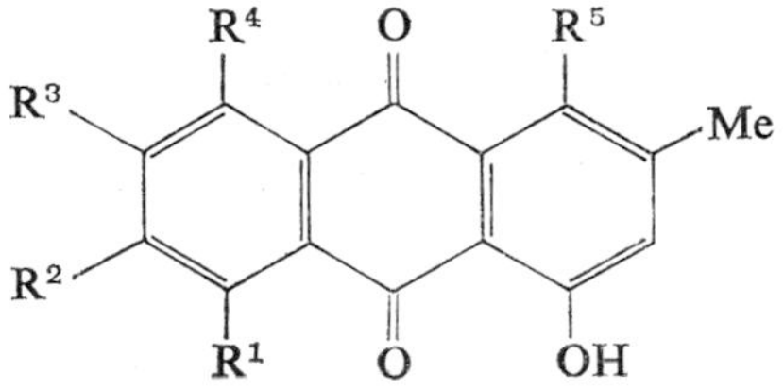

	R^1	R^2	R^3	R^4	R^5	
a	H	H	H	H	H	pachybasin *Pachybasidium candidum*,[322] *Trichoderma viride*[323]
b	OH	H	H	H	H	chrysophanol *P. islandicum*,[324] and other Fungi Imperfecti
c	OH	H	OH	H	H	emodin L^{352} *Cortinarius sanguineus*,[325] Fungi Imperfecti, higher plants
d	OMe	H	OH	H	H	emodin-5-methyl ether (questin) *P. frequentans*[326]
e	OH	H	OMe	H	H	physcion Fungi Imperfecti, lichens, higher plants[327]
f	OH	H	OMe	OH	H	xanthorin *Xanthoria elegans*[328]
g	OH	OH	OMe	H	H	dermoglaucin *Cortinarius sanguineus*[329]

222
(cont.)

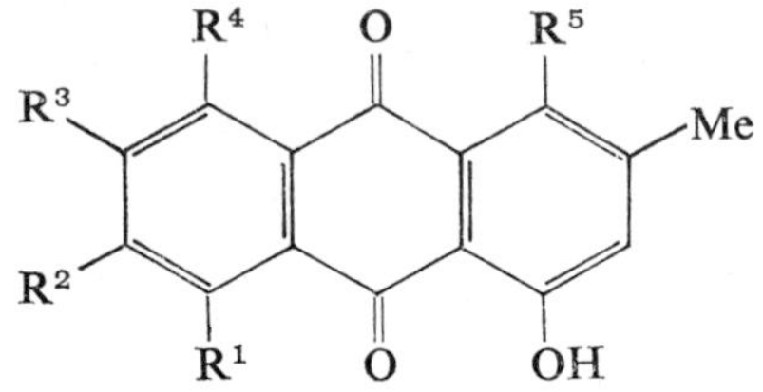

	R^1	R^2	R^3	R^4	R^5		
h	OH	OH	OMe	OH	H	dermocybin *Cortinarius sanguineus*[325,329]	
i	OH	Cl	OH	H	H	AO-1 *Anaptychia obscurata*,[330] *A. fumigatus*[331]	
j	OH	Cl	OMe	H	H	fragillin *Sphaerophorus* spp.[332]	
k	OH	Cl	OH	Cl	H	AO-2 *Anaptychia obscurata*[330]	
l	OH	Cl	OH	OH	H	papulosin, 6-chloro-8-hydroxyemodin *Lasallia papulosa*,[333,334] *Valaria rubricosa*[334]	
m	OH	Cl	OH	H	OH	6-chloro-1-hydroxyemodin *Lasallia papulosa*,[334] *Valaria rubricosa*[334]	
n	OH	H	H	OH	H	helminthosporin *Helminthosporium* spp.[335]	L[354]
o	OH	H	H	H	OH	islandicin *P. islandicum*[336]	L[353,357]
p	OH	H	H	OH	OH	cynodontin *Helminthosporium* spp.[337]	L[355]
q	OH	Me	H	OH	OH	1,4,5,8-tetrahydroxy-2,6-dimethyl- anthraquinone *Curvularia* spp.[365]	
r	OH	H	OH	H	OH	catenarin *Helminthosporium* spp., other Fungi Imperfecti[338]	
s	OH	H	OMe	H	OH	erythroglaucin *A. glaucus*[339]	

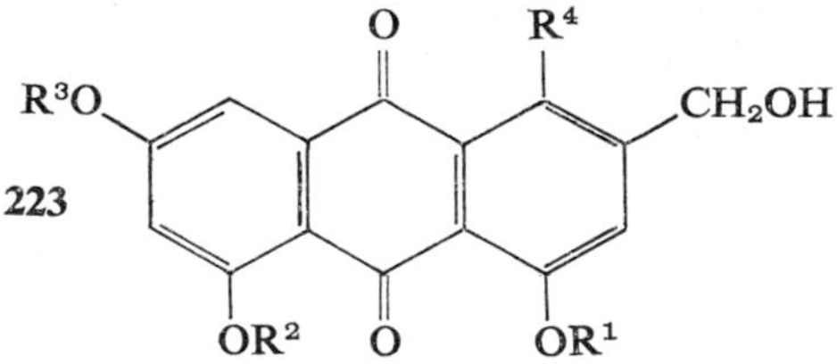

223

	R^1	R^2	R^3	R^4	
a	H	H	H	H	ω-hydroxyemodin *Penicillium* spp.[340]
b	H	Me	H	H	ω-hydroxyemodin-5-methyl ether *P. frequentans*[326]
c	H	H	Me	H	teloschistin *Teloschiste flavians*[341]
d	Me	H	H	H	roseopurpurin *P. roseo-purpureum*[342]
e	H	H	H	OH	tritisporin *Helminthosporium tritici-vulgaris*[338]

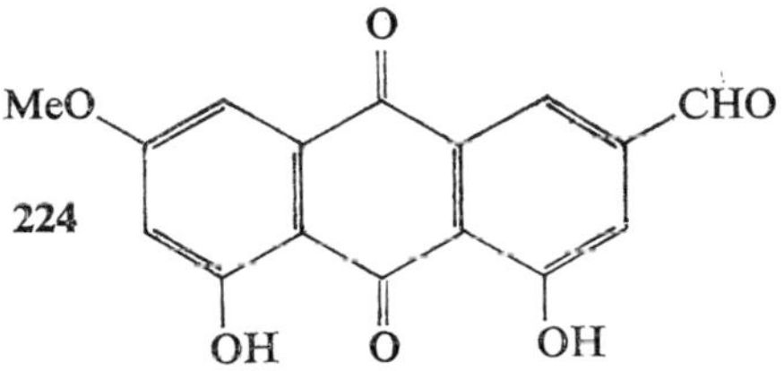

224

fallacinal
Xanthoria fallax[343]

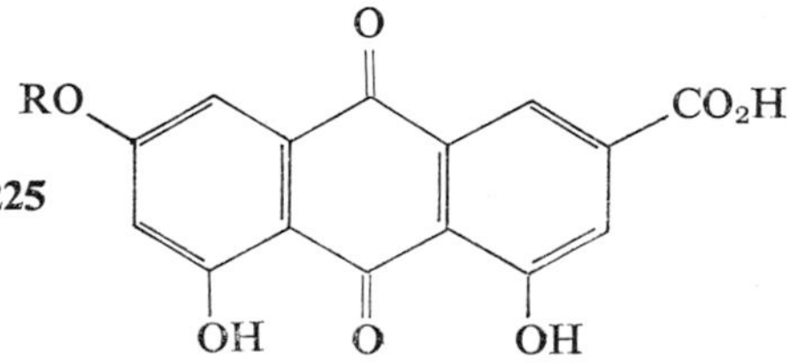

225

a: R = H, emodic acid
      P. cyclopium[340]

b: R = Me, parietinic acid
      Xanthoria parietina[367]

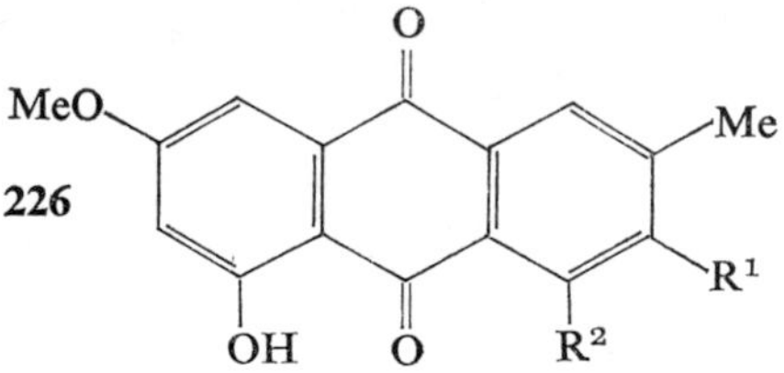

226

a: $R^1 = R^2 = H$, methylxanthopurpurin-7-methyl ether
   Alternaria solani[346]

b: $R^1 = OH, R^2 = H$, macrosporin
   A. porri,[345] *A. solani*[346]

c: $R^1 = R^2 = OH$, 7-methoxy-2-methyl-3,4,5-trihydroxyanthraquinone
   A. solani[346]

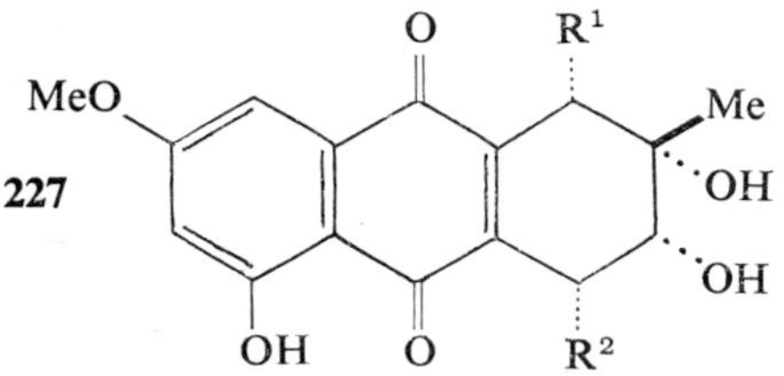

227

a: $R^1 = R^2 = H$, altersolanol B
   A. solani[346]

b: $R^1 = H, R^2 = OH$, bostrycin
   Bostrychonema alpestre[347]

c: $R^1 = R^2 = OH$, altersolanol A
   A. solani[346]

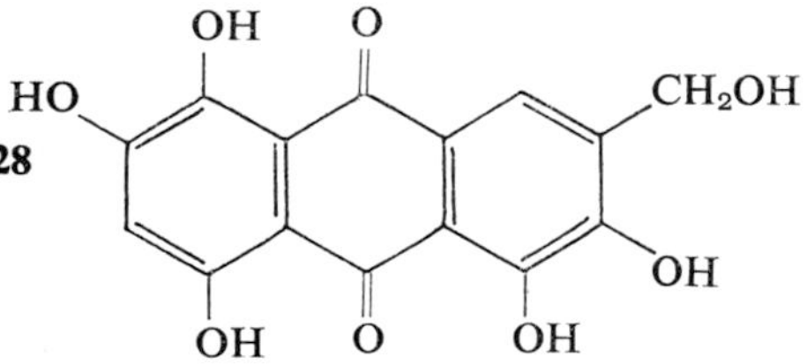

228

asperthecin
Aspergillus spp.[348]

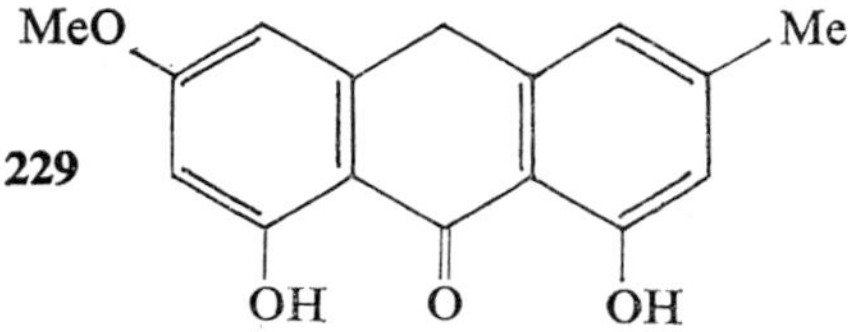

229 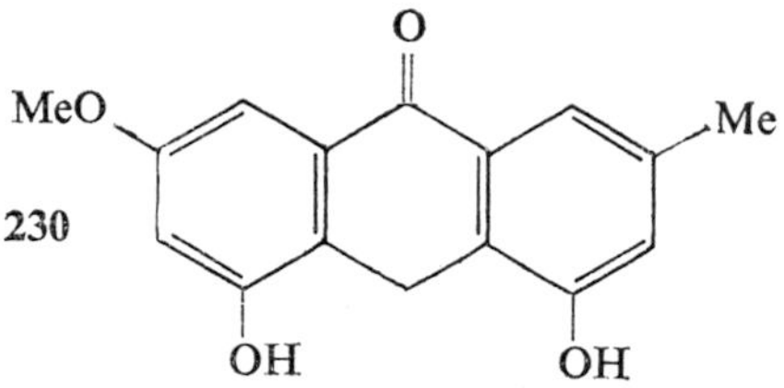   **230**

physcion anthrones A and B
A. glaucus[339]

231

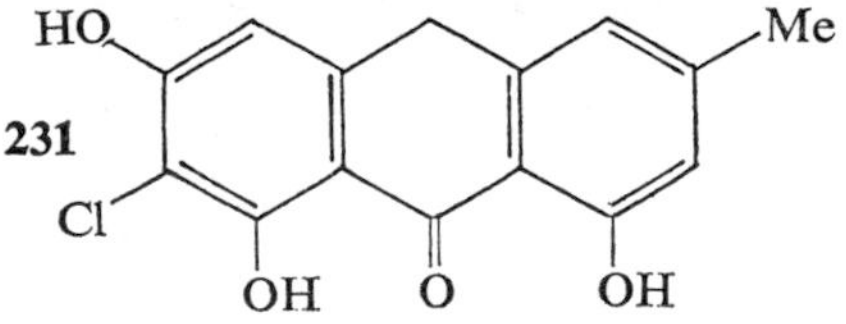

2-chloro-1,3,8-trihydroxy-6-methylanthrone
A. fumigatus[331]

232

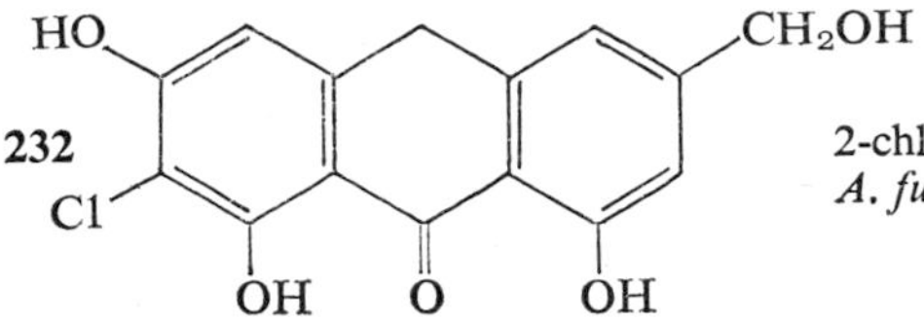

2-chloro-1,3,8-trihydroxy-6-hydroxymethylanthrone
A. fumigatus[331]

233

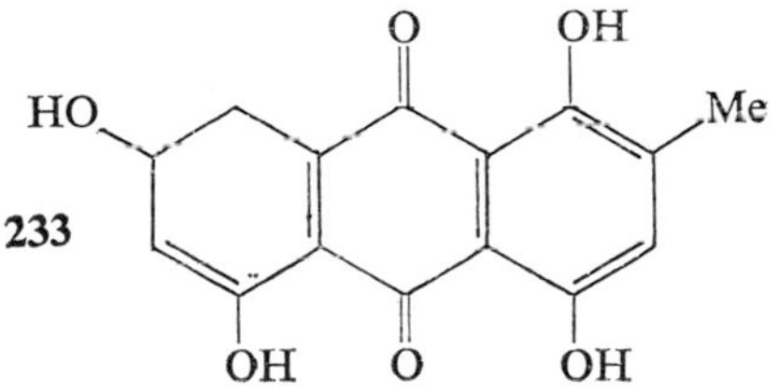

dihydrocatenarin
P. islandicum[349]

234

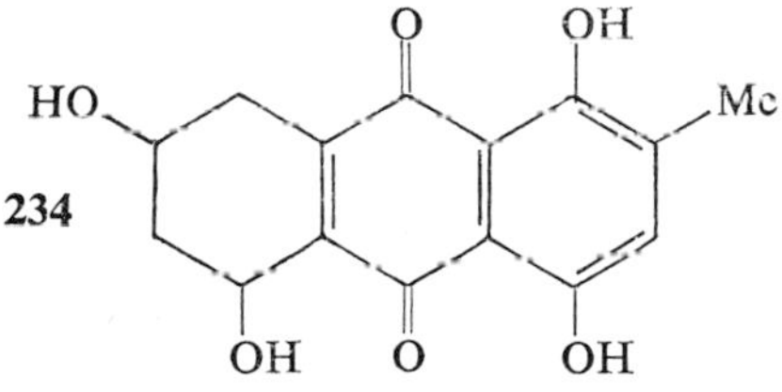

tetrahydrocatenarin
P. islandicum[340]

The anthraquinones have been isolated from Fungi Imperfecti, Basidiomycetes, lichens and higher plants, and some have been obtained from more
than one of these sources. For example emodin (**222c**) was first isolated in
1858 from rhubarb[351] and occurs either in the free state or as a glycoside in
many higher plants; the first isolation from a fungus was from the toadstool
Cortinarius sanguineus[325] and it has since been isolated from several Fungi

Imperfecti. *P. islandicum* is a particularly rich source of anthraquinones and their dimers.

The polyketide origin of emodin,[352] islandicin,[353] helminthosporin[354] and cynodontin[355] has been demonstrated with [1-^{14}C]acetate. Gatenbeck[353] has also used [1-^{14}C,^{18}O]acetate to study the origin of the oxygen atoms of islandicin and emodin† and found, as expected, that three oxygen atoms of islandicin and four oxygen atoms of emodin are derived by reduction of the carbonyl group of acetate. These experiments do not answer the question of whether the anthraquinones are formed from a single polyketide chain

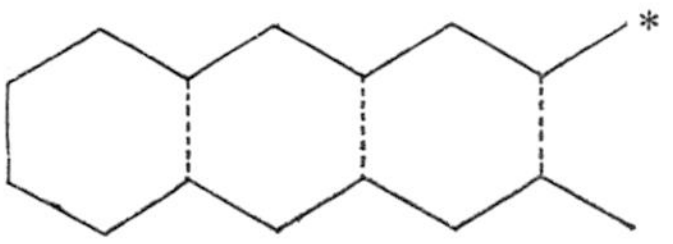

235a 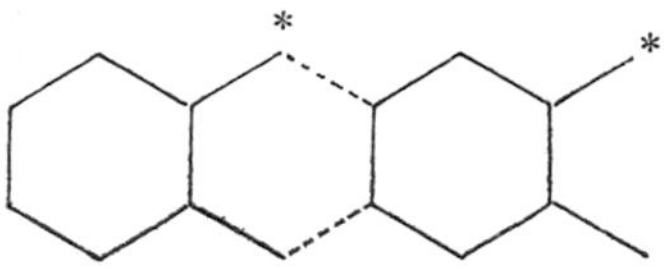   235b

(as in **235a**) or from two preformed benzene derivatives (as in **235b**); the latter process has been suggested on several occasions[356] and is analagous to the "two-chain theory" for the heptaketides (p. 156). The problem has been resolved in the case of islandicin by experiments with [^{14}C]malonate[357] which show conclusively that only one "starter" acetate unit (marked with an asterisk) is present, as in **235a,** and not two, as required by **235b.** A similar result has been obtained with the dimeric compound rugulosin (**254b**) (see below).

Chrysophanol is formed by the polyketide route in higher plants,[358] and this is probably true for related compounds such as emodin and physcion. On the other hand some anthraquinones of higher plants are formed by a different route, involving condensation of mevalonate with a naphthalenic precursor (Scheme 31).[359]

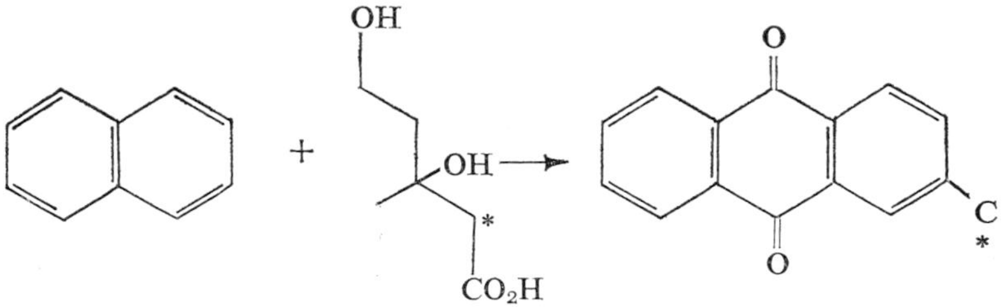

Scheme 31: Non-polyketide biosynthesis of some anthraquinones in higher plants

† The emodin was obtained by cleavage of the more readily accessible dimer, skyrin (**239**) (see below).

The "parent" anthraquinone is endocrocin (221a) which retains all the oxygen atoms of the polyketide precursor and has had only the second quinone oxygen introduced. With the exception of clavorubin (221b) the remaining anthraquinones all lack the carboxyl group of endocrocin. In general the oxygenation pattern reflects the polyketide origin of the compounds, hydroxyl (or methoxyl) groups being present at position 4 (two exceptions are discussed below) and usually at positions 5 and 7. Extra oxygenation can occur at positions 1,6 and 8, and halogen has been found at positions 6 and 8. The methyl group may be oxidized to alcohol, aldehyde or

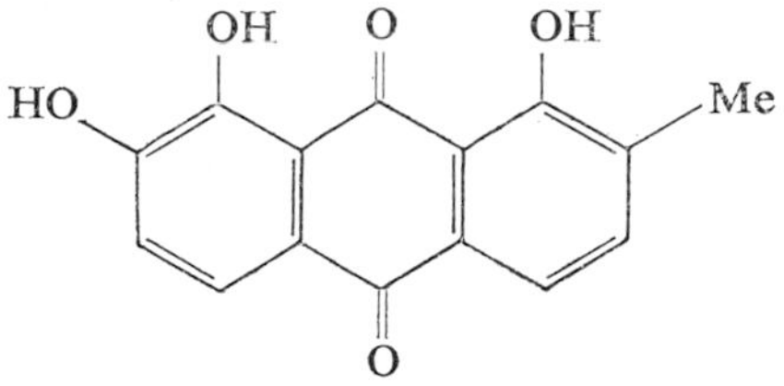

236

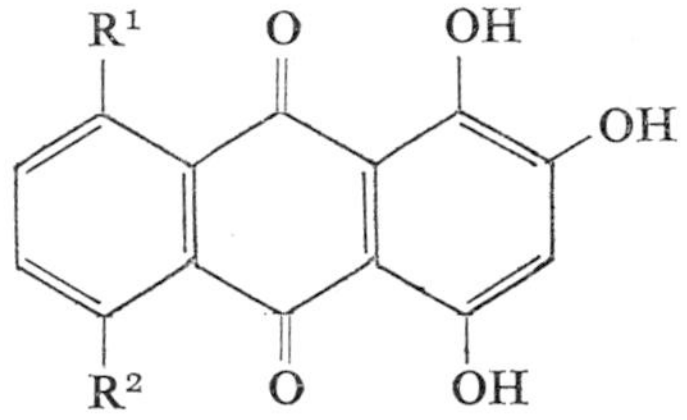

237   (R^1 or $R^2 = CO_2H$)

acid. Compound (222q) is the only example where C-methylation, common among other polyketides, seems to have occurred. Although anthrones should be the precursors of anthraquinones, until the recent isolation of the chloro compounds (231) and (232), the physcion anthrones (229) and (230) were the only monomeric examples of the class from fungi. The anthrone (230) must be derived from, rather than be a precursor of, physcion (222e).

Methylxanthopurpurin-7-methyl ether (226a) and macrosporin (226b) are the only apparently polyketide-derived anthraquinones which lack a hydroxyl group at positions 4. This could result from hydration of the aromatic system to give, say, bostrycin (227b), followed by dehydration with loss of hydroxyl from position 4. This could also account for the presence of a 3-hydroxyl group (another uncommon feature) in macrosporin, and the co-occurrence of 226a and 226b with the altersolanols (227a) and (227c) suggests a biosynthetic relationship between the two series.

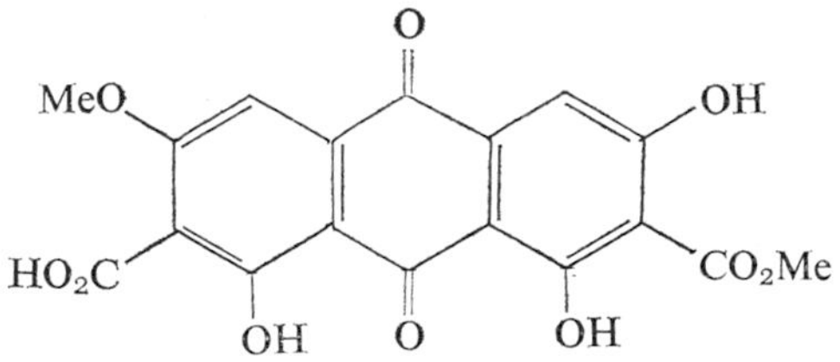

238

Finally, there are three compounds for which anthraquinone structures have been proposed which do not fit into the polyketide pattern discussed above; in each case the structure has now been shown to be incorrect or doubtful. Synthesis of structure (236) proposed[360] for cladofulvin has shown it to be incorrect;[361] structure (237) proposed[362] for boletol has been questioned;[363] and rhodocladonic acid, for which structure (238) was proposed,[364] has been shown to be a naphthoquinone (see p. 147).

B.6.a.ii. *Dimeric anthraquinones*

A feature of Fungi Imperfecti, which has been studied particularly in the case of *Penicillium islandicum,* is their ability to form dimeric anthraquinones, presumably by oxidative coupling. The simplest example is skyrin (239),

239

the dimer of emodin, and, as a result of a detailed study of the metabolites of *P. islandicum* by Shibata and his colleagues, the compounds formed by similar coupling of emodin (222c), chrysophanol (222b), islandicin (222o) and catenarin (222r) in all possible combinations are now known;[366] they are listed in Table 1. The compounds are all optically active and give the same (positive Cotton) O.R.D. curves.

Table 1: The dimeric anthraquinones

	Name	Monomers
239	Skyrin	$(emodin)_2$
240	Dianhydrorugulosin	$(chrysophanol)_2$
241	Iridoskyrin	$(islandicin)_2$
242	Dicatenarin	$(catenarin)_2$
243	Roseoskyrin	chrysophanol + islandicin
244	Auroskyrin	chrysophanol + emodin
245	Rhodoislandin A	chrysophanol + catenarin
246	Rhodoislandin B	emodin + islandicin
247	Punicoskyrin	islandicin + catenarin
248	Aurantioskyrin	emodin + catenarin
249	Oxyskyrin	emodin + ω-hydroxyemodin (223a)
250	Skyrinol	$(\omega$-hydroxyemodin$)_2$

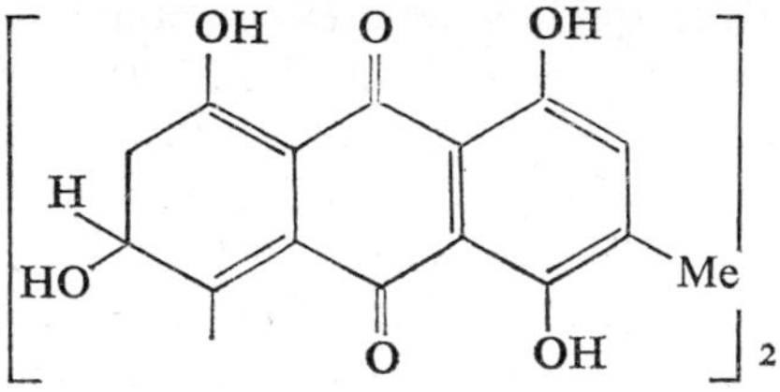

251

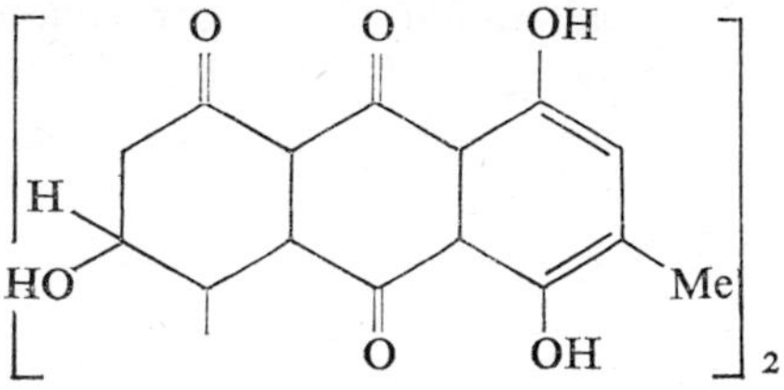

252

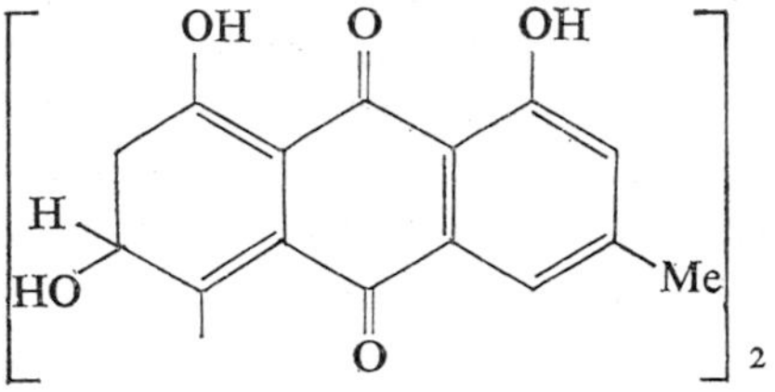

253

254

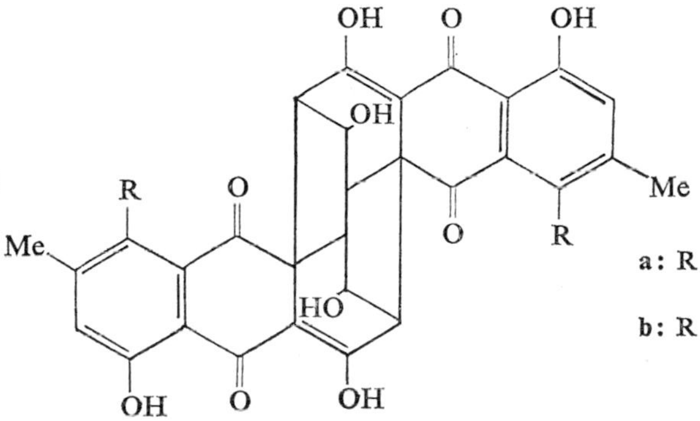

a: R = OH, luteoskyrin
        P. islandicum[368]
b: R = H, rugulosin
        P. rugulosus[368]

255

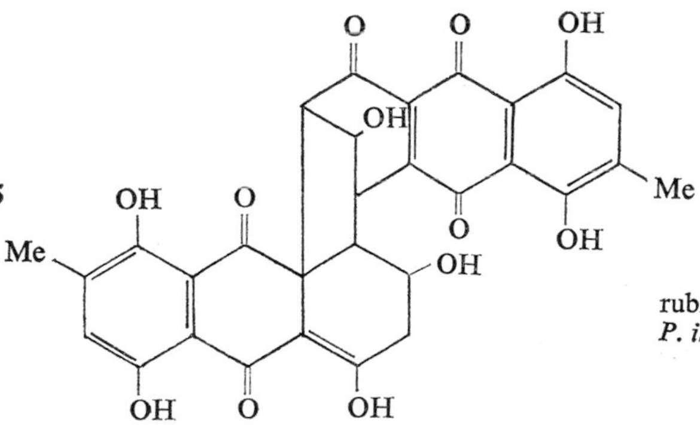

rubroskyrin
P. islandicum[368]

In addition to the bisanthraquinones there are three coupled compounds—luteoskyrin and rubroskyrin from *P. islandicum* and rugulosin from *P. rugulosum*—which were long believed to have the partially reduced bis-anthraquinone structures (251) to (253) respectively, but whose structures have now been revised to (254a), (255) and (254b) respectively.[368]

The acetate origin of skyrin (239)[352,353] and the acetate–malonate origin of rugulosin (254b)[369] have been confirmed. In the latter experiments two acetate "starter" units were detected, demonstrating (a) that the dimeric compounds are formed from two monomers and (b) that each monomer is derived from a single polyketide chain (see islandicin above). The biosynthetic relationship between the monomers, the dimers, and the modified dimers has been studied[370] by incubating *P. islandicum* for a short time with [1-^{14}C] acetate and plotting specific activity against time for islandicin, skyrin (239), rubroskyrin (255), and iridoskyrin (241). Analysis of the results showed (1) that skyrin was the first to appear (!) and that islandicin is not a precursor of iridoskyrin and (2) that neither rubroskyrin nor iridoskyrin is a precursor of the other. In other words, the various anthraquinone metabolites are formed in parallel and not in sequence. The finding[371] that mutation of *P. islandicum* failed to block selectively any of the anthraquinones is in accord with this. The problem has been discussed by Shibata[294] and by Bu'Lock,[372] and it seems likely that the interconversions occur at an earlier stage in the biosynthesis, possibly at the anthrone level.

B.6.a.iii *Dimeric Anthrones*

256   penicilliopsin
Penicilliopsis clavariaeformis[373]

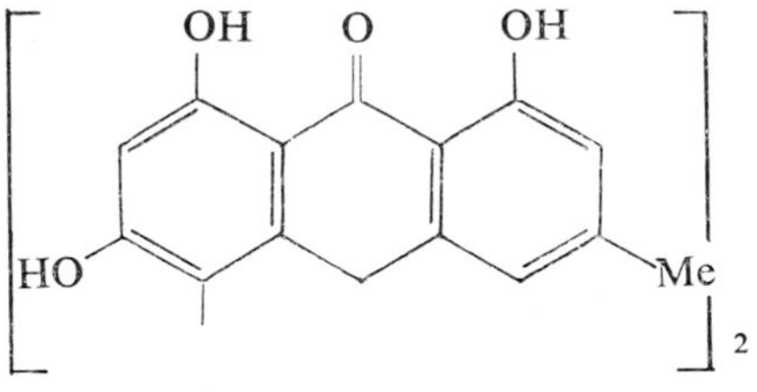

257   flavoobscurin A
Anaptychia obscurata[374]

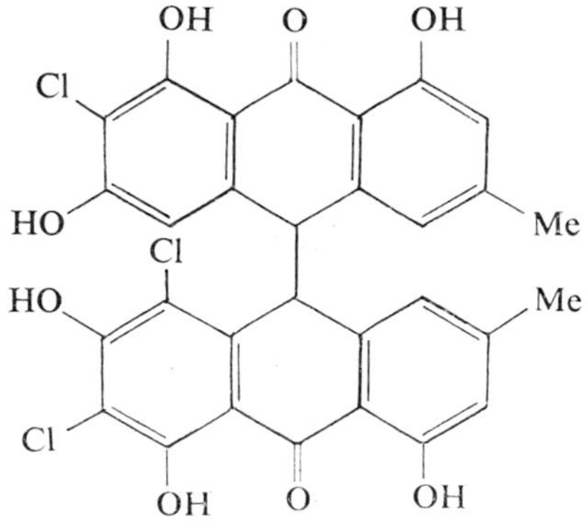

258   flavoobscurins B$_1$ and B$_2$
Anaptychia obscurata[374]

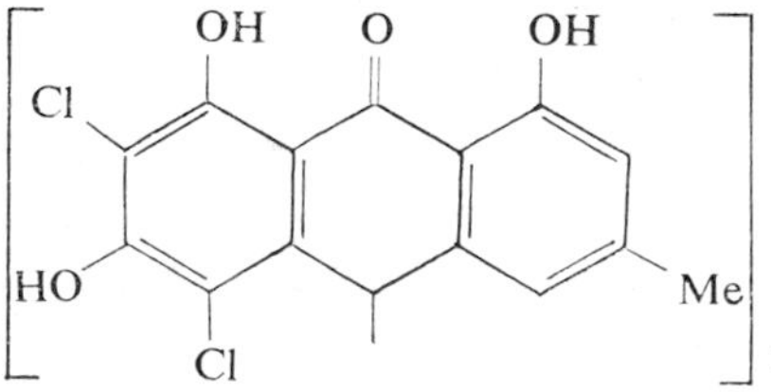

259   flavomannin
P. wortmanni [375]

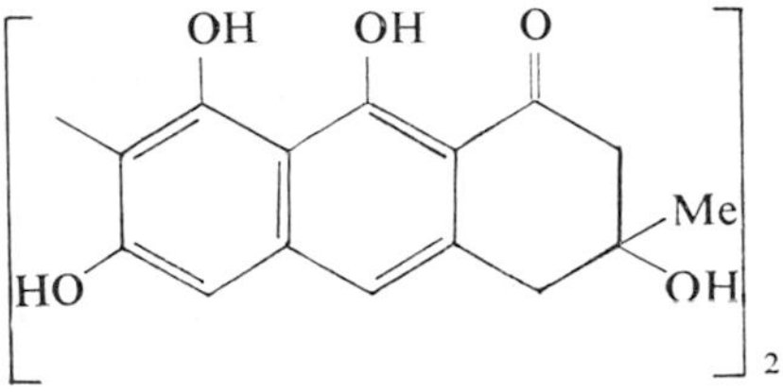

The flavoobscurins are coupled products of the anthrones corresponding to the quinones AO–1 (**222i**) and AO–2 (**222k**); flavoobscurins B$_1$ and B$_2$ are stereoisomers. Flavomannin (**259**) is a hydrated anthrone dimer.

B.6.a.iv. *The Ergochromes*[376]

Ergot, the sclerotia produced by *Claviceps purpurea* when growing on rye, contains a mixture of pigments from which the first crystalline product was isolated in 1877. Only in recent years, however, has the resolution of the mixture into its components permitted rational structural work to be carried out. An excellent review of the early work on the compounds has been given by Eglinton *et al.*[377] The pigments are 2,2'- or 4,4'- dimers of the monomeric structures of Table 2, three 4,4'-compounds and seven 2,2'-compounds being known. Franck and his colleagues[378] use the nomenclature "ergochrome AA (4,4')" etc., for the compounds, which are listed in Table 2; thus ergoflavin, the first compound to have its structure established by chemical[379a] and X-ray analysis,[379b] becomes ergochrome CC (2,2'). Earlier, the compounds had been termed ergochrysins, secalonic acids and chrysergonic acids (see, e.g. refs. 380, 381). The ergochromes are also produced when *C. purpurea* is grown on liquid media.[382] Recently ergochrome AA (4,4') (secalonic acid A) has been isolated from the lichen *Parmelia entotheiochroa*.[383]

The co-occurrence of the ergochromes and the anthraquinones (**221a**) and (**221b**) suggested that the ergochromes might arise by cleavage of an

Table 2: The ergochromes

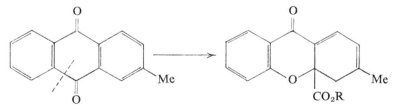

260	ergochrome AA (4,4′)	
261	ergochrome BB (4,4′)	
262	ergochrome CC (2,2′)	
263	ergochrome AB (4,4′)	L^{384}
264	ergochrome AC (2,2′)	

265	ergochrome BC (2,2′)
266	ergochrome AD (2,2′)
267	ergochrome BD (2,2′)
268	ergochrome CD (2,2′)
269	ergochrome DD (2,2′)

anthraquinone or related compound, as indicated in Scheme 32. Support for this hypothesis has been provided by Franck and Hüper[384] who showed that ergochrome AB (4,4′) (secalonic acid C) produced in the presence of [2-^{14}C]acetate had the expected distribution of radioactivity. More recently

Scheme 32: The derivation of the ergochromes from an anthraquinone precursor

Franck and his colleagues[385] have shown that ergochromes produced in the presence of [1-^{14}C]acetate, [2-^{14}C]acetate, or [2-^{3}H$_2$]acetate have labelling patterns consistent with their derivation via endocrocin, and that radioactivity from labelled emodin is incorporated into the ergochromes.

B.6.a.v. *Sulochrin and Related Compounds*

70

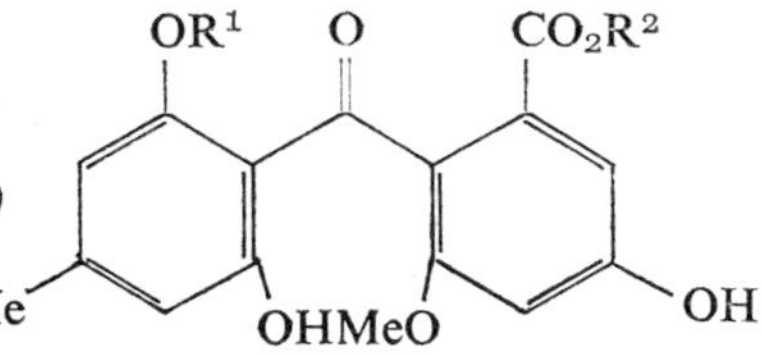

OR¹ O CO₂R² / Me / OHMeO / OH

a: $R^1 = H, R^2 = Me$, sulochrin $L^{398-401}$
   Widespread among Fungi Imperfecti
b: $R^1 = R^2 = Me$, monomethylsulochrin
   A. fumigatus[386]
c: $R^1 = R^2 = H$, desmethylsulochrin (Compound G)
   Oospora sulphurea-ochracea[387]

1

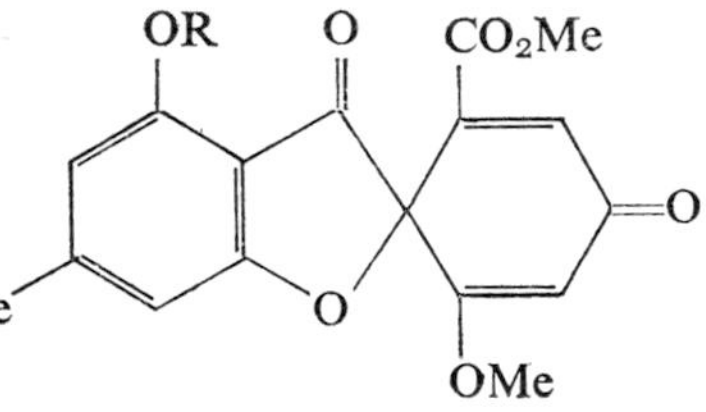

OR O CO₂Me / Me / O / OMe

a: $R = H$, (−)-dechlorogeodin (Compound E)
   Oospora sulphurea-ochracea[387]
   (+)-dechlorogeodin L^{400}
   P. frequentans[326]
b: $R = Me$, trypacidin
   A. fumigatus[386,388]

2

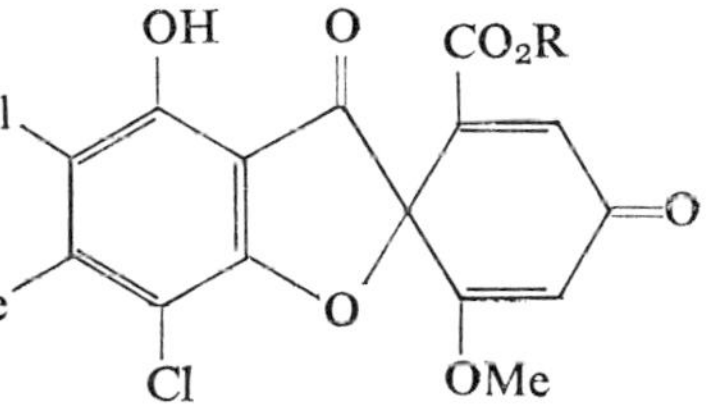

OH O CO₂R / Cl / Me / Cl / O / OMe

a: $R = H$, erdin
   A. terreus[389]
b: $R = Me$, geodin
   A. terreus[389]

3

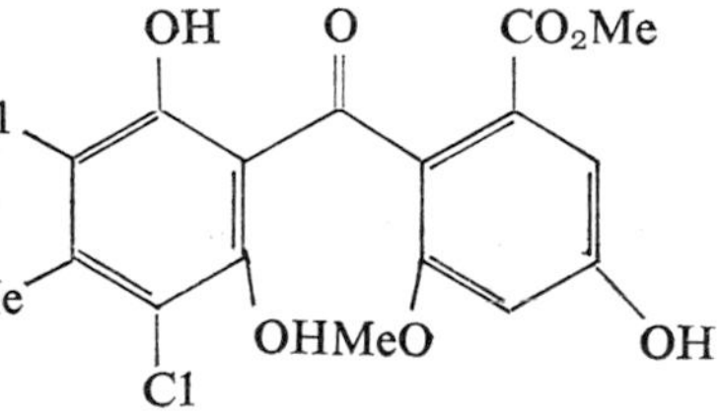

OH O CO₂Me / Cl / Me / Cl / OHMeO / OH

dihydrogeodin
A. terreus[390]

274

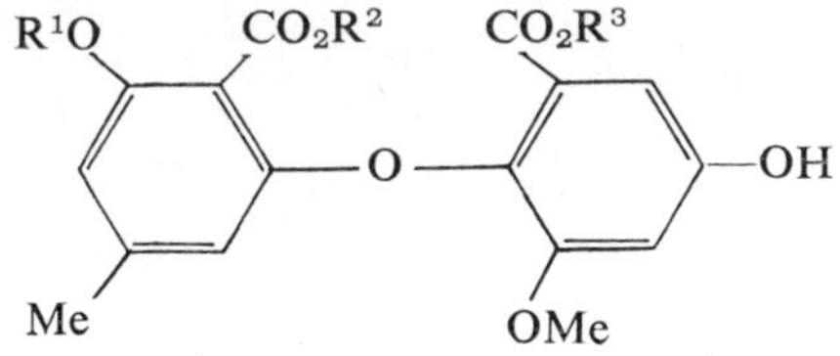

	R^1	R^2	R^3	
a	H	H	H	monomethylosoic acid (Compound F)
b	H	H	Me	dimethylosoic acid (Compound A, asterric acid)
c	Ac	H	Me	Compound C
d	H	Me	Me	Compound D (methyl asterrate)
				Oospora sulphurea-ochracea[387]

275

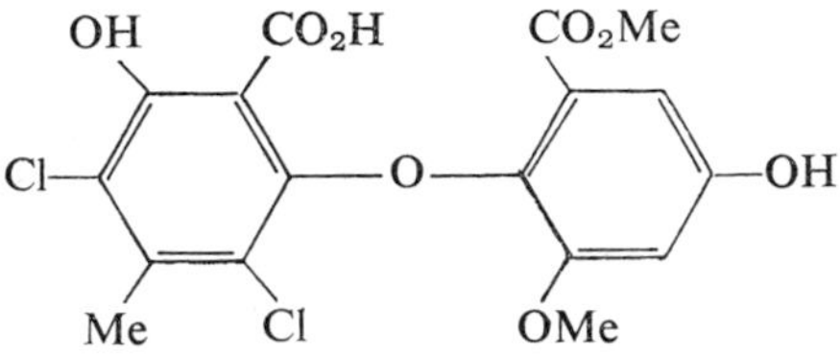

geodin hydrate
A. terreus[75]

276

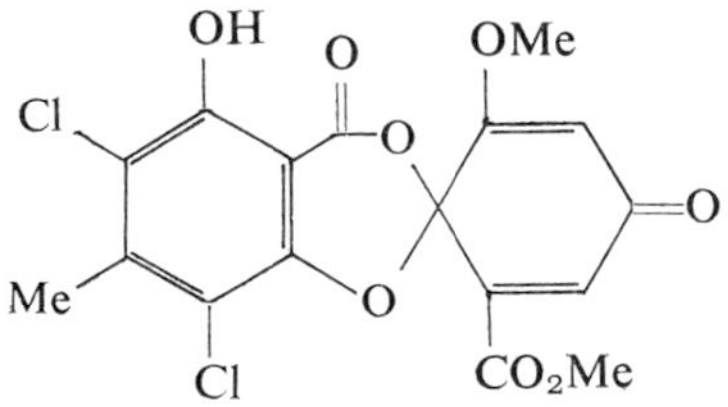

geodoxin
A. terreus[391]

277

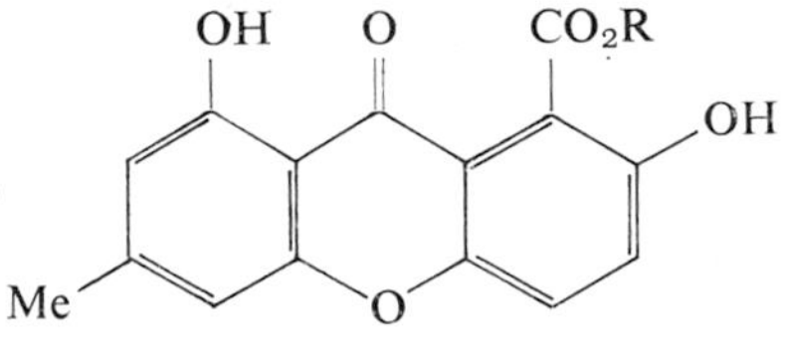

a: R = Me, pinselin
b: R = H, pinselic acid
        A. amarum[392]

278

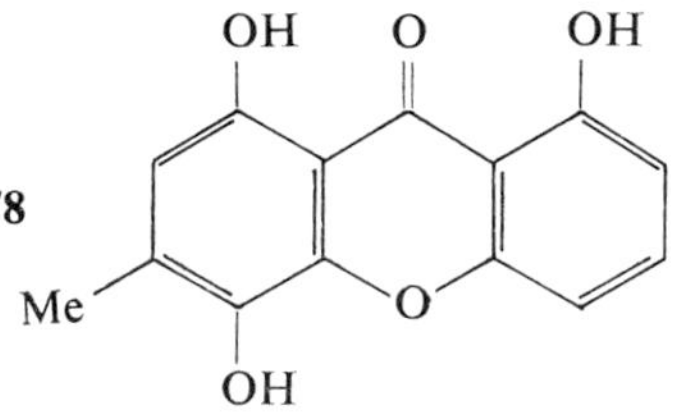

ravenelin
Helminthosporium raveneli[393]

Most of the compounds fall into one of two closely related series distinguished by the presence or absence of chlorine. The compounds containing chlorine have been isolated only from *A. terreus* and the dechloro-compounds were isolated first from *Oospora sulphurea-ochracea*, though sulochrin[390] and asterric acid (274b)[394] have since been obtained from *A. terreus*.

The relationship between the benzophenones and the spiro-compounds is the same as that between griseofulvin and the griseophenones (see p. 144), and the diphenyl ethers (274) and (275) and geodoxin (276) might represent further transformations of the spiro-compounds as in Scheme 33,[395] the three steps of which have been simulated *in vitro*.[394,395] Mutation experiments[390] supported this sequence for the dechloro-compounds and indicated

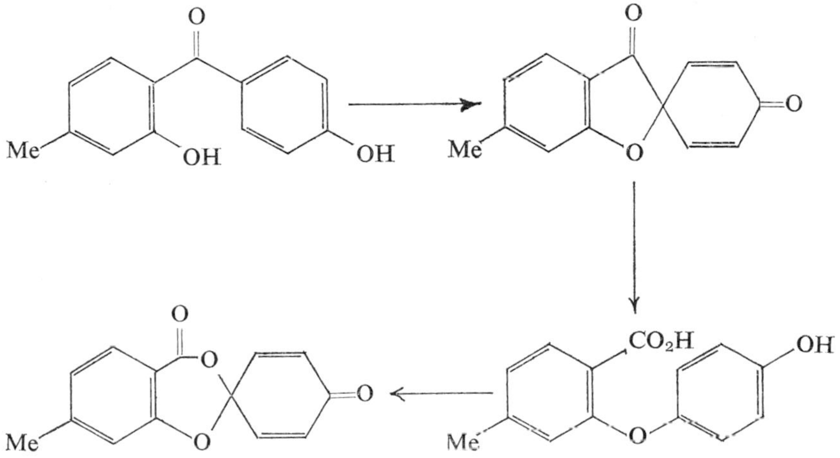

Scheme 33: Possible biosynthetic relationship between the *A. terreus* metabolites

that geodin hydrate (275) and geodoxin (276) are formed from geodin. However, no mutant could be isolated which produced dihydrogeodin (273) but not geodin, suggesting that the latter arises by chlorination of bis-dechlorogeodin rather than by oxidation of dihydrogeodin. Rhodes *et al.*[396] have produced evidence which supports this sequence, so that the chlorination steps in the biosyntheses of geodin and griseofulvin seem to occur at different stages. Although *O*-methylation is common among benzophenones, both in this series and in the griseofulvin series (see p. 142) the 2- and 4'- hydroxyl groups which are necessary for the phenolic coupling to occur are always free. This would be consistent with methylation at the grisan level though the above evidence appears to rule this out.

While the inter-relationship of the metabolites seems clear, there is still controversy over the origin of their carbon skeletons, which cannot be derived by folding of a straight octaketide chain. Two alternatives have been suggested. In the first, due to Gatenbeck,[397] sulochrin might arise by oxidative cleavage of emodin-5-methyl ether (Scheme 34); support for this postulate was provided by the co-occurrence of emodin-5-methyl ether and sulochrin

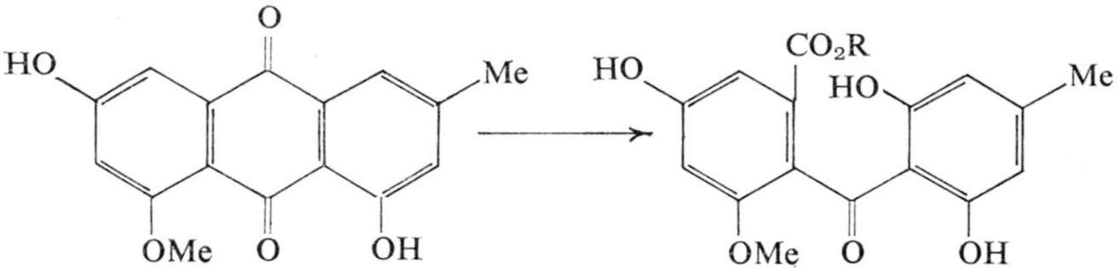

Scheme 34: Formation of sulochrin from emodin-5-methyl ether

in *P. frequentans*.[326] The second alternative, first discussed by Tatum,[356a] is that sulochrin-type compounds might arise by condensation of two aromatic compounds (Scheme 35) (and might in fact be precursors of, rather than degradation products of, anthraquinones, i.e. intermediates in the "two-

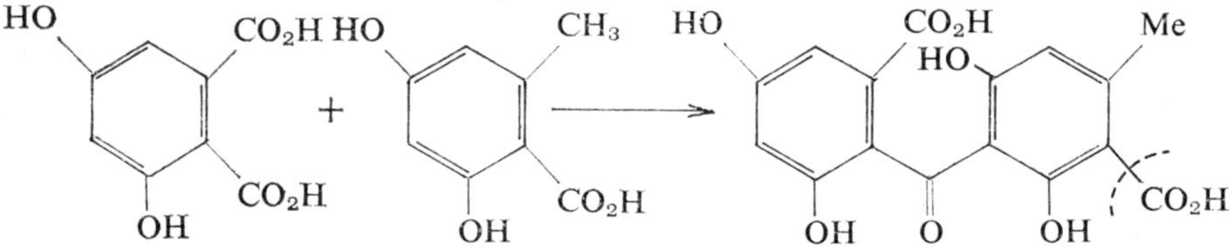

Scheme 35: Formation of sulochrin from benzene precursors

chain" theory discussed above p. 162). The isolation of orsellinic and 6-methylsalicylic acids from mutants of *A. terreus* which did not produce sulochrin has been argued in support of this theory.[390]

As in the case of anthraquinone biosynthesis, the two alternatives should be distinguishable by experiments with labelled malonate. Hassall and his colleagues have undertaken such experiments and have produced results[398] purporting to show that two acetate starter units are involved in sulochrin biosynthesis in accord with the route of Scheme 35. However, careful examination of their results does not bear out their conclusion. In further experiments, using [2-^{14}C]acetate,[399] Hassall has found that the carboxyl carbon atom of sulochrin contains appreciably higher radioactivity than is present in the atoms of the ring to which it is attached, suggesting that this carbon is derived from an acetate "starter" unit and supporting the "two-unit" process of Scheme 35 (without providing any evidence as to the nature of the two intermediates).

On the other hand, two groups have produced evidence in favour of the derivation of sulochrin from a single polyketide chain via an anthraquinone. First, Misconi and Stickings[400] used [^{14}C]malonate to study the biosynthesis of questin (222d) and (+)-dechlorogeodin (271a) in *P. frequentans* and found that both compounds are derived from one acetyl and seven malonyl units, in accord with Gatenbeck's results with islandicin[357] and supporting the mechanism of Scheme 34 for the biosynthesis of sulochrin-type compounds. Secondly, Gatenbeck[401] found that emodin (222c) and its 5-methyl ether (222d) can act as precursors of sulochrin in *P. frequentans*, that in the presence of the metabolic inhibitor aminopterin production of sulochrin is diminished and anthraquinones accumulate, and that emodin and sulochrin are precursors of geodin in *A. terreus*.

Thus the origin of the carbon skeletons of these compounds is in doubt and it may be that their classification with the anthraquinones is not valid.

B.6.b. *Compounds of Type*

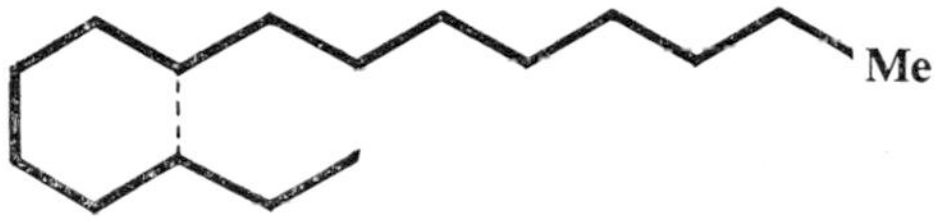

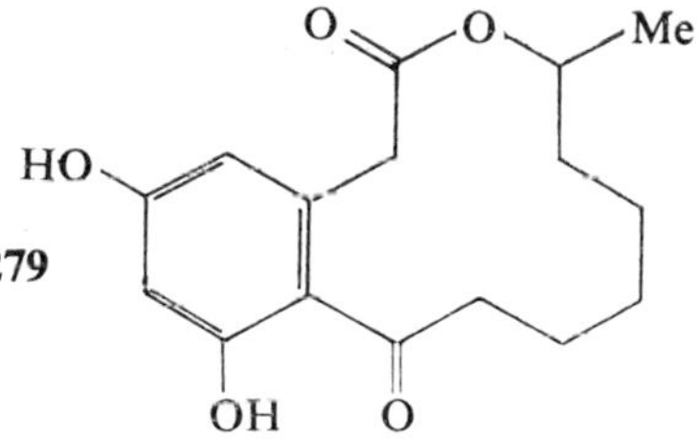

curvularin L^{404}
Curvularia spp.[403]

279

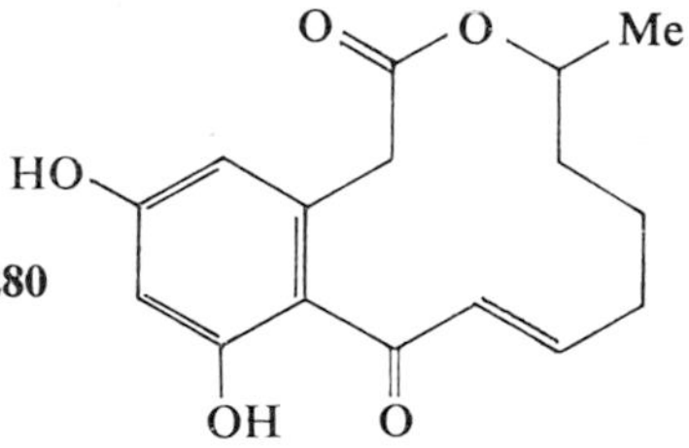

dehydrocurvularin (*Curvularia* compound II)
Curvularia spp.[403]

280

Curvularin is derived from acetate in the expected manner.

B.6.c. *Compounds of Type*

281 AcO

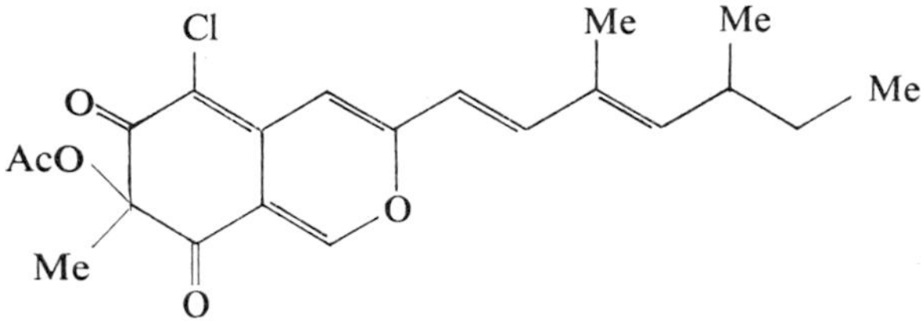

sclerotiorin L^{186}
P. sclerotiorum[405]
7-*epi*-sclerotiorin
P. hirayamae[406,407]

282

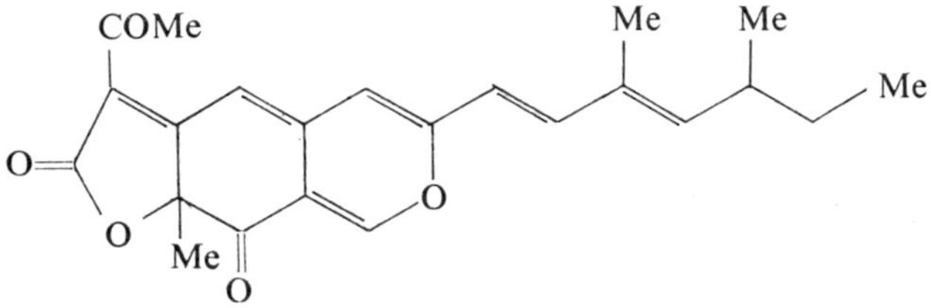

rotiorin $L^{238,40}$
P. sclerotiorum[405]

283

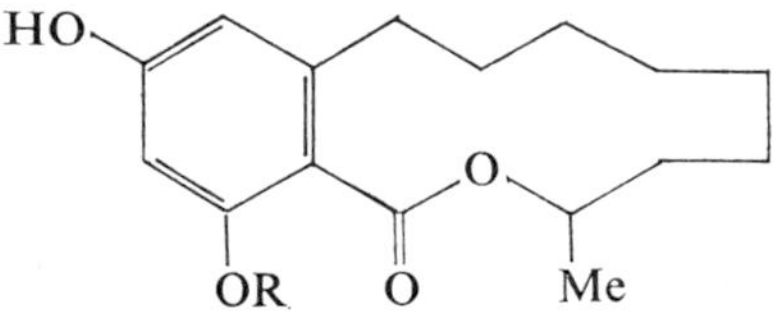

a: R = Me,   lasiodiplodin

b: R = H,   desmethyl- lasiodiplodin
             Lasiodiplodia theobromae[167]

Sclerotiorin and rotiorin are members of the azaphilone group (see p. 138). A metabolite of *P. hirayamae* was at first thought to be the (−)-antipode of sclerotiorin[406] but is in fact 7-*epi*-sclerotiorin.[407] Acetate and formate are incorporated into sclerotiorin[186] and rotiorin[238] in the expected manner, but experiments with malonate[408] revealed the interesting fact that the acetoacetyl residue of rotiorin is derived from two acetate residues without intervention of malonate, as in the biosynthesis of mevalonic acid. The side-chains of rubropunctatin and monascorubrin (p. 139) are also derived in an unusual manner. The three "extra" methyl groups of sclerotiorin are transferred from methionine with retention of their hydrogen atoms.[409]

The resorcylic acid lactone lasiodiplodin (**283**) is formally derived by an alternative cyclization of the precursor of curvularin (**279**). Monorden (**288**) and zearalenone (289) are examples of this cyclization in the nonaketide series.

B.6.d. *Compounds of Type*

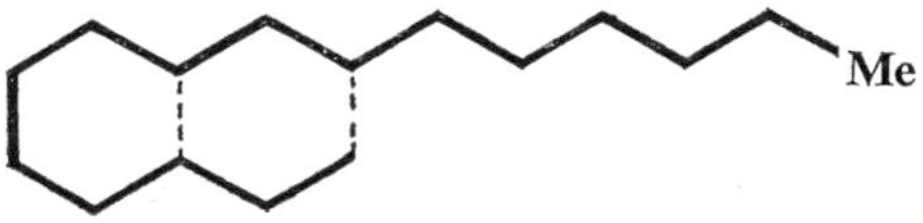

284

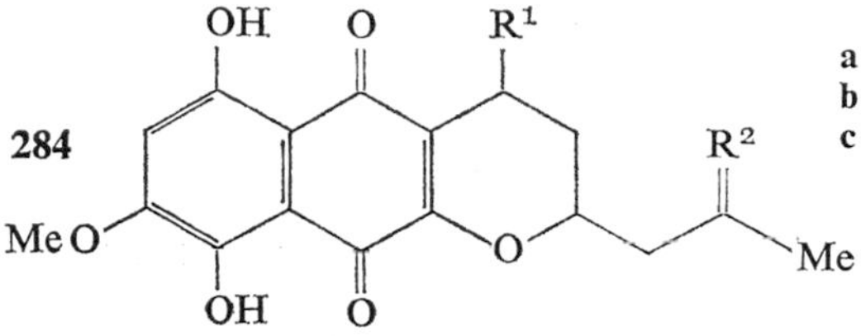

a: R^1 = OH, R^2 = O, erythrostominone
b: R^1 = H, R^2 = O, deoxyerythrostominone
c: R^1 = H, R^2 = H,OH, deoxyerythrostominol
Gnomonia erythrostoma[410]

B.6.e. *Compounds of Type*

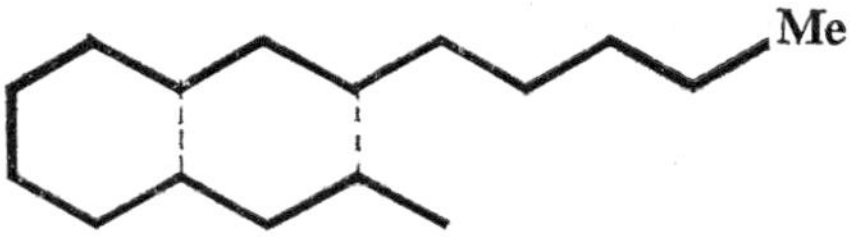

285

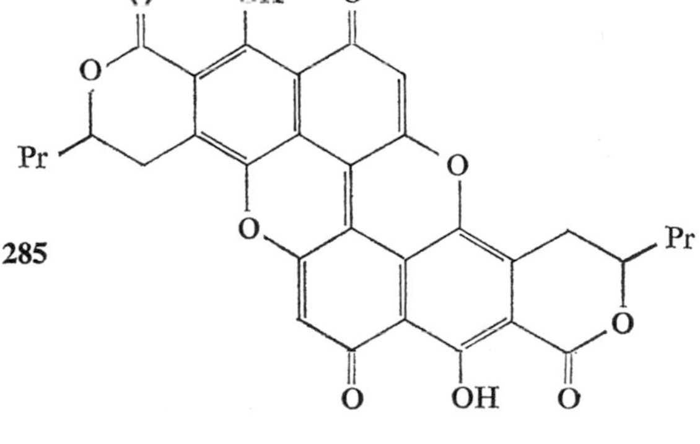

xylindein
Chlorociboria aeruginosa[411]

B.6.f. *Compounds of Type*

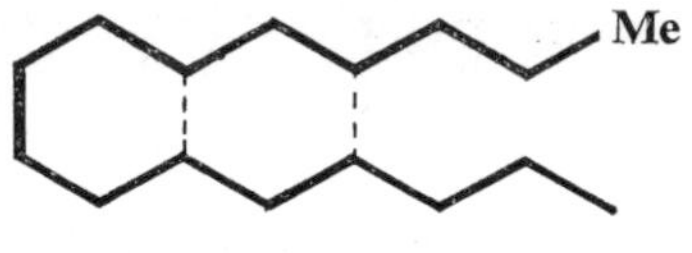

286

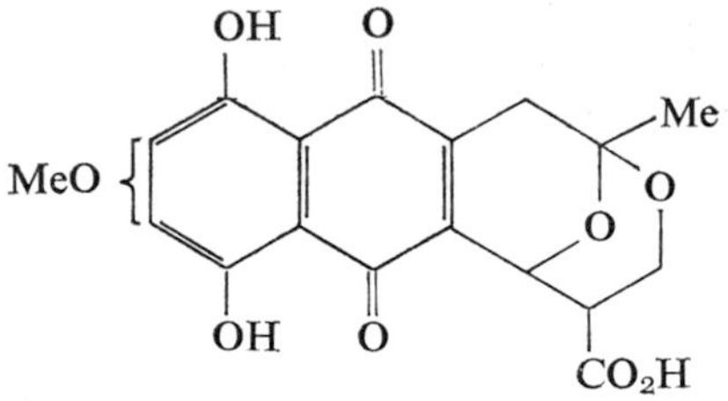

marticin
Fusarium spp.[412]

B.7. NONAKETIDES

There are relatively few nonaketides. The largest group comprises the aflatoxins and related compounds, though there is still some controversy over the origin of these compounds, and it is possible that they are misplaced here. The tetracyclines, though not strictly nonaketides, are also most conveniently discussed in this section.

B.7.a. *Open-chain Compounds*

287

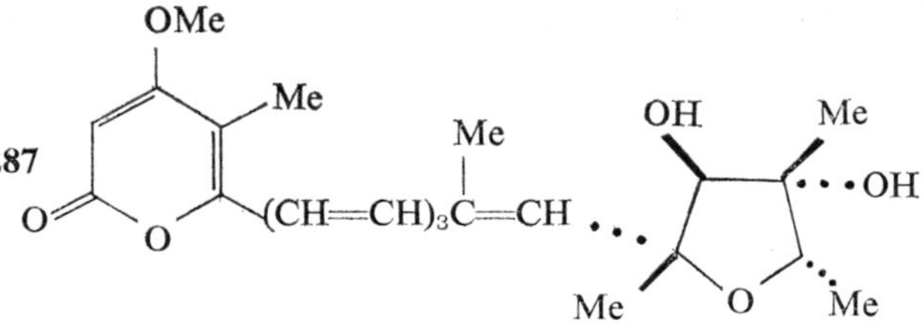

citreoviridin
Penicillium spp.[413]

B.7.b. *Compounds of Type*

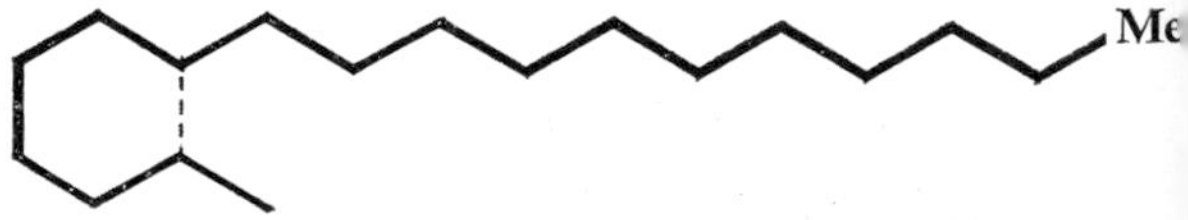

288

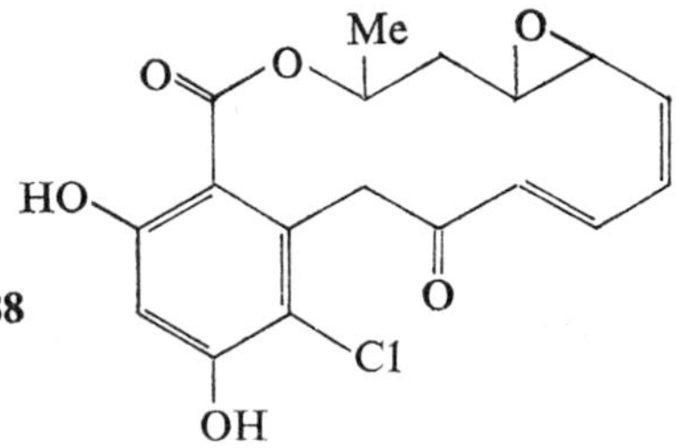

monorden (radicicol)
Monosporium bonorden,[414]
Nectria radicicola[415]

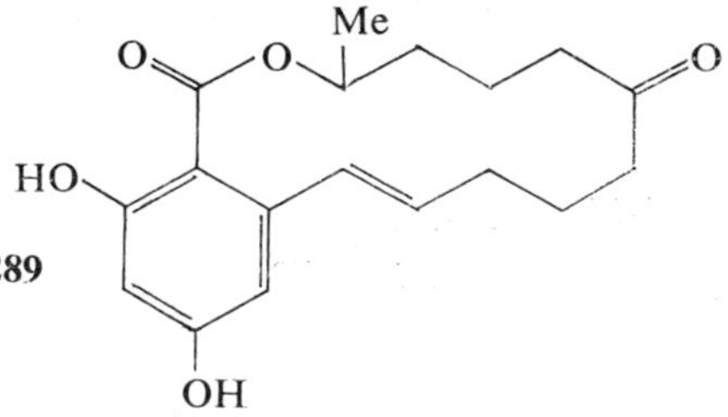

zearalenone
Gibberella zea,[416]
  Fusarium moniliforme[417]

289

The resorcylic acid lactones (288) and (289) are related to lasiodiplodin (283). Zearalenone is the toxin responsible for genital abnormalities in farm animals feeding on mouldy grain; it has anabolic and uterotrophic activity.

B.7.c. *Compounds of Type*

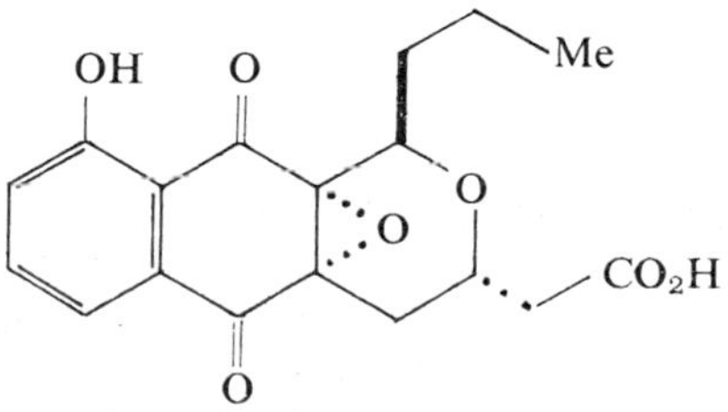

290

frenolicin
Streptomyces fradiae[418]

B.7.d. *Compounds of Type*

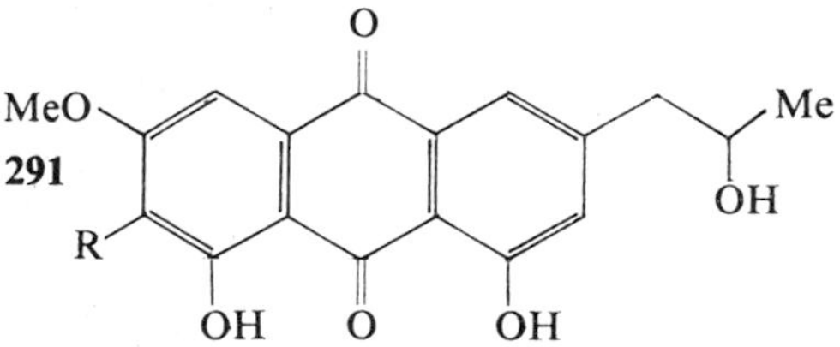

291

a: R = H, nalgiovensin
b: R = Cl, nalgiolaxin
  P. nalgiovensis[419]

B.7.e. *The Aflatoxins and Related Compounds*

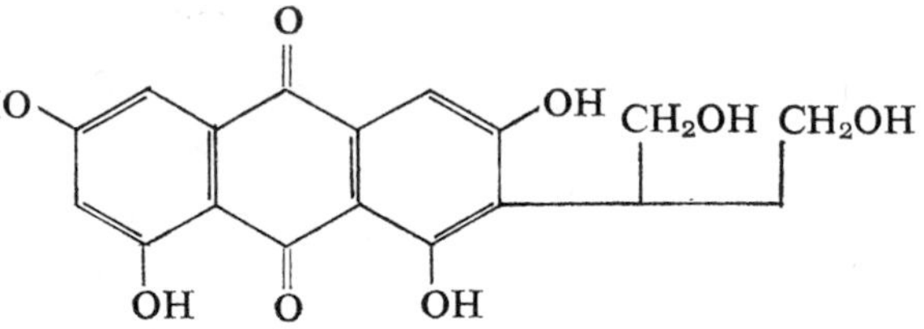

292

HO — ... — OH CH₂OH CH₂OH

versiconol
A. versicolor[420]

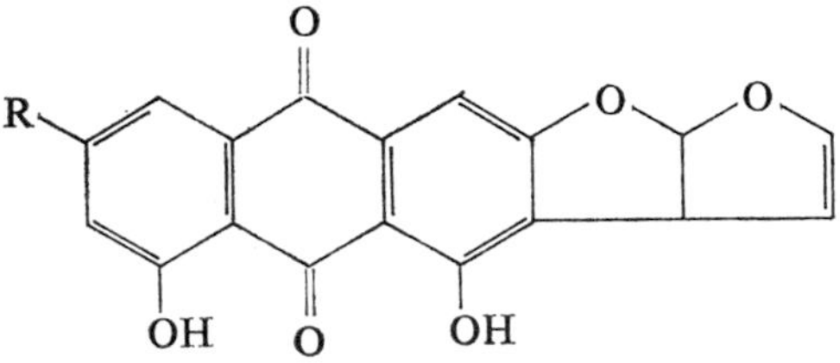

293

a: R = OH, versicolorin A
    A. versicolor[421]

b: R = H, deoxyversicolorin A
    A. versicolor[422]

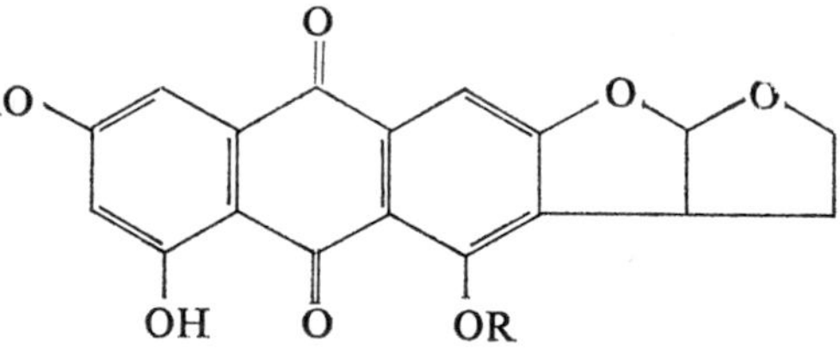

294

a: R = H, versicolorin B
    (versicolorin C = racemate
    A. versicolor,[421] *A. flavus*[402]

b: R = Me, aversin
    A. versicolor[423]

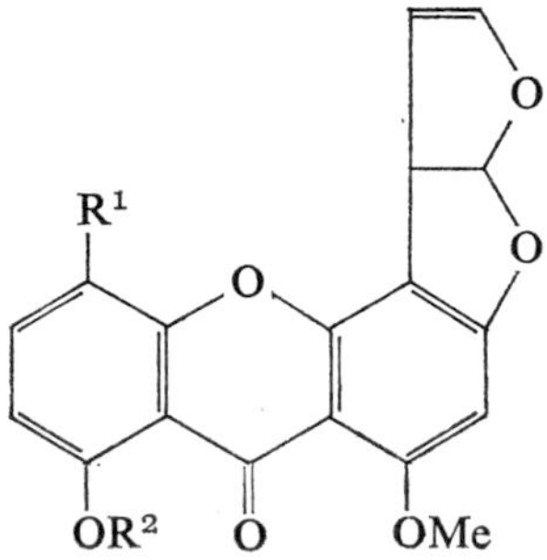

295

a: R¹ = R² = H, sterigmatocystin
    A. versicolor[424]

b: R¹ = OMe, R² = H, 5-methoxysterig-
    matocystin
    A. versicolor[423]

c: R¹ = H, R² = Me, O-methylsterig-
    matocystin
    A. flavus[425]

296

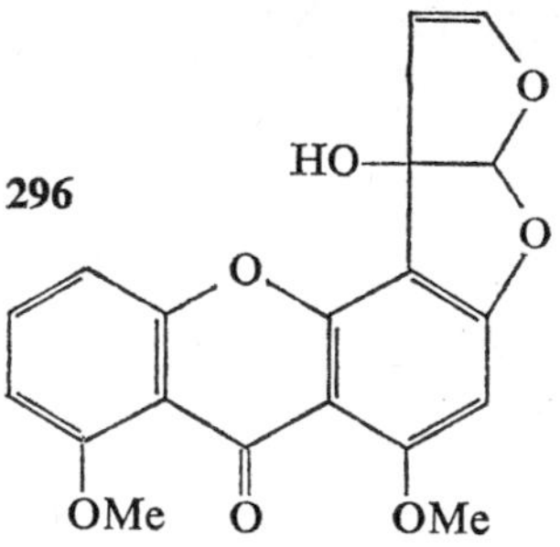

aspertoxin (3-hydroxy-6,7-
     dimethoxydifuroxanthone)
A. flavus[426]

297

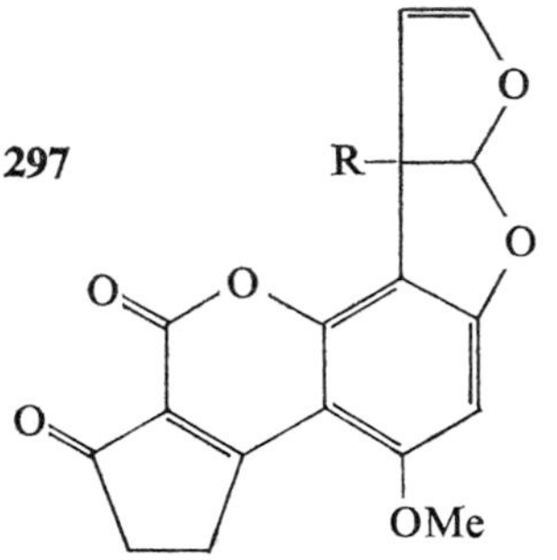

a: R = H, aflatoxin B$_1$ L[437a]
        A. flavus,[427] *P. puberulum*[428]

b: R = OH, aflatoxin M$_1$
        A. flavús[429]

298

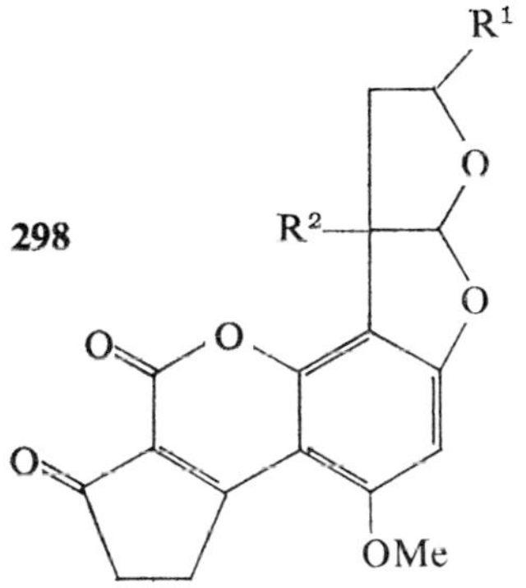

a: R^1 = R^2 = H, aflatoxin B$_2$
        A. flavus[427a,430]

b: R^1 = H, R^2 = OH, aflatoxin M$_2$
        A. flavus[429]

c: R^1 = OH, R^2 = H, aflatoxin B$_{2a}$
        A. flavus[431]

299

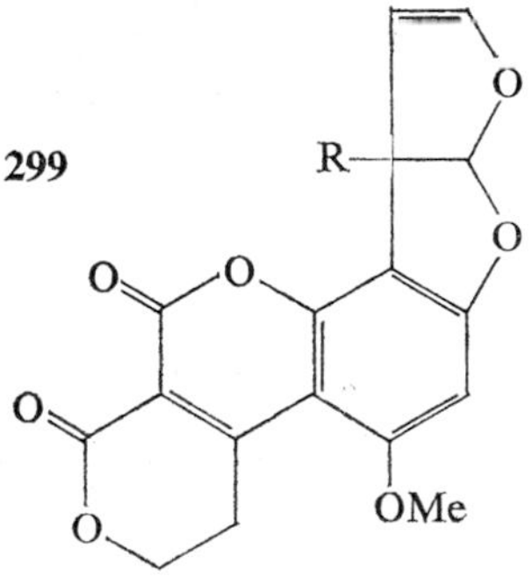

a: R = H, aflatoxin G$_1$
        A. flavus,[427] *P. puberulum*[428]

b: R = OH, aflatoxin GM$_1$
        A. flavus[402]

300

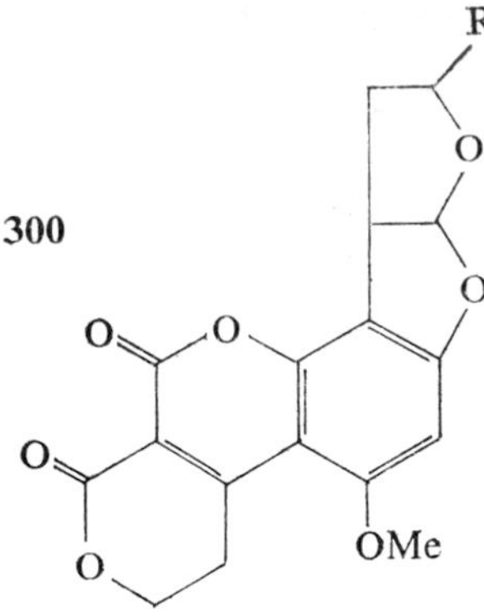

a: R = H, aflatoxin G_2
  A. flavus[427a]
b: R = OH, aflatoxin G_{2a}
  A. flavus[431]

301

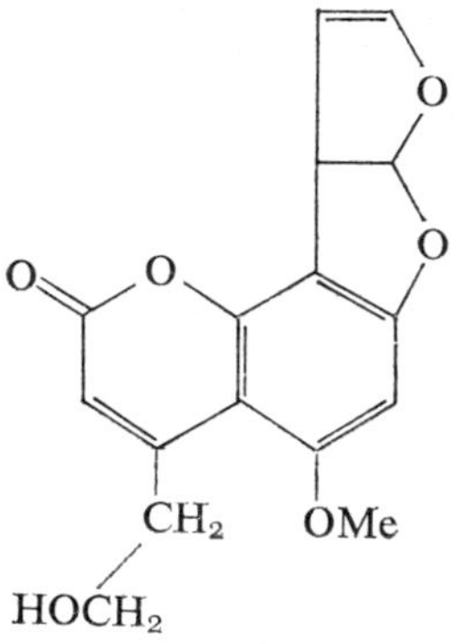

aflatoxin B_3
A. flavus[402]

At first glance, the only feature common to all the compounds discussed in this section is the bisfurano-system. It is not immediately apparent how this system arises in nature, nor indeed how the rest of the sterigmatocystin and aflatoxin molecules are formed. It is only with the recent biosynthetic studies, discussed below, by Büchi's group that the possible nonaketide origin of all the compounds has come to be recognised.

Originally all the compounds were isolated from *A. flavus* or *A. versicolor* (the claim[428] that aflatoxins B_1 and G_1 are produced by *P. puberulum* has been questioned[432]) and it is indicative of a common biosynthetic origin that sterigmatocystin derivatives have been obtained from both sources. The first of the compounds to be isolated was sterigmatocystin (**295a**), the correct structure for which was proposed in 1962. Subsequently 5-methoxysterigmatocystin (**295b**)† was isolated from *A. versicolor* along with aversin (**294b**). The alternative structure (**302**) for aversin could not be excluded[328] but in view of its close relationship to the versicolorins, the configuration of which

† This compound was originally thought to be 6-methoxysterigmatocystin.

has been established by degradation to the anthraquinone (303), structure
(294b) seems the more likely. Versiconol (292) appears to be converted to
versicolorin A by replacement cultures of *A. versicolor*.[420] *A. versicolor* also
produces a series of pigments clearly derivable from a straight C_{20} chain
(see p. 186).

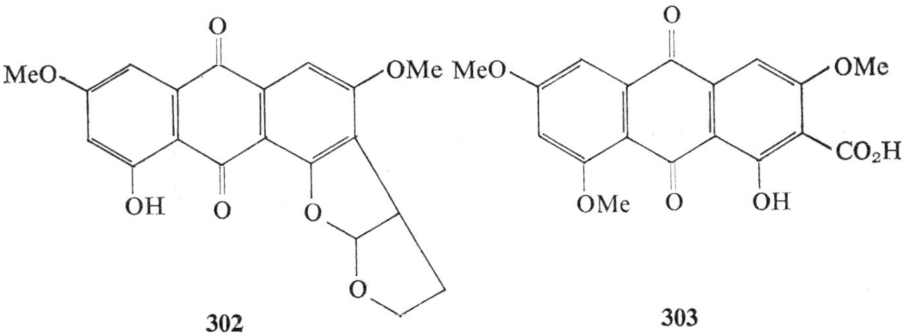

302   303

The aflatoxins are of considerable economic significance as a result of their
toxicity[433] to animals fed with mouldy groundnut meal. The first aflatoxins
to be isolated were called B and G,[434] the letters referring to the colours—
blue and green respectively—of the fluorescence produced by the compounds.
Subsequently[427a] aflatoxin B was shown to contain a small quantity of its
dihydro derivative and the two compounds were re-named B_1 and B_2 respectively; at the same time, the dihydro derivative of aflatoxin G was
obtained. More recently aflatoxins M_1 and M_2 (so designated because they
were first detected in milk from cows fed on groundnut meal containing
aflatoxins) and aflatoxins B_{2a} and G_{2a} have been isolated; the latter compounds may be artefacts since they are formed from B_1 and G_1 respectively

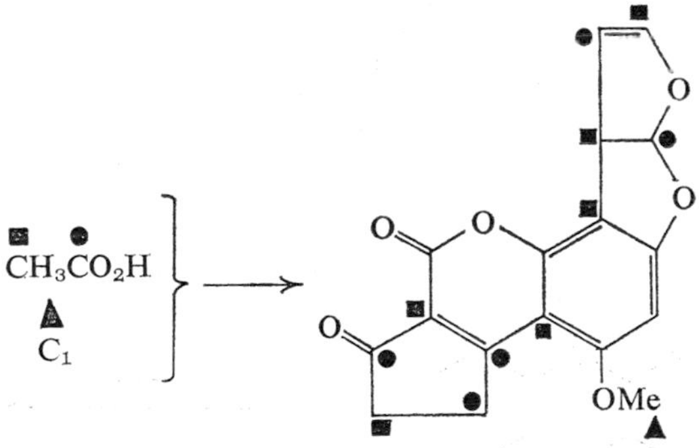

Scheme 36: The incorporation of radioactivity into the aflatoxins[437]

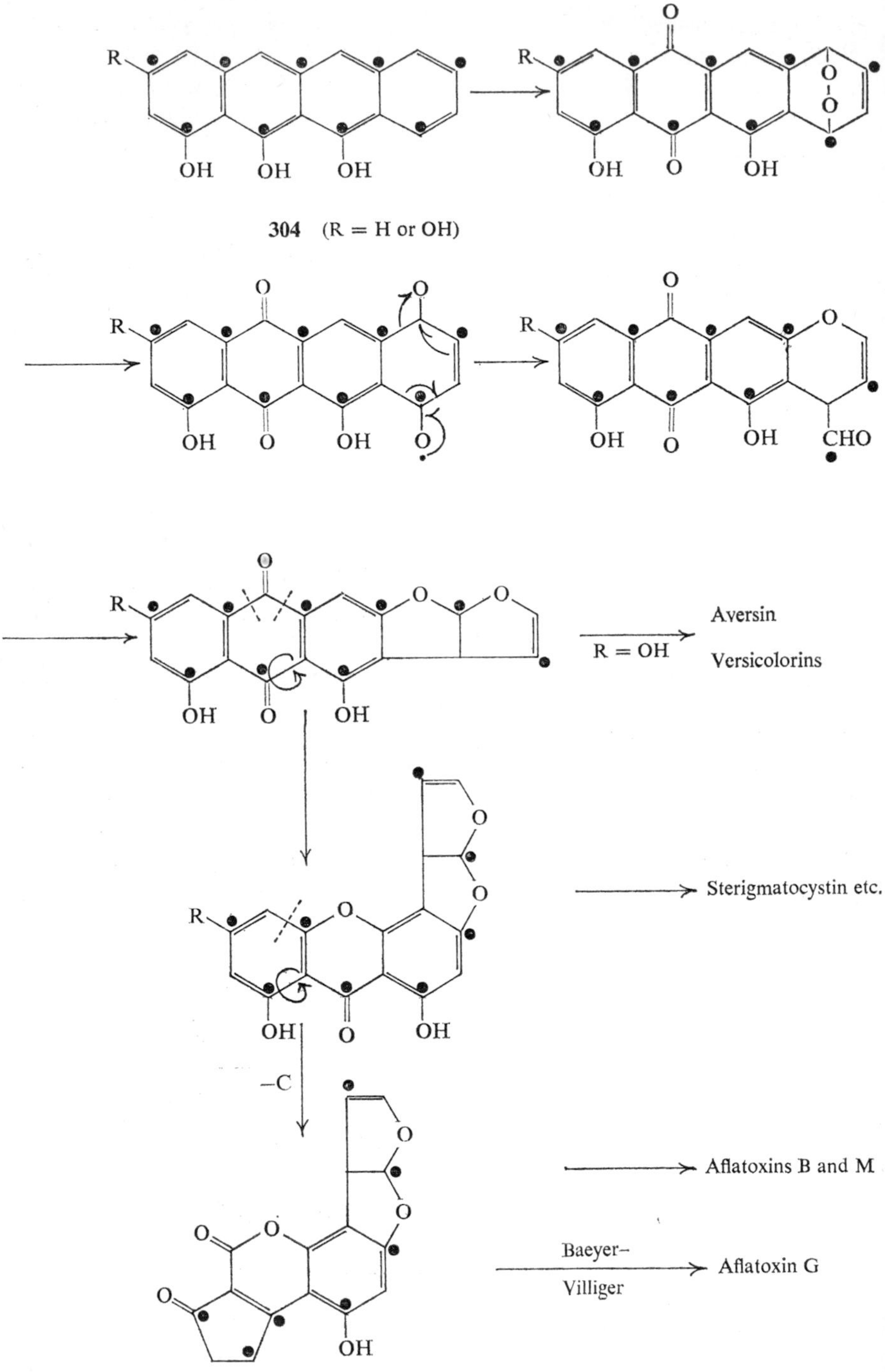

304   (R = H or OH)

Aversin

$\xrightarrow{R = OH}$

Versicolorins

Sterigmatocystin etc.

Aflatoxins B and M

$\xrightarrow[\text{Villiger}]{\text{Baeyer--}}$ Aflatoxin G

Scheme 37: Suggested biosynthesis of the bisfurans[437]

under acidic conditions. The structures of aflatoxins B_1 and G_1 were derived by Büchi and his collaborators[427b] and were confirmed[435] by X-ray analysis; the absolute configurations of the molecules has been deduced.[436]

The biosynthesis of aflatoxin B_1 has been studied by Büchi and his co-workers[437a] who were able, as a result of elegant degradative experiments, to establish the labelling pattern of Scheme 36. They interpreted[437b] this labelling pattern in terms of the sequence of Scheme 37, which also interrelates all the bisfurano-compounds discussed in this section. Büchi points out that the correct labelling pattern can also be derived starting from benzanthracene derivatives rather than from the naphthacene (**304**).

Holker and Mulheirn[438] find that incorporation of [1-^{14}C]acetate into sterigmatocystin gives a distribution of radioactivity similar to that observed in the aflatoxins by Büchi *et al*. However, they find that the carbon atoms of the furano-system are significantly less active than those of the xanthone system, suggesting a biosynthesis from two acetate-derived chains rather than from a single chain as suggested by Büchi (Scheme 37).

B.7.f. *The Tetracyclines*

The tetracyclines (**305**) are broad-spectrum antibiotics produced by *Streptomyces* spp. As a result of their commercial importance their bio-synthesis has received greater attention than that of any other polyketide and a detailed biosynthetic pathway has been elucidated by tracer experiments, by precursor experiments, and above all by mutation experiments. The subject has been extensively reviewed, most recently by McCormick,[439] and the present state of our knowledge is illustrated in Scheme 38 for 7-chlorotetracycline; the other tetracyclines probably arise by variations of this sequence.

	R^1	R^2	R^3	
a	H	H	H	6-demethyltetracycline
b	Cl	H	H	7-chloro-6-demethyltetracycline
c	H	Me	H	tetracycline
d	H	Me	OH	5-hydroxytetracycline (oxytetracycline, terramycin)
e	Cl	Me	H	7-chlorotetracycline (aureomycin)

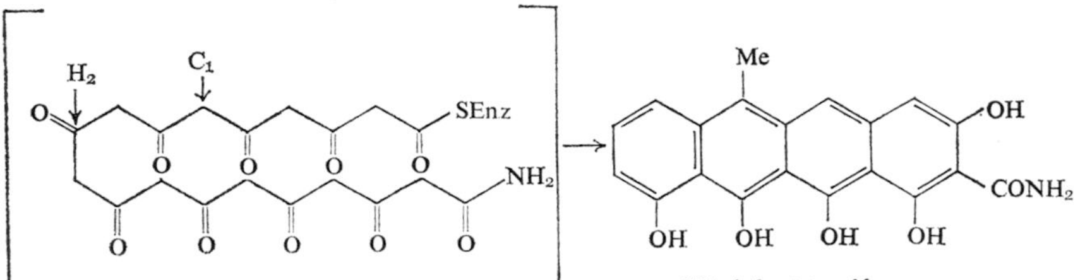

6-Methylpretetramid

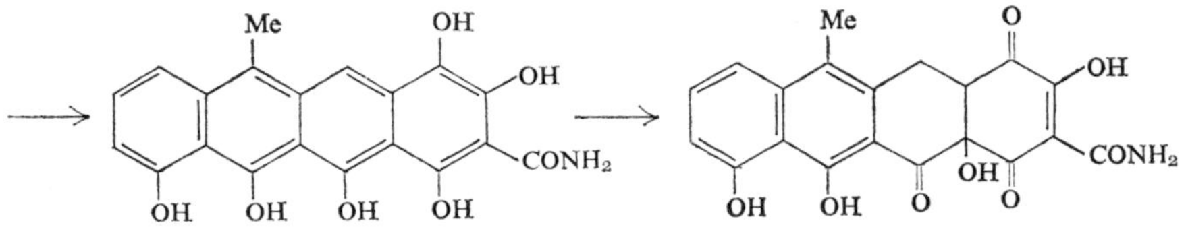

4-Hydroxy-6-methylpretetramid

4-Ketodedimethylaminoanhydro-
tetracycline

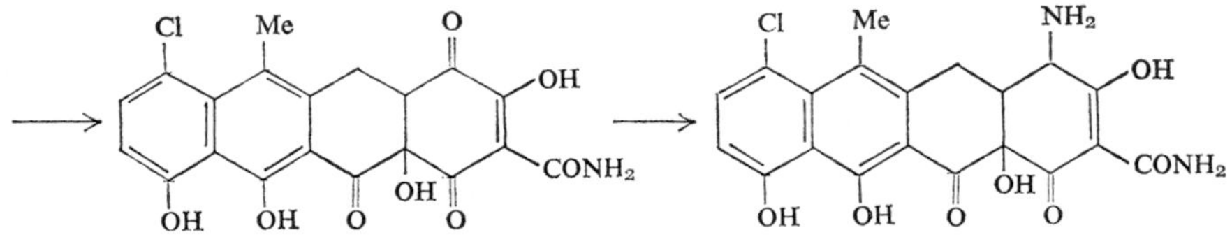

4-Ketodedimethylaminoanhydro-
7-chlorotetracycline

7-Chloro-4-aminodedimethyl-
aminoanhydrotetracycline

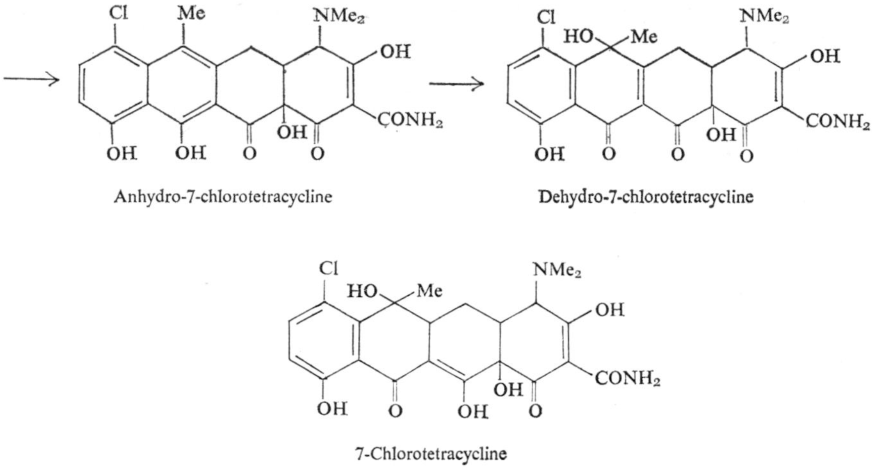

Anhydro-7-chlorotetracycline

Dehydro-7-chlorotetracycline

7-Chlorotetracycline

Scheme 38: The biosynthesis of the tetracyclines

Tetracycline biosynthesis illustrates several features frequently encountered in polyketide biosynthesis—cyclization of the precursor to give an aromatic system which undergoes modification, loss of an oxygen atom before or during aromatization, introduction of a methyl group at a prearomatic stage, and introduction of chlorine into an aromatic compound. An unusual feature is the replacement of the normal acetyl coenzyme A "starter" by, probably, malonamoyl coenzyme A, thus accounting for the early introduction of the carboxamide group. In the related 2-acetyl-decarboxamidotetracycline (306) the malonamoyl residue appears to have been replaced by an acetoacetyl residue.

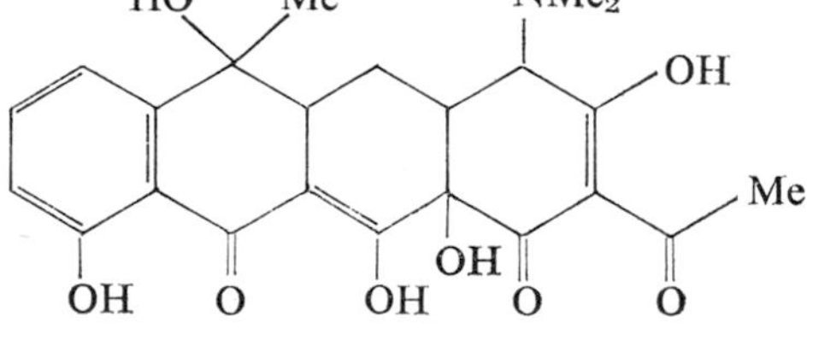

306

In principle, the tetracyclines could equally well be derived from a polyketide folded as in (307), but the isolation of protetrone (308),[440] and the

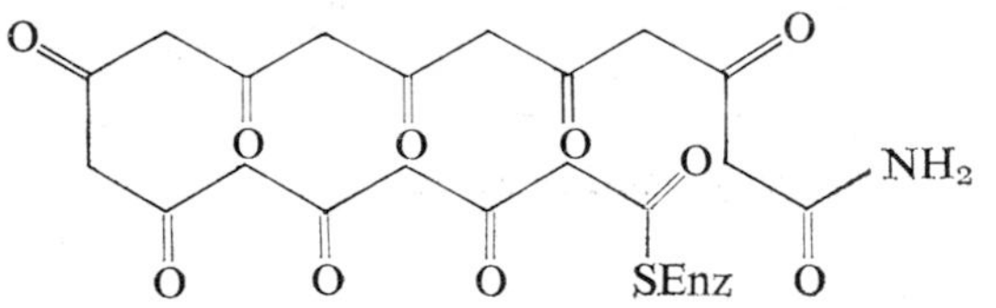

307 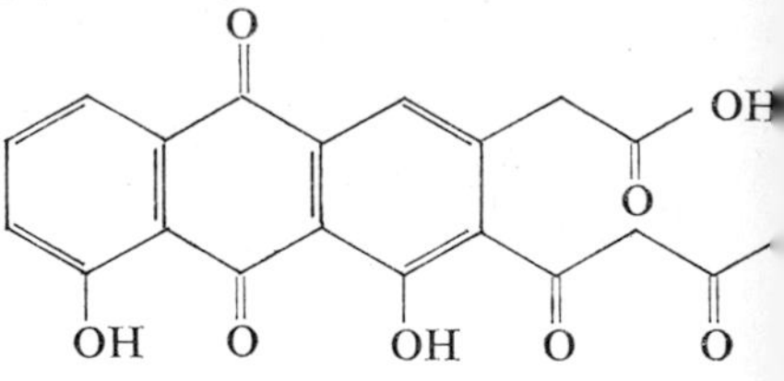   308

methylanthrone (309)[441] from blocked mutants of *S. aureofaciens* provides
evidence for folding as in Scheme 38.

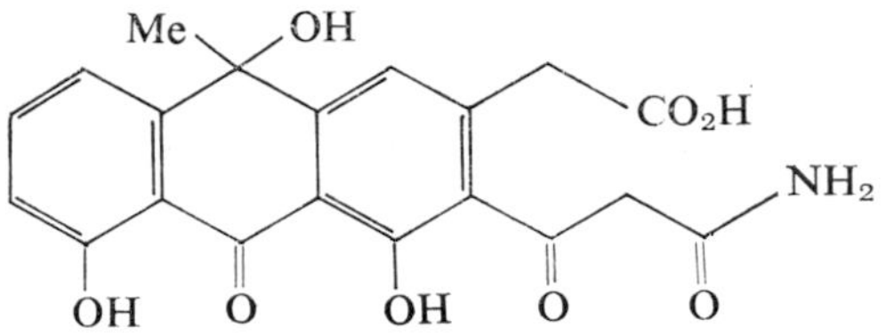

309

B.8. DECAKETIDES

The fungal metabolites derived from an acetyl unit and nine malonyl
units are all closely related and are discussed in section (a). In addition to
these we shall also mention a few *Streptomyces* products which appear to
be decaketides, and the anthracyclines, in which the acetate "starter" unit is
replaced by propionate. A decaketide chain is involved in the biosynthesis of
erythroskyrin (p. 287).

B.8.a. *Compounds of Type*

310

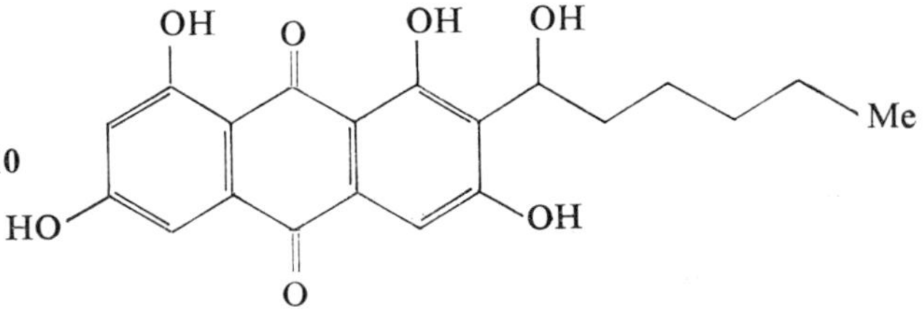

averantin
Aspergillus versicol

311

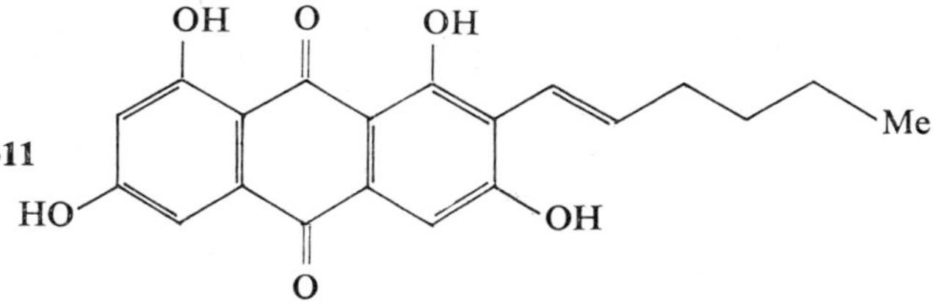

OH O OH

HO

O

=Me

averythrin
A. versicolor[443]

312

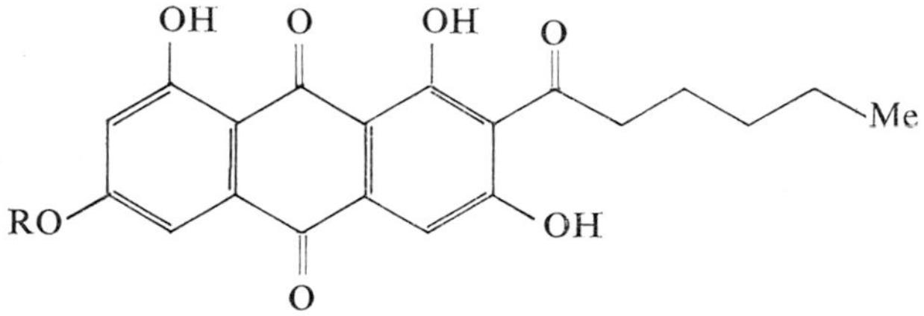

OH O OH O

RO

OH

O

=Me

a: R = H, norsolorinic acid
   Solorinia crocea,[444] *A. versicolor*[445]

b: R = Me, solorinic acid
   S. crocea[446]

313

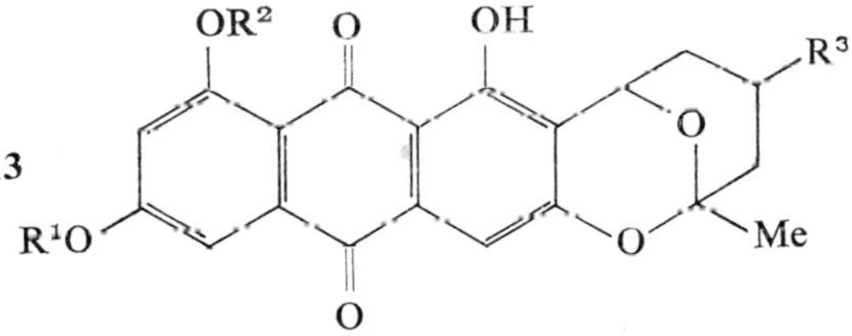

OR² O OH

R¹O

O

O

R³

Me

O

	R¹	R²	R³	
a	H	H	H	averufin *A. versicolor*[447]
b	H	Me	H	averufin methyl ether *A. versicolor*[447b]
c	Me	Me	H	averufin dimethyl ether *A. versicolor*[447b]
d	H	H	OH	hydroxyaverufin *A. versicolor*[448]

314

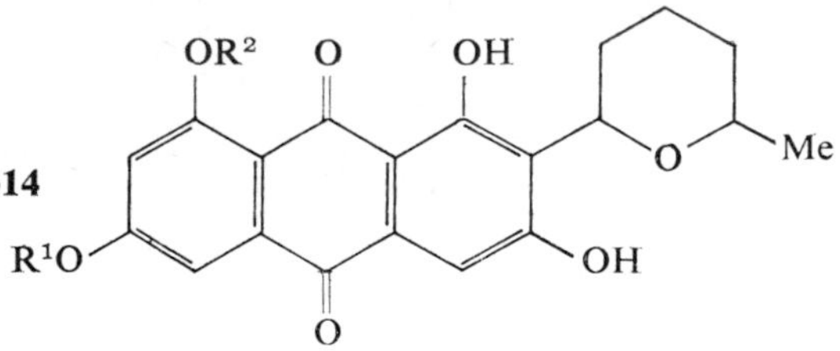

a: $R^1 = R^2 = H$,   *A. versicolor,* [447b] *A. flavus*[402]
b: $R^1 = H, R^2 = Me$,   *A. versicolor*[447b]
c: $R^1 = R^2 = Me$,   *A. versicolor*[447b]

Several of the closely related series of compounds from *A. versicolor* were obtained from u.v. mutants during a biosynthetic investigation.[447b] The isolation of solorinic acid (**312b**) and norsolorinic acid (**312a**) from the lichen *Solorinia crocea* provides an example of the close relationship between lichen and fungal products.

B.8.b. *Possible Decaketides from Streptomyces spp.*

315

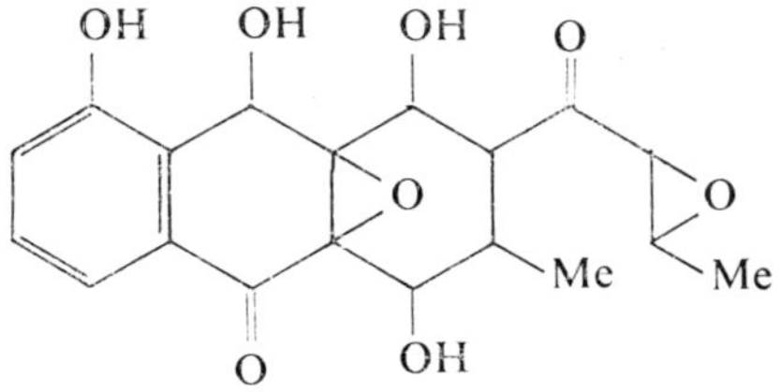

cervicarcin
S. ogaensis[449]

316

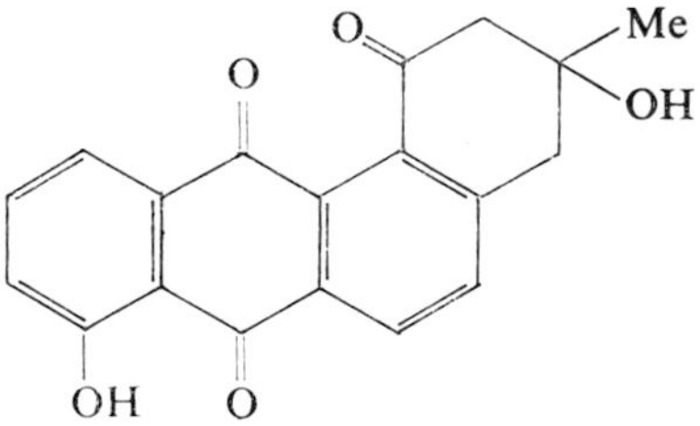

tetrangulol
Streptomyces rimosus[450]

317

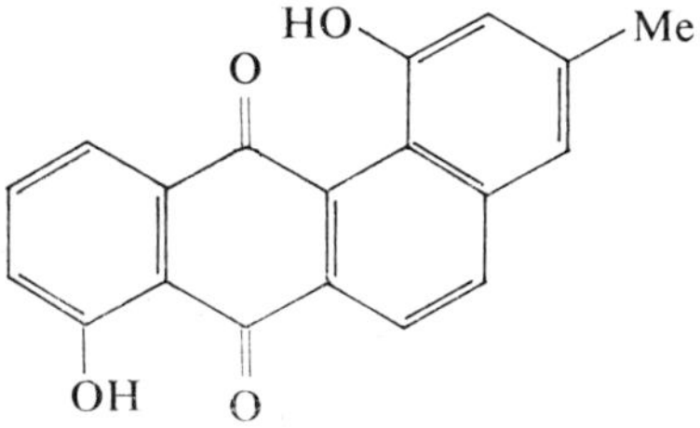

tetrangomycin
S. rimosus[450]

318 HO,H

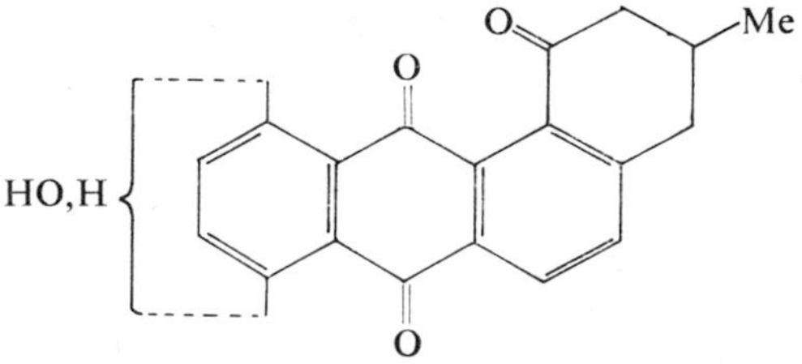

ochramycinone
Streptomyces spp.[451]

319

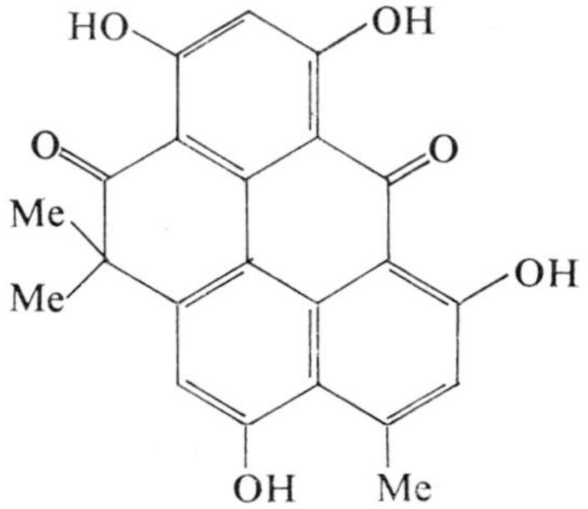

resistomycin (X-340)
Streptomyces resistomycificus[452]

The classification of these compounds is speculative. Cervicarcin could arise as in **320**, and tetrangulol, tetrangomycin, and ochramycinone as in **321**. The introduction of two C_1-units onto the same carbon as in **322** leading to resistomycin (other foldings of the chain are possible[452b]) has not been observed in fungal products but occurs in higher plants.[453]

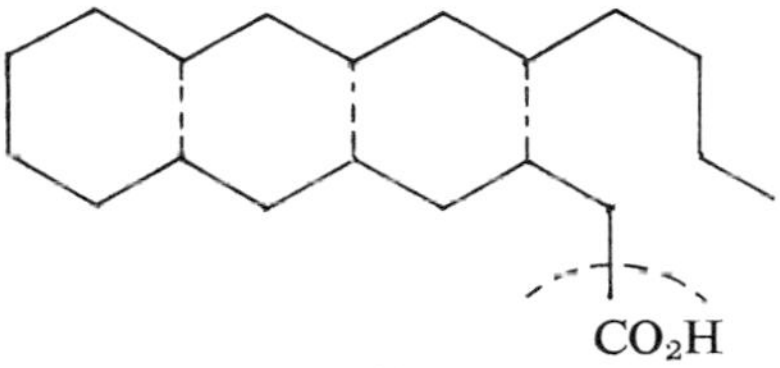

320

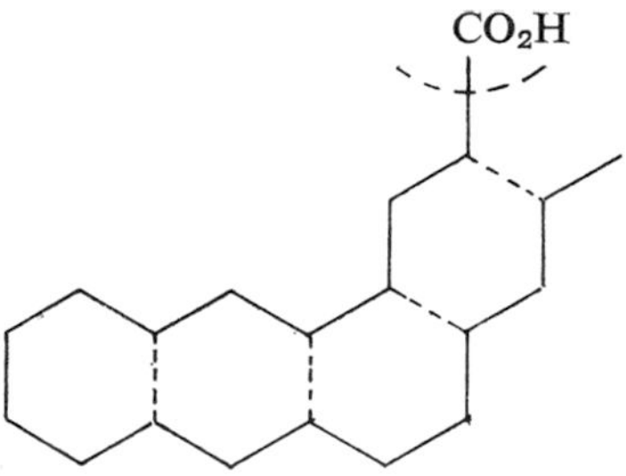

321

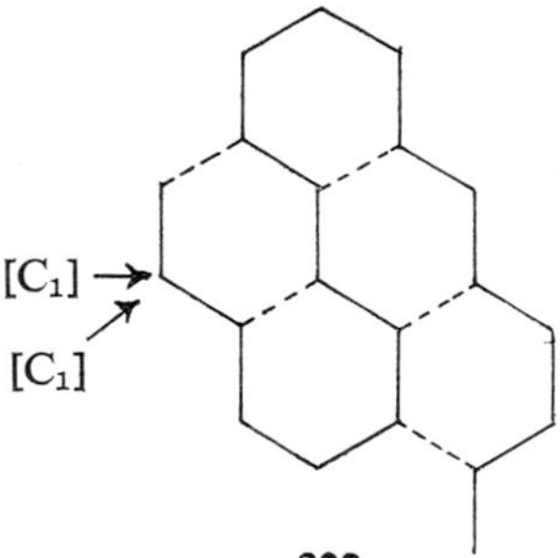

322

B.8.c. *The Anthracyclines*

The anthracyclines are a group of antibiotic glycosides produced by *Streptomyces* spp. The aglycones, whose chemistry has been reviewed,[454] consist of an anthraquinone chromophore with a fourth (saturated) ring fused on, and from a biosynthetic point of view three main structural variations may be discerned. These are illustrated by ε-pyrromycinone (rutilantinone) (323),[455] β-rhodomycinone (324),[456] and daunomycinone (325).[457] Pyrromycinone has been shown[458] to arise by condensation of a propionyl with nine acetyl (malonyl) units as shown in Scheme 39 (cf. homoorsellinic

323

324

325

acid, p. 76). β-Rhodomycinone is clearly derived by decarboxylation of a pyrromycinone-type precursor, but the presence of an oxygen atom on the

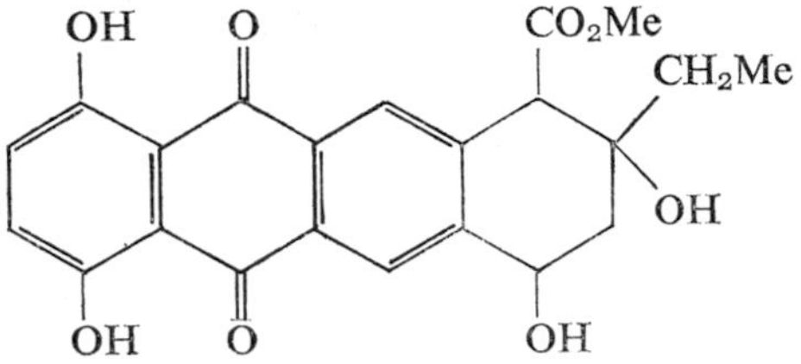

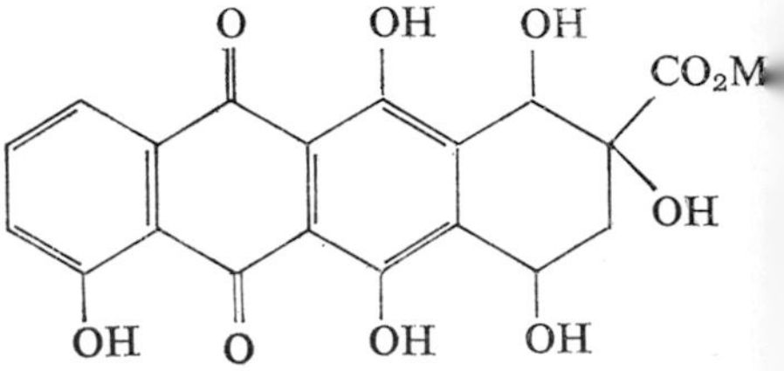

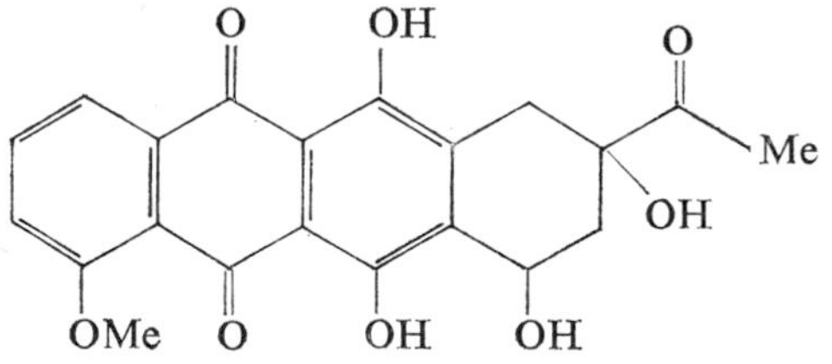

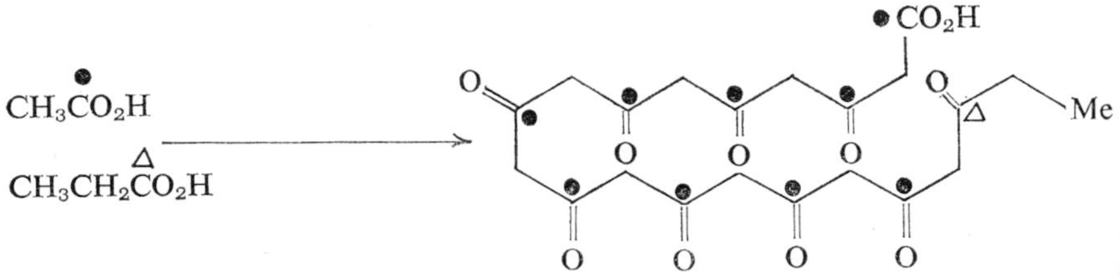

Scheme 39: The incorporation of acetate and propionate into ε-pyrromycinone

side-chain of daunomycinone raises the possibility that it is a true decaketide (**326**); however, the oxygenation pattern, especially the presence of oxygen at position 9, makes it more likely that daunomycinone arises by the same route as the other anthracyclines.

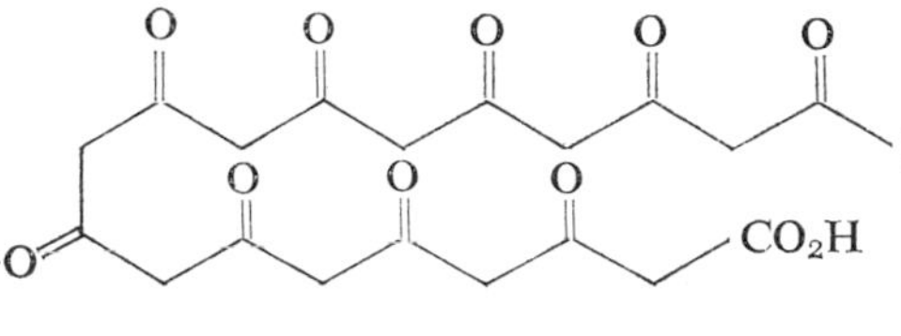

326

B.9. COMPOUNDS FORMED FROM MORE THAN TEN "C₂-UNITS"

Siphulin (**327**), from the lichen *Siphula ceratites*,[459] is the only obviously polyketide-derived compound which can be formed from a chain of more than twenty carbons; it may be formed by condensation of an octanoyl "starter" with eight malonate units.

The only other compounds with chains longer than twenty carbon atoms are the polyene antifungal antibiotics produced by *Streptomyces* spp. These compounds, some of which occur as glycosides with amino-sugars, contain a macrolide ring and a chromophore of four to seven double bonds. They are exemplified by lagosin[460] and fungichromin[461] which have the same gross structure (**328a**) (though it is not clear whether the compounds are identical), filipin (**328b**),[462] and pimaracin (**329**).[463] Compounds (**328**) and (**329**) possess chains of twenty-six and twenty-eight carbons respectively. The methyl group of lagosin may be derived from propionate rather than from the C_1-pool since similar groups in nystatin (the full structure of which is not yet known) are known to arise from propionate.[464]

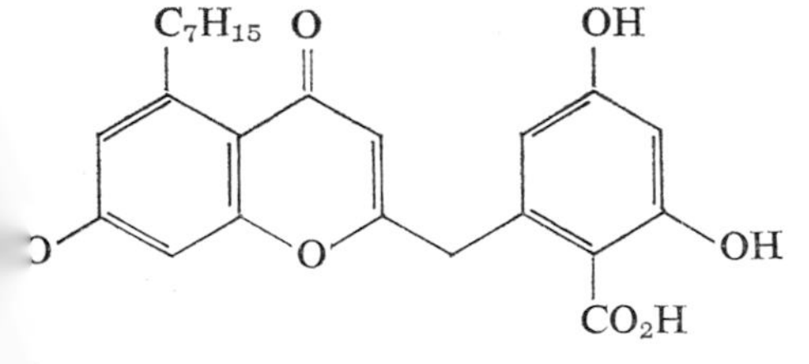

327

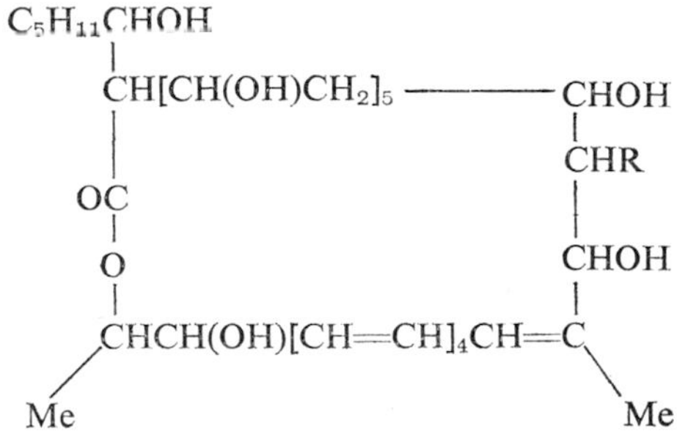

328   a: R = OH
       b: R = H

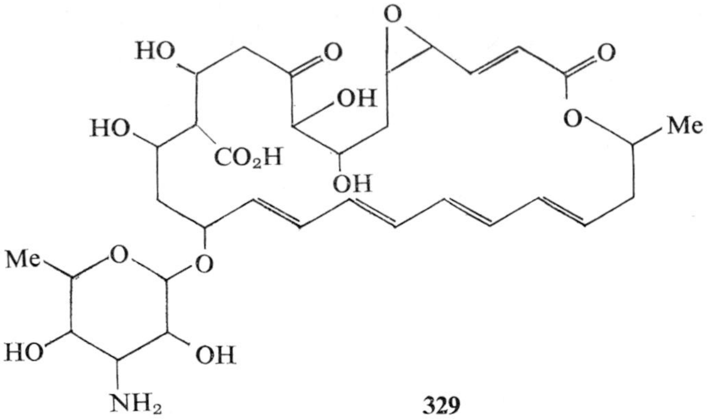

329

B.10. MISCELLANEOUS COMPOUNDS DERIVED FROM ACETATE

Included here are compounds derived from acetate whose mode of assembly is not yet sufficiently clear to permit their classification in the previous sections.

330

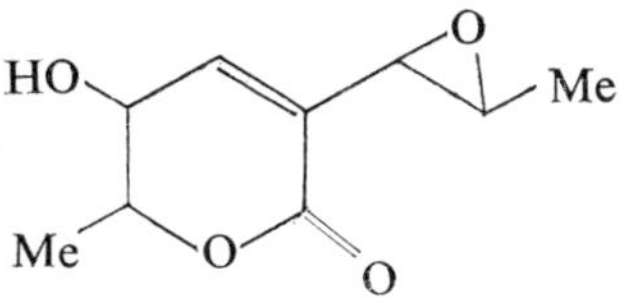

3-epoxypropyl-5,6-dihydro-5-hydroxy- L^4
   6-methylpyrone
Aspergillus melleus[465]

331

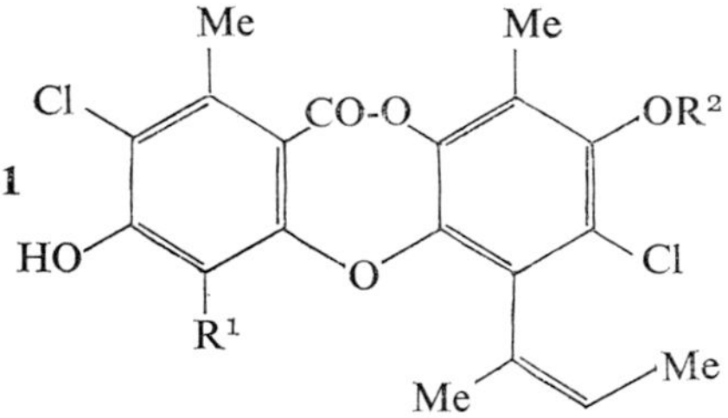

a: R^1 = Cl, R^2 = H, nornidulin (ustin) L^4
b: R^1 = Cl, R^2 = Me, nidulin
c: R^1 = R^2 = H, dechloronornidulin
                A. nidulans[467]

332

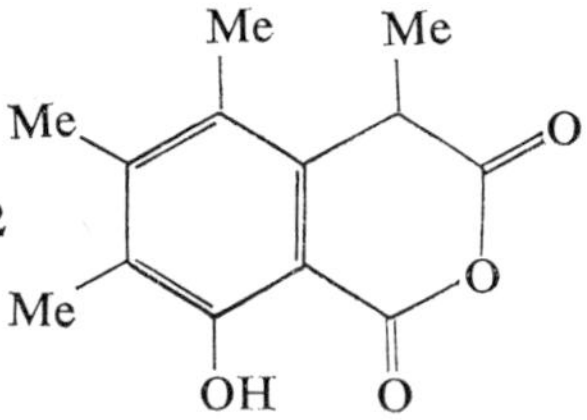

sclerin L^4
Sclerotinia libertiana,[469] *S. sclerotiorum*[470]

333

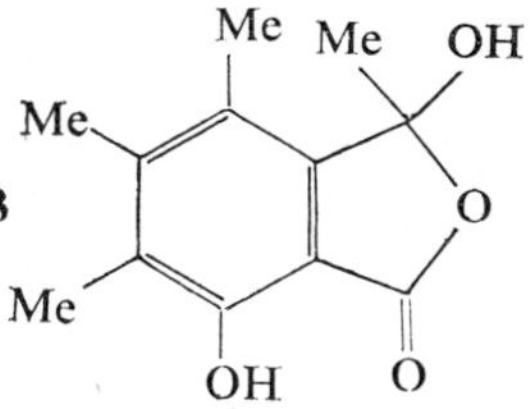

sclerolide
S. libertiana[472]

334

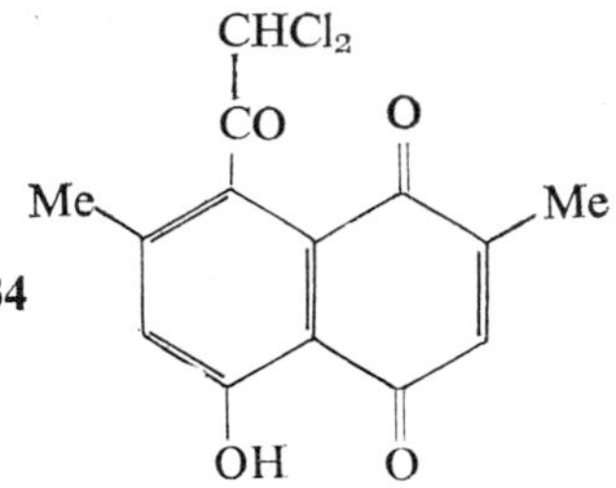

mollisin   *L*[474]
Mollisia caesia, M. gallens[473]

335

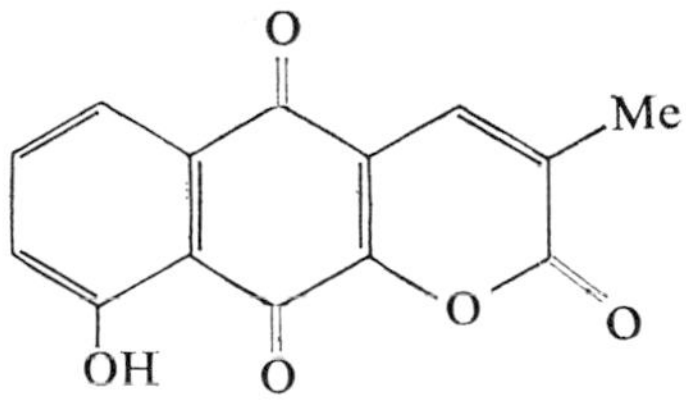

lambertellin
Lambertella hicoriae,[475] *L. corni-maris*[476]

336

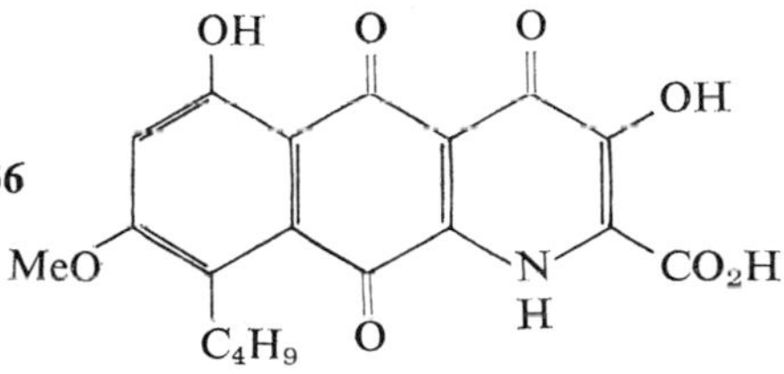

phomazarin   *L*[355]
Phoma terrestris[477]

Incorporation of labelled acetate into the epoxypyrone (**330**) gives the labelling pattern shown in formula (**337**). This distribution of radioactivity, especially the linkage of two carbon atoms derived from the methyl group

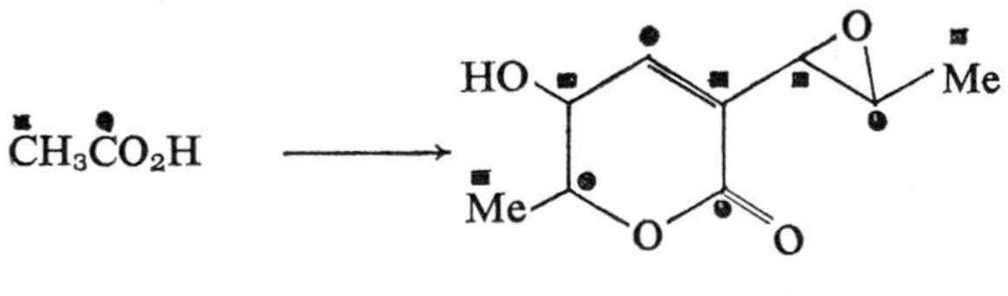

CH_3CO_2H ⟶

337

of acetate, is difficult to interpret in terms of the normal processes of polyketide biosynthesis.

Nidulin (**331**) is the only depsidone to have been isolated from a fungus. Ring A of nidulin is a simple orsellinic acid derivative but ring B is unusual. The incorporation of [1-^{14}C]acetate and [U-^{14}C]isoleucine into nidulin[468] is

Scheme 40: Suggested route for the biosynthesis of nidulin

consistent with the formation of ring B from a C_5 acid, derived from isoleucine, and three acetate (malonate) units with decarboxylation (Scheme 40); the incorporation of C_1 precursors was not studied.

The structure of the plant-growth promotor sclerin (**332**) cannot be accounted for by "normal" biosynthetic pathways. Kubota *et al.*[471] have studied the incorporation of [1-^{14}C]acetate, [2-^{14}C]acetate and [^{14}C]formate into sclerin and explain their results in terms of condensation of two polyketide chains (route *a* in Scheme 41). Sclerin is a cometabolite of sclerotinins A and B (**118**) which are obviously related to citrinin (**120**) and the other

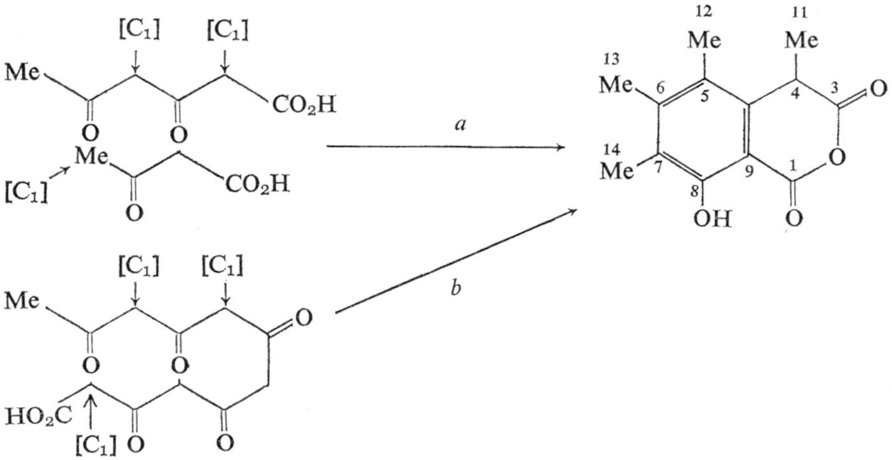

Scheme 41: Alternative biosyntheses for sclerin

isocoumarins (see p. 116), and the authors suggest that the sclerotinins, and also citrinin (and presumably all the isocoumarins?) might also be derived from two chains (Scheme 42). However, these two-chain schemes involve two processes which are almost without precedent in polyketide biosynthesis— the introduction of a methyl group onto the methyl group of a polyketide chain (a quinone-methide intermediate can be invoked to explain the only other example of this, see barnol, p. 100), and an aldol condensation onto the methyl group of a polyketide chain (this *might* be involved in the formation

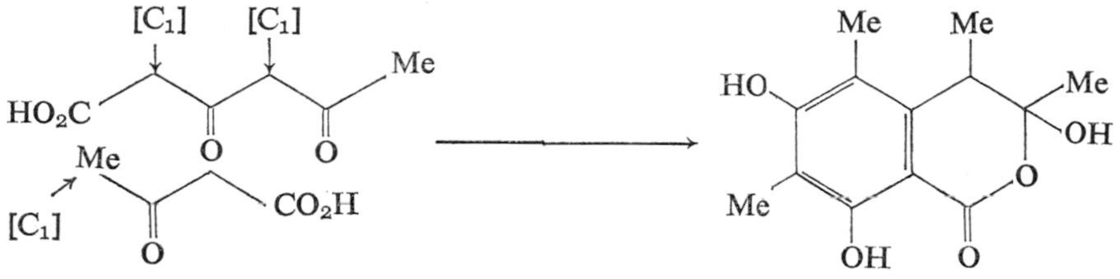

Scheme 42: A two-chain derivation of sclerotinin A (**338**)

of the C_{10} naphthalene compounds, but no direct evidence is available, see p. 100).

In the degradation of the labelled sclerin only carbon atoms 1,3,4 and 11 were isolated unambiguously. The remaining carbon atoms were isolated by Kuhn–Roth degradations, a notoriously unreliable technique when applied to compounds containing more than one aromatic methyl group, and some of the experimental errors are large; moreover some of the theoretical values appear to have been miscalculated or misquoted. However, the radioactivities of C-1, C-3, C-4 and C-11 and of three degradation products are consistent with the presence in sclerin of three carbon atoms from the C_1-pool, and five carbons from each of C-1 and C-2 of acetate, as required by the two-chain hypothesis. In view of the experimental errors involved, alternative *b* of Scheme 41, which involves processes for which there are clearly established precedents and which represents an alternative folding of the chain leading to the sclerotinins, should not be completely dismissed.

On the other hand, mollisin (**334**) probably does arise from two chains since Bentley and Gatenbeck[474] have shown that neither methionine nor mevalonate is incorporated into the molecule. The results of experiments with [1-^{14}C]acetate and [1,3-^{14}C]malonate, using Kuhn–Roth degradation to isolate carbons 11 and 12 and 2 and 7, are consistent with a biosynthesis of type (**339**) or (**340**), though (**340**) was preferred since it involves chlorination at a more active position. No specific labelling was observed in the malonate experiment, and although [2-^{14}C]acetate and [2-^{14}C]malonate were incorporated into mollisin the radioactive product was not degraded. The latter was an unfortunate omission since, for a two-chain derivation of type (339) or (340), the Kuhn–Roth acetic acid from these experiments should have

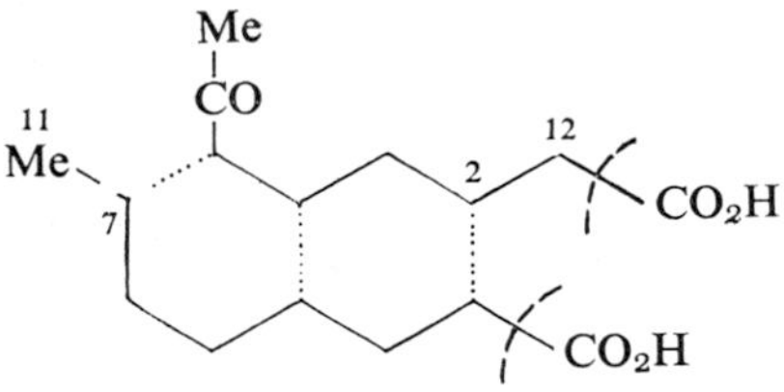

339 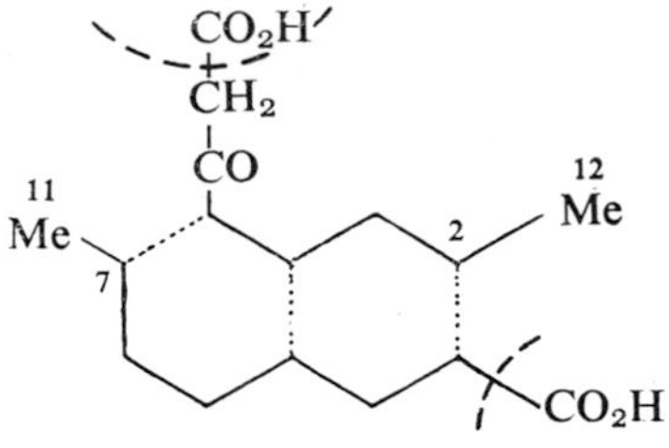   340

contained 25% of the radioactivity of mollisin, (as opposed to 33% in the experiment with [1-^{14}C]acetate).

Structure (335) for lambertellin has been confirmed by synthesis[478] (an isomeric structure had been proposed[476]). Lambertellin might arise by degradation of an anthraquinone intermediate, e.g. chrysophanol (341) though incorporation of acetate has yet to be demonstrated.

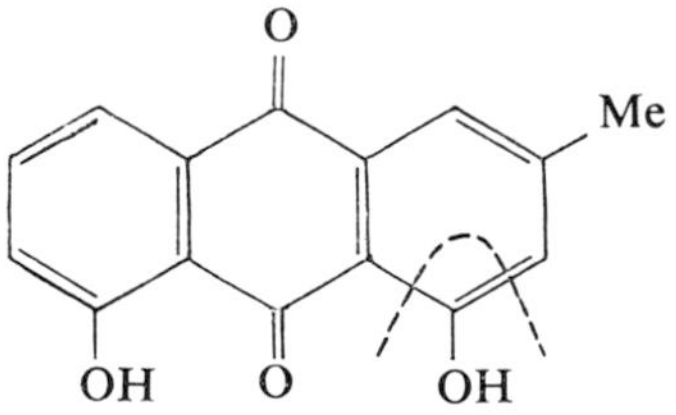

341

Birch et al.[355] showed that phomazarin (336) is biosynthesized from at least eight acetate units and that the carboxyl group arises from the methyl group of acetate. At that time, phomazarin was believed to be isomeric with (336), but subsequently Birch et al.[477b] proposed structure (336) and re-interpreted their labelling data in terms of (342).

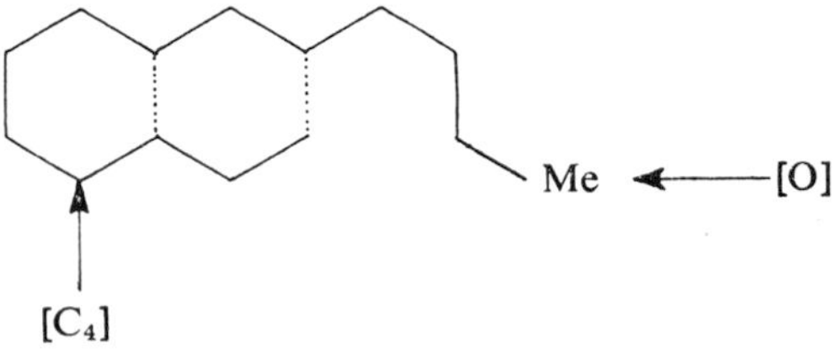

342

C. SUMMARY

Perusal of this chapter enables us to form a picture of the frequency of occurrence of the various chain-lengths and types of cyclization among the fungal polyketides, and certain conclusions may be drawn which are discussed below. It should be borne in mind, however, that the picture is partly man-made and that we are seeing the distribution not of the polyketides of fungal origin, but of those *which have so far been isolated*. The picture is therefore

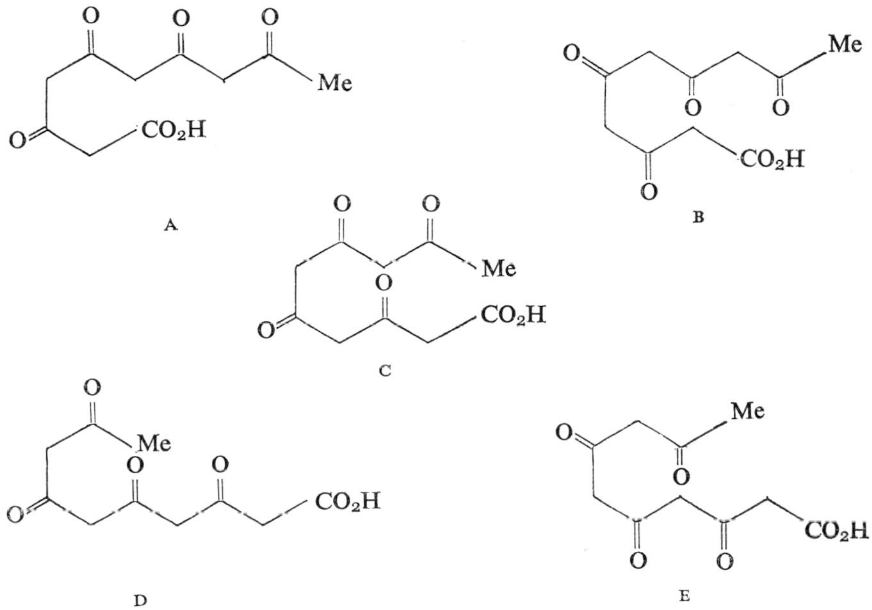

Scheme 43: The theoretically possible foldings of a pentaketide chain; cyclizations of types D and E have not been observed

somewhat distorted for two reasons. Firstly, certain classes of compound have a greater tendency to crystallize than others, and until the recent advent of techniques such as thin layer chromatography, n.m.r. spectroscopy and mass spectrometry, non-crystalline products have presented serious problems of purity determination and have therefore been largely ignored; more non-crystalline compounds are now being examined and their structures may well exemplify some of the presently missing types. Secondly, the isolation of one or more examples of a class has often led to a systematic search for related compounds, perhaps in connection with biosynthetic studies, so that the size of a class may reflect the intensity of the search for its members rather than their frequency of occurrence.

C.1. DISTRIBUTION OF CHAIN-LENGTHS

Tetra-, penta-, hepta- and octaketides are numerous, while tri-, hexa-, nona- and decaketides are less common. There is a sharp cutout at the decaketide level, and only one true polyketide derived from a chain longer than C_{20} is known.

C.2. TYPES OF CYCLIZATION

For each polyketide chain most, and in some cases all, possible cyclizations (to give six-membered rings) are exemplified, with one notable restriction. Uncyclized residues from the methyl ends of polyketide chains are never shorter than residues from the carboxyl end of the chain. This is illustrated in Scheme 43 for folding of the chain derived from five "C_2-units"; compounds derived from foldings of types A to C are known, but compounds derived from types D and E are not. This restriction is presumably a result of the mechanism of formation of aromatic compounds from polyketide precursors.

C.3. OXYGENATION PATTERNS

We have seen that in general the oxygen atoms of the polyketide-derived compounds occupy positions corresponding to the keto-groups of the polyketide precursor but that these oxygen atoms may be lost and "extra" oxygen atoms introduced. However, one general "rule" may be discerned— if a compound is the result of a cyclization involving the methylene group α- to the terminal carboxyl group of a polyketide chain (Scheme 44) then the oxygen β- to the carboxyl group is *always* retained. This, too, may be a result of the mechanism of aromatization of the polyketide precursors.

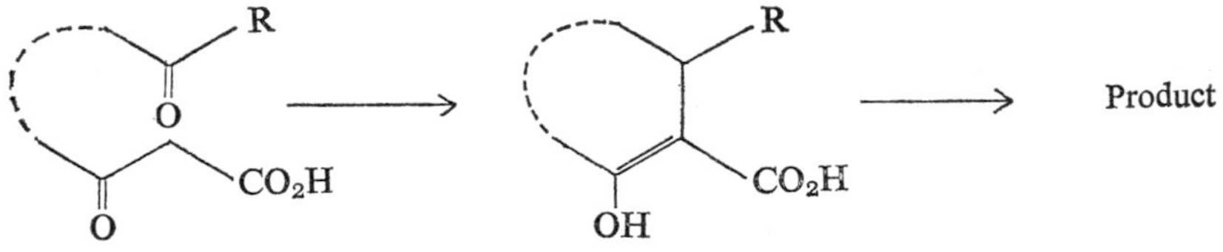

Scheme 44

REFERENCES

1. See, for example, J. H. Richards and J. B. Hendrickson, "The Biosynthesis of Steroids, Terpenes, and Acetogenins". Benjamin, New York, 1964; J. D. Bu'Lock, "The Biosynthesis of Natural Products", Ch.2. McGraw-Hill, London, 1965.
2. A. J. Birch, *Science*, 1967, **156**, 202.
3. K. Mosbach, *Acta Chem. Scand.*, 1964, **18**, 1591.
4. See, for example, Z. Vaněk and J. Majer. *In* "Antibiotics Vol. II. Biosynthesis", (D. Gottlieb and P. D. Shaw, eds.), p. 154. Springer-Verlag, New York, 1967.
5. W. I. Taylor and A. R. Battersby (eds.), "Oxidative Coupling of Phenols". Edward Arnold, London, 1967.
6. W. B. Turner, unpublished result.
7. R. J. Light, *Arch. Biochem. Biophys.*, 1965, **112**, 163.
8. J. E. Nixon, G. R. Putz and J. W. Porter, *J. Biol. Chem.*, 1968, **243**, 5471; M. Yalpani, K. Willecke and F. Lynen, *Eur. J. Biochem.*, 1969, **8**, 495.
9. G. M. Gaucher and M. G. Shepherd, *Biochem. Biophys. Res. Commun.*, 1968, **32**, 664.
10. F. Lynen and M. Tada, *Angew. Chem.*, 1961, **73**, 513; R. J. Light, *J. Biol. Chem.*, 1967, **242**, 1880.
11. E. W. Bassett and S. W. Tanenbaum, *Biochim. Biophys. Acta*, 1960, **40**, 535.
12. E. W. Bassett and S. W. Tanenbaum, *Biochim. Biophys. Acta*, 1962, **59**, 524.
13. J. D. Bu'Lock, "Essays in Biosynthesis and Microbial Development", p. 19. Wiley, New York, 1967.
14. S. Gatenbeck, P. O. Eriksson and Y. Hansson, *Acta Chem. Scand.*, 1969, **23**, 699.
15. T. E. Acker, P. E. Brenneisen and S. W. Tanenbaum, *J. Amer. Chem. Soc.*, 1966, **88**, 834.
16. R. Bentley and P. M. Zwitkowits, *J. Amer. Chem. Soc.*, 1967, **89**, 676
17. T. M. Harris, C. M. Harris and R. J. Light, *Biochim. Biophys. Acta*, 1966, **121**, 420.
18. (a) S. Nozoe, K. Hirai, K. Tsuda, K. Ishibashi, M. Shirasaka and J. F. Grove, *Tetrahedron Lett.*, 1965, 4675; (b) J. F. Grove, *J. Chem. Soc.*, 1964, 3234.
19. Z. Kis, P. Furger and H. P. Sigg., *Experientia*, 1969, **25**, 123.
20. A. D. Argoudelis and J. F. Zieserl, *Tetrahedron Lett.*, 1966, p. 1969.
21. R. H. Evans, G. A. Ellestad and M. P. Kunstmann, *Tetrahedron Lett.*, 1969, 1791.
22. J. F. Grove, R. N. Speake and G. Ward, *J. Chem. Soc. C*, 1966, 230; J. MacMillan and R. J. Pryce, *Tetrahedron Lett.*, 1968, 5497; J. W. Powell and W. B. Whalley, *J. Chem. Soc. C*, 1969, 911.
23. A. J. Birch, R. A. Massy-Westropp and C. J. Moye, *Aust. J. Chem.*, 1955, **8**, 539.
24. J. D. Bu'Lock, H. M. Smalley and G. N. Smith, *J. Biol. Chem.*, 1962, **237**, 1778.
25. E. W. Bassett and S. W. Tanenbaum, *Experientia*, 1958, **14**, 38.
26. S. Gatenbeck, *Acta Chem. Scand.*, 1957, **11**, 555.
27. S. Gatenbeck, *Acta Chem. Scand.*, 1958, **12**, 1985.

28. For references see ref. 77b.
29. S. Gatenbeck and K. Mosbach, *Acta Chem. Scand.*, 1959, **13**, 1561.
30. K. Mosbach, *Acta Chem. Scand.*, 1960, **14**, 457.
31. J. H. Birkinshaw and A. Gowlland, *Biochem. J.*, 1962, **84**, 342.
32. J. A. Ballantine, C. H. Hassall and B. D. Jones, *Phytochemistry*, 1968, **7**, 1529.
33. J. A. Ballantine, C. H. Hassall and G. Jones, *J. Chem. Soc.*, 1965, 4672.
34. J. A. Ballantine, C. H. Hassall, B. D. Jones and G. Jones, *Phytochemistry*, 1967, **6**, 1157.
35. L. Reio, *J. Chromatography*, 1958, **1**, 338; K. Mosbach, *Z. Naturforsch. B*, 1959, **14**, 69.
36. D. Hess, *Z. Naturforsch. B*, 1959, **14**, 345.
37. R. Bentley and J. G. Keil, *J. Biol. Chem.*, 1962, **237**, 867.
38. N. M. Packter, *Biochem J.*, 1965, **97**, 321.
39. M. C. Rebstock, *Arch. Biochem. Biophys.*, 1964, **104**, 156.
40. P. Simonart and A. Wiaux, *Bull. Soc. Chim. Biol.*, 1959, **41**, 537.
41. G. Pettersson, *Acta Chem. Scand.*, 1964, **18**, 1202.
42. (a) H. Raistrick and P. Simonart, *Biochem. J.*, 1933, **27**, 628; (b) A. Brack, *Helv. Chim. Acta*, 1947, **30**, 1; (c) J. Barta and R. Mecir, *Experientia*, 1948, **4**, 277.
43. A. Closse, R. Mauli and H. P. Sigg, *Helv. Chim. Acta*, 1966, **49**, 204.
44. S. Gatenbeck and I. Lönnroth, *Acta Chem. Scand.*, 1962, **16**, 2298.
45. W. K. Anslow and H. Raistrick, *Biochem. J.*, 1938, **32**, 687.
46. N. M. Packter and J. Glover, *Biochim. Biophys. Acta*, 1965, **100**, 50.
47. N. M. Packter, *Biochem. J.*, 1965, **97**, 321.
48. N. M. Packter, *Biochem. J.*, 1966, **98**, 353.
49. G. Pettersson, *Acta Chem. Scand.*, 1963, **17**, 1323.
50. G. Pettersson, *Acta Chem. Scand.*, 1963, **17**, 1771.
51. G. Pettersson, *Acta Chem. Scand.*, 1964, **18**, 335.
52. G. Pettersson, *Acta Chem. Scand.*, 1964, **18**, 1202.
53. G. Pettersson, *Acta Chem. Scand.*, 1965, **19**, 543.
54. W. K. Anslow and H. Raistrick, *Biochem J.*, 1938, **32**, 2288.
55. D. F. Jones, unpublished result.
56. G. Pettersson, *Acta Chem. Scand.*, 1965, **19**, 1016.
57. M. Anchel, A. Hervey, F. Kavanagh, J. Polatnick and W. J. Robbins, *Proc. Nat. Acad. Sci. U.S.*, 1948, **34**, 498.
58. A. J. Birch, "CIBA Foundation Symposium on Quinones in Electron Transport", p. 233. London, 1961.
59. G. Pettersson, *Acta Chem. Scand.*, 1966, **20**, 45.
60. G. Pettersson, *Acta Chem. Scand.*, 1966, **20**, 151.
61. N. M. Packter, *Abstr. 3rd Meet. Fed. Europ. Biochem. Soc.*, *Warsaw*, 1966, F270; *Biochem. J.*, 1969, **114**, 369.
62. J. C. Sheehan, W. B. Lawson and R. J. Gaul, *J. Amer. Chem. Soc.*, 1958, **80**, 5536.
63. G. Read and L. C. Vining, *Chem. Commun.*, 1968, 935; G. Read, D. W. S. Westlake and L. C. Vining, *Can. J. Biochem.*, 1969, **47**, 1071.
64. M. W. Miller, *Abstr. 154th A.C.S. Meeting*, *Chicago*, 1967, S118.
65. Y. Yamamoto, K. Nitta, K. Tango and S. Saito, *Chem. Pharm. Bull. (Tokyo)*, 1965, **13**, 935.
66. G. Pettersson, *Acta Chem. Scand.*, 1965, **19**, 414.
67. N. M. Packter and M. W. Steward, *Biochem. J.*, 1967, **102**, 122.

68. G. Pettersson, *Acta Chem. Scand.*, 1965, **19**, 2013.
69. K. Mosbach and U. Ehrensvärd, *Biochem. Biophys. Res. Commun.*, 1966, **22**, 145.
70. G. Pettersson, *Acta Chem. Scand.*, 1964, **18**, 1428, 1839.
71. G. Pettersson, *Acta Chem. Scand.*, 1966, **20**, 657.
72. P. Simonart and H. Verachtert, *Bull. Soc. Chim. Biol.*, 1967, **49**, 543.
73. G. Pettersson, *Acta Chem. Scand.*, 1966, **20**, 45.
74. Y. Yamamoto, K. Nitta, Y. Terashima, J. Ishikawa and N. Watanabe, *Chem. Pharm. Bull. (Tokyo)*, 1965, **13**, 1009.
75. R. F. Curtis, P. C. Harries, C. H. Hassall and J. D. Levi, *Biochem. J.*, 1964, **90**, 43.
76. H. Raistrick and P. Rudman, *Biochem. J.*, 1956, **63**, 395.
77. G. Pettersson, *Acta Chem. Scand.*, 1965, **19**, (a) p. 35; (b) p. 1724.
78. I. Ljungcrantz and K. Mosbach, *Acta Chem. Scand.*, 1964, **18**, 638.
79. I. Ljungcrantz and K. Mosbach, *Biochim. Biophys. Acta*, 1964, **86**, 203; *Physiol. Plantarum*, 1965, **18**, 1.
80. J. H. Birkinshaw, H. Raistrick, D. J. Ross and C. E. Stickings, *Biochem. J.*, 1952, **50**, 610; L. A. Duncanson, J. F. Grove and J. Zealley, *J. Chem. Soc.*, 1953, 3637.
81. A. J. Birch and M. Kocor, *J. Chem. Soc.*, 1960, 866.
82. W. W. Andres, M. P. Kunstmann and L. A. Mitscher, *Experientia*, 1967, **23**, 703.
83. S. Naito and Y. Kaneko, *Tetrahedron Lett.*, 1969, 4675.
84. H. Raistrick and D. J. Ross, *Biochem. J.*, 1952, **50**, 635.
85. A. J. Birch and E. Pride, *J. Chem. Soc.*, 1962, 370.
86. J. H. Birkinshaw, P. Chaplen and R. Lahoz-Oliver, *Biochem. J.*, 1957, **67**, 155.
87. A. N. Starratt, *Can. J. Chem.*, 1968, **46**, 767.
88. P. W. Brian, P. J. Curtis and H. G. Hemming, *J. Gen. Microbiol.*, 1948, **2**, 341.
89. R. F. Curtis, P. C. Harries, C. H. Hassall and J. D. Levi, *J. Chem. Soc. C*, 1966, 168.
90. P. W. Brian, P. J. Curtis, S. R. Howland, E. G. Jeffreys and H. Raudnitz, *Experientia*, 1951, **7**, 266; J. F. Grove, *J. Chem. Soc. C*, 1966, 985.
91. A. J. Birch, R. I. Fryer and H. Smith, *Proc. Chem. Soc.*, 1958, 343.
92. R. Bentley and W. V. Lavate, *J. Biol. Chem.*, 1965, **240**, 532.
93. G. Pettersson, *Acta Chem. Scand.*, 1965, **19**, 1827.
94. G. Pettersson, *Acta Chem. Scand.*, 1964, **18**, 2303.
95. C.-K. Wat, A. Tse, R. J. Bandoni and G. H. N. Towers, *Phytochemistry*, 1968, **7**, 2177.
96. G. Pettersson, *Acta Chem. Scand.*, 1965, **19**, 414.
97. M. W. Steward and N. M. Packter, *Biochem. J.*, 1968, **109**, 1.
98. M. Lenfant, G. Farrugia and E. Lederer, *C.R. Acad. Sci. Paris Ser. C*, 1969, **268**, 1986.
99. S. H. El Basyouni and L. C. Vining, *Can. J. Biochem.*, 1966, **44**, 557.
100. A. J. Birch and R. I. Fryer, *Aust. J. Chem.*, 1969, **22**, 1319.
101. T. Posternak, *Helv. Chim. Acta*, 1938, **21**, 1326; E. A. H. Friedheim, *Helv. Chim. Acta*, 1938, **21**, 1464.
102. B. Akermark, H. Erdtman and C. A. Wachtmeister, *Acta Chem., Scand.*, 1959, **13**, 1855.
103. J. Santesson, *Acta Chem. Scand.*, 1967, **21**, 1111; S. Huneck, K. Schreiber, G. Snatzke and P. Trška, *Z. Naturforsch. B*, 1970, **25**, 265.
104. C. A. Wachtmeister, *Acta Chem. Scand.*, 1956, **10**, 1404.

105. S. Shibata, *Chem. Pharm. Bull.* (*Tokyo*), 1957, **5**, 488.
106. E. Charollais, S. Fliszar and T. Posternak, *Arch. Sci. Geneva*, 1963, **16**, 474,
107. O. F. Black and C. L. Alsberg, *U.S. Dep. Agr., Bur. Plant. Ind. Bull.*, No. 270, 1913.
108. J. H. Birkinshaw, A. E. Oxford and H. Raistrick, *Biochem. J.*, 1936, **30**, 394.
109. A. J. Birch, G. E. Blance and H. Smith, *J. Chem. Soc.*, 1958, 4582.
110. K. Mosbach, *Acta Chem. Scand.*, 1960, **14**, 457.
111. R. B. Woodward and G. Singh, *J. Amer. Chem. Soc.*, 1949, **71**, 758; *Experientia*, 1950, **6**, 238.
112. J. H. Birkinshaw, *Ann. Rev. Biochem.*, 1953, **22**, 371.
113. J. D. Bu'Lock and A. J. Ryan, *Proc. Chem. Soc.*, 1958, 222.
114. E. W. Bassett and S. W. Tanenbaum, *Biochim. Biophys. Acta*, 1958, **28**, 247.
115. E. W. Bassett and S. W. Tanenbaum, *Biochim. Biophys. Acta*, 1960, **40**, 535.
116. A. I. Scott and M. Yalpani, *Chem. Commun.*, 1967, 945.
117. S. Huneck, "Lichen Substances". *In* "Progress in Phytochemistry", Vol. I. Interscience, London, 1968.
118. M. Yamazaki, M. Matsuo and S. Shibata, *Chem. Pharm. Bull.* (*Tokyo*), 1965, **13**, 1015.
119. K. Mosbach, *Acta Chem. Scand.*, 1964, **18**, 329.
120. M. Yamazaki and S. Shibata, *Chem. Pharm. Bull* (*Tokyo*), 1966, **14**, 96.
121. T. Komiya and S. Shibata, *Chem. Pharm. Bull.* (*Tokyo*), 1969, **17**, 1305.
122. F. Bergel, A. L. Morrison, A. R. Moss and H. Rinderknecht, *J. Chem. Soc.*, 1944, 415; C. H. Hassall and A. R. Todd, *J. Chem. Soc.*, 1947, 611.
123. D. C. Allport and J. D. Bu'Lock, *J. Chem. Soc.*, 1960, 654.
124. See, for example, Y. Asahina and S. Shibata, "Chemistry of Lichen Substances". Japan Society for the Promotion of Science, Tokyo, 1954.
125. S. Shibata and H. Taguchi, *Tetrahedron Lett.*, 1967, 4867.
126. S. Gatenbeck and U. Brunsberg, *Acta Chem. Scand.*, 1966, **20**, 2334.
127. H. Taguchi, U. Sankawa and S. Shibata, *Chem. Pharm. Bull.* (*Tokyo*), 1969, **17**, 2054.
128. J. H. Birkinshaw, A. R. Chambers and H. Raistrick, *Biochem. J.*, 1942, **36**, 242.
129. R. Bentley, *J. Biol. Chem.*, 1963, **238**, 1895.
130. I. G. Andrew and W. Segal, *J. Chem. Soc.*, 1964, 607.
131. J. H. Birkinshaw and H. Raistrick, *Biochem. J.*, 1932, **26**, 441.
132. L. D. Ferretti and J. H. Richards, *Proc. Nat. Acad. Sci. U.S.*, 1960, **46**, 1438.
133. W. Segal, *J. Chem. Soc.*, 1959, 2847.
134. M. J. S. Dewar, *Nature*, 1945, **155**, 50.
135. R. E. Corbett, A. W. Johnson and A. R. Todd, *J. Chem. Soc.*, 1950, 6.
136. G. Aulin-Erdtman, *Acta Chem. Scand.*, 1951, **5**, 301.
137. P. V. Divekar, P. E. Brenneisen and S. W. Tanenbaum, *Biochim. Biophys. Acta*, 1961, **50**, 588.
138. R. Bentley and C. P. Thiessen, *J. Biol. Chem.*, 1963, **238**, 3811.
139. S. W. Tanenbaum and E. W. Bassett, *Biochim. Biophys. Acta*, 1962, **59**, 524.
140. R. Robinson, *Chem. and Ind.*, 1951, 12.
141. T. R. Seshadri, *J. Sci. Ind. Res* (*India*), 1955, **14B**, 248.
142. R. Bentley, J. A. Ghaphery and J. G. Keil, *Arch. Biochem. Biophys.*, 1965, **111**, 80.
143. R. Bentley and P. M. Zwitkowits, *J. Amer. Chem. Soc.*, 1967, **89**, 681.
144. G. S. Marx and S. W. Tanenbaum, *J. Amer. Chem. Soc.*, 1968, **90**, 5302.
145. R. Bentley and I. M. Campbell. *In* "Comprehensive Biochemistry" (M. Florkin and E. H. Stotz, eds.) Vol. 20, p. 453. Elsevier, London, 1968.

146. J. H. Richards and J. B. Hendrickson, "The Biosynthesis of Steroids, Terpenes and Acetogenins", p. 169. Benjamin, New York, 1964.

147. J. H. Birkinshaw, A. Bracken and E. N. Morgan, *Biochem. J.*, 1948, **42**, xxxix; J. H. Birkinshaw, H. Raistrick and D. J. Ross, *Biochem. J.*, 1952, **50**, 630.

148. A. J. Birch, R. J. English, R. A. Massy-Westropp and H. Smith, *J. Chem. Soc.*, 1958, 369.

149. Unpublished result cited by A. J. Birch, *Proc. Chem. Soc.*, 1962, 10.

150. A. J. Birch, *Ann. Rev. Plant Physiol.*, 1968, **19**, 321.

151. I. M. Campbell, C. H. Calzadilla and N. J. McCorkindale, *Tetrahedron Lett.*, 1966, 5107.

152. G. Jaureguiberry, G. Farrugia-Fougerouse, H. Audier and E. Lederer, *C.R. Acad. Sci. Paris, Ser. C*, 1964, **259**, 3108.

153. P. W. Clutterbuck, A. E. Oxford, H. Raistrick and G. Smith, *Biochem. J.*, 1932, **26**, 1441; A. E. Oxford and H. Raistrick, *Biochem. J.*, 1933, **27**, 634.

154. A. E. Oxford and H. Raistrick, *Biochem. J.*, 1932, **26**, 1902.

155. S. Gatenbeck, *Acta Chem. Scand.*, 1958, **12**, 1985.

156. H. Raistrick and C. E. Stickings, *Biochem. J.*, 1951, **48**, 53.

157. V. C. Vora, *J. Sci. Ind. Res. (India)*, 1954, **13B**, 842.

158. F. H. Stodola, C. Cabot and C. R. Benjamin, *Biochem. J.*, 1964, **93**, 92.

159. S. W. Tanenbaum, S. G. Agarwal, T. Williams and R. G. Pitcher, *Tetrahedron Lett.*, 1970, 2377.

160. E. Nishikawa, *J. Agr. Chem. Soc. Japan*, 1933, **9**, 772.

161. E. L. Patterson, W. W. Andres and N. Bohonos, *Experientia*, 1966, **22**, 209.

162. J. D. Bu'Lock and G. N. Smith, unpublished.

163. G. Bendz, *Ark. Kemi*, 1959, **14**, 511.

164. R. F. Curtis, *Experientia*, 1968, **24**, 1187; A. Stoessl, *Biochem. Biophys. Res. Commun.*, 1969, **35**, 186.

165. (a) P. W. Brian, H. G. Hemming, J. S. Moffatt and C. H. Unwin, *Trans. Brit. Mycol. Soc.*, 1953, **36**, 243; (b) A. J. Birch, L. Loh, A. Pelter, J. H. Birkinshaw, P. Chaplen, A. H. Manchanda and M. Riano-Martin, *Tetrahedron Lett.*, 1965, 29; *Aust. J. Chem.*, 1969, **22**, 1933.

166. R. Aue, R. Mauli and H.P. Sigg, *Experientia*, 1966, **22**, 575.

167. D. C. Aldridge, S. Galt, D. Giles and W. B. Turner, *J. Chem. Soc.*, in the press.

168. B. F. Burrows, unpublished.

169. L. A. Mitscher, W. W. Andres and W. McCrae, *Experientia*, 1964, **20**, 258.

170. D. Giles and W. B. Turner, *J. Chem. Soc. C*, 1969, 2187.

171. W. J. McGahren and L. A. Mitscher, *J. Org. Chem.*, 1968, **33**, 1577.

172. M. E. Munk, D. B. Nelson, F. J. Antosz, D. L. Herald and T. H. Haskell, *J. Amer. Chem. Soc.*, 1968, **90**, 1087.

173. A. Ballio, S. Barcellona and B. Santurbano, *Tetrahedron Lett.*, 1966, 3723.

174. I. Yamamoto, Y. Yamamoto and K. Nitta, *Agr. Biol. Chem. (Tokyo)*, 1961, **25**, 405; K. Nitta, C. Takura, I. Yamamoto and Y. Yamamoto, *Agr. Biol. Chem. (Tokyo)*, 1963, **27**, 813.

175. K. Nitta, Y. Yamamoto, T. Inoue and T. Hyodo, *Chem. Pharm. Bull. (Tokyo)*, 1966, **14**, 363.

176. K. Nitta, Y. Yamamoto, I. Yamamoto and S. Yamatodani, *Agr. Biol. Chem. (Tokyo)*, 1963, **27**, 822.

177. I. Yamamoto, K. Nitta and Y. Yamamoto, *Agr. Biol. Chem. (Tokyo)*, 1962, **26**, 486; K. Nitta, J. Imai, I. Yamamoto and Y. Yamamoto, *Agr. Biol. Chem. (Tokyo)*, 1963, **27**, 817.

178. D. C. Aldridge, J. F. Grove and W. B. Turner, *J. Chem. Soc C*, 1966, 126.

179. K. J. van der Merwe, P. S. Steyn and L. Fourie, *J. Chem. Soc.*, 1965, 7083.
180. W. van Walbeek, P. M. Scott, J. Harwig and J. W. Lawrence, *Can. J. Microbiol.*, 1969, **15**, 1281.
181. T. Sassa, H. Aoki, M. Namiki and K. Munakata, *Agr. Biol. Chem.* (*Tokyo*), 1968, **32**, 1432.
182. R. F. Curtis, C. H. Hassall and M. Nazar, *J. Chem. Soc. C*, 1968, 85.
183. C. H. Hassall and D. W. Jones, *J. Chem. Soc.*, 1962, 4189.
184. A. C. Hetherington and H. Raistrick, *Phil. Trans. Roy. Soc.* (*London*), 1931, **220B**, 269.
185. A. J. Ewart, *Ann. Bot.*, 1933, **47**, 913.
186. A. J. Birch, P. Fitton, E. Pride, A. J. Ryan, H. Smith and W. B. Whalley, *J. Chem. Soc.*, 1958, 4576.
187. E. Schwenk, G. J. Alexander, A. M. Gold and D. F. Stevens, *J. Biol. Chem.*, 1958, **233**, 1211.
188. O. R. Rodig, L. C. Ellis and I. T. Glover, *Biochemistry*, 1966, **5**, 2458.
189. W. B. Whalley. *In* "Progress in Organic Chemistry" (J. W. Cook, ed.), Vol. 4, p. 72. Butterworth, London.
190. A. J. Birch, J. J. Wright, F. Gager, L.Mo and A. Pelter, *Aust. J. Chem.*, 1967, **22**, 2429.
191. H. Raistrick and G. Smith, *Biochem. J.*, 1935, **29**, 606; J. F. Grove, *J. Chem. Soc.*, 1954, 4693; D. H. R. Barton and E. Miller, *J. Chem. Soc.*, 1955, 1028.
192. W. J. McGahren, J. H. van den Hende and L. A. Mitscher, *J. Amer. Chem. Soc.*, 1969, **91**, 157.
193. A. J. Birch, A. Cassera and A. R. Jones, *Chem. Commun.*, 1965, 167.
194. C. J. Moye and S. Sternhell, *Aust. J. Chem.*, 1966, **19**, 2107; A. W. Burgstahler, T. B. Lewis and M. C. Abdel-Rahman, *J. Org. Chem.*, 1966, **31**, 3516.
195. A. Kamal, N. Ahmad, M. Ali Khan and I. H. Qureshi, *Tetrahedron*, 1962, **18**, 433.
196. A. Kamal, M. Ali Khan and A. Ali Qureshi, *Tetrahedron*, 1963, **19**, 111.
197. A. Kamal and M. A. Sandhu, *Tetrahedron Lett*, 1963, 611.
198. A. Kamal, A. Ali Qureshi and A. Ahmad, *Tetrahedron*, 1965, **21**, 1411.
199. A. Kamal, A. Ali Qureshi, M. Ali Khan and F. M. Khan, *Tetrahedron*, 1963, **19**, 117.
200. A. Ali Qureshi, R. W. Rickards and A. Kamal, *Tetrahedron*, 1967, **23**, 3801.
201. S. E. Michael, *Biochem. J.*, 1948, **43**, 528.
202. A. J. Birch, A. J. Ryan, J. Schofield and H. Smith, *J. Chem. Soc.*, 1965, 1231.
203. D. C. Allport and J. D. Bu'Lock, *J. Chem. Soc.*, 1960, 654.
204. M. Arshad, J. P. Devlin, W. D. Ollis and R. E. Wheeler, *Chem. Commun.*, 1968, 154.
205. (a) K. Soviar, O. Motl, A. Samek and J. Smolikova, *Tetrahedron Lett.*, 1967, 2277; (b) D. J. Aberhart, K. H. Overton and S. Huneck, *J. Chem. Soc. C*, 1969, 704.
206. C. H. Fox and S. Huneck, *Phytochemistry*, 1969, **8**, 1301.
207. D. C. Allport and J. D. Bu'Lock, *J. Chem. Soc.*, 1958, 4090.
208. J. M. Anderson and J. Murray, *Chem. and Ind.*, 1956, 376.
209. J. E. Davies, F. E. King and J. C. Roberts, *Chem. and Ind.*, 1954, 1110.
210. (a) H. Nishikawa, *Agr. Biol. Chem.* (*Tokyo*), 1962, **26**, 696; (b) S. Natori, Y. Inouye and H. Nishikawa, *Chem. Pharm. Bull.* (*Tokyo*), 1967, **15**, 380.
211. R. Howe and R. H. Moore, *Experientia*, 1969, **25**, 474.
212. N. N. Gerber and B. Wieclawek, *J. Org. Chem.*, 1966, **31**, 1496.
213. G. Bendz, *Acta Chem. Scand.*, 1951, **5**, 489.

214. K. Suzuki, T. Sassa, H. Tanaka, H. Aoki and M. Namiki, *Agr. Biol. Chem. (Tokyo)*, 1968, **32,** 1471.
215. G. Read and L. C. Vining, *Chem. and Ind.*, 1963, 1239; G. Read, A. Rashid and L. C. Vining, *J. Chem. Soc. C*, 1969, 2059.
216. See, for example, E. Leistner and M. H. Zenk, *Z. Naturforsch. B*, 1968, **23,** 259.
217. J. D. Bu'Lock. "Essays in Biosynthesis and Microbial Development", pp. 7 ff. and 27 ff. Wiley, New York, 1967.
218. M. Barbetta, G. Casnati and A. Ricca, *Rend. Ist. Lombardo Sci. Lett. A*, 1967, **101,** 75.
219. P. V. Divekar, H. Raistrick, T. A. Dobson and L. C. Vining, *Can. J. Chem.*, 1965, **43,** 1835.
220. A. G. McInnes, D. G. Smith, L. C. Vining and J. L. C. Wright, *Chem. Commun.*, 1968, 1669; J. L. C. Wright, D. G. Smith, A. G. McInnes, L. C. Vining and D. W. S. Westlake, *Can. J. Biochem.*, 1969, **47,** 945.
221. S. Takeuchi, H. Yonehara and H. Shoji, *J. Antibiotics (Tokyo)*, 1964, **17,** 267; S. Takeuchi, H. Yonehara and H. Umezawa, *J. Antibiotics (Tokyo)*, 1959, **12,** 195; S. Takeuchi and H. Yonehara, *J. Antibiotics (Tokyo)*, 1969, **22,** 179.
222. B. F. Burrows, *Chem. Commun.*, 1967, 597.
223. D. J. Cram and M. Tishler, *J. Amer. Chem. Soc.*, 1948, **70,** 4238.
224. J. F. Grove, *J. Chem. Soc.*, 1952, 4056; J. R. Bartels-Keith, *J. Chem. Soc.*, 1960, 860; 1662.
225. N. Tanake and H. Umezawa, *J. Antibiotics (Tokyo)*, 1962, **15,** 189.
226. H. Umezawa, *J. Antibiotics (Tokyo)*, 1962, **15,** 228.
227. W. B. Turner, *J. Chem. Soc.*, 1961, 522.
228. D. J. Cram and M. Tishler, *J. Amer. Chem. Soc.*, 1948, **70,** 4238; D. J. Cram, *J. Amer. Chem. Soc.*, 1948, **70,** 4240.
229. H. Iizuka and M. Iida, *Nature*, 1962, **196,** 681.
230. E. Hardegger, W. Rieder, A. Walser and F. Kugler, *Helv. Chim. Acta*, 1966, **49,** 1283.
231. I. Iwai and H. Mishima, *Chem. and Ind.*, 1965, 186.
232. G. Büchi, J. D. White and G. N. Wogan, *J. Amer. Chem. Soc.*, 1965, **87,** 3484.
233. R. Locci, L. Merlini, G. Hasini and J. Rogers Locci, *Giorn. Microbiol.*, 1966, **15,** 93.
234. E. J. Haws, J. S. E. Holker, A. Kelly, A. D. G. Powell and A. Robertson, *J. Chem. Soc.*, 1959, 3598; E. J. Haws and J. S. E. Holker, *J. Chem. Soc.*, 1961, 3820.
235. B. C. Fielding, E. J. Haws, J. S. E. Holker, A. D. G. Powell, A. Robertson, D. N. Stanway and W. B. Whalley, *Tetrahedron Lett.*, 1960, No. 5 p. 24, and references cited.
236. B. C. Fielding, J. S. E. Holker, D. F. Jones, A. D. G. Powell, K. W. Richmond, A. Robertson and W. B. Whalley, *J. Chem. Soc.*, 1961, 4579.
237. W. B. Whalley, *In* "Proceedings of the Symposium on the Chemistry and Biochemistry of Fungi and Yeasts, Dublin, 1963", p. 565. Butterworths, London.
238. (a) J. R. Hadfield, J. S. E. Holker and D. N. Stanway, *J. Chem. Soc. C*, 1967, 751; (b) M. Kurono, K. Nakanishi, K. Shindo and M. Tada, *Chem. Pharm. Bull. (Tokyo)*, 1963, **11,** 359; A. J. Birch, A. Cassera, P. Fitton, J. S. E. Holker, H. Smith, G. A. Thompson and W. B. Whalley, *J. Chem. Soc.*, 1962, 3583.
239. S. Iwasaki, S. Nozoe, S. Okuda, Z. Sato and T. Kozaka, *Tetrahedron Lett.*, 1969, 3977.

240. J. H. Birkinshaw and H. Raistrick, *Biochem. J.*, 1936, **30**, 801; A. Bracken, A. Pocker and H. Raistrick, *Biochem. J.*, 1954, **57**, 587; K. Bowden, B. Lythgoe and D. J. S. Marsden, *J. Chem. Soc.*, 1959, 1662.
241. P. J. Curtis, H. G. Hemming and W. K. Smith, *Nature*, 1951, **167**, 557; J. F. Grove and B. K. Tidd, *Chem. and Ind.*, 1963, 412; H. P. Sigg, *Helv. Chim. Acta*, 1963, **46**, 1061.
242. A. J. Birch and M. Kocor, *J. Chem. Soc.*, 1960, 866; P. Chaplen and R. Thomas, *Biochem. J.*, 1960, **77**, 91.
243. D. C. Aldridge and W. B. Turner, *J. Chem. Soc. C*, in the press.
244. P. W. Brian, P. J. Curtis, H. G. Hemming and G. L. F. Norris, *Trans. Brit. Mycol. Soc.*, 1957, **40**, 369; J. F. W. McOmie, A. B. Turner and M. S. Tute, *J. Chem. Soc. C*, 1966, 1608.
245. S. W. Tanenbaum and S. Nakajima, *Biochemistry*, 1969, **8**, 4622, 4626.
246. H. Raistrick, R. Robinson and A. R. Todd, *J. Chem. Soc.*, 1937, 80; A. Quilico, C. Cardani and L. Panizzi, *Atti. Accad. Nazl. Lincei, Rend.*, *Classe Sci. Fis., Mat. Nat.*, 1953, **14**, 358 (*Chem. Abstr.*, 1954, **48**, 10663); A. Quilico, C. Cardani and G. S. d'Alcontres, *Gazz. Chim. Ital.*, 1953, **83**, 754.
247. A. J. Birch, A. J. Ryan, J. Schofield and H. Smith, *J. Chem. Soc.*, 1965, 1231.
248. A. E. Oxford, H. Raistrick and P. Simonart, *Biochem. J.*, 1939, **33**, 240; for other references see the review articles, ref. 264.
249. A. J. Birch, R. A. Massy-Westropp, R. W. Rickards and H. Smith, *J. Chem. Soc.*, 1958, 360.
250. A. Rhodes, G. A. Somerfield and M. P. McGonagle, *Biochem. J.*, 1963, **88**, 349.
251. J. MacMillan, *J. Chem. Soc.*, 1954, 2585.
252. J. MacMillan, *J. Chem. Soc.*, 1953, 1697.
253. W. J. McMaster, A. I. Scott and S. Trippett, *J. Chem. Soc.*, 1960, 4628.
254. A. Rhodes, B. Boothroyd, M. P. McGonagle and G. A. Somerfield *Biochem. J.*, 1961, **81**, 28.
255. J. Santesson, *Acta Chem. Scand.*, 1968, **22**, 1698.
256. Y. Asahina and H. Nogami, *Bull. Chem. Soc. Japan*, 1942, **17**, 202.
257. J. Santesson, *Ark. Kemi*, 1969, **30**, 461.
258. J. Santesson, *Ark. Kemi*, 1969, **30**, 455.
259. J. Santesson, *Ark. Kemi*, 1969, **30**, 449.
260. J. Santesson, *Ark. Kemi*, 1969, **31**, 57.
261. J. Santesson and C. A. Wachtmeister, *Ark. Kemi*, 1969, **30**, 445.
262. J. Santesson, *Ark. Kemi*, 1969, **31**, 121.
263. S. Huneck and J. Santesson, *Z. Naturforsch. B*, 1969, **24**, 756.
264. P. W. Brian, *Trans. Brit. Mycol. Soc.*, 1960, **43**, 1; A. Rhodes, *Progr. Ind. Microbiol.*, 1963, **4**, 167; J. F. Grove, *Quart. Rev.*, 1963, **17**, 1; *Fortschr. Chem. Org. Naturstoffe*, 1964, **22**, 203.
265. P. W. Brian, P. J. Curtis and H. G. Hemming, *Trans. Brit. Mycol. Soc.*, 1946, **29**, 173.
266. J. MacMillan, *J. Chem. Soc.*, 1959, 1823.
267. W. A. C. Brown and G. A. Sim, *J. Chem. Soc.*, 1963, 1050.
268. D. H. R. Barton and I. Cohen, *Festschrift A. Stoll*, p. 117. Birkhäuser, Basel, 1956.
269. A. J. Birch, A Cassera and R. W. Rickards, *Chem. and Ind.*, 1961, 792.
270. S. Okuda, H. Isaka, M. Iida, Y. Minemura, H. Iizuka and K. Tsuda, *J. Pharm. Soc. Japan*, 1967, **87**, 1003.
271. T. Kametani, S. Hibino and S. Takano, *Chem. Commun.*, 1969, 131.

272. H. R. V. Arnstein and A. H. Cook, *J. Chem. Soc.*, 1947, 1021; H. W. Ruelius and A. Gauhe, *Ann. Chem.*, 1950, **569**, 38; 1951, **570**, 121.

273. S. Gatenbeck and R. Bentley, *Biochem. J.*, 1965, **94**, 478.

274. G. P. Arsenault, *Tetrahedron*, 1968, **24**, 4745.

275. B. E. Cross, P. L. Myers and G. R. B. Webster, *J. Chem. Soc. C*, 1970, 930.

276. F. A. Cajori, T. T. Otani and M. A. Hamilton, *J. Biol. Chem.*, 1954, **208**, 107.

277. G. P. Arsenault, *Tetrahedron*, 1965, 4033.

278. J. C. Roberts and C. W. H. Warren, *J. Chem. Soc.*, 1955, 2992.

279. H. Kern and S. Naef-Roth, *Phytopathol. Z.*, 1967, **60**, 316.

280. U. Weiss, H. Ziffer, T. J. Batterham, M. Blumer, W. H. L. Hackeng, H. Copier and C. A. Salemink, *Can. J. Microbiol.*, 1965, **11**, 57; R. J. J. Ch. Lousberg, C. A. Salemink, U. Weiss and T. J. Batterham, *J. Chem. Soc. C*, 1969, 1219.

281. Ching-Tan Chen, K. Nakanishi and S. Natori, *Chem. Pharm. Bull.* (*Tokyo*), 1966, **14**, 1434.

282. F. Blank, A. S. Ng and G. Just, *Can. J. Chem.*, 1966, **44**, 2873; 1969, **47**, 1223.

283. J. N. Ashley, B. C. Hobbs and H. Raistrick, *Biochem. J.*, 1937, **31**, 385; R. P. Mull and F. F. Nord, *Arch. Biochem. Biophys.*, 1944, **4**, 419.

284. O. L. Galmarini and F. H. Stodola, *J. Org. Chem.*, 1965, **30**, 112.

285. N. A. Lund, A. Robertson and W. B. Whalley, *J. Chem. Soc.*, 1953, 2434.

286. P. M. Baker and E. Bullock, *Can. J. Chem.*, 1969, **47**, 2421.

287. S. Shibata, A. Ohta and Y. Ogihara, *Chem. Pharm. Bull.* (*Tokyo*), 1963, **11**, 1174, 1179; S. Shibata and Y. Ogihara, *Chem. Pharm. Bull.* (*Tokyo*), 1963, **11**, 1576.

288. G. Tertzakian, R. H. Haskins, G. P. Slater and L. R. Nesbitt, *Proc. Chem. Soc.*, 1964, 195.

289. J. N. Ashley, B. C. Hobbs and H. Raistrick, *Biochem. J.*, 1937, **31**, 385.

290. G. R. Birchall, K. Bowden, U. Weiss and W. B. Whalley, *J. Chem. Soc., C*, 1966, 2237.

291. T. Takeda, E. Morishita and S. Shibata, *Chem. Pharm. Bull.* (*Tokyo*), 1968, **16**, 2213.

292. H. Tanaka, P.-L. Wang, O. Yamada and T. Tamura, *Agr. Biol. Chem.* (*Tokyo*), 1966, **30**, 107; P.-L. Wang and H. Tanaka, *Agr. Biol. Chem.* (*Tokyo*), 1966, **30**, 683.

293. S. Shibata, E. Morishita, T. Takeda and K. Sakata, *Tetrahedron Lett.*, 1966, 4855; P. M. Baker and J. C. Roberts, *J. Chem. Soc. C*, 1966, 2234; J. S. Gray, G. C. J. Martin and W. Rigby, *J. Chem. Soc. C*, 1967, 2580.

294. S. Shibata, *Chem. In Britain*, 1967, 110.

295. (a) G. Just, W. C. Day and F. Blank, *Can. J. Chem.*, 1963, **41**, 74; (b) J. C. Wirth, T. E. Beesley and S. R. Anand, *Phytochemistry*, 1965, **4**, 505.

296. H. Raistrick, C. E. Stickings and R. Thomas, *Biochem. J.*, 1953, **55**, 421.

297. G. G. Freeman, *Phytochemistry*, 1966, **5**, 719.

298. J. C. Overeem and A. van Dijkman, *Rec. Trav. Chim. Pays-Bas*, 1968, **87**, 940.

299. R. Thomas, *Biochem. J.*, 1961, **78**, 748.

300. S. Gatenbeck and S. Hermodsson, *Acta Chem. Scand.*, 1965, **19**, 65.

301. S. Sjoland and S. Gatenbeck, *Acta Chem. Scand.*, 1966, **20**, 1053.

302. K. G. Neill and H. Raistrick, *Biochem. J.*, 1957, **65**, 166.

303. N. Narasimhachari, K. S. Gopalkrishnan, R. H. Haskins and L. C. Vining, *Can. J. Microbiol.*, 1963, **9**, 134.

304. R. E. Harman, J. Cason, F. H. Stodola and A. L. Adkins, *J. Org. Chem.*, 1955, **20**, 1260.

305. J. A. Galarraga, K. G. Neill and H. Raistrick, *Biochem. J.*, 1955, **61**, 456.
306. N. Narasimhachari and L. C. Vining, *Can. J. Chem.*, 1963, **41**, 641.
307. R. Thomas, *Biochem. J.*, 1961, **78**, 807.
308. S. Shibata, Y. Ogihara, N. Tokutake and O. Tanaka, *Tetrahedron Lett.*, 1965, 1287.
309. U. Sankawa, H. Taguchi, Y. Ogihara and S. Shibata, *Tetrahedron Lett.*, 1966, 2883.
310. Y. Ogihara, Y. Iitaka and S. Shibata, *Tetrahedron Lett.*, 1965, 1289.
311. I. C. Paul, G. A. Sim and G. A. Morrison, *Proc. Chem. Soc.*, 1962, 352.
312. D. H. R. Barton, P. de Mayo, G. A. Morrison and H. Raistrick, *Tetrahedron*, 1959, **6**, 48.
313. W. B. Turner, unpublished result.
314. A. C. Hetherington and H. Raistrick, *Phil. Trans. Roy. Soc. (London)*, 1931, **B220,** 209; A. Robertson, W. B. Whalley and J. Yates, *J. Chem. Soc.*, 1951, 2013.
315. A. E. Oxford, H. Raistrick and P. Simonart, *Biochem. J.*, 1935, **29**, 1102; F. M. Dean, R. A. Eade, R. Moubasher and A. Robertson, *J. Chem. Soc.*, 1957, 3497.
316. K. Mosbach and S. Gatenbeck, *Biochem. Biophys. Res. Commun.*, 1963, **11,** 166.
317. D. F. Jones, Ph.D. Thesis, Liverpool, 1960.
318. Z. Vaněk and M. Soucek, *Folia Microbiol.*, 1962, **7**, 262.
319. Y. Asahina and F. Fuzikawa, *Chem. Ber.*, 1935, **68B**, 1558.
320. B. Franck and T. Reschke, *Chem. Ber.*, 1960, **93**, 347.
321. B. Franck and I. Zimmer, *Chem. Ber.*, 1965, **98**, 1514.
322. S. Shibata and M. Takido, *Pharm. Bull. (Tokyo)*, 1955, **3**, 156.
323. G. P. Slater, R. H. Haskins, L. R. Hogge and L. R. Nesbitt, *Can. J. Chem.*, 1967, **45**, 92.
324. B. H. Howard and H. Raistrick, *Biochem. J.*, 1950, **46**, 49.
325. F. Kögl and J. J. Postowsky, *Ann. Chem.*, 1925, **444**, 1.
326. C. E. Stickings and A. Mahmoodian, *Chem. and Ind.*, 1962, 1718; *Biochem. J.*, 1964, **92**, 369.
327. R. H. Thomson, "Naturally Occurring Quinones", p. 196. Butterworth, London, 1957.
328. W. Steglich, W. Lösel and W. Reininger, *Tetrahedron Lett.*, 1967, 4719.
329. W. Steglich and V. Austel, *Tetrahedron Lett.*, 1966, 3077.
330. I. Yosioka, H. Yamauchi, K. Morimoto and I. Kitagawa, *Tetrahedron Lett.*, 1968, 1149.
331. Y. Yamamoto, N. Kiriyama and S. Arahata, *Chem. Pharm. Bull. (Tokyo)*, 1968, **16**, 304.
332. T. Bruun, D. P. Hollis and R. Ryhage, *Acta Chem. Scand.*, 1965, **19**, 839.
333. C. H. Fox, W. S. G. Maass and T. P. Forrest, *Tetrahedron Lett.*, 1969, 919.
334. G. Bohman, *Acta Chem. Scand.*, 1969, **23**, 2241.
335. H. Raistrick, R. Robinson and A. R. Todd, *J. Chem. Soc.*, 1933, 488.
336. B. H. Howard and H. Raistrick, *Biochem. J.*, 1949, **44**, 227.
337. H. Raistrick, R. Robinson and A. R. Todd, *Biochem. J.*, 1933, **27**, 1170.
338. H. Raistrick, R. Robinson and A. R. Todd, *Biochem. J.*, 1934, **28**, 559.
339. J. N. Ashley, H. Raistrick and T. Richards, *Biochem. J.*, 1939, **33**, 1291.
340. W. K. Anslow, J. Breen and H. Raistrick, *Biochem. J.*, 1940, **34**, 159.
341. T. R. Seshadri and S. S. Subramanian, *Proc. Indian Acad. Sci.*, 1949, **30A**, 67.
342. T. Posternak, *Helv. Chim. Acta*, 1940, **23**, 1046; H. G. Hind, *Biochem. J.*, 1940, **34**, 67, 577.
343. T. Murakami, *Pharm. Bull. (Tokyo)*, 1956, **4**, 298.

344. W. Escherich, *Biochem. Z.*, 1958, **330**, 73.
345. R. Suemitsu, T. Matsui and M. Hiura, *Bull. Agr. Chem. Soc. Japan*, 1957, **21**, 1; R. Suemitsu, M. Nakajima and M. Hiura, *Agr. Biol. Chem. (Tokyo)*, 1961, **25**, 100.
346. A. Stoessl, *Chem. Commun.*, 1967, 307; *Can. J. Chem.*, 1969, **47**, 767, 777.
347. T. Noda, T. Take, M. Otani, K. Miyauchi, J. Abe and T. Watanabé, *Tetrahedron Lett.*, 1968, 6087; A. Takenaka, A. Furusaki, T. Watanabé, T. Noda, T. Take, T. Watanabé and J. Abe, *Tetrahedron Lett.*, 1968, 6091; T. Noda, T. Take, and J. Abe, *Tetrahedron*, 1970, **26**, 1339.
348. J. H. Birkinshaw and R. Gourlay, *Biochem. J.*, 1961, **81**, 618.
349. J. D. Bu'Lock and J. R. Smith, *J. Chem. Soc. C*, 1968, 309.
350. B. H. Howard and H. Raistrick, *Biochem. J.*, 1954, **56**, 56; S. Shibata, T. Ikekawa and T. Kishi, *Chem. Pharm. Bull. (Tokyo)*, 1960, **8**, 889.
351. W. de la Rue and H. Muller, *J. Chem. Soc.*, 1858, **10**, 304.
352. S. Gatenbeck, *Acta Chem. Scand.*, 1958, **12**, 1211.
353. S. Gatenbeck, *Acta Chem. Scand.*, 1960, **14**, 296.
354. A. J. Birch, A. J. Ryan and H. Smith, *J. Chem. Soc.*, 1958, 4773.
355. A. J. Birch, R. I. Frycr, P. J. Thomson and H. Smith, *Nature*, 1961, **190**, 441.
356. (a) E. L. Tatum, *Ann. Rev. Biochem.*, 1944, **13**, 667; (b) K. Aghoramurthy and T. R. Seshadri, *J. Sci. Ind. Res (India)*, 1954, **13A**, 114; (c) H. Raistrick, *Acta Chem. Fenn.*, 1950, **10A**, 237.
357. S. Gatenbeck, *Acta Chem. Scand.*, 1962, **16**, 1053.
358. E. Leistner and M. H. Zenk, *Chem. Commun.*, 1969, 210.
359. E. Leistner and M. H. Zenk, *Tetrahedron Lett.*, 1968, 1395.
360. G. Agosti, J. H. Birkinshaw and P. Chaplen, *Biochem. J.*, 1962, **85**, 528.
361. V. M. Chari, S. Neelakantan and T. R. Seshadri, *Tetrahedron Lett.*, 1967, 999.
362. F. Kögl and W. B. Deijs, *Ann. Chem.*, 1935, **515**, 10, 23.
363. P. C. Beaumont, R. L. Edwards and G. C. Elsworthy, *J. Chem. Soc. C*, 1968, 2968; W. Steglich, W. Furtner and A. Prox, *Z. Naturforsch. B*, 1968, **23**, 1044.
364. S. Shibata, *J. Pharm. Soc. Japan*, 1941, **61**, 320.
365. F. Bohlmann, W. Luders and W. Plettner, *Arch. Pharm.*, 1961, **294**, 521; R. G. Coombe, J. J. Jacobs and T. R. Watson, *Aust. J. Chem.*, 1968, **21**, 783.
366. Y. Ogihara, N. Kobayashi and S. Shibata, *Tetrahedron Lett.*, 1968, 1881.
367. A. E. Oxford and H. Raistrick, *Biochem. J.*, 1940, **34**, 790; S. Shibata, J. Shoji, A. Ohta and M. Watanabe, *Pharm. Bull. (Tokyo)*, 1957, **5**, 380.
368. U. Sankawa, Y. Iitaka, N. Kobayashi, Y. Ogihara and S. Shibata, *Tetrahedron Lett.*, 1968, 6135.
369. S. Shibata and T. Ikekawa, *Chem. Pharm. Bull. (Tokyo)*, 1963, **11**, 368.
370. S. Gatenbeck, *Acta Chem. Scand.*, 1960, **14**, 102.
371. S. Gatenbeck and P. Barbesgård, *Acta Chem. Scand.*, 1960, **14**, 230.
372. J. D. Bu'Lock, "The Biosynthesis of Natural Products", p. 36. McGraw-Hill, London, 1965.
373. A. E. Oxford and H. Raistrick, *Biochem. J.*, 1940, **34**, 790; H. Brockmann and H. Eggers, *Angew. Chem.*, 1955, **67**, 706.
374. I. Yosioka, H. Yamauchi, K. Morimoto and I. Kitagawa, *Tetrahedron Lett.*, 1968, 3749.
375. J. Atherton, B. W. Bycroft, J. C. Roberts, P. Roffey and M. E. Wilcox, *J. Chem. Soc. C*, 1968, 2560.
376. B. Franck, *Angew. Chem. Intern. Ed.*, 1969, **8**, 251.
377. G. Eglinton, F. E. King, G. Lloyd, J. W. Loder, J. R. Marshall, A. Robertson and W. B. Whalley, *J. Chem. Soc.*, 1958, 1833.

378. B. Franck, G. Baumann and U. Ohnsorge, *Tetrahedron Lett.*, 1965, 2031.
379. (a) J. W. Apsimon, J. A. Corran, N. G. Creasey, K. Y. Sim and W. B. Whalley, *Proc. Chem. Soc.*, 1963, 209; *J. Chem. Soc.*, 1965, 4130; (b) J. D. M. Asher, A. T. McPhail, J. M. Robertson, J. V. Silverton and G. A. Sim, *Proc. Chem. Soc.*, 1963, 210.
380. D. J. Aberhart, Y. S. Chen, P. de Mayo and J. B. Stothers, *Tetrahedron*, 1965, 21, 1417.
381. J. W. Apsimon, J. A. Corran, N. G. Creasey, W. Marlow, W. B. Whalley and K. Y. Sim, *J. Chem. Soc.*, 1965, 4144.
382. D. Gröger, *Planta Med.*, 1960, 8, 430.
383. I. Yosioka, T. Nakanishi, S. Izumi and I. Kitagawa, *Chem. Pharm. Bull. (Tokyo)*, 1968, 16, 2090.
384. B. Franck and F. Hüper, *Angew. Chem. Intern. Ed.*, 1966, 5, 728.
385. B. Franck, F. Hüper, D. Gröger, and D. Erge, *Chem. Ber.*, 1968, 101, 1954; D. Gröger, D. Erge, B. Franck, U. Ohnsorge, H. Flasch and F. Hüper, *Chem. Ber.*, 1968, 101, 1970.
386. W. B. Turner, *J. Chem. Soc.*, 1965, 6658.
387. S. Natori and H. Nishikawa, *Chem. Pharm. Bull. (Tokyo)*, 1962, 10, 117.
388. J. Balan, A Kjaer, S. Kovac and R. H. Shapiro, *Acta Chem. Scand.*, 1965, 19, 528.
389. H. Raistrick and G. Smith, *Biochem. J.*, 1936, 30, 1315; D. H. R. Barton and A. I. Scott, *J. Chem. Soc.*, 1958, 1767.
390. R. F. Curtis, P. C. Harries, C. H. Hassall and J. D. Levi, *Biochem. J.*, 1964, 90, 43.
391. C. H. Hassall and T. C. McMorris, *J. Chem. Soc.*, 1959, 2831.
392. H. Munakata, *J. Agr. Biol. Chem. (Japan)*, 1943, 19, 343; *J. Biochem. (Japan)*, 1953, 40, 451.
393. H. Raistrick, R. Robinson and D. E. White, *Biochem. J.*, 1936, 30, 1303.
394. C. H. Hassall and J. R. Lewis, *J. Chem. Soc.*, 1961, 2312.
395. R. F. Curtis, C. H. Hassall and D. W. Jones, *Chem. and Ind.*, 1959, 1283.
396. A. Rhodes, M. P. McGonagle and G. A. Somerfield, *Chem. and Ind.*, 1962, 611.
397. S. Gatenbeck, *Svensk Kem. Tidskr.*, 1960, 72, 188.
398. R. F. Curtis, P. C. Harries, C. H. Hassall, J. D. Levi and D. M. Phillips, *J. Chem. Soc. C*, 1966, 168.
399. R. F. Curtis, C. H. Hassall and R. K. Pike, *J. Chem. Soc. C*, 1968, 1807.
400. L. Y. Misconi and C. E. Stickings, *5th International Symp. Chem. Nat. Prods.*, *London*, 1968, Abstr. C75.
401. S. Gatenbeck and L. Malmström, *Acta Chem. Scand.*, 1969, 23, 3493.
402. J. G. Heathcote and M. F. Dutton, *Tetrahedron*, 1969, 25, 1497.
403. O. C. Musgrave, *J. Chem. Soc.*, 1956, 4301; H. D. Munro, O. C. Musgrave and R. Templeton, *J. Chem. Soc. C*, 1967, 947.
404. A. J. Birch, O. C. Musgrave, R. W. Rickards and H. Smith, *J. Chem. Soc. C*, 1959, 3146.
405. F. M. Dean, J. Staunton and W. B. Whalley, *J. Chem. Soc. C*, 1959, 3004; J. S. E. Holker, W. J. Ross, J. Staunton and W. B. Whalley, *J. Chem. Soc. C*, 1962, 4150.
406. S. Udagawa, *Chem. Pharm. Bull. (Tokyo)*, 1963, 11, 366.
407. E. M. Gregory and W. B. Turner *Chem. and Ind.*, 1963, 1625.
408. J. S. E. Holker, J. Staunton and W. B. Whalley, *J. Chem. Soc.*, 1964, 16.
409. G. Jaureguiberry, M. Lenfant, B. C. Das and E. Lederer, *Tetrahedron*, 1966, suppl. 8, Part I, 27.

410. B. E. Cross, M. N. Edinberry and W. B. Turner, *Chem. Commun.*, 1970, 209.

411. G. M. Blackburn, A. H. Neilson and Lord Todd, *Proc. Chem. Soc.*, 1962, 327; G. M. Blackburn, D. E. U. Ekong, A. H. Neilson and Lord Todd, *Chimia*, 1965, **19**, 208.

412. H. Kern and S. Naef-Roth, *Phytopathol. Z.*, 1965, **53**, 45.

413. N. Sakabe, T. Goto and Y. Hirata, *Tetrahedron Lett.*, 1964, 1825.

414. F. McCapra, A. I. Scott, P. Delmotte, J. Delmotte-Plaquee and N. S. Bhacca, *Tetrahedron Lett.*, 1964, 869.

415. R. N. Mirrington, E. Ritchie, C. W. Shoppee, W. C. Taylor and S. Sternhell, *Tetrahedron Lett.*, 1964, 365.

416. M. Stob, R. S. Baldwin, J. Tuite, F. N. Andrews and K. G. Gillette, *Nature*, 1962, **196**, 1318; W. H. Urry, H. L. Wehrmeister, E. B. Hodge and P. H. Hidy, *Tetrahedron Lett.*, 1966, 3109.

417. C. J. Mirocha, C. M. Christensen and G. H. Nelson, *Appl. Microbiol.*, 1969, **17**, 482.

418. G. A. Ellestad, H. A. Whaley and E. L. Patterson, *J. Amer. Chem. Soc.*, 1966, **88**, 4109.

419. H. Raistrick and J. Ziffer, *Biochem. J.*, 1951, **49**, 563; A. J. Birch and R. A. Massy-Westropp, *J. Chem. Soc.*, 1957, 2215; A. J. Birch and K. S. J. Stapleford, *J. Chem. Soc.*, 1967, 2570.

420. Y. Hatsuda, T. Hamasaki, M. Ishida and S. Yoshikawa, *Agr. Biol. Chem. (Tokyo)*, 1969, **33**, 131.

421. T. Hamasaki, Y. Hatsuda, N. Terashima and M. Renbutsu, *Agr. Biol. Chem. (Tokyo)*, 1967, **31**, 11.

422. J. S. E. Holker, personal communcation.

423. E. Bullock, D. Kirkaldy, J. C. Roberts and J. G. Underwood, *J. Chem. Soc.*, 1963, 829; J. S. E. Holker and S. A. Kagal, *Chem. Commun.*, 1968, 1574.

424. Y. Hatsuda and S. Kuyama, *J. Agr. Chem. Soc. Japan*, 1954, **28**, 989; J. E. Davies, J. C. Roberts and S. C. Wallwork, *Chem. and Ind.*, 1956, 178; J. H. Birkinshaw and I. M. M. Hammady, *Biochem. J.*, 1957, **65**, 162; E. Bullock, J. C. Roberts and J. G. Underwood, *J. Chem. Soc.*, 1962 4179.

425. H. J. Burkhardt and J. Forgacs, *Tetrahedron*, 1968, **24**, 717.

426. A. C. Waiss, M. Wiley, D. R. Black and R. E. Lundin, *Tetrahedron Lett.*, 1968, 3207; J. V. Rodricks, E. Lustig, A. D. Campbell and L. Stoloff, *Tetrahedron Lett.*, 1968, 2975.

427. (a) R. D. Hartley, B. F. Nesbitt and J. O'Kelley, *Nature*, 1963, **198**, 1056; (b) T. Asao, G. Büchi, M. M. Abdel-Kader, S. B. Chang, E. L. Wick and G. N. Wogan, *J. Amer. Chem. Soc.*, 1963, **85**, 1706; 1965, **87**, 882; (c) K. J. van der Merwe, L. Fourie and de B. Scott, *Chem. and Ind.*, 1963, 1660.

428. F. A. Hodges, J. R. Zust, H. R. Smith, A. A. Nelson, B. H. Armbrecht and A. D. Campbell, *Science*, 1964, **145**, 1439.

429. C. W. Holzapfel, P. S. Steyn and I. F. H. Purchase, *Tetrahedron Lett.*, 1966, 2799.

430. D. A. van Dorp, A. S. M. van der Zijden, R. K. Beerthuis, S. Sparreboom, W. O. Ord, K. de Jong and R. Keunung, *Rec. Trav. Chim.*, 1963, **82**, 587.

431. M. F. Dutton and J. G. Heathcote, *Chem. and Ind.*, 1968, 418.

432. B. J. Wilson, T. C. Campbell, A. W. Hayes and R. T. Hamilton, *Appl. Microbiol.*, 1968, **16**, 819.

433. R. Allcroft and R. B. A. Carnaghan, *Chem. and Ind.*, 1963, 50; K. Sargeant, R. B. A. Carnaghan and R. Allcroft, *Chem. and Ind.*, 1963, 53.

434. B. F. Nesbitt, J. O'Kelley, K. Sargeant and A. Sheridan, *Nature*, 1962, **195**, 1062.
435. K. K. Cheung and G. A. Sim, *Nature*, 1964, **201**, 1185; T. C. van Soest and A. F. Peerdman, *Koninkl. Ned. Akad. Wetenschap. Proc. Ser. B*, 1964, **67**, 469.
436. S. Brechbuhler, G. Büchi and G. Milne, *J. Org. Chem.*, 1967, **32**, 2641.
437. M. Biollaz, G. Büchi and G. Milne, *J. Amer. Chem. Soc.*, 1968, **90**, (a) 5017; (b) 5019; *J. Amer. Chem. Soc.*, 1970, **92**, 1035.
438. J. S. E. Holker and L. J. Mulheirn, *Chem. Commun.*, 1968, 1576.
439. J. R. D. McCormick. In "Antibiotics. Vol. II. Biosynthesis" (D. Gottlieb and P. D. Shaw, eds.), p. 113. Springer-Verlag, New York, 1967.
440. J. R. D. McCormick and E. R. Jensen, *J. Amer. Chem. Soc.*, 1968, **90**, 7126.
441. J. R. D. McCormick, E. R. Jensen, N. H. Arnold, H. S. Corey, U. H. Joachim, S. Johnson, P. A. Miller and N. O. Sjolander, *J. Amer. Chem. Soc.*, 1968, **90**, 7127.
442. J. H. Birkinshaw and I. M. M. Hammady, *Biochem. J.*, 1957, **65**, 162; J. H. Birkinshaw, J. C. Roberts and P. Roffey, *J. Chem. Soc. C*, 1966, 855.
443. J. C. Roberts and P. Roffey, *J. Chem. Soc.*, 1965, 3666.
444. H. A. Anderson, R. H. Thomson and J. W. Wells, *J. Chem. Soc. C*, 1966, 1727.
445. T. Hamasaki, M. Renbutsu and Y. Hatsuda, *Agr. Biol. Chem. (Tokyo)*, 1967, **31**, 1513.
446. G. Koller and H. Russ, *Monatsh. Chem.*, 1937, **70**, 54.
447. (a) D. F. G. Pusey and J. C. Roberts, *J. Chem. Soc.*, 1963, 3542; P. Roffey and M. V. Sargent, *Chem. Commun.*, 1966, 913; P. Roffey, M. V. Sargent and J. A. Knight, *J. Chem. Soc. C*, 1967, 2328; (b) J. S. E. Holker, S. A. Kagal, L. J. Mulheirn and P. M. White, *Chem. Commun.*, 1966, 911.
448. J. S. E. Holker, personal communication.
449. S. Marumo, K. Sagaki and S. Suzuki, *J. Amer. Chem. Soc.*, 1964, **86**, 4507.
450. M. P. Kunstmann and L. A. Mitscher, *J. Org. Chem.*, 1966, **31**, 2920.
451. J. H. Bowie and A. W. Johnson, *Tetrahedron Lett.*, 1967, 1449.
452. (a) H. Brockmann and T. Reschke, *Tetrahedron Lett.*, 1968, 3167; (b) N A. Bailey, C. P. Falshaw, W. D. Ollis, M. Watanabe, M. M. Dhar, A. W. Khan and V. C. Vora, *Chem. Commun.*, 1968, 374.
453. T. A. Geissman. In "Biosynthesis of Natural Products" (P. Bernfeld, ed.), 2nd edition, p. 782, Pergamon, London, 1967.
454. W. D. Ollis and I. O. Sutherland. "Recent Developments in the Chemistry of Natural Phenolic Compounds", p. 212. Pergamon, London, 1961.
455. W. D. Ollis, I. O. Sutherland and J. J. Gordon, *Tetrahedron Lett.*, 1959, **16**, 17; H. Brockmann, H. Brockmann Jr., J. J. Gordon, W. Keller-Schierlein, W. Lenk, W. D. Ollis, V. Prelog and I. O. Sutherland, *Tetrahedron Lett.* 1960, **8**, 25.
456. H. Brockmann, P. Boldt and J. Niemeyer, *Chem. Ber.*, 1963, **96**, 1356.
457. F. Arcamone, G. Franceschi, P. Orezzi and S. Penco, *Tetrahedron Lett.*, 1968, 3349; F. Arcamone, G. Cassinelli, G. Franceschi and P. Orezzi, *Tetrahedron Lett.*, 1968, 3353.
458. W. D. Ollis, I. O. Sutherland, R. C. Codner, J. J. Gordon and G. A. Miller, *Proc. Chem. Soc.*, 1960, 347.
459. T. Bruun, *Acta Chem. Scand.*, 1965, **19**, 1677.
460. M. L. Dhar, V. Thaller and M. C. Whiting, *J. Chem. Soc.*, 1964, 842.
461. A. C. Cope, R. K. Bly, E. P. Burrows, O. J. Ceder, E. Ciganek, B. T. Gillis, R. F. Porter and H. E. Johnson, *J. Amer. Chem. Soc.*, 1962, **84**, 2170.

462. O. Ceder and R. Ryhage, *Acta Chem. Scand.*, 1964, **18**, 558.

463. O. Ceder, *Acta Chem. Scand.*, 1964, **18**, 126.

464. A. J. Birch. *In* "Antibiotics, Vol. II, Biosynthesis" (D. Gottlieb and P. D. Shaw, eds.), p. 228. Springer-Verlag, New York, 1967.

465. S. D. Mills and W. B. Turner, *J. Chem. Soc.*, 1967, 2242.

466. J. Staunton, private communication.

467. F. M. Dean, J. C. Roberts and A. Robertson, *J. Chem. Soc.*, 1954, 1432; F. M. Dean, D. S. Deohra, A. D. T. Erni, D. W. Hughes and J. C. Roberts, *J. Chem. Soc.*, 1960, 4829.

468. W. F. Beach and J. H. Richards, *J. Org. Chem.*, 1963, **28**, 2746.

469. Y. Satomura and A. Sato, *Agr. Biol. Chem.* (*Tokyo*), 1965, **29**, 337; T. Tokoroyama, T. Kamikawa and T. Kubota, *Tetrahedron*, 1968, **24**, 2345.

470. T. Sassa, H. Aoki, M. Namiki and K. Munakata, *Agr. Biol. Chem.* (*Tokyo*), 1968, **32**, 1432.

471. T. Kubota, T. Tokoroyama, S. Oi and Y. Satomura, *Tetrahedron Lett.*, 1969, 631.

472. T. Kubota, T. Tokoroyama, T. Kamikawa and Y. Satomura, *Tetrahedron Lett.*, 1966, 5205.

473. G. J. M. van der Kerk and J. C. Overeem, *Rec. Trav. Chim.*, 1957, **76**, 425; 1964, **83**, 995.

474. R. Bentley and S. Gatenbeck, *Biochemistry*, 1965, **4**, 1150.

475. T. Sproston, H. Tomlinson, G. E. Milo and A. Jones, *Phytopathol.*, 1962, **52**, 735; T. B. Tjio, T. Sproston and H. Tomlinson, *Lloydia*, 1965, **28**, 359.

476. J. J. Armstrong and W. B. Turner, *J. Chem. Soc.*, 1965, 5927.

477. (a) F. Kögl and J. Sparenburg, *Rec. Trav. Chim.*, 1940, **59**, 1180; (b) A. J. Birch, D. N. Butler and R. W. Rickards, *Tetrahedron Lett.*, 1964, 1853.

478. P. M. Brown, V. Krishnamoorthy, J. W. Mathieson and R. H. Thomson, *J. Chem. Soc. C*, 1970, 109.

CHAPTER 6

Terpenes and Steroids

THE TERPENES and steroids are widely distributed in nature and those of
fungal origin form only a small fraction of the known examples. In spite of
their relatively small number the fungal terpenes and steroids are significant
for two reasons. First, many of the terpenes produced by fungi are of types
not encountered elsewhere. Secondly, the convenience of fungi for biosynthetic
experiments has led to detailed studies of the biosynthesis of some of the
compounds, and the results from these experiments have implications for the
biosynthesis of compounds from other sources.

Some steroids have hormonal activities essential for mammals, and it has
recently become apparent that terpenoid compounds also have hormonal
roles in the development of fungi. The problem has been studied particularly
in the Phycomycetes, and antheridiol and the trisporic acids have been
recognized as sex hormones, while sirenin is responsible for the attraction
of male gametes to female gametes in *Allomyces*. These compounds are
discussed in the appropriate parts of this chapter, and their biological effects
have been reviewed.[1] Exogenous steroids, especially cholesterol, have been
shown to affect growth and sexual reproduction of *Pythium* and *Phytoph-
thora*.[2] The gibberellins are plant-growth hormones, but have no identified
role in fungi.

A. ISOPRENOID BIOSYNTHESIS

The biosynthesis of terpenes and steroids has been extensively reviewed,[4]
and here we shall emphasize the more recent developments. As in the case of
the polyketides, the concepts of terpenoid biosynthesis were developed by
organic chemists as a means of correlating the structures of a wide range of
natural products and as an aid in the determination of the structures of new
compounds. But because of the wide distribution of terpenoid compounds
and the essential role of some steroids in mammalian systems, some aspects
of terpenoid biosynthesis, especially the steps leading via squalene to the
steroids, have been extensively studied at the enzyme level (a comprehensive
account of this aspect has recently appeared[3]). On the other hand, the
biosynthesis of most of the species-specific terpenes is still discussed in terms
of chemically reasonable cyclizations of the acyclic intermediates, though in a

few cases the details of the processes are being elucidated with the aid of mevalonate specifically labelled with tritium.

A.1. THE FORMATION OF ISOPENTENYL PYROPHOSPHATE

The "isoprene unit" of the chemist's "isoprene rule" is now known to be isopentenyl pyrophosphate, formed from acetate via mevalonate (Scheme 1). We saw in Chapter 4 that the first step of fatty acid biosynthesis is the

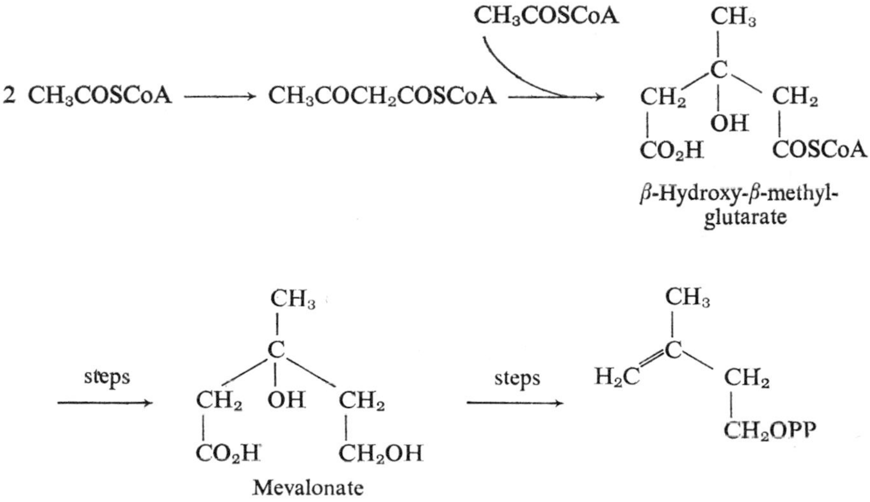

Scheme 1: The biosynthesis of isopentenyl pyrophosphate from acetate via mevalonate

condensation of protein-bound acetate and malonate to yield protein-bound acetoacetate, and noted that the intervention of malonate makes the process energetically favourable. For the biosynthesis of mevalonate, acetoacetate is formed by condensation of two molecules of acetyl coenzyme A, one of the acetyl residues probably being transferred to an enzyme thiol group before condensation. The equilibrium is shifted in favour of condensation by reaction of acetoacetyl coenzyme A with a third molecule of acetyl coenzyme A to give β-hydroxy-β-methylglutarate which is reduced to mevalonate. The direct formation of an acetoacetyl residue from acetyl residues is encountered in the biosynthesis of the polyketide-derived secondary metabolite rotiorin (p. 174).

A.2. THE FORMATION OF THE ACYCLIC TERPENES

Isopentenyl pyrophosphate isomerizes to dimethylallyl pyrophosphate which condenses with a second molecule of isopentenyl pyrophosphate to

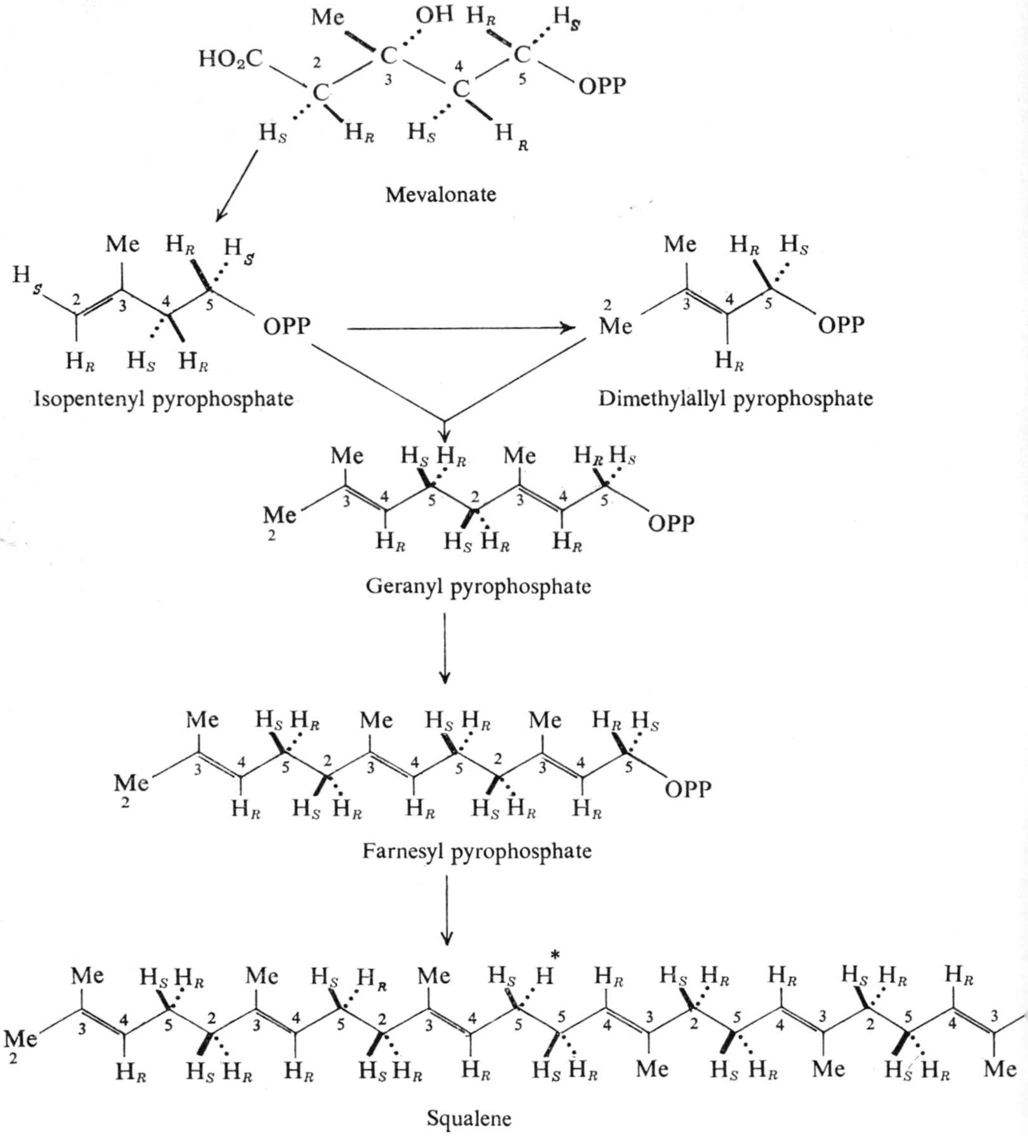

Scheme 2: The stereochemistry of the formation of the acyclic terpenes. The numbers refer to the carbon atoms of mevalonate and the subscripts S and R to the configuration of the protons in mevalonate; H* is the proton transferred from NADPH during squalene formation.

give geranyl pyrophosphate, the precursor of the monoterpenes. Further condensations with isopentenyl pyrophosphate yield (1) farnesyl pyrophosphate, the precursor of the sesquiterpenes and, via squalene, of the triterpenes, (2) geranylgeranyl pyrophosphate, the precursor of the diterpenes and of the carotenoids, (3) geranylfarnesyl pyrophosphate, the precursor of the sesterterpenes and (4) chains containing up to ten C_5 units which are present in the essential cell components plastoquinone, ubiquinone and vitamin K. Polymerization of isopentenyl pyrophosphate with the formation of *cis*- rather than the usual *trans*-double bonds is involved in the biosynthesis of rubber.

An important recent development has been the elucidation by Cornforth and Popjak[5] of the stereochemistry of the reactions. Their results are summarized in Scheme 2, from which it will be seen that (1) in the formation of isopentenyl pyrophosphate the *R*-proton of C-2 becomes *trans* to the methyl group, (2) in the formation of dimethylallyl pyrophosphate the methyl group *trans* to the methylene group is derived from C-2 of mevalonate and the *S*-proton is lost from C-4 (this process is repeated during the subsequent condensations of isopentenyl pyrophosphate with the allylic pyrophosphates), (3) at each condensation step, inversion occurs at C-5, and (4) during the linking of two farnesyl residues to form squalene the terminal *R*-hydrogen of one of them is replaced by hydrogen from NADPH.

The work of Cornforth and Popjak has not only provided information on the mechanisms of the primary steps of isoprenoid biosynthesis but has also made available new techniques for the study of the changes which the primary products undergo to give secondary metabolites. These techniques are being used increasingly to study the biosynthesis of fungal metabolites (see, for example, the gibberellins, rosenonolactone, pleuromutilin and the ophiobolins).

A.3. THE MECHANISM OF SQUALENE FORMATION

Several mechanisms for the condensation of two molecules of farnesyl pyrophosphate to give squalene have been discussed[6] and chemical analogies have been devised for one of them.[7] However, a new intermediate in squalene biosynthesis has recently been isolated by Rilling,[8] and should this prove to be an obligatory precursor of squalene some of the postulated mechanisms will be excluded.

The new precursor, which has been named presqualene,[9] was first obtained by incubation of a yeast extract, capable of squalene biosynthesis, with geranyl pyrophosphate and isopentenyl pyrophosphate *in the absence of NADPH*. When a similar experiment was carried out with farnesyl pyrophosphate labelled with ^{14}C and ^{32}P (1:1), presqualene with a $^{14}C : ^{32}P$ ratio of $1:0.56$ was obtained. Incubation of radioactive presqualene with the yeast extract *in the presence of NADPH* gave radioactive squalene. Popjak *et al.*[10] have repeated Rilling's experiments and were able to obtain pre-squalene using a yeast preparation but not using rat or pig liver microsomal preparations. Structures (**1**)[9] and (**2**)[10] have been independently proposed for presqualene.

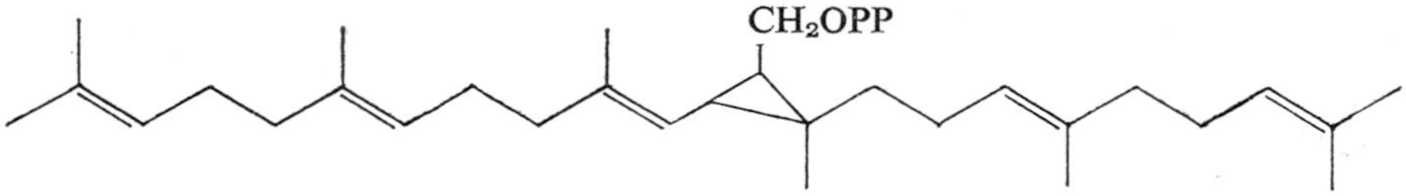

1

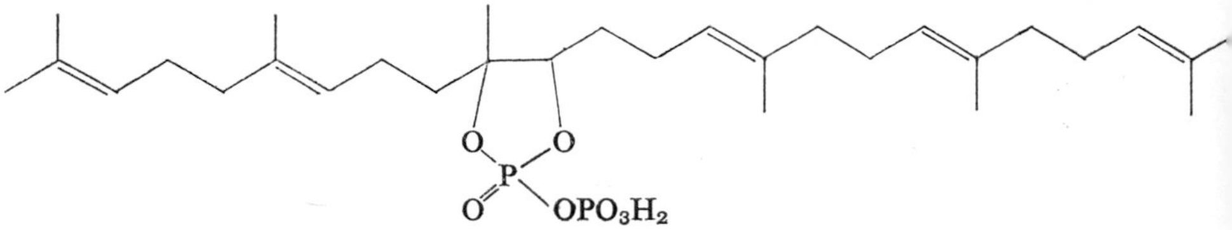

2

B. THE TERPENES AND STEROIDS OF FUNGI

B.1. MONOTERPENES

Mycelianamide contains a geranyl side-chain, and the side-chain of mycophenolic acid may have been derived from geraniol, but no simple monoterpene has been isolated from a fungus. Whether this reflects the

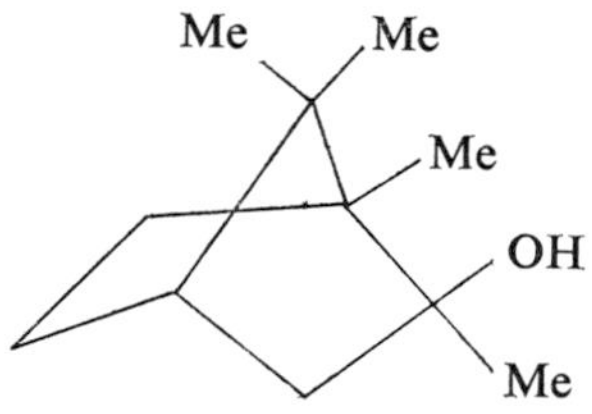

3

absence of such compounds or failure to detect those which are present remains to be seen, but Sprecher[11] was unable to detect monoterpenes in the steam-volatile fractions of fungi of the genera *Ceratocystis*, *Penicillium* and *Fusarium*, or of the mycobiont of the lichen *Evernia prunastri*. 2-*exo*-Hydroxy-2-methylbornane (**3**), recently obtained from three *Streptomyces* spp.,[12] is the only monoterpene-like compound to be isolated from bacteria, though it may result from degradation of a sesquiterpene.

B.2. SESQUITERPENES

In the Pfizer Handbook of Microbial Metabolites[13] only three fungal metabolites were listed as sesquiterpenes, but in the nine years since its publication over thirty fungal metabolites, many of them isolated much earlier, have been recognized as sesquiterpenes. The compounds have been isolated mainly from Fungi Imperfecti and Basidiomycetes (both from fruiting bodies and from mycelial cultures).

B.2.a. *The Trichothecanes*

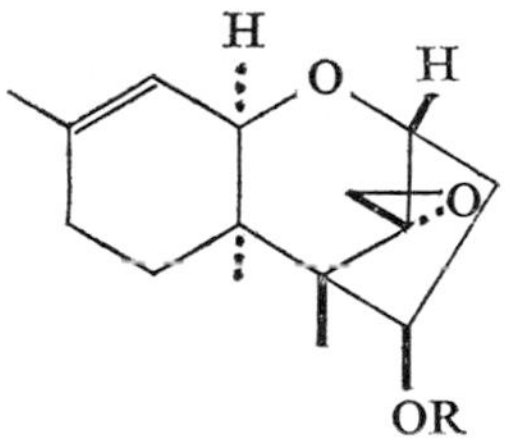

a: R = Ac, trichodermin
        Trichoderma sp.[14]

b: R = H, trichodermol, roridin C
        Myrothecium roridum[15]

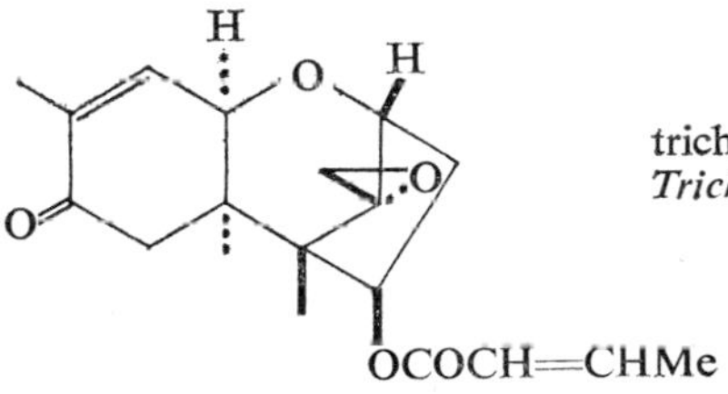

trichothecin L[17,18]
Trichothecium roseum[16]

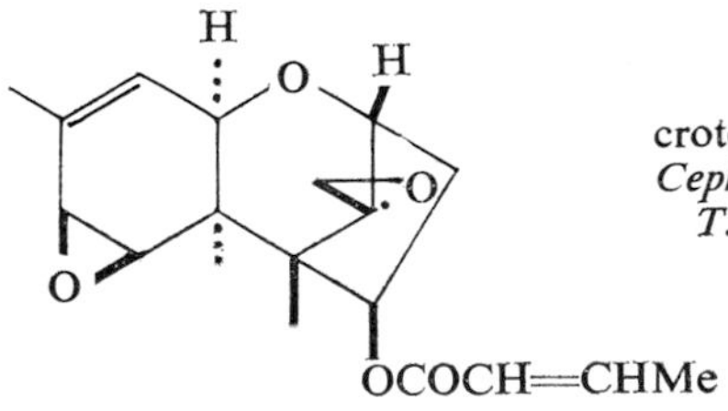

crotocin (antibiotic T)
Cephalosporium crotocinigenum,[20]
        T. roseum[21]

7

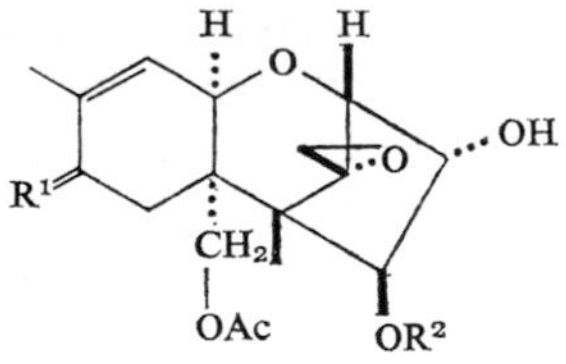

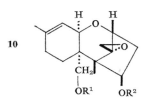

a: $R^1 = H_2$, $R^2 = Ac$, diacetoxyscirpenol
  Fusarium scirpi (= *F. equiseti*)[22,23]

b: $R^1 = H$, $OCOCH_2CHMe_2$, $R^2 = Ac$, toxin T-2
  F. tricinctum,[24] *Trichothecium lignorum*[25]

c: $R^1 = H$, $OCOCH_2CHMe_2$, $R^2 = H$, toxin HT-2
  F. tricinetum[25]

8

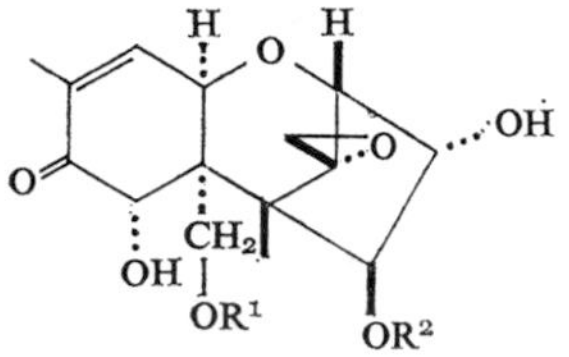

a: $R^1 = R^2 = H$, nivalenol
  F. nivale[26]

b: $R^1 = H$, $R^2 = Ac$, nivalenol monoacetate, fusarenon
  F. nivale[26]

c: $R^1 = R^2 = Ac$, 4β,15-diacetoxy-3α,7α-dihydroxy-12,13-epoxytrichothec-9-en-8-one
  F. scirpi[22,23c,23f]

9

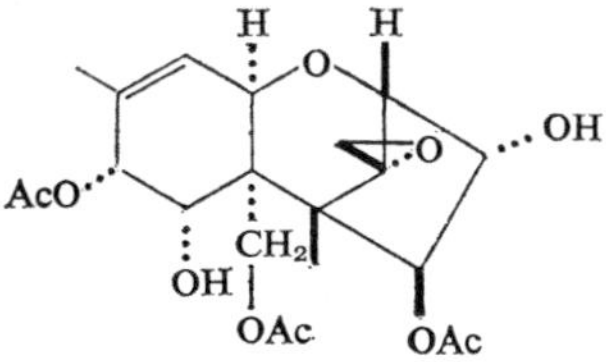

4β,8α,15-triacetoxy-12,13-epoxytrichothec-9-ene-3α,7α-diol
F. scirpi[27]

10

a: $R^1 = R^2 = H$, verrucarol
b: $R^1 = R^2 = Ac$, diacetoxyverrucarol
  Myrothecium sp.[28]

a: $R^1 = R^2 = H$, verrucarol

b: $R^1 = R^2 = Ac$, diacetoxyverrucarol
         Myrothecium sp.[28]

c: $R^1, R^2 =$ $\overset{\text{OH}}{\underset{\text{H}}{\vdots}}\;\overset{\text{H}}{\underset{\text{Me}}{\vdots}}$

c: $R^1, R^2 =$ $-COC-CCH_2CH_2OCOCH\overset{t}{=}CHCH\overset{c}{=}CHCO-$, verrucarin A
               $\underset{H}{|}\;\underset{Me}{|}$ *M. verrucaria*[15,29]

d: $R^1, R^2 =$ $-COCH\overset{\displaystyle O}{\overbrace{}}CCH_2CH_2OCOCH\overset{t}{=}CHCH\overset{c}{=}CHCO-$, verrucarin B
                       $\underset{Me}{|}$ *M. verrucaria*[15,30]

e: $R^1, R^2 =$ $-COCH=CCH_2CH_2OCOCH\overset{t}{=}CHCH\overset{c}{=}CHCO-$, verrucarin J
                 $\underset{Me}{|}$ *M. verrucaria*[15,31]

f: $R^1, R^2 =$ $-CO\overset{2'}{C}OCCH_2CH_2OCOCH\overset{t}{=}CHCH\overset{c}{=}CHCO-$,
                 $\underset{Me}{|}$ 2'-dehydroverrucarin A
                                      M. verrucaria[32]

g: $R^1, R^2 =$ $-COC-CCH_2CH_2OCHCH\overset{t}{=}CHCH\overset{c}{=}CHCO-$, roridin A
           $\underset{H}{|}\;\underset{Me}{|}$ $\underset{CH(OH)Me}{|}$ *M. roridum*[15,33]

h: $R^1, R^2 =$ $-COCH\overset{\displaystyle O}{\overbrace{}}CCH_2CH_2OCHCH\overset{t}{=}CHCH\overset{c}{=}CHCO-$, roridin D
              $\underset{Me}{|}$ $\underset{CH(OH)Me}{|}$ *M. roridum*[15,34]

i: $R^1, R^2 =$ $-COCH=CCH_2CH_2OCHCH\overset{t}{=}CHCH\overset{c}{=}CHCO-$, roridin E
            $\underset{Me}{|}$ $\underset{CH(OH)Me}{|}$ *M. roridum*[15,40]

The largest class of fungal sesquiterpenes is based on the trichothecane skeleton (**11**).[35] All the known examples have a 12,13-epoxide ring, a 9,10-double bond, and from one to four hydroxyl groups which are usually acylated. In the verrucarins (**10c**)–(**10f**) and the roridins (**10g**)–(**10i**) a macrocyclic

11

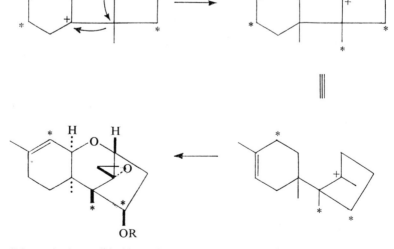

Scheme 3: A possible biosynthetic route from farnesyl pyrosphosphate to trichodermin and the other trichothecanes

ring has been formed between two of the hydroxyl groups of the trichothecane moiety, verrucarol (10a); the verrucarins and roridins have been independently isolated, either pure or as mixtures, by several groups.[36] It has been suggested[23e,27] that nivalenol (8a) and its monacetate (8b) are artefacts formed from the diacetate (8c) during isolation. The epoxytrichothec-9-enes are cytotoxic, fungitoxic, phytotoxic and vesicant.

The correct structure for the sequiterpene skeleton of the trichothecanes was established by X-ray analysis[37] of the p-bromobenzoate of trichodermol (= roridin C) (4b) and by chemical correlation of trichodermol with trichothecin[14] and with verrucarol,[14,38] which in turn was related to diacetoxyscirpenol.[23b] The full structure of verrucarin A has been confirmed by X-ray analysis.[39]

Trichothecin produced in the presence of [2-^{14}C]mevalonate has the labelling pattern shown in Scheme 3,[17] which can arise by cyclization

14

of farnesyl pyrophosphate to a bisabolene intermediate [shown as γ-bisabolene (13) in Scheme 3, though it could equally well be β-bisabolene (14)] followed by two 1,2-methyl migrations. The double bond in the cyclohexane ring of bisabolene has to rearrange in order to give the labelling pattern observed in trichothecin.

All-trans-[1-^{14}C]farnesyl pyrophosphate is incorporated into trichothecin,[18] and is believed to be a precursor of all the sesquiterpenes, but a double bond has to isomerize to the cis form at some stage in the cyclization of farnesol to give the bisabolenes (Scheme 3) and other cyclic sesquiterpenes (see below) (cf. also the cyclization of geranylfarnesol to give the ophiobolins, p. 252). Further evidence for the intermediacy of an all-trans compound is provided by the incorporation of 5R-[5-^{3}H]mevalonate into trichothecin with retention of tritium[19] (in the biosynthesis of rubber, polymerization of isopentenyl pyrophosphate gives cis double bonds with loss of the 5R-hydrogen of mevalonate).

B.2.b. *Helminthosporal and Related Compounds*

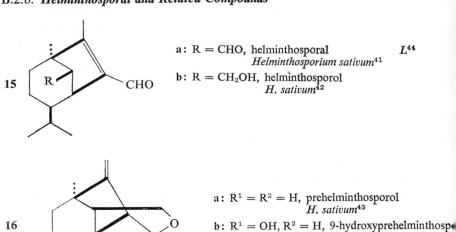

15

a: R = CHO, helminthosporal   *L*[44]
       Helminthosporium sativum[41]

b: R = CH₂OH, helminthosporol
       H. sativum[42]

16

a: R¹ = R² = H, prehelminthosporol
       H. sativum[43]

b: R¹ = OH, R² = H, 9-hydroxyprehelminthosp⋯
       Cochliobolus sativus[45]

17

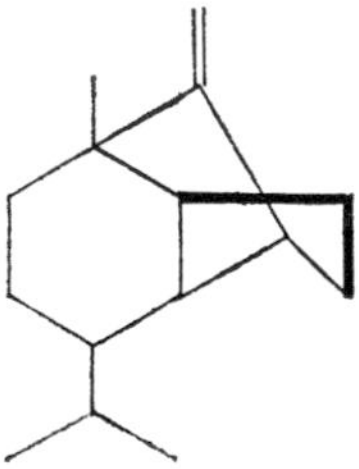

sativene
H. sativum[46]

It is now known that helminthosporal is an artefact of the isolation procedure,[43] being derived from the acetal (**18a**) present in the crude extract. The acetal (**16**, R¹ = H, R² = Et) was also isolated, and it was found that if

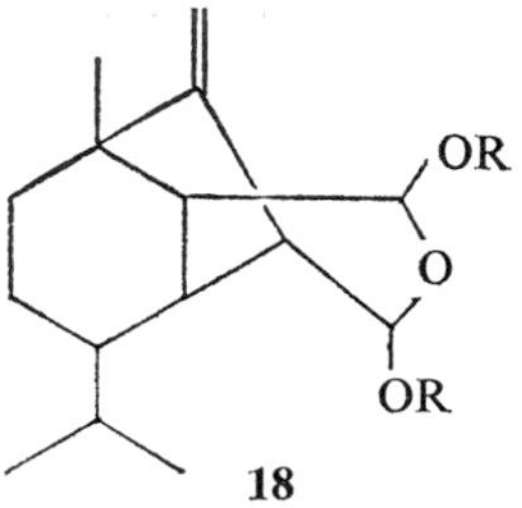

18

a: R = Et
b: R = H

extracts were prepared without the use of ethanol, prehelminthosporol (16a) was obtained. The corresponding "prehelminthosporal" (18b) could not be detected and it was suggested that the helminthosporal precursor, giving the acetal (18a) in the presence of ethanol, might be an unstable dimeric acetal. The isolation of helminthosporol (15b)[42] involved heating the broth at 100° for 15 min., so that helminthosporol may well be an artefact of this procedure. Prehelminthosporol and 9-hydroxyprehelminthosporol occur as mixtures of epimers at the acetal carbon.[45]

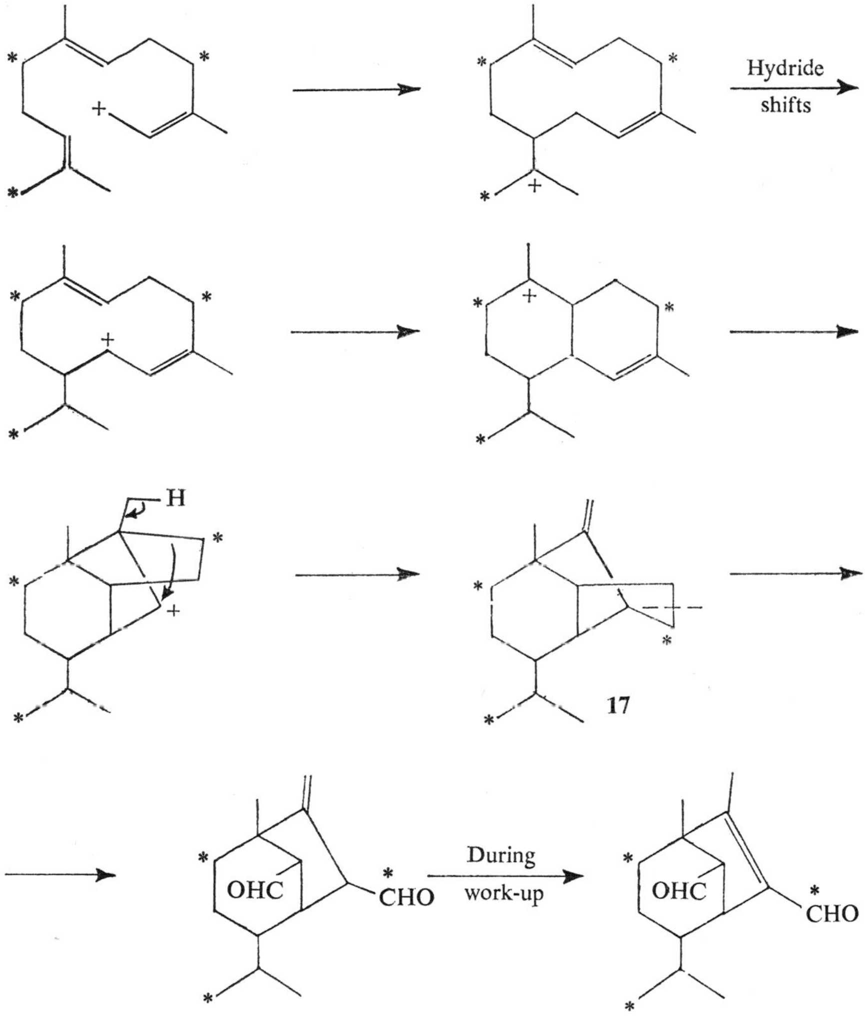

Scheme 4: Possible biosynthetic sequence from farnesol to helminthosporal

Degradation of helminthosporal produced in the presence of [2-^{14}C]-mevalonate showed that about one third of the radioactivity was located at the aldehyde carbon, consistent with a biosynthetic sequence such as that of Scheme 4 in which sativene (**17**) is an intermediate.

B.2.c. *Sesquiterpenes Derivable from "Protoilludane"*

19

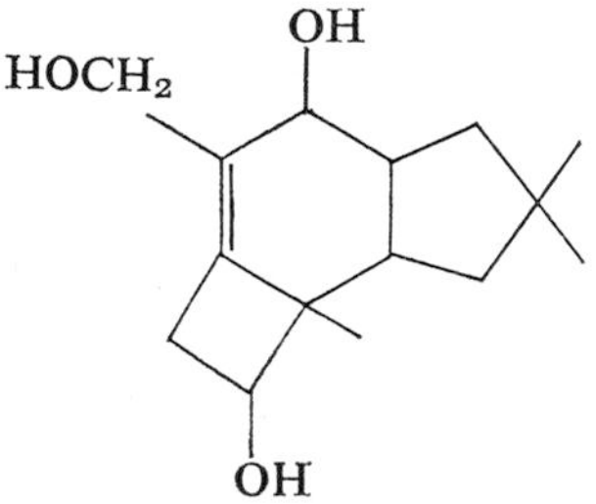

illudol
Clitocybe illudens[47,48]

20

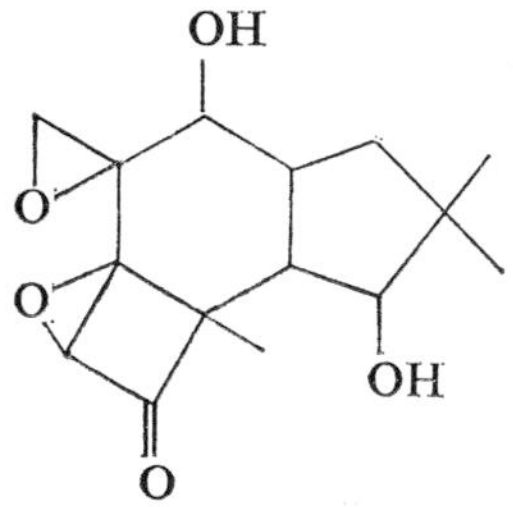

coriolin
Coriolus consors[49]

21

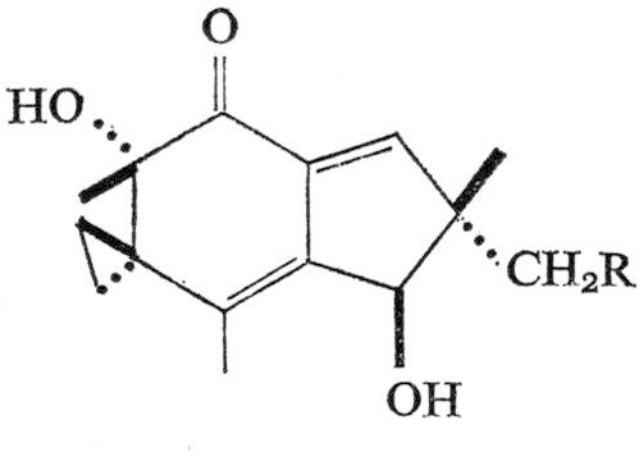

a: R = H, illudin M
    Clitocybe illudens[47, 50]

b: R = OH, illudin S, lampterol
    C. illudens[47,50],
    Lampteromyces japonicus[51]

22

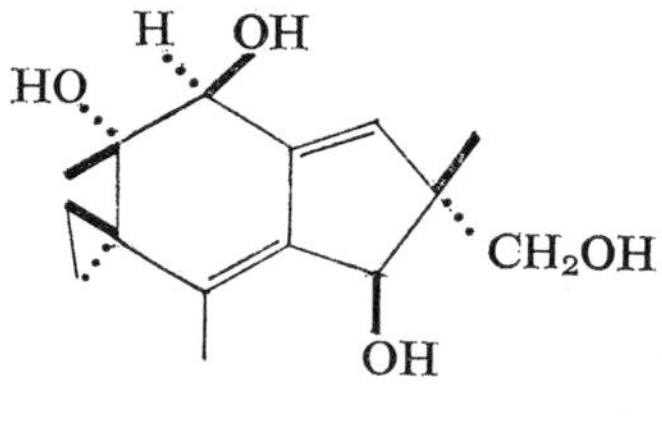

dihydroilludin S
L. japonicus[52]

23

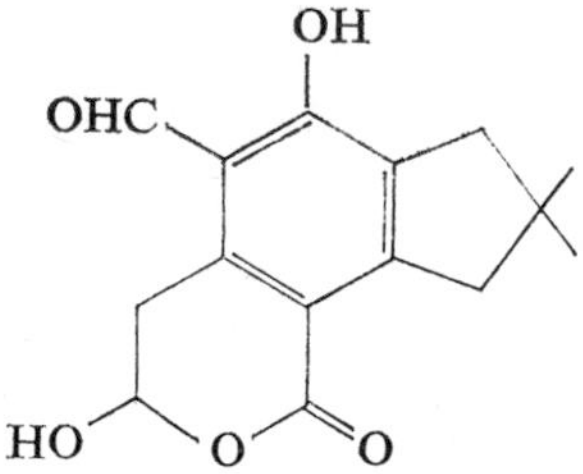

illudalic acid
C. illudens[53]

24

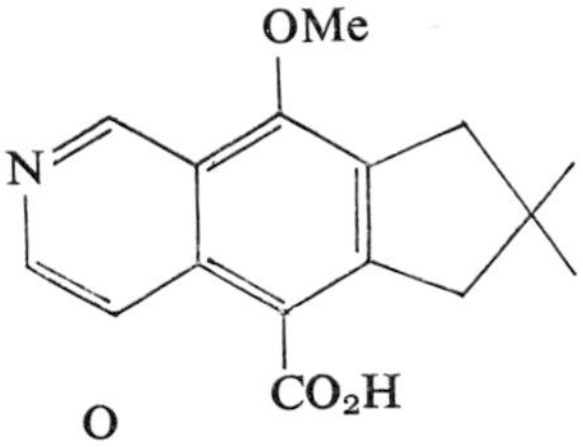

illudinine
C. illudens[53]

25

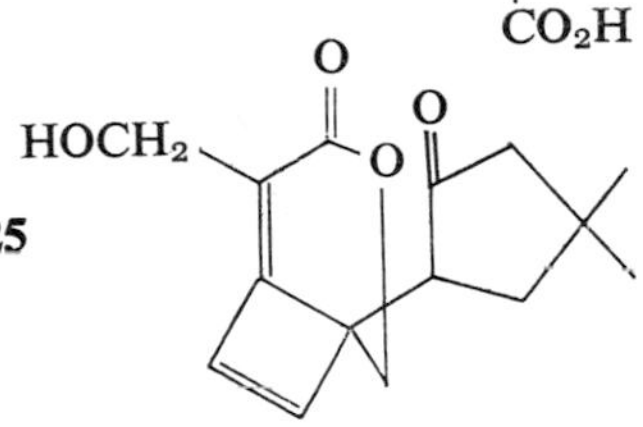

fomannosin
Fomes annosus[54]

26

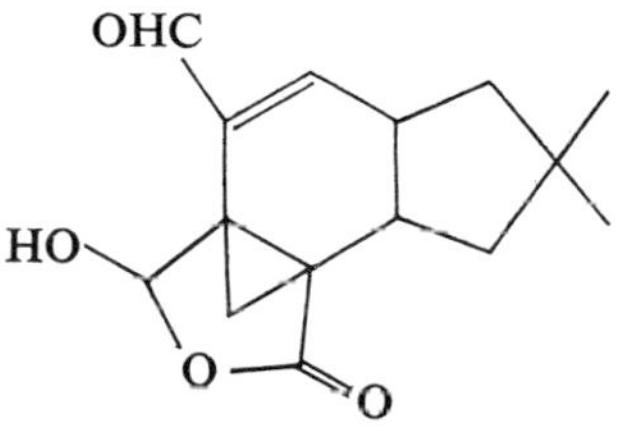

marasmic acid L[56]
Marasmius conigenus[55,56]

27

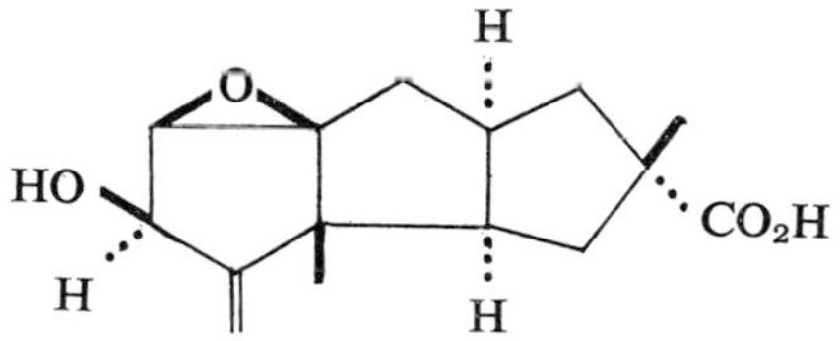

hirsutic acid C
Stereum hirsutum[57,58]

These compounds, which are all products of Basidiomycetes are formally derivable from a "protoilludane" intermediate (**29,** Scheme 5) which has the carbon skeleton of illudol. Protoilludane can arise from farnesol via humulene (**28**). Alternative cyclizations of humulene, not involving the ion (**29**) have been proposed[58] for the formation of hirsutic acid C. The incorporation of [2-^{14}C]mevalonate into marasmic acid has been studied.[56] Since the location

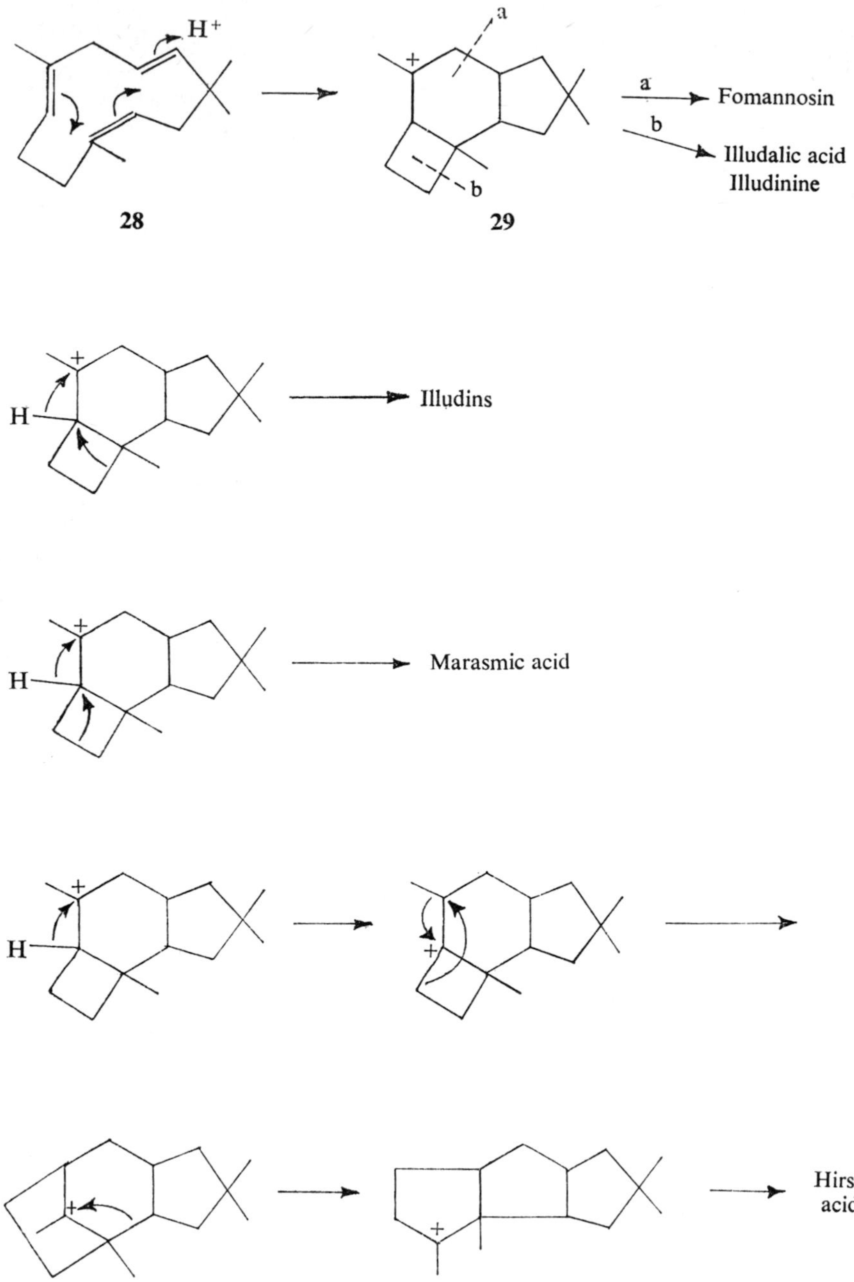

Scheme 5: Formal derivation of fungal sesquiterpenes from a "protoilludane" intermediate (29)

of the radioactivity in marasmic acid depended upon Kuhn–Roth oxidations the results are not conclusive, though they are in fair agreement with theoretical prediction.

Fomannosin is phytotoxic, illudins M and S and marasmic acid are antibacterial, and hirsutic acid C is an inactive precursor of an antibacterial compound of unknown structure. Illudins M and S co-occurred with an "inactive compound"[47] which was subsequently named illudol,[48] though the melting points of the "inactive compound" and illudol differ markedly. Illudoic acid, another metabolite of *C. illudens*, has been tentatively assigned

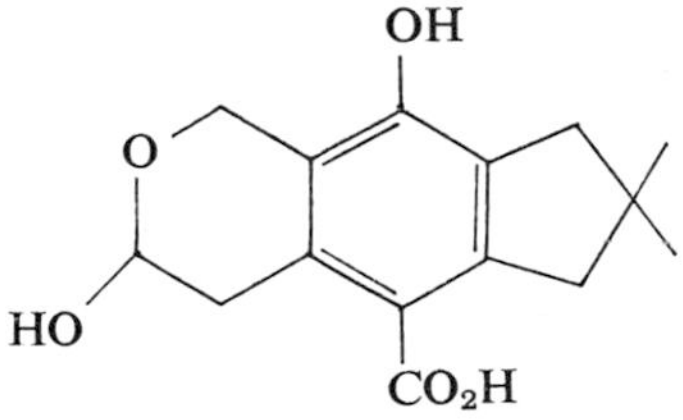

30

structure (**30**).[53] The structures of illudin S,[51] fomannosin,[54] and hirsutic acid C[58] have been confirmed by X-ray analysis.

B.2.d. *Miscellaneous Sesquiterpenes*

31

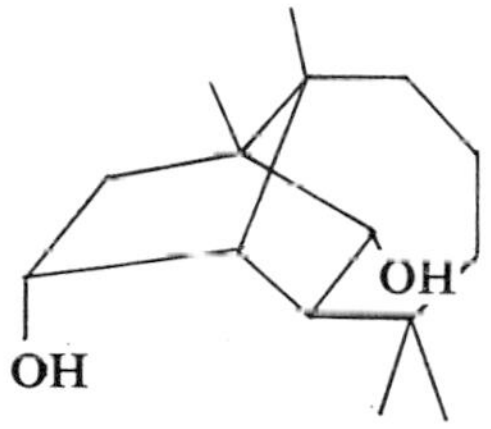

culmorin
Fusarium culmorum[59,60]

32

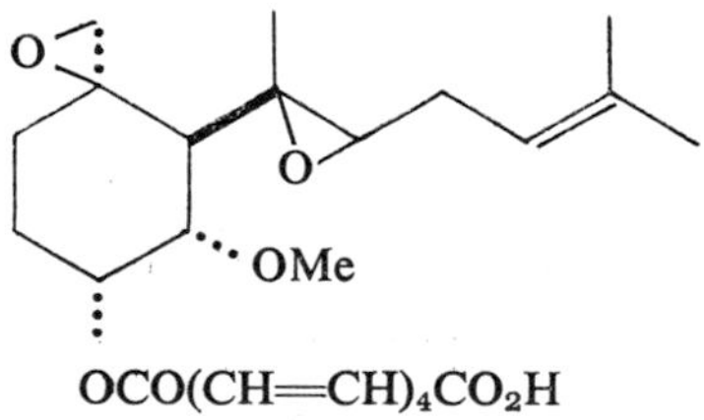

fumagillin   *L*[62]
Aspergillus fumigatus[61]

$OCO(CH\!=\!CH)_4CO_2H$

33 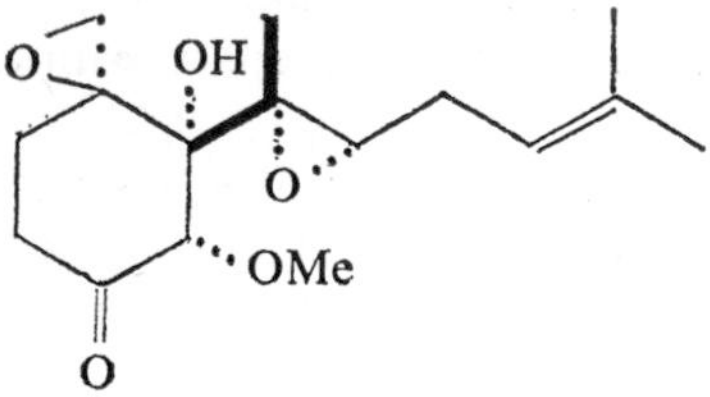
O
OH
O
OMe
O

ovalicin, graphinone
Pseudorotium ovalis[63a]
Graphium sp.[63b]

34 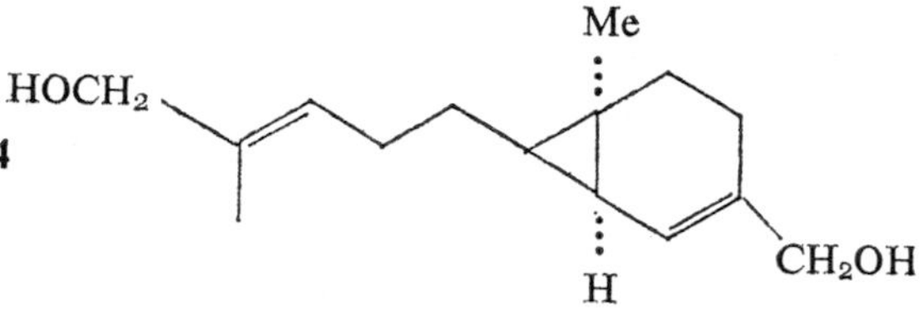
Me
$HOCH_2$
CH_2OH
H

sirenin
Allomyces spp.[64]

35 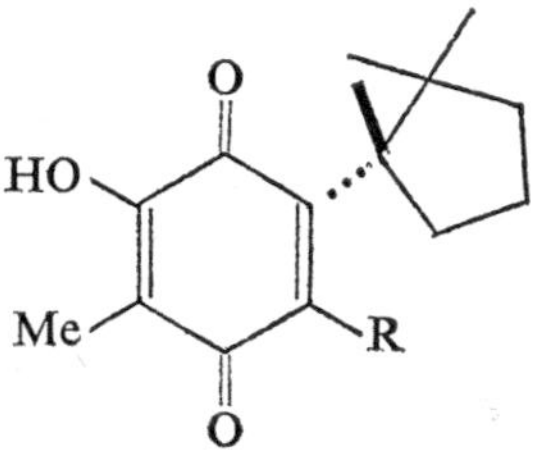
O
HO
Me
R
O

a: R = OH, helicobasidin *L*[66,71]
Helicobasidium mompa[65,66]
b: R = H, deoxyhelicobasidin
H. mompa[67]

36 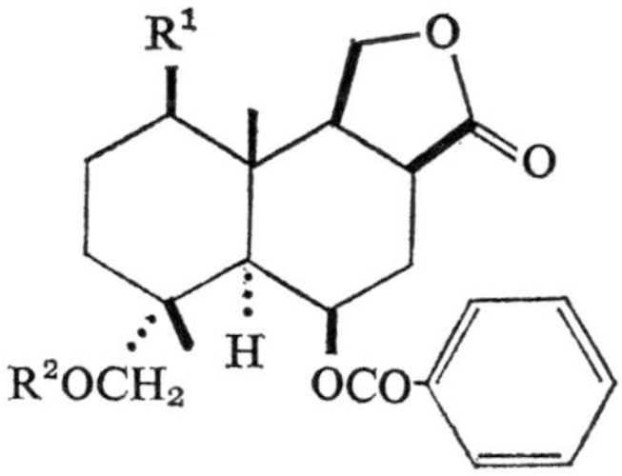
R^1
O
O
H
R^2OCH_2
OCO

a: R^1 = OH, R^2 = Ac, pebrolide
b: R^1 = H, R^2 = Ac, 1-deoxypebrolide
c: R^1 = OH, R^2 = H, desacetylpebrolide
Penicillium brevi-compactum

37 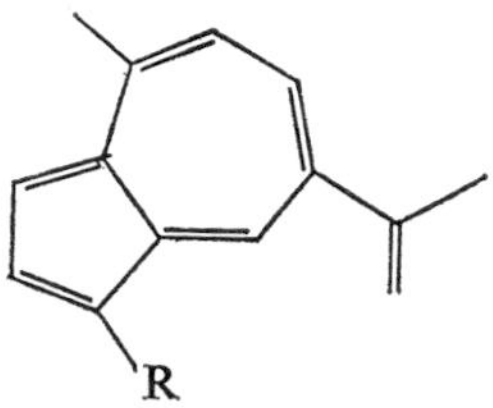
R

a: R = CHO, lactaroviolin
Lactarius deliciosus[68]
b: R = Me, lactarazulene
L. deliciosus[69]

38

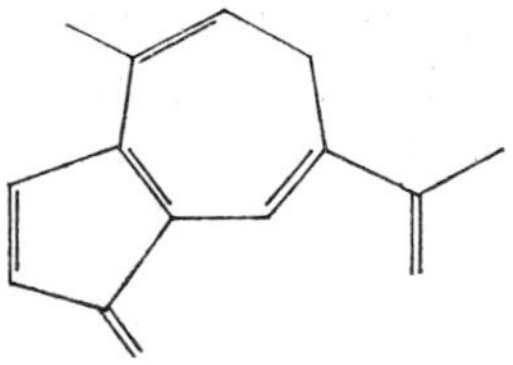

lactarofulvene
L. deliciosus[70]

The skeleton of culmorin is enantiomeric with that of longiborneol, and helicobasidin has the skeleton of a postulated intermediate between bisabo-lene and the trichothecanes (see Scheme 3). The sesquiterpene nature of helicobasidin has been confirmed;[66,71] it is the only example of a mevalonate-derived quinone to have been isolated from a fungus though several such compounds have been obtained from higher plants. Pebrolide has the drimane skeleton; it is possible that antibiotics LL-Z1271 (**54**) are also derived from this skeleton (see p. 238).

The alcohol moiety of fumagillin is derived from three mevalonate units,[62] and a possible biosynthetic route from farnesyl pyrophosphate is shown in Scheme 6.

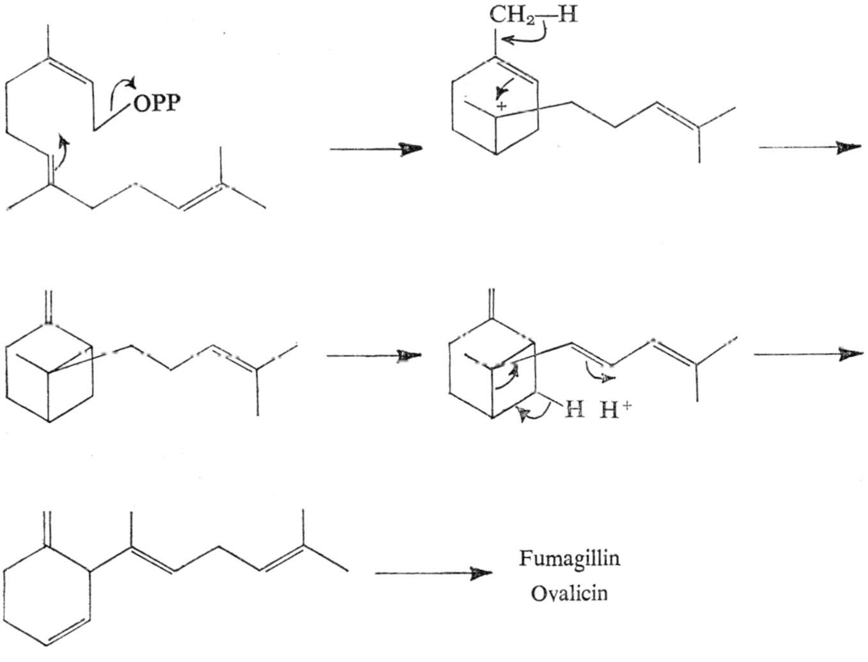

Fumagillin
Ovalicin

Scheme 6: Possible biosynthetic sequence from farnesyl pyrophosphate (written in the *cis* form) to fumagillin and ovalicin

The gametes of the aquatic organism *Allomyces* are motile, and the male gametes are attracted to the female gametes. The attraction is mediated by sirenin (**34**), which is secreted by the female gametes.

B.2.e. *Triprenyl Phenols*

39

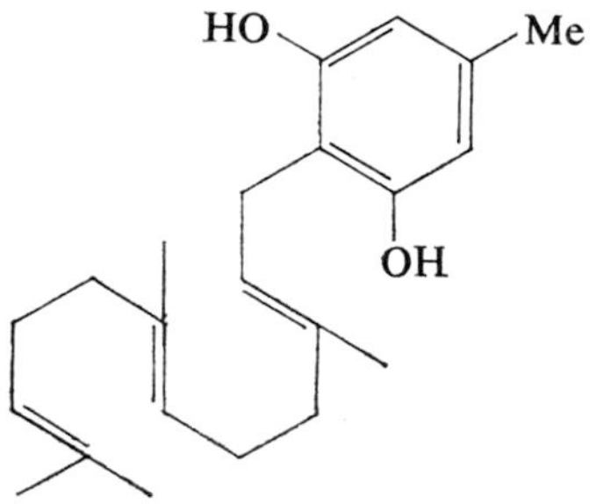

grifolin
Grifola confluens[72]

40

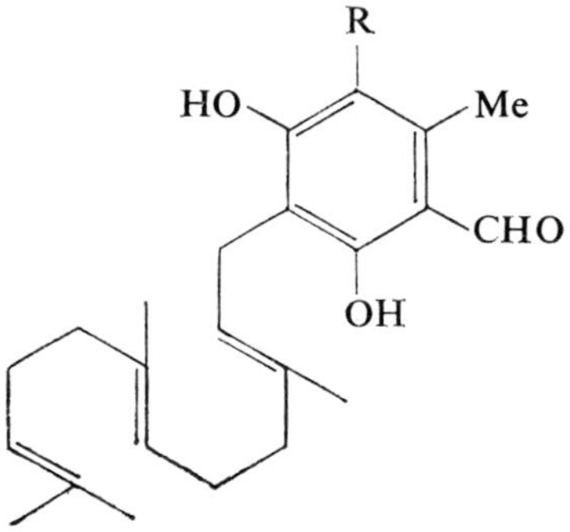

a: R = Cl, antibiotic LL-Z1272α
b: R = H, antibiotic LL-Z1272β
        Fusarium sp.[73]

41

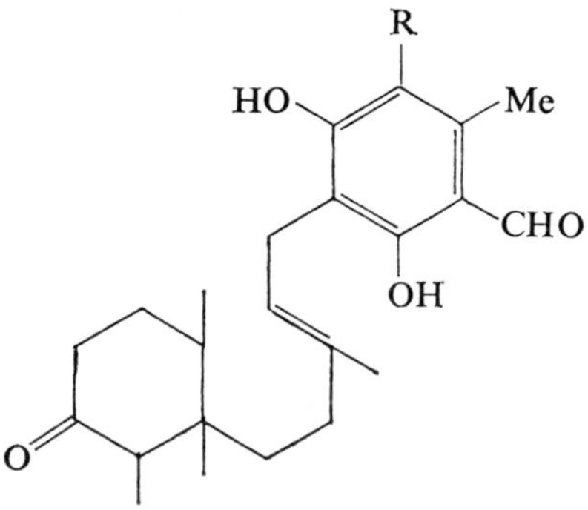

a: R = Cl, antibiotic LL-Z1272δ
b: R = H, antibiotic LL-Z1272ε
        Fusarium sp.[73]

42

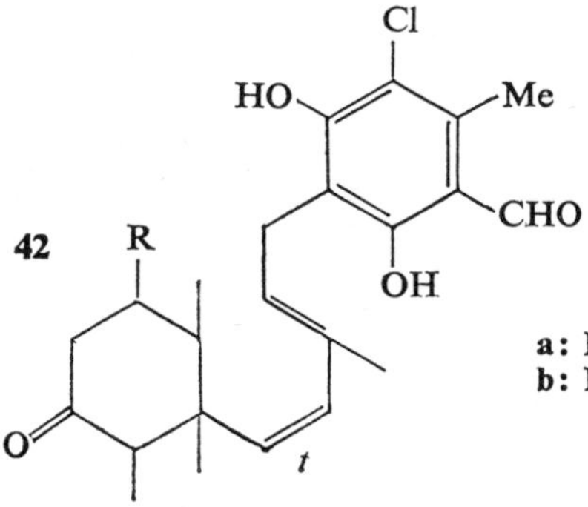

a: R = H, antibiotic LL-Z1272γ, ascochlorin
b: R = OAc, antibiotic LL-Z1272ζ
Fusarium sp,[73] Ascochyta viciae[74]

43

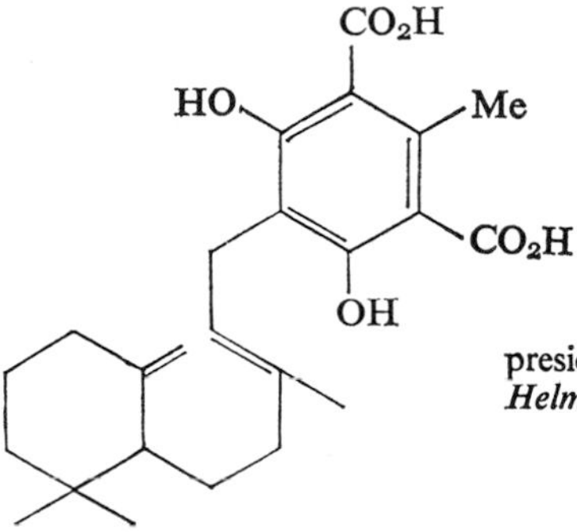

presiccanochromenic acid
Helminthosporium siccans[75]

44

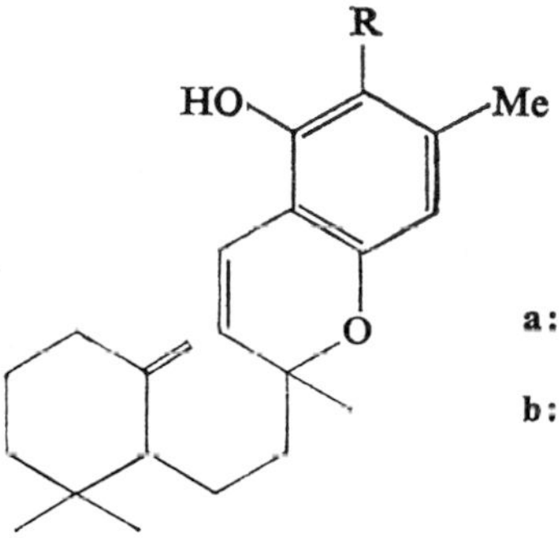

a: R = CO₂H, siccanochromenic acid
H. siccans[75]

b: R = H, siccanochromene A
H. siccans[76]

45

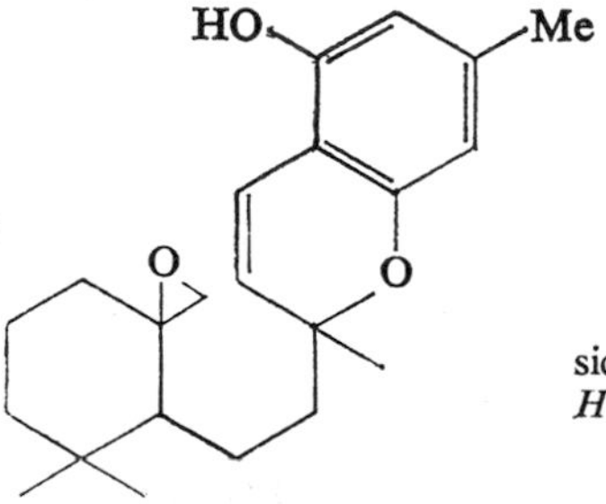

siccanochromene B
H. siccans[76]

46

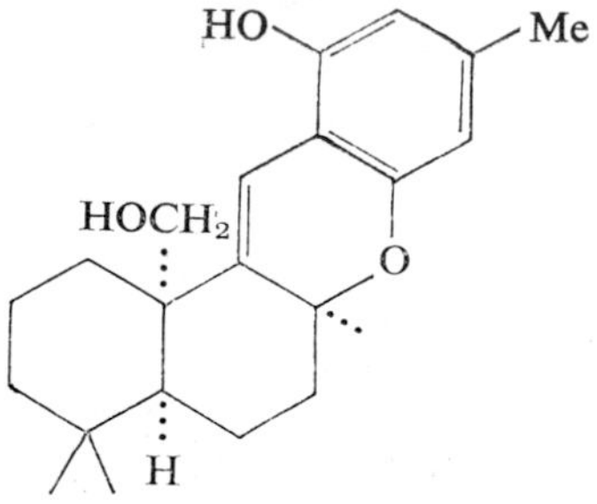

siccanochromene E
H. siccans[75]

47

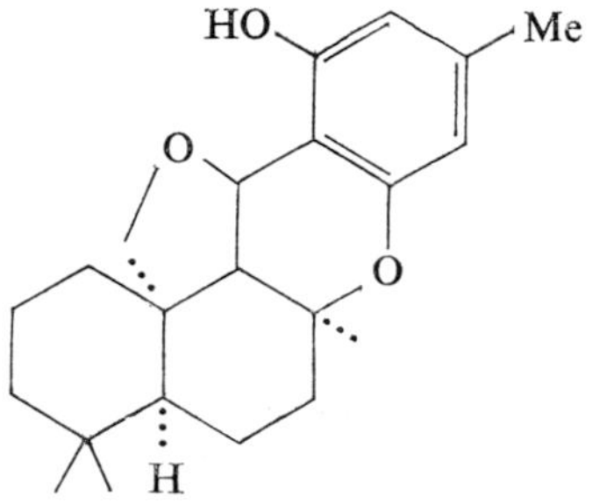

siccanin
H. siccans[77]

48

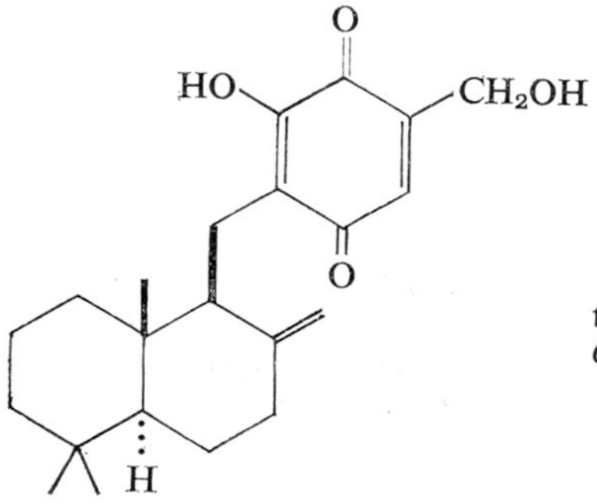

tauranin
Oospora aurantia[78]

In these compounds a sesquiterpene unit is attached to a polyketide-derived aromatic nucleus of the orsellinic acid/orcinol type. The co-occurrence of acyclic and cyclic compounds in the same series suggests that the cyclizations occur after introduction of a farnesyl side-chain onto the aromatic system. In the formation of compounds (**41**) and (**42**) an extra methyl group migration has occurred to give a skeleton not encountered previously; the keto-group in these compounds may result from initial formation of an

epoxide as in squalene cyclization (p. 253). The *cis* ring fusion of compounds
(**46**) and (**47**) is unusual; compound (**48**) has the more normal *trans* configuration, cf. pebrolide (**36**). The chromene ring system of compounds
(**44**), (**45**) and (**46**) is analogous to that of mycochromenic acid (p. 115).

B.3. DITERPENES

Diterpenes appear to be of limited distribution among the fungi, having
so far been isolated from only six species (a geranylgeranyl side-chain is
present in bovinone, p. 356). Most diterpenes, and all those from fungi, can
be derived from intermediates possessing the labdane (**49**) or pimarane (**50**)
skeletons. The cyclization of geranylgeranyl pyrophosphate can give rise to
two enantiomeric series of compounds—the "normal" series, whose stereo-

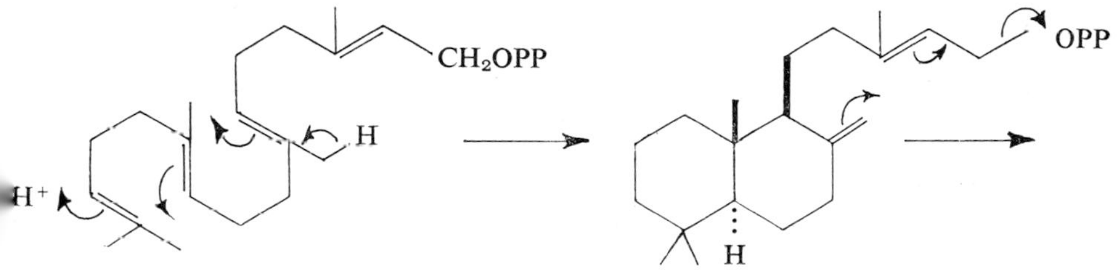

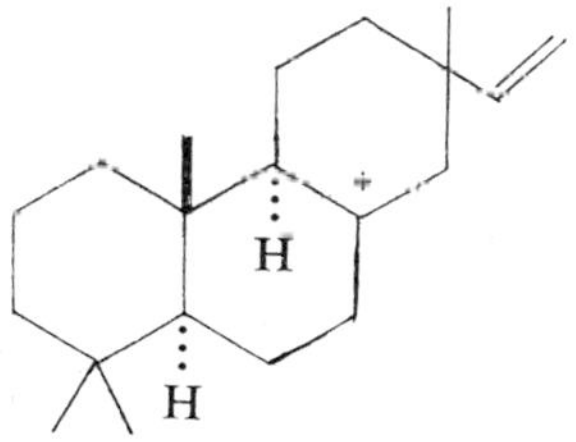

Scheme 7: The formation of the labdane (**49**) and pimarane (**50**) skeletons from
geranylgeranyl pyrophosphate

chemistry is illustrated in Scheme 7, and the "antipodal" series. Examples of
both types have been obtained from fungi, and in the normal series both
configurations at C-13 of the pimarane skeleton are known.

B.3.a. *Miscellaneous Diterpenes*

51

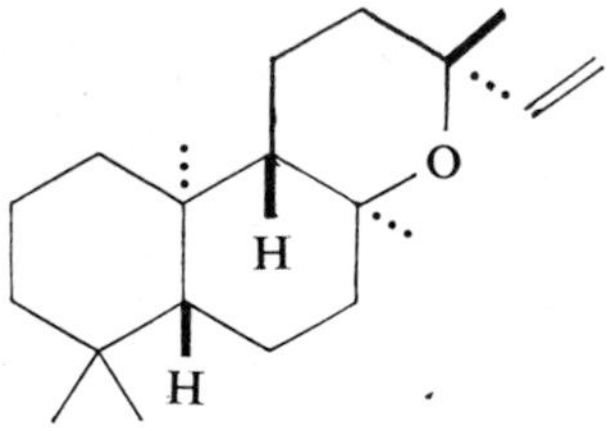

13-*epi*-(−)-manoyl oxide
Gibberella fujikuroi[79]

52

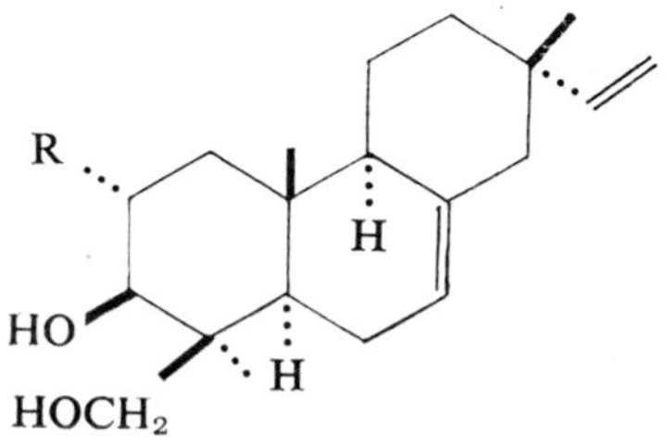

a: R = OH, virescenol A
b: R = H, virescenol B
        Oospora virescens[80]

HO
HOCH₂

53

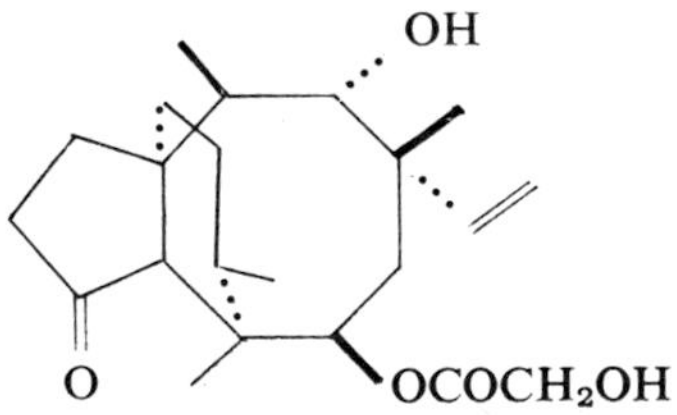

pleuromutilin   L[83,84]
Pleurotus mutilus,[81] *P. passeckerianus*,[81]
Drosophila subatrata[82]

OCOCH₂OH

54

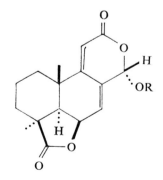

a: R = Me, antibiotic LL-Z1271α
b: R = H, antibiotic LL-Z1271γ
        Acrostalagmus sp.[85]

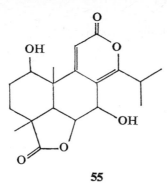

55

Scheme 8: A biosynthesis of pleuromutilin which accounts for the observed distribution
of radioactivity (•) from [2-¹⁴C]mevalonate[83]

A

B

[O]

[O]

53

13-*epi*-(−)-Manoyl oxide (**51**) has an antipodal labdane skeleton and the virescenols (**52**), which occur as glycosides, have normal pimarane skeletons. Antibiotics LL-Z1271α (**54a**) and LL-Z1271γ (**54b**) are the only known C_{16} terpenes. Their resemblance to, for example, nagilactone A (**55**) suggests that they are degraded diterpenes, but they could arise by introduction of a C_1-unit onto the sesquiterpenoid drimane skeleton.

The distribution of radioactivity in the antibiotic pleuromutilin (**53**) produced in the presence of [2-^{14}C]mevalonate and of [1-^{14}C]acetate is consistent with the biogenesis of Scheme 8.[83,84] Several details of the process have been established[83] with [^{3}H]mevalonate. The stereochemistry of the ion A precludes a synchronous migration of the proton and the methyl group, and the ion B is neutralized by solvolysis accompanied by a 1,5-transannular hydride shift.

B.3.b. *Rosenonolactone and its Congeners*

56

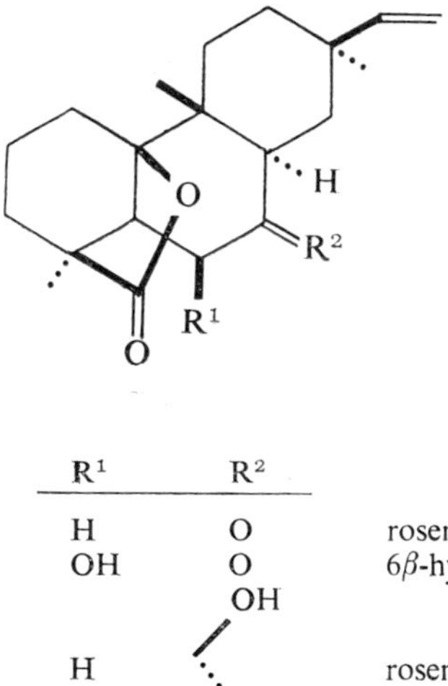

	R^1	R^2		
a	H	O	rosenonolactone (rosein I)[87,89,90a]	*L*[91,92]
b	OH	O	6β-hydroxyrosenonolactone[93]	
c	H	OH / H	rosenololactone[21]	
d	H	H_2	7-deoxyrosenonolactone[94]	
e	OH	H_2	rosololactone (rosein II, rosonolactone)[87,89,90b]	

57

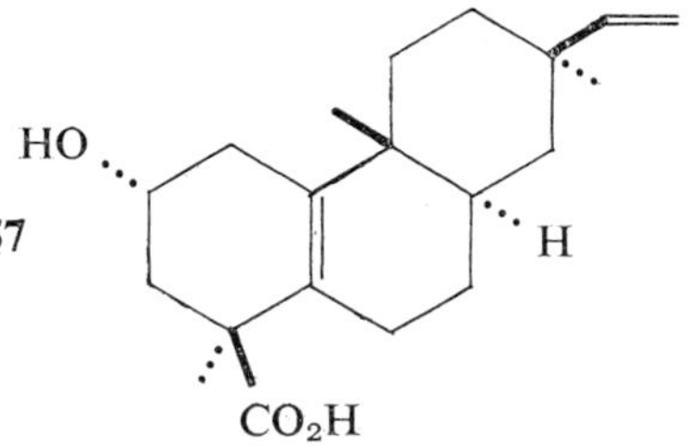

isorosenolic acid[95]

HO

CO_2H

These compounds are metabolites of *Trichothecium roseum*, which also produces the sesquiterpenes trichothecin and crotocin. Rosenonolactone **(56a)** is also produced by *T. luteum* and *T. cytosporium*, and rosololactone **(56e)** by *T. cytosporium*.[88]

Rosenonolactone **(56a)** and rosololactone **(56e)** were first isolated[87] as roseins I and II and subsequently studied[89] as rosenonolactone and rosono-lactone. The structure of rosenonolactone was established by Robertson and his co-workers[90a] who also studied the chemistry of rosonolactone, giving it its present name of rosololactone.[90b] The relative[96] and absolute[94] stereochemistry of rosenonolactone apart from the configuration at C-13 was established by chemical and O.R.D. experiments, and the full structure of rosololactone was derived by X-ray analysis of its dibromide;[97] since rosololactone has been related chemically to rosenonolactone[97] the full absolute stereochemistry of both compounds is known.

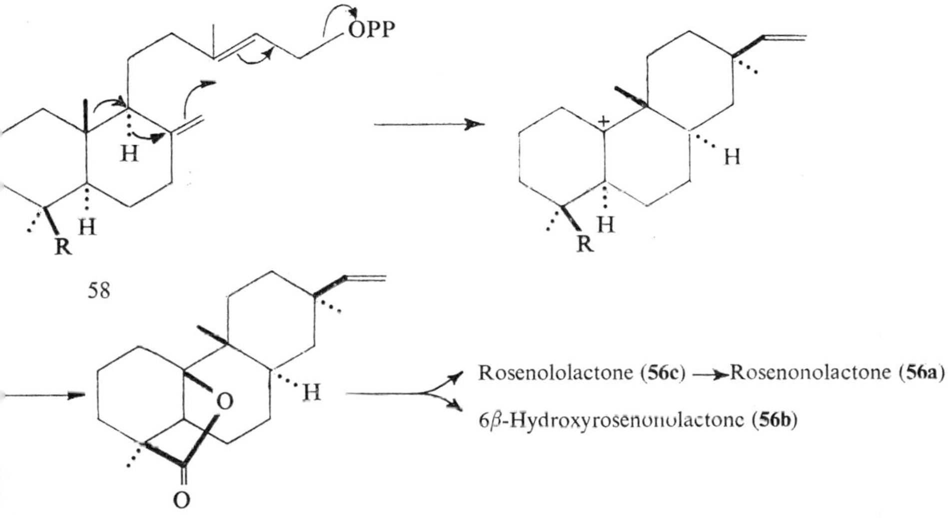

Rosenololactone **(56c)** ⟶ Rosenonolactone **(56a)**

6β-Hydroxyrosenonolactone **(56b)**

7-Deoxyrosenonolactone **(56d)**

Scheme 9: The biosynthesis of rosenonolactone and its congeners

The diterpenoid nature of rosenonolactone (Scheme 9) has been confirmed by labelling experiments with [1-^{14}C]acetate,[91] [2-^{14}C]mevalonate[91,92] and [1-^{3}H]geranylgeraniol.[98] Support for a labdane intermediate is provided[99] by the specific incorporation of the pyrophosphate **(58, R = Me)** into rosen-onolactone. Incorporation of 4R-[4-^{3}H,2-^{14}C]mevalonate into rosenonolac-tone proceeded with retention of four tritium atoms, three of which were located at C-5, C-8, and C-18.[98] This confirms the migration of hydrogen from C-9 to C-8 and excludes a $\Delta^{5,10}$-intermediate. [5-^{3}H$_2$, 2-^{14}C]Mevalonate gave rosenonolactone with eight tritium atoms two of which are at C-6,[98] excluding a $\Delta^{5,6}$-intermediate (*a priori* the 5-hydrogen could have migrated to C-10 with formation of a 5,6-double bond and then migrated back to C-5 during lactone formation). Finally, [2-^{3}H$_2$, 2-^{14}C]mevalonate gave rosenono-lactone with six tritium atoms of which two are at C-1, excluding a $\Delta^{1,10}$-intermediate. Since the migrating methyl group is *cis* to the lactone ring, a concerted process is unlikely and it has been suggested that an α-oriented bond might be formed between C-10 and an enzyme which is displaced during lactone formation. Incorporation experiments[98,100] show that 7-deoxyrosenonolactone is a precursor of the oxygenated compounds.

B.3.c. *The Gibberellins and Kauranes*

In the early stages of a disease of rice plants caused by *Gibberella fujikuroi* (the perfect stage of *Fusarium moniliforme*) the infected plants grow more rapidly than normal. Filtrates from cultures of *G. fujikuroi* were found to produce a similar growth-promoting effect and a crystalline active product,

59   **60**

later shown to be a mixture of gibberellins, was isolated. The compounds assumed a wider significance with the discovery that gibberellic acid and other gibberellins (many of which have not been detected in *G. fujikuroi*) are endogenous plant hormones. The scientific and commercial importance of gibberellic acid promoted a detailed examination of the products of *G. fujikuroi* by Cross and his colleagues at the Akers Research Laboratories of I.C.I. As a result of this search, many new compounds based on the gibbane (**59**) and the kaurane (**60**) skeletons were isolated and the biosynthetic relationship between the compounds was established. The history, the chemistry, and the biological properties of the gibberellins have been reviewed.[101]

B.3.c.i. *The C$_{19}$ gibbanes*

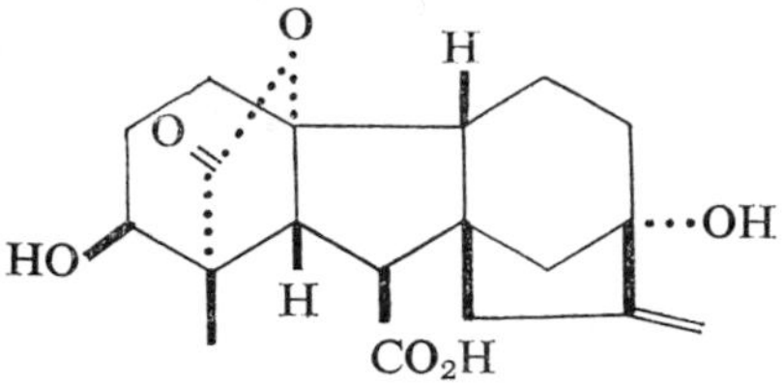

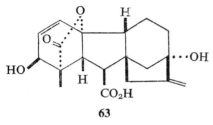

61

Gibberellin A$_1$[101]

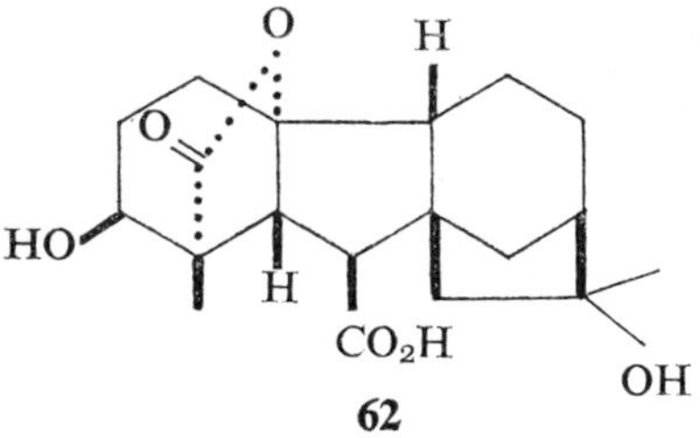

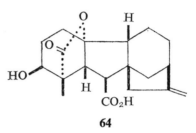

62

Gibberellin A$_2$[101]

63

Gibberellin A$_3$(gibberellic acid)[101]

64

Gibberellin A$_4$[101]

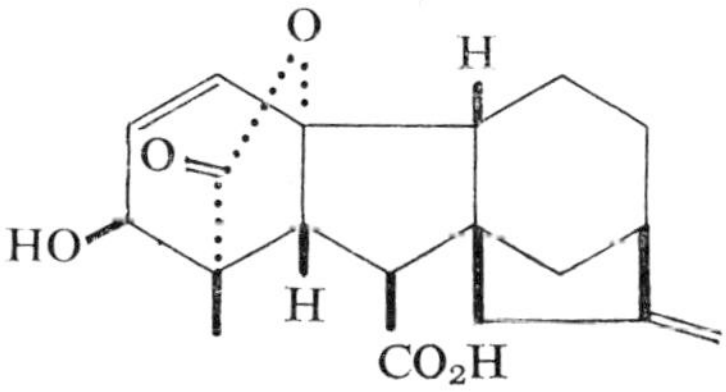

65

Gibberellin A$_7$[102]

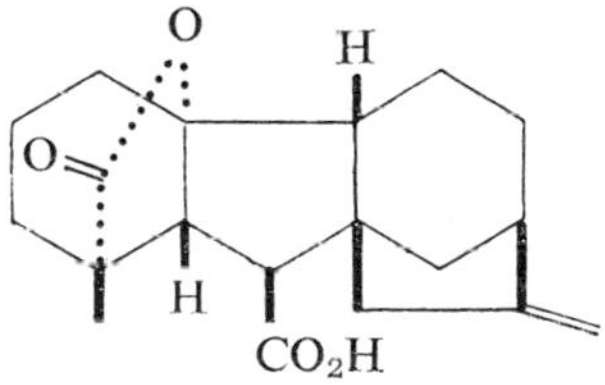

66

Gibberellin A$_9$[102]

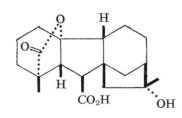

67

Gibberellin A$_{10}$[103]

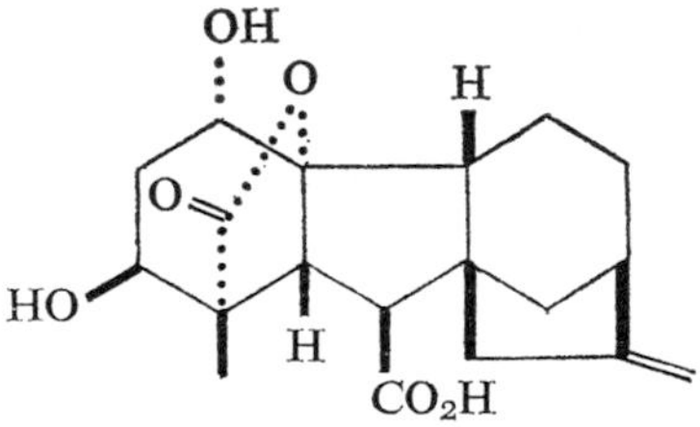

68

Gibberellin A$_{16}$[104]

B.3.c.ii *The C$_{20}$ gibbanes*

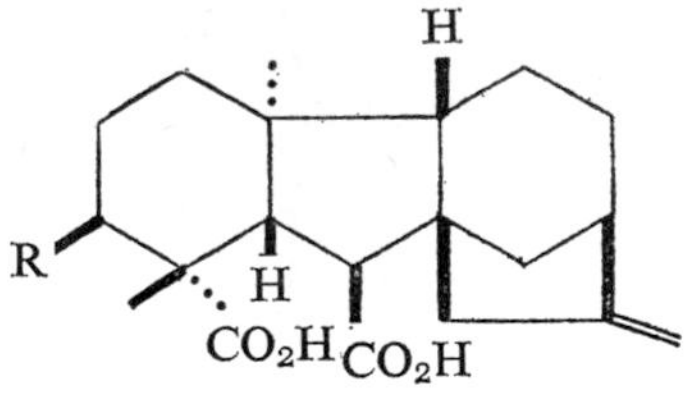

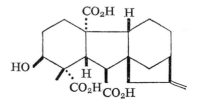

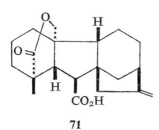

69

a: R = H, gibberellin A$_{12}$[105]
b: R = OH, gibberellin A$_{14}$[106]

70

Gibberellin A$_{13}$[107]

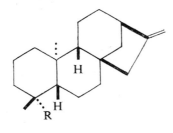

71

Gibberellin A$_{15}$[108]

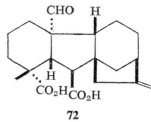

72

Gibberellin A$_{24}$[109]

B.3.c.iii. *The Kauranes*

73

a: R = Me, (−)-kaurene[77]
b: R = CO$_2$H, (−)-kaurenoic acid[122]

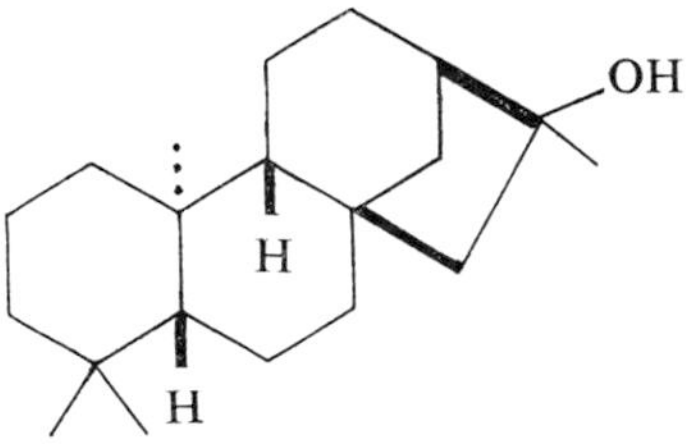

74

(−)-kauranol[110]†

† Also isolated from the lichen *Ramalina tumidula*[114]

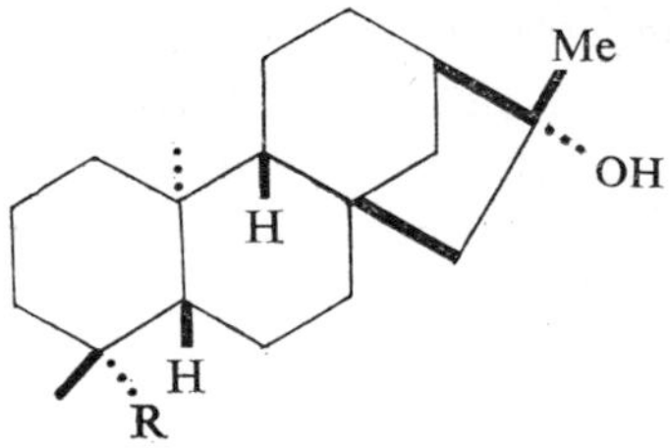

75

a : R = CHO, 16α-hydroxy-(−)-kauran-19-al[111]
b : R = CO$_2$H 16α-hydroxy-(−)-kauran-19-oic acid[111]

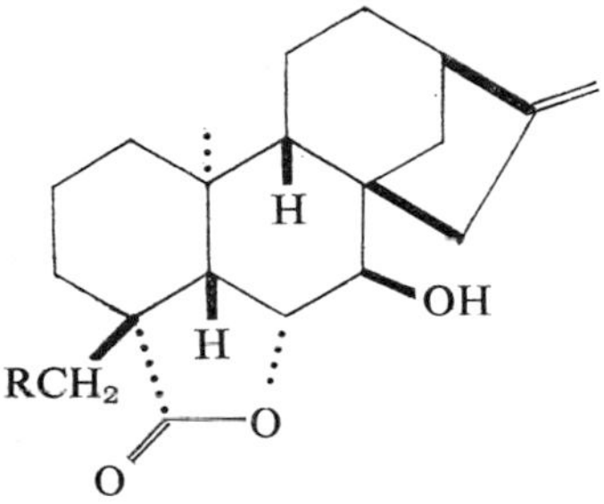

76

a : R = H, 7-hydroxykaurenolide[110]
b : R = OH, 7,18-dihydroxykaurenolide[112]

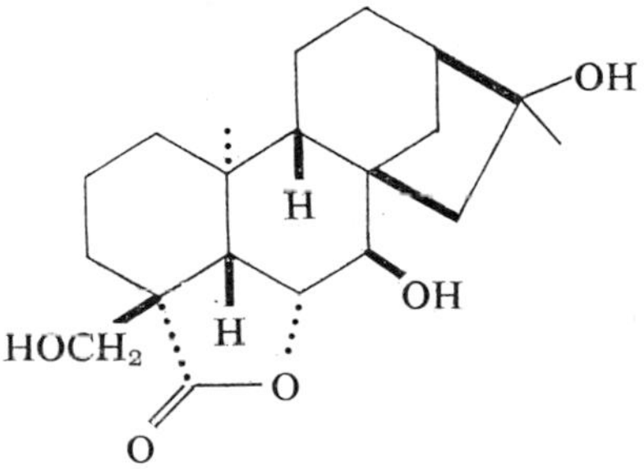

77

7,16,18-trihydroxykauranolide[112]

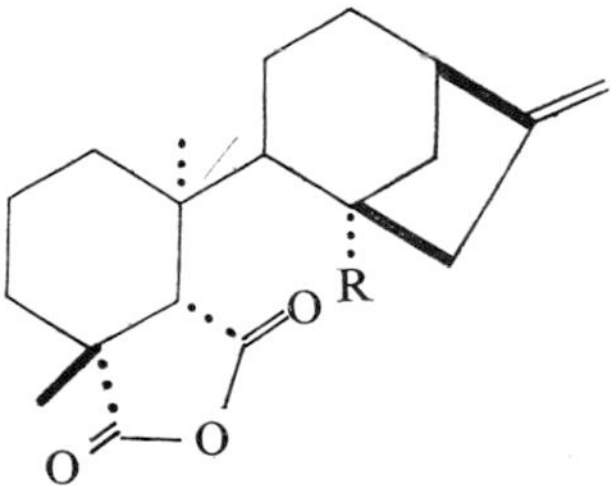

78

a : R = CHO, fujenal[113]
b : R = CO$_2$H, fujenoic acid[113]

B.3.c.iv. *The Biosynthesis of the Kauranes and Gibberellins*

The incorporation of radioactive acetate and mevalonate into gibberellic acid gave the labelling pattern shown in Scheme 10,[115] consistent with the formation of gibberellic acid from geranylgeraniol via (−)-kaurene. The

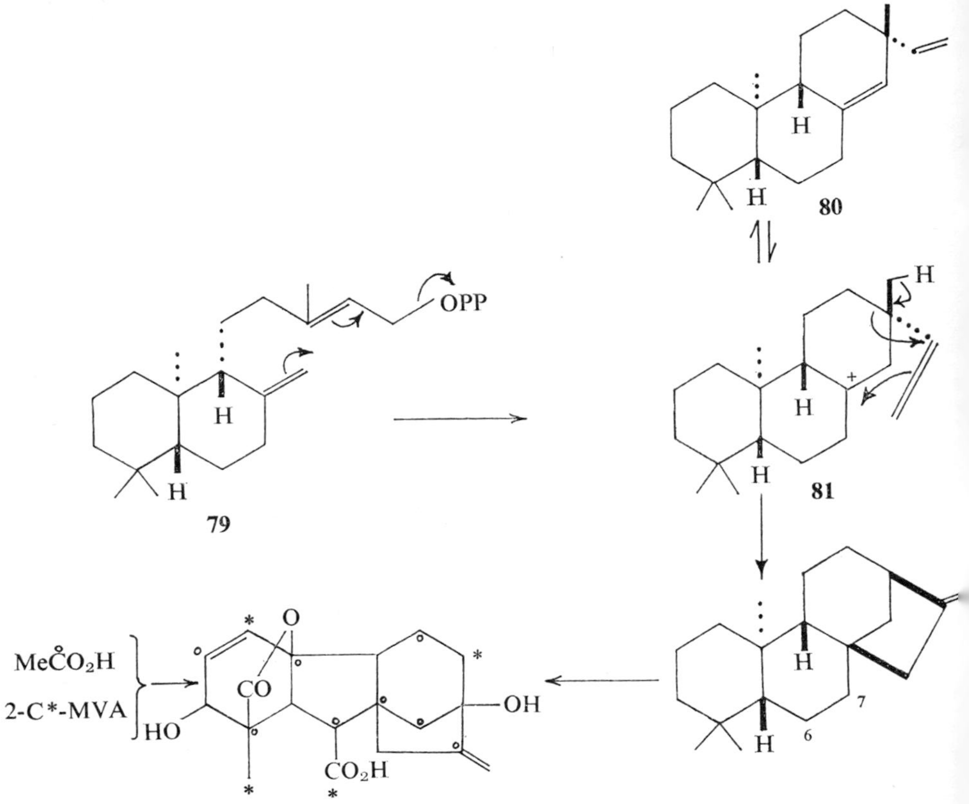

OPP

MeCO₂H
MeĊO₂H

2-C*-MVA

HO

ĊO₂H

OH

79

80

81

6

7

Scheme 10: The incorporation of acetate and mevalonate into gibberellic acid and the biosynthesis of kaurene

results also confirm the mechanism shown in Scheme 10 (originally proposed by Wenkert[116]) for the formation of the tetracyclic skeleton from a tricyclic precursor, and show that C-7 of kaurene becomes the 10-carboxyl group of gibberellic acid. The subsequent isolation of (−)-kaurene from *G. fujikuroi* and the specific incorporation of labelled (−)-kaurene into gibberellic acid[117] confirmed the role of (−)-kaurene as a precursor of the gibberellins.

Biosynthesis of (−)-kaurene

(−)-Labda-8,13-dien-15-yl pyrophosphate (copalyl pyrophosphate) (**79**), the enantiomer of the precursor (**49**) of rosenonolactone, is specifically incorporated into (−)-kaurene.[118] The corresponding alcohol has not been obtained from cultures of *G. fujikuroi* (though the related compound (51) has) but a soluble enzyme system capable of converting geranylgeranyl pyrophosphate or mevalonate to the pyrophosphate (79) and of converting this to (−)-kaurene has been obtained from mycelium of *G. fujikuroi*.[119]

Experiments with $4R$-[4-^{3}H,2-^{14}C]mevalonate and with [2-^{3}H$_2$,2-^{14}C]-mevalonate exclude the intervention of intermediates possessing a $\Delta^{7,8}$ or $\Delta^{8,9}$ double bond, but the $\Delta^{8,14}$-diene (80) is specifically incorporated into gibberellic acid,[118] presumably via a kaurene derivative. (An earlier report[120] of the incorporation of the diene (80) into gibberellic acid was subsequently withdrawn.[121]) The very low incorporation of the diene (80) suggests that it is not an intermediate in the biosynthesis of (−)-kaurene but that it can serve as a precursor by equilibration with the ion (81).

Conversion of (−)-kaurene to the gibbanes

(−)-Kauren-19-ol (73, $R = CH_2OH$),[123a,124,125] (−)-kaurenal (73, $R =$ CHO)[126] and (−)-kaurenoic acid (73b)[127] are incorporated into gibberellic acid by *G. fujikuroi* suggesting that the first stage in the conversion of (−)-kaurene to the gibbanes is the stepwise oxidation of the 4α-methyl group. 7β-Hydroxykaurenoic acid (82) and the aldehyde (83) have been detected in cultures of *G. fujikuroi* by dilution analysis after feeding radio-active (−)-kaurene,[120] and are incorporated into gibberellic acid.[128,129] The ring contraction of (82) to give (83) requires a suitable leaving group at C-6; since tritium derived from $5R$-[5-^{3}H]mevalonate is lost during this process, the leaving group has the 6β-stereochemistry.[130] This is in accord with the earlier observation that 7β-hydroxykaurenolide (76a), the corresponding hydroxy-acid, and 7α-hydroxykaurenolide, all of which possess a 6α-sub-stituent, are not incorporated into gibberellic acid[123b] [7-hydroxykaurenolide (76a) was incorporated into 7,18-dihydroxykaurenolide (76b)].

Biosynthetic relationships among the gibbanes

Of the C_{20} gibbanes, gibberellin A_{13} (70) is not incorporated into gibberellic acid, gibberellin A_{14} (69b) is efficiently incorporated, and gibberellin A_{12} (69a) is only poorly incorporated though the corresponding diol is efficiently incorporated.[129] The last result suggests that gibberellin A_{12} is at the wrong oxidation level and that the corresponding 10-aldehyde (83) is the true intermediate for gibberellin A_{14} and gibberellic acid.

Gibberellin A_9 is converted to gibberellin A_{10} but not to gibberellic acid,[117] and a mixture of gibberellins A_4 and A_7 is converted to a mixture of gib-berellins A_1 and A_3.[125] The likely interrelationships of the gibbanes is sum-marized in Scheme 12, though there is no direct evidence for several of the proposed steps.

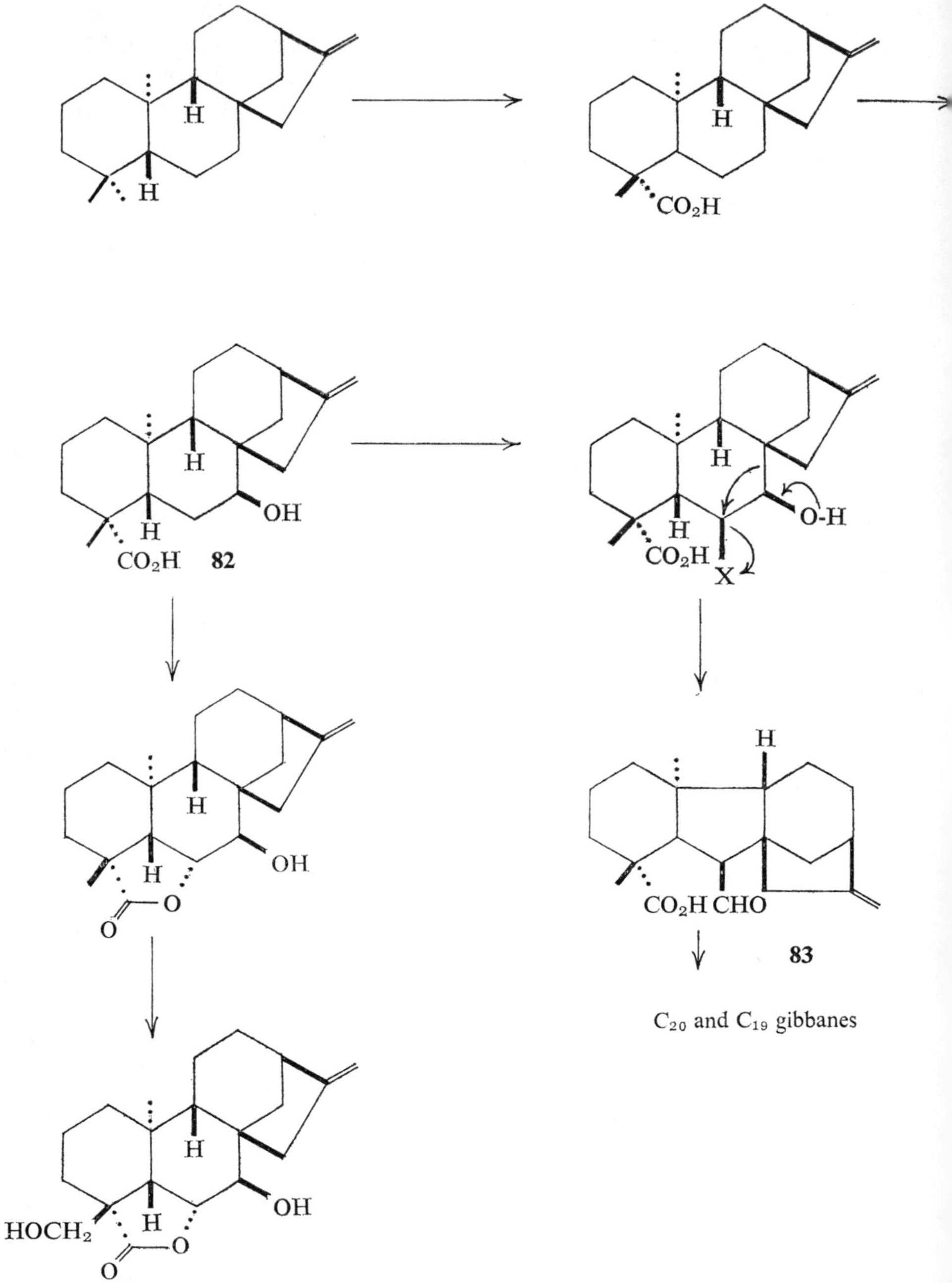

Scheme 11: Possible biosynthetic routes from kaurene to the kaurenolides and the gibbanes

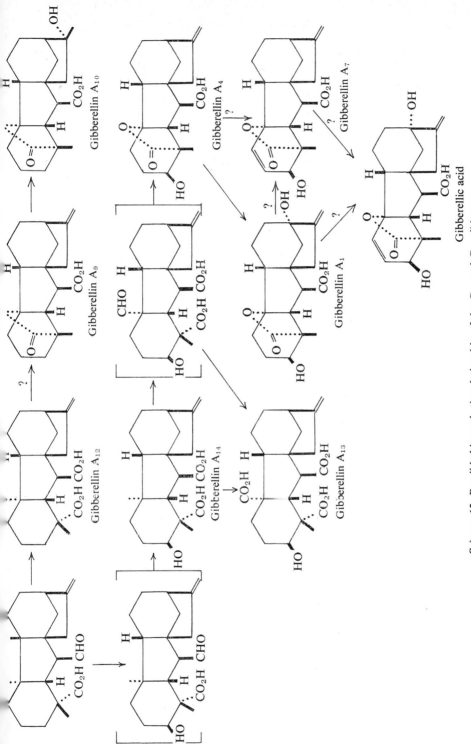

Scheme 12: Possible biosynthetic relationship of the C_{20} and C_{20} gibbanes

The incorporation of the tritium-labelled mevalonates into gibberellic acid provides information on several of the steps. No $\Delta^{4a,4b}$, $\Delta^{4a,4}$ or $\Delta^{4a,10a}$ double bond is involved in the formation of the lactone ring,[118] suggesting that the loss of the methyl group might occur by oxidation to the aldehyde level followed by a Baeyer–Villiger oxidation to give first a formate and then a hydroxyl group. This would also explain the non-incorporation of gibberellin A_{13} into gibberellic acid. The introduction of the 3-hydroxyl group proceeds with retention of configuration,[118] and the formation of the ring A double bond involves *cis* elimination of hydrogen from the α-face of a saturated gibberellin.[130]

B.4. SESTERTERPENES

The sesterterpenes are C_{25} compounds derived from geranylfarnesyl pyrophosphate. The first compound shown to belong to the class was

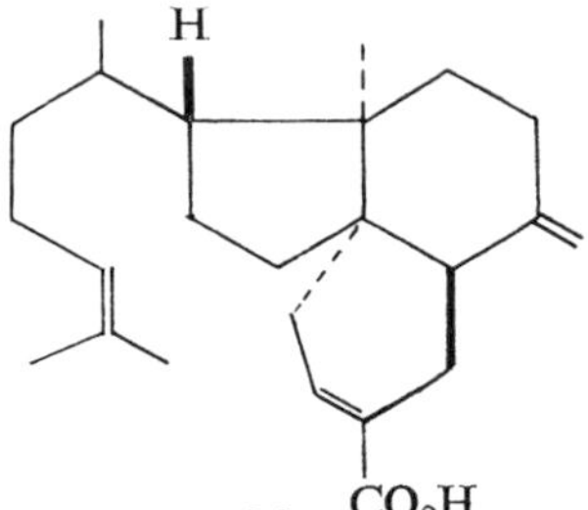

84 CO₂H   85

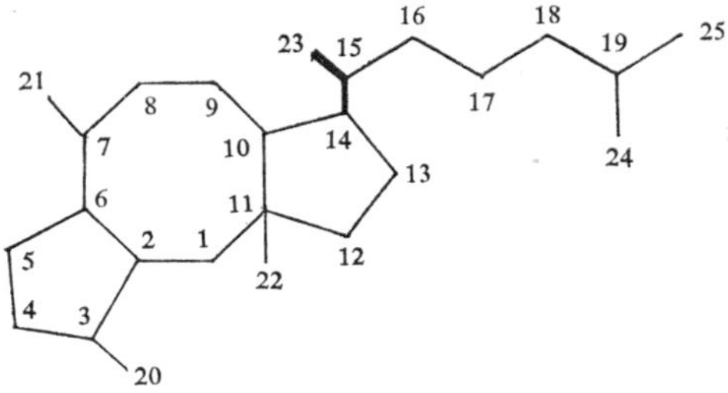

gascardic acid (**84**), isolated from secretions of the insect *Gascardia madagascariensis* (D. Arigoni, Chemical Society Autumn Meeting, Nottingham, 1965). The remaining members of the class, obtained from insects[131] and from fungi (see below), are based on the ophiobolane skeleton (**85**).[132]

B.4.a. *The Fungal Sesterterpenes*

86

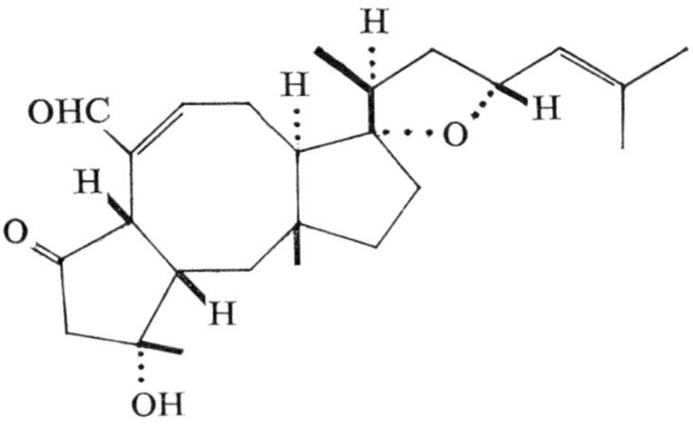

ophiobolin A (ophiobolin,   L[145,147]
cochliobolin, ophiobalin,
cochliobolin A)

Cochliobolus (= *Ophiobolus*)
  miyabeanus[133,134]
  *Helminthosporium turcicum, H.
  zizaniae, H. panici-miliacei,
  H. leersii*[135]

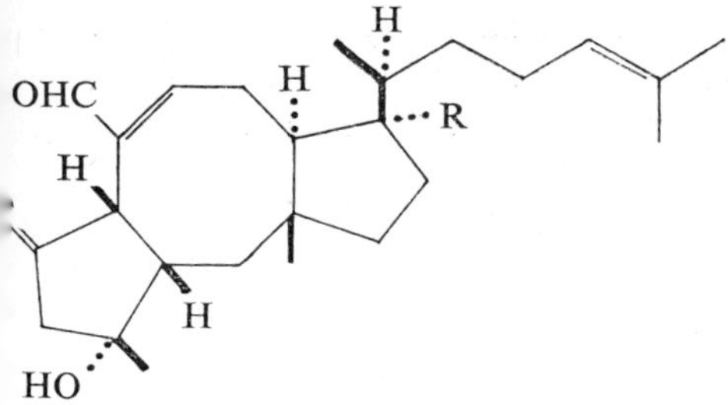

a: R = OH, ophiobolin B (zizanin, $L^{145,147}$
     zizanin B, ophiobolosin A,
     cochliobolin B)
b: R = H, ophiobolin C (zizanin A) $L^{149,150}$
     H. turcicum, H. zizaniae,[135]
     Cochliobolus miyabeanus,
     C. heterostrophus

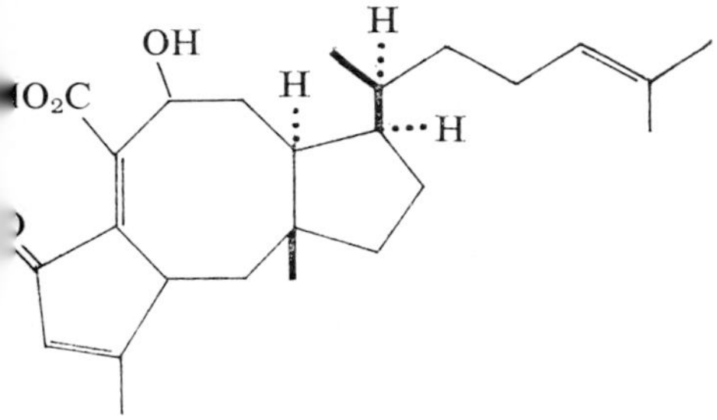

ophiobolin D (cephalonic acid)
Cephalosporium caerulens[136]

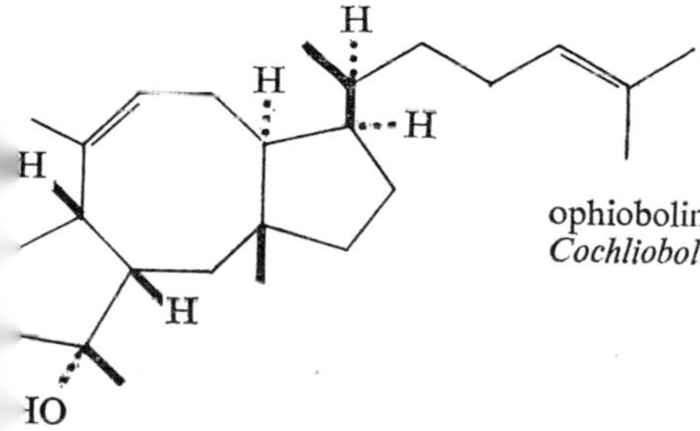

ophiobolin F L^{146}
Cochliobolus heterostrophus[137]

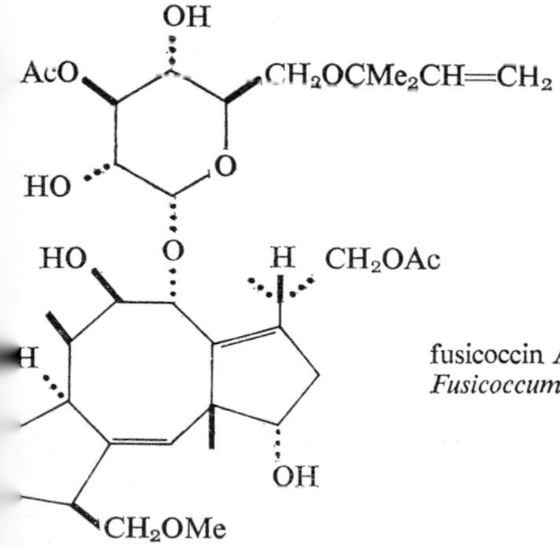

fusicoccin A
Fusicoccum amygdali[138]

Ophiobolins A,B,C and F are phytotoxins produced by a related group of plant pathogens (*Cochliobolus miyabeanus* is the perfect stage of *Helminthosporium oryzae*). The compounds were isolated independently under a variety of names which have now been rationalized.[132] The structure of ophiobolin A was derived[136] by X-ray analysis of a bromomethoxy-derivative and those of ophiobolin B[139] and ophiobolin C[140] by relating them to ophiobolin A. Ophiobolin D (cephalonic acid) is a weakly antibacterial metabolite of *Cephalosporium caerulens*, an organism which produces helvolic acid (p. 264) (high yields of ophiobolin D were obtained from one fermentation which could not be repeated, a circumstance familiar to those who have worked with fungal metabolites!), and its structure was determined by X-ray analysis,[141] and by correlation with ophiobolin C.[142] The structure of fusicoccin A was determined independently by two groups.[143,144]

91

It was immediately apparent[136] that the ophiobolane skeleton could arise by cyclization of the then unknown geranylfarnesyl pyrophosphate and this has been confirmed by labelling experiments with [2-^{14}C]mevalonate, which gave ophiobolin A with the distribution of radioactivity shown in (**91**),[145] and by the conversion of synthetic all-*trans*-geranylfarnesyl pyrophosphate into ophiobolin F by a cell-free preparation of *C. heterostrophus*. [146] Geranylfarnesol has been isolated from insect wax[131] and geranylnerolidol from *C. heterostrophus*.[137]

Ophiobolin C is converted to ophiobolin B by *C. heterostrophus*,[147] and ophiobolin B to ophiobolin A by *C. heterostrophus*[147] and by *C. miyabeanus*.[145] Ophiobolin F was suggested[116] as a precursor of ophiobolin C before its isolation.

Incorporation of ^{18}O from atmospheric oxygen occurs at C-14 (as expected for an oxygenase reaction) but not at C-3.[147,148] Incorporation of [2-^{3}H$_2$]-mevalonate into ophiobolins C and A revealed two interesting, and unexpected, facts (Scheme 13)—there is a 1,5-hydride shift from C-8 to C-15 during the cyclization process, and one atom of tritium is lost from C-24 during the formation of ophiobolin A from ophiobolin C. The stereochemistry of the

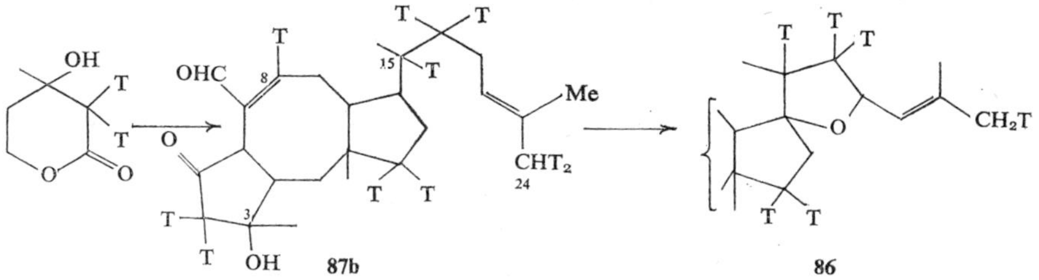

Scheme 13: The incorporation of [2-^{3}H$_2$]mevalonate into ophiobolin C (**87b**) and ophiobolin A (**86**)[148]

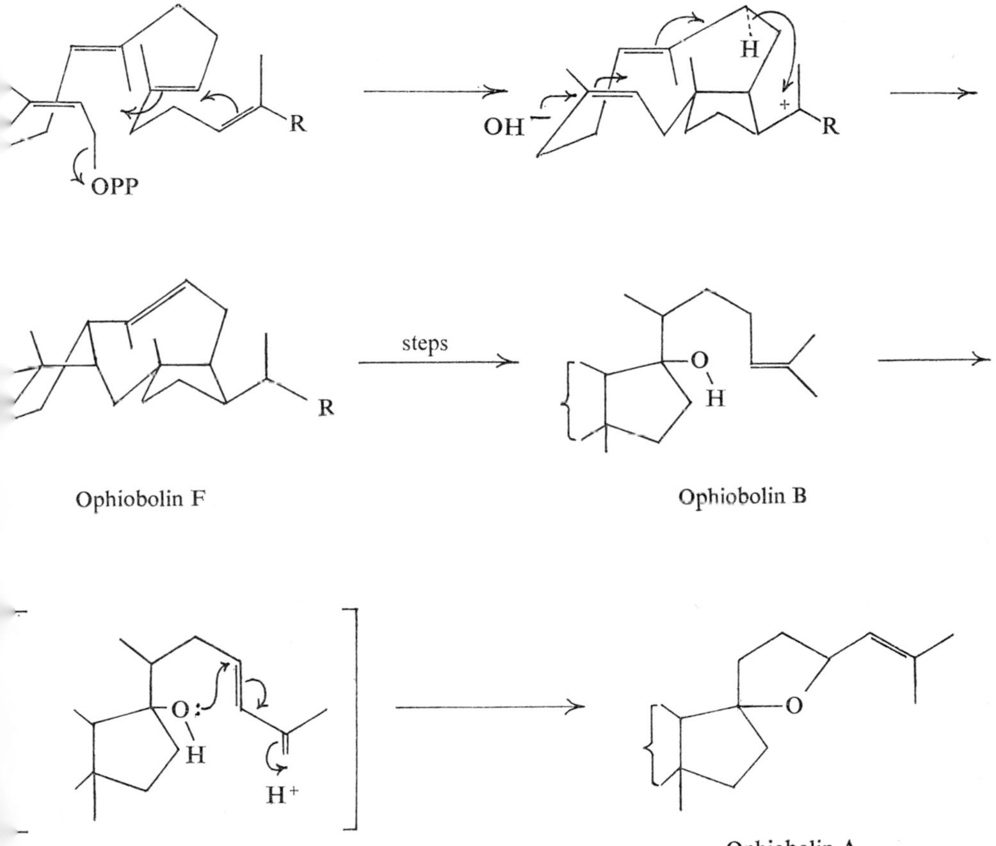

Ophiobolin F

Ophiobolin B

Ophiobolin A

Scheme 14: Possible mechanism for the biosynthesis of the ophiobolins from geranylfarnesyl pyrophosphate

1,5-hydride shift has been studied[149] using 2R- and 2S-[2-³H]mevalonate; only the product from the 2R-precursor possessed tritium at C-15, i.e. the 8β rather than the 8α hydrogen migrates. The results of these experiments with tritium-labelled precursors are accommodated by a sequence such as that shown in Scheme 14.

The incorporation, noted above, of all-*trans*-geranylfarnesyl pyrophosphate into ophiobolin F is in accord with the results of incorporation experiments with 4R-[150,151] and 4S-[2-¹⁴C,4-³H]mevalonate[150] into the ophiobolins; the 4R isomer retains, and the 4S isomer loses, all its tritium during the formation of ophiobolin C, the normal result during terpene biosynthesis (p. 216).

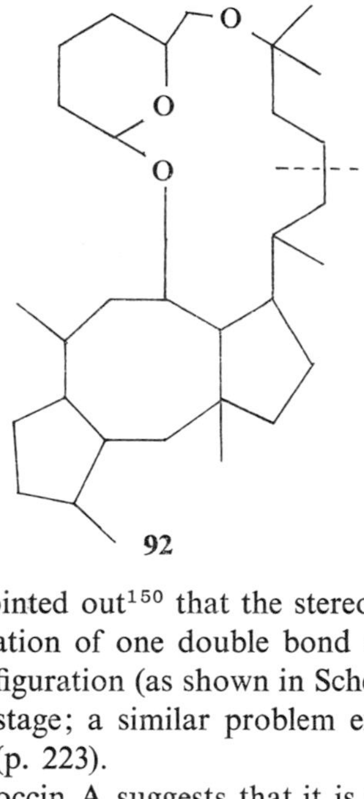

92

However, it has been pointed out[150] that the stereochemistry of the ophiobolins requires isomerization of one double bond of geranylfarnesyl pyrophosphate to the *cis*-configuration (as shown in Scheme 14) or an isomerization at the monocyclic stage; a similar problem exists in the biosynthesis of many sesquiterpenes (p. 223).

The structure of fusicoccin A suggests that it is a sesterterpene in which the side-chain has been linked to the nucleus through a sugar residue and then cleaved as shown in (92). On the other hand, fusicoccin may be a diterpene, and its resemblance to the sesterterpenes coincidental.

B.5. TRITERPENES AND STEROIDS

B.5.a. *Biosynthesis*

B.5.a.i. *The Cyclization of Squalene*

The role of squalene as an obligatory precursor of the steroids and triterpenes is firmly established and the mechanism of its cyclization in a variety of conformations has been extensively discussed.[4] The process is now known to be initiated by the formation of squalene oxide,[152] and is illustrated in Scheme 15 by the formation of lanosterol (94) and hence zymosterol (95). The transformation of lanosterol to zymosterol involves the stepwise loss of three methyl groups by oxidation and decarboxylation; the order in which the methyl groups are lost depends upon the biological system.

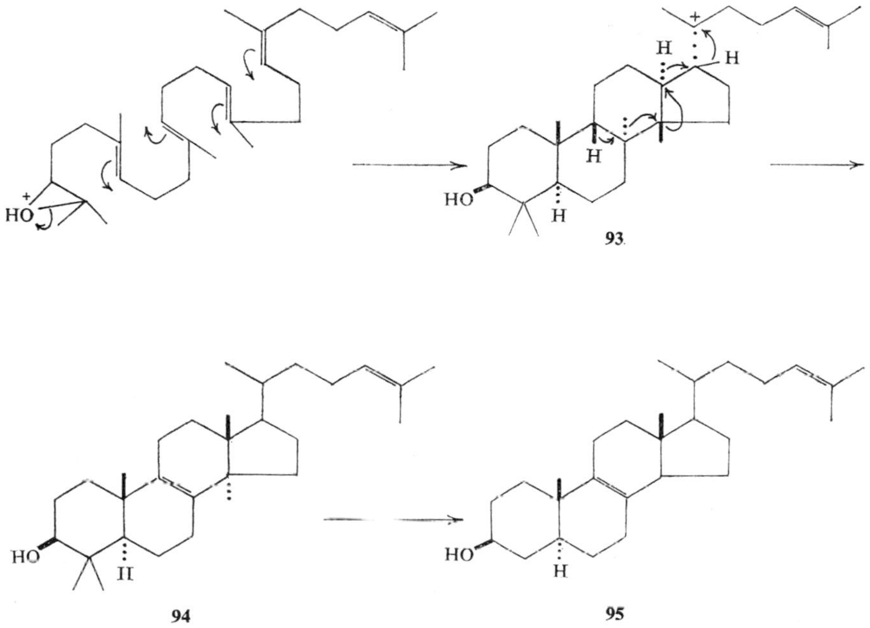

Scheme 15: The cyclization of squalene oxide to give the tetracyclic triterpenes and hence the sterols

Early support for the involvement of squalene in the biosynthesis of fungal terpenoids was provided by studies of the incorporation of acetate into ergosterol (103) in yeast[153] and into eburicoic acid (118a) in *Polyporus sulphureus*.[154] Moreover, [1-³H₂]farnesyl pyrophosphate gives eburicoic acid with tritium localized at C-11 and C-12.[155] Finally, squalene oxide is specifically incorporated into fusidic acid (p. 266) and into lanosterol by a cell-free extract of *Phycomyces blakesleeanus*.[156]

B.5.a.ii. *The C_{28} and C_{31} Compounds*

A feature of the fungal triterpenes and steroids is the frequent occurrence of an extra methyl or methylene group at C-24 to give either C_{28} compounds, e.g. ergosterol (**103**), or C_{31} compounds, e.g. eburicoic acid (**118a**). The incorporation of formate into ergosterol[157] and into eburicoic acid,[158] and of the methyl group of methionine[159] or of *S*-adenosylmethionine[160] into ergosterol, shows that the extra carbon atom is derived from the C_1-pool (this was the first demonstration of transmethylation from methionine onto

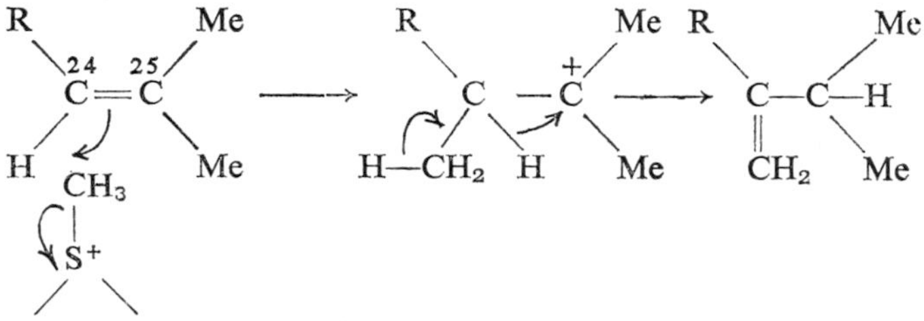

Scheme 16: The mechanism of the introduction of the "extra" carbon at C-24 of ergosterol and eburicoic acid

carbon). A purified *S*-adenosylmethionine:Δ^{24}-sterol methyl transferase, which will convert zymosterol (**95**) to fecosterol (**102**) in stoichiometric amount, has been obtained from yeast.[161] During the biosynthesis of ergosterol one of the hydrogen atoms of the methyl group of methionine is lost[162] and a hydrogen atom migrates from C-24 to C-25 of the side-chain.[163,164] These facts can be accommodated by the mechanism of Scheme 16. Support for a 24-methylene intermediate is provided by the incorporation of labelled 24-methylenedihydrolanosterol (**121a**) into ergosterol in yeasts[165,166] and into eburicoic acid in *P. sulphureus*.[165] The steps in the conversion of the 24-methylene side-chain to that of ergosterol and the timing of the introduction of the methylene group are discussed in Section B.5.b.ii below.

B.5.b. *Fungal Triterpenes and Steroids*

The compounds are classified according to their carbon skeletons. Thus we shall deal with (i) the C_{27} sterols, (ii) the C_{28} sterols related to ergosterol (C_{28} sterols resulting from retention of one of the methyl groups of lanosterol are discussed in Section B.5.b.iv), (iii) the C_{30} tetracyclic triterpenes, (iv) the C_{31} tetracyclic triterpenes, (v) miscellaneous triterpenes and steroids, (vi) the fusidanes and (vii) viridin and related compounds. The last two groups have skeletons of types so far encountered only in fungal products.

B.5.b.i. C_{27} *Compounds*

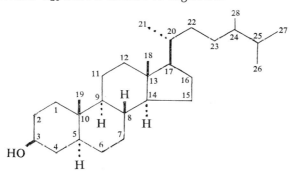

95

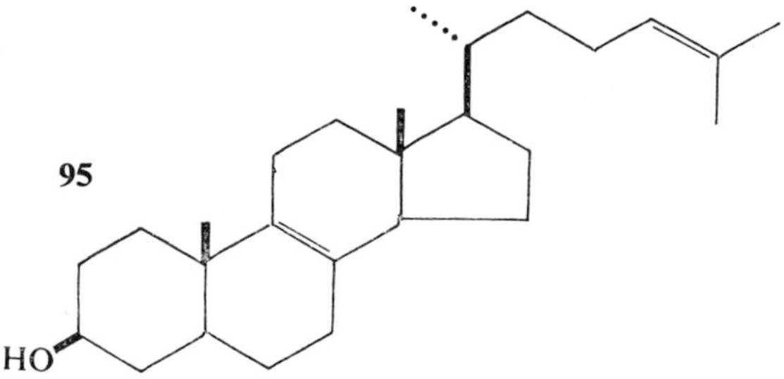

zymosterol
Yeast[167,171]

96

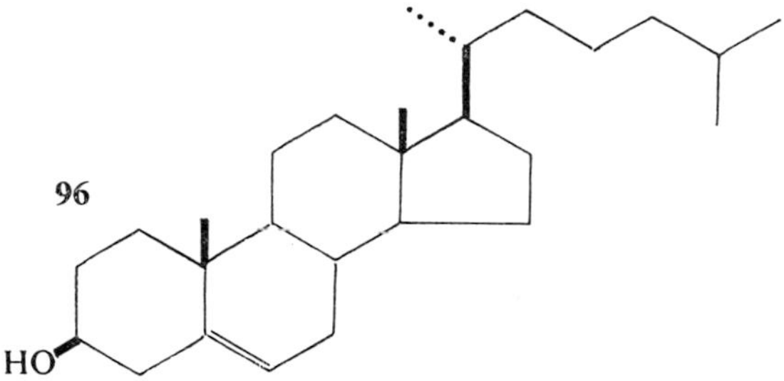

cholesterol
Penicillium funiculosum[168]

B.5.b.ii. C_{28} *Sterols Related to Ergosterol*

		Position of double bonds	
97	fungisterol	7	widespread (see ref. 169)
98	22,23-dihydroergosterol	5,7	yeasts[170,171]
99	5,6-dihydroergosterol	7,22	yeasts, *Claviceps purpurea*,[172,173] *Fomes applanatus*[174] *Polyporus pinicola*[169]
100	episterol	7,24(28)	yeasts[170]
101	ascosterol	8,23	yeasts[170,171]
102	fecosterol	8,24(28)	yeasts[170,171]
103	ergosterol	5,7,22	widespread
104	ergosta-5,7,24(28)-trien-3-ol	5,7,24(28)	*Phycomyces blakesleeanus*[175]
105	14-dehydroergosterol	5,7,14,22	*Aspergillus niger*[176]
106	24(28)-dehydroergosterol	5,7,22,24(28)	yeasts[177]

107

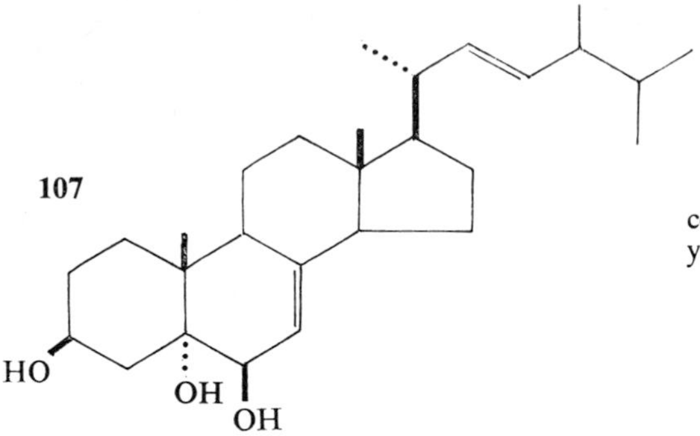

cerevisterol
yeast, *A. phalloides*[178,179]

108

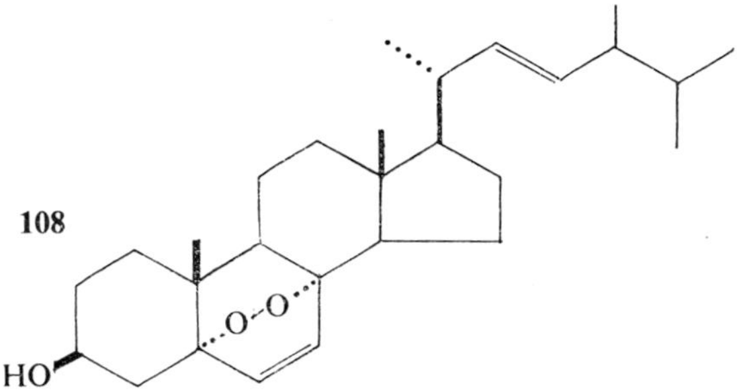

ergosterol peroxide
Aspergillus fumigatus,[180] and others

109

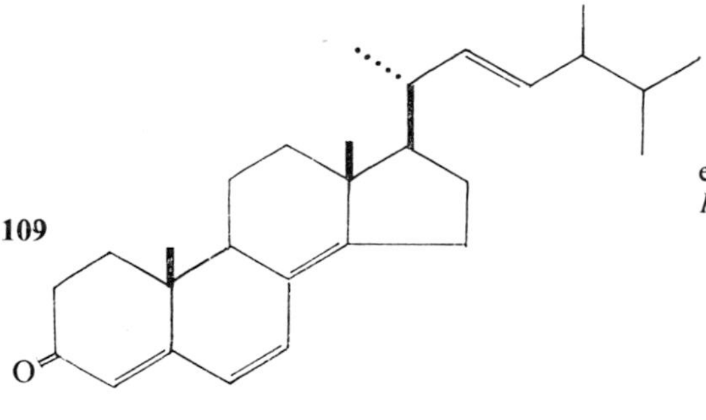

ergosta-4,6,8(14),22-tetraen-3-one
Fomes officinalis[181]

110

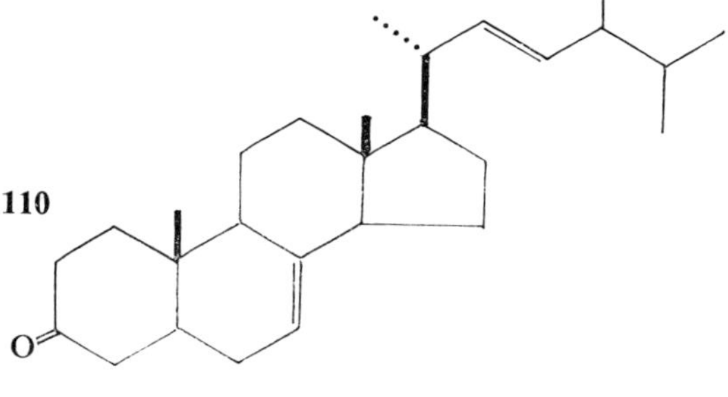

ergosta-7,22-dien-3-one
F. fomentarius[182]

Ergosterol (**103**) was first isolated from ergot (*Claviceps purpurea*) and has since been found in a wide variety of fungi. It is particularly abundant in yeasts, where it often occurs with related minor components. Ergosterol peroxide, too, is frequently obtained from fungi, and it has been suggested that it is an artefact.[183]

The biosynthesis of ergosterol in yeasts has been extensively studied. We have already discussed the origin of C-28, and two other problems merit comment—the timing of the introduction of C-28 in relation to the loss of the methyl groups of lanosterol, and the elaboration of the side-chain and of ring B.

The frequent occurrence of C_{31} compounds and the observation[184] that squalene and lanosterol, but not zymosterol, are converted into ergosterol by cell-free yeast homogenates suggest that methylation of lanosterol occurs before loss of the methyl groups, and this is supported by the incorporation, referred to above, of 24-methylenedihydrolanosterol (**121a**) into ergosterol. On the other hand, Katsuki and Bloch[185] have obtained results with cell-free extracts which suggest that in the biosynthesis of ergosterol, methylation occurs at the C_{29} level, and it may be that more than one pathway is available between lanosterol and ergosterol. That this is so is also suggested by the observation[186] that 4α-methyl-24-methylene-24,25-dihydrozymosterol (**124**), 4α-methylzymosterol, and obtusifoliol (**111**) are all converted to ergosterol by growing yeast cultures.

111   112

The incorporation of 24-methyldihydrolanosterol and of ergost-7-en-3β-ol (**112**) into ergosterol shows that the presence of the 24-methylene group is not necessary for the introduction of the 22,23-double bond,[166] though a compound with the 22,24(28)-diene system is a more likely natural intermediate.[175,185] The formation of the 22,23-double bond involves a *cis* elimination of hydrogen atoms,[187] the stereochemistry of the process in

A. fumigatus being opposite to that involved in the biosynthesis of porifera-sterol in the phytoflagellate, *Ochromonas malhamensis*.

Goulston and Mercer[175] have shown that episterol (**100**) is converted to ergosterol in *Phycomyces blakesleeanus*, probably via the triene (**104**), and Akhtar and his colleagues have shown that ergosta-8,14-dien-3β-ol is converted to ergosterol,[188] and that the formation of the 5,6-double bond involves a *cis* loss of hydrogen and does not proceed by hydroxylation and dehydration.[189]

B.5.b.iii. *The C$_{30}$ Tetracyclic Triterpenes*

113

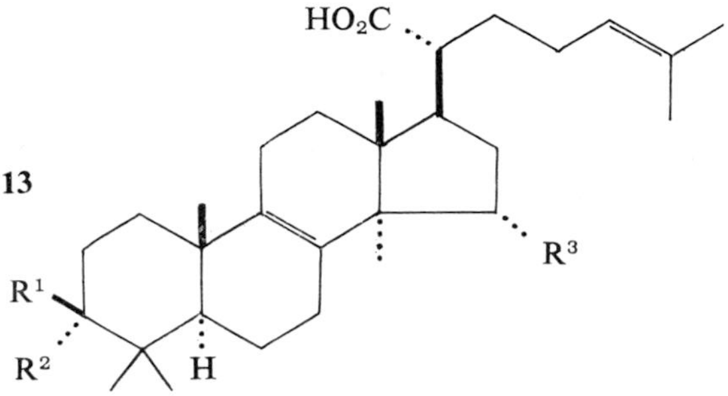

	R^1	R^2	R^3	
a	H	OH	H	3α-hydroxylanosta-8,24-dien-21-oic acid *Polyporus pinicola*[190]
b	OH	H	H	3β-hydroxylanosta-8,24-dien-21-oic acid *Trametes odorata*[191]
c		=O	H	pinicolic acid A *Polyporus pinicola*[192]
d	OH	H	OH	15α-hydroxytrametenolic acid *Lenzites trabea*[193]

114

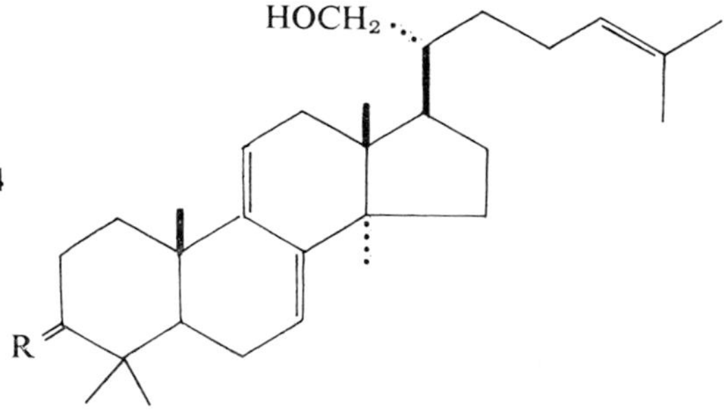

a: R = H, β-OH, lanosta-7,9(11),24-triene-3β,21-diol
b: R = O, 21-hydroxylanosta-7,9(11),24-triene-3-one
  Polyporus pinicola[169]

115

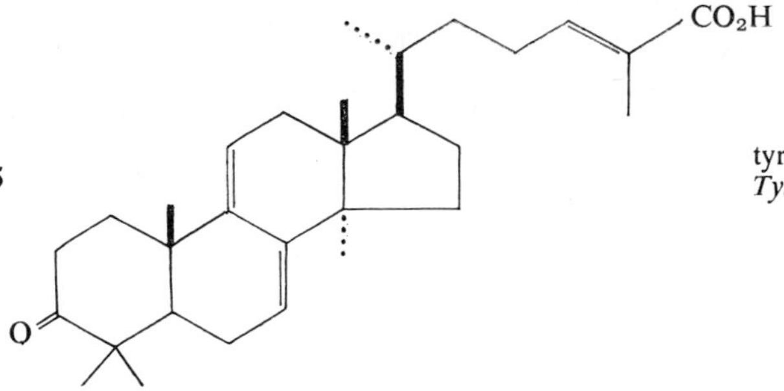

tyromycic acid
Tyromyces albidus[194]

116

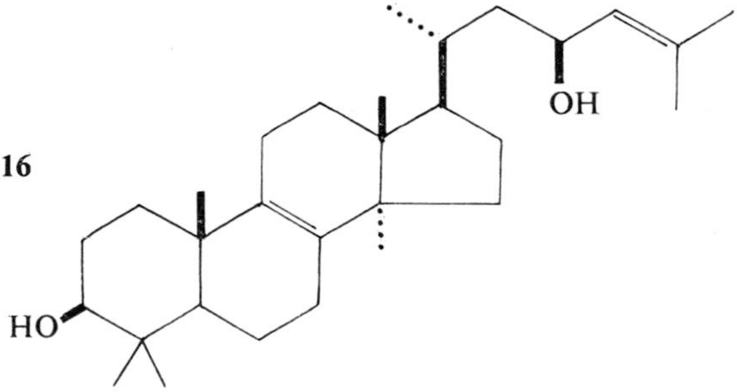

23S-hydroxylanosterol
Scleroderma aurantium[195]

117

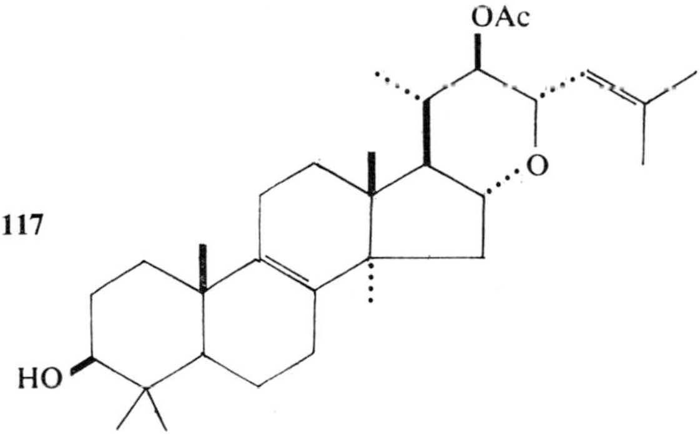

echinodol
Echinodontium tinctorium[196]

The fungal C_{30} tetracyclic triterpenes have been obtained only from Basidiomycetes, and are all based on the lanostane skeleton.

B.5.b.iv. *The C_{31} Tetracyclic Triterpenes*

118

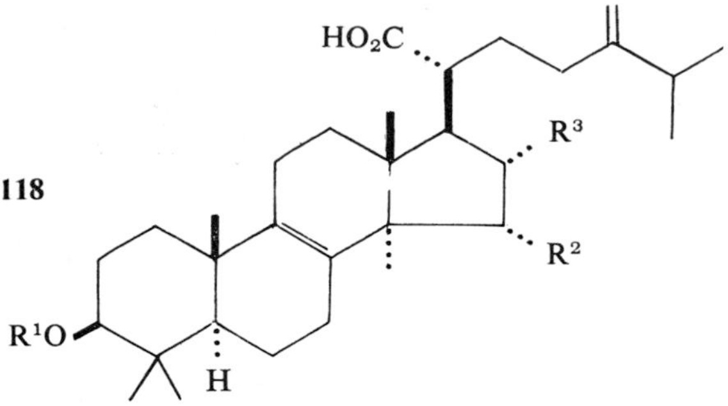

	R^1	R^2	R^3	
a	H	H	H	eburicoic acid *Polyporus anthracophilus, P. eucalyptorum, P. sulphureus, Fomes officinalis, Lentinus dactyloides,*[197] *Poria cocos, Polyporus hispidus,*[198] *Oospora astringens*[199]
b	Ac	H	H	3-*O*-acetyleburicoic acid *Polyporus anthracophilus*[197a]
c	H	H	OH	tumulosic acid *P. tumulosus, P. australiensis, P. betulinus, Poria cocos*[198]
d	H	OH	H	sulphurenic acid *Polyporus sulphureus*[200]

119

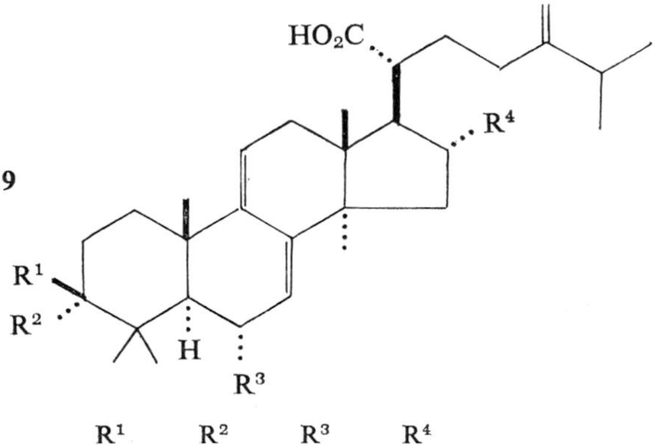

	R^1	R^2	R^3	R^4	
a	=O		H	OH	polyporenic acid C *Polyporus betulinus, P. benzoinus*[201]
b	=O		OH	OH	6α-hydroxypolyporenic acid C *Trametes feei*[202]
c	OH	H	H	H	dehydroeburicoic acid *Fomes officinalis, Lentinus dactyloides*[198]
d	OH	H	H	OH	dehydrotumulosic acid *Polyporus tumulosus, P. australiensis, P. betulinus, Poria cocos*[198]

120

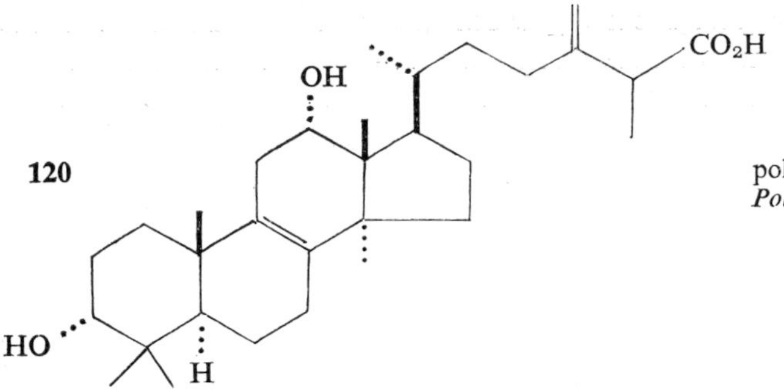

polyporenic acid A
Polyporus betulinus[203]

121

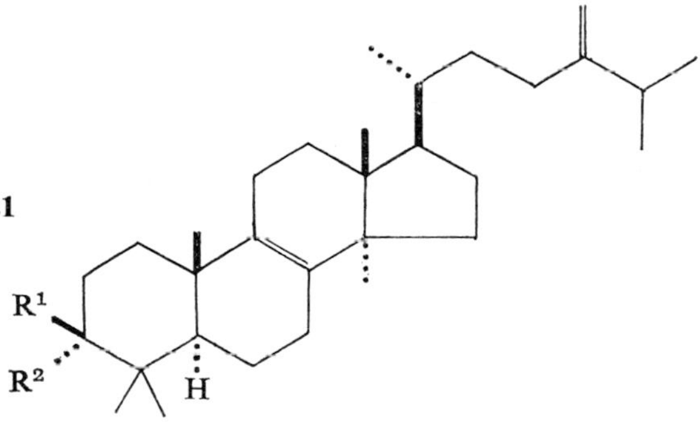

a: $R^1 = OH, R^2 = H$, 24-methylenedihydrolanosterol
    Phycomyces blakesleeanus[204]

b: $R^1 = H, R^2 = OH$, eburicol
    Fomes officinalis[205]

122

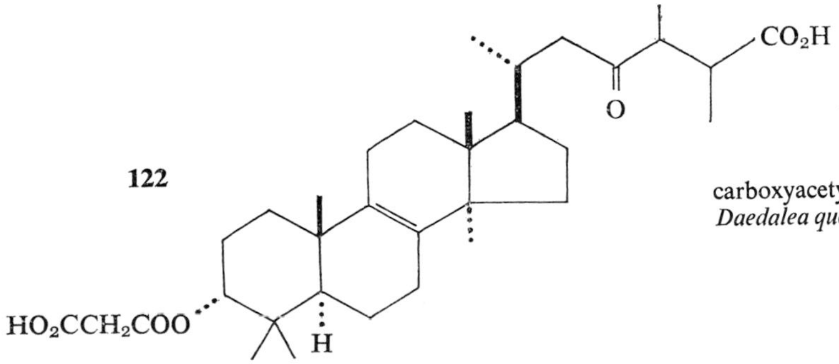

carboxyacetylquercinic acid
Daedalea quercinus[206]

Like the C_{30} compounds, the C_{31} tetracyclic triterpenes are based on the lanostane skeleton and, with the exception of 24-methylenedihydrolanosterol (**121a**), have been obtained from Basidiomycetes. Pachymic acid is the *O*-acetate of polyporenic acid B[207] which has been shown to be a mixture of tumulosic and dehydrotumulosic acids.[198]

The first isolation of the tetracyclic triterpenes from the fruiting bodies of Basidiomycetes involved a saponification step. Re-examination[208] of the products of *Polyporus betulinus* without saponification has shown that the major triterpene, polyporenic acid A (**120**) occurs as its 3α-esters with acetic, malonic, caproic and β-hydroxy-β-methylglutaric acids; the esters of the dibasic acids occur partly as mixed esters with methanol.

B.5.b.v. *Miscellaneous Triterpenes and Steroids*

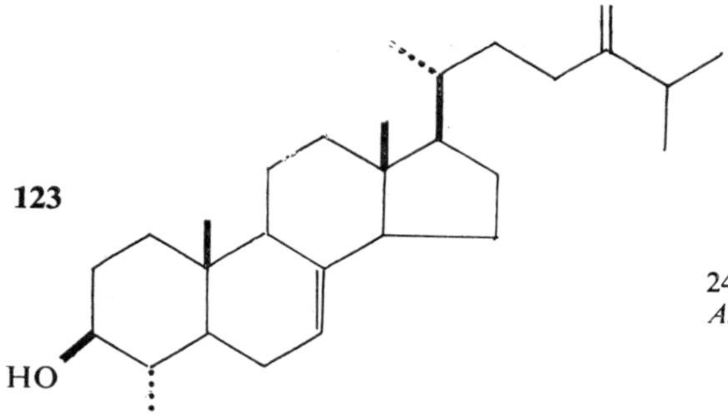

24-methylenelophenol
Aspergillus fumigatus[204]

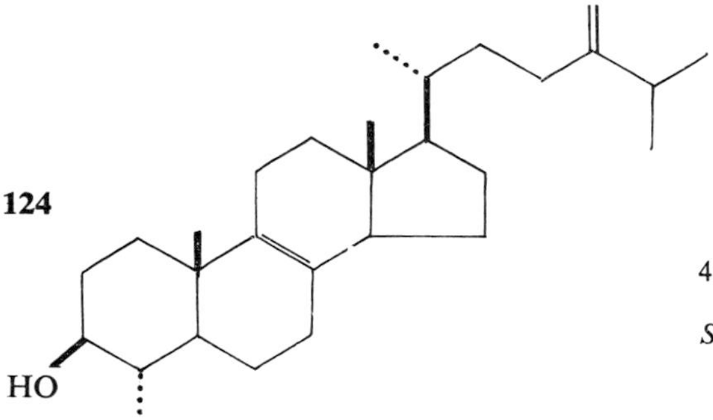

4α-methyl-24-methylene-
24,25-dihydrozymosterol
Saccharomyces cerevisiae[186]

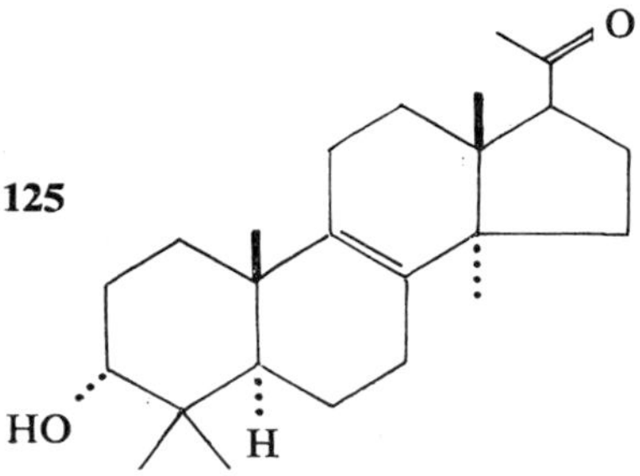

3α-hydroxy-4,4,14α-trimethyl-
Δ^8-5α-pregnen-20-one
Fomes officinalis[205]

126

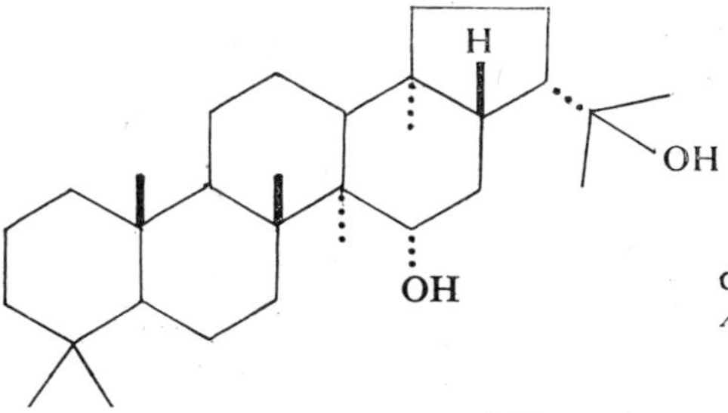

dustanin
Aspergillus sp.[209]

127

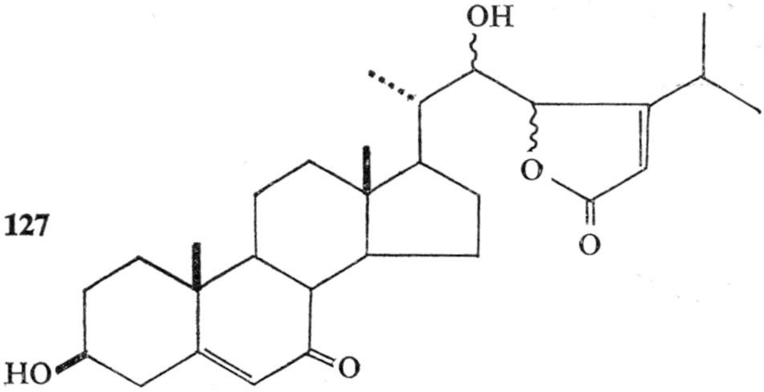

antheridiol
Achlya bisexualis[210]

Dustanin (**126**) is the only pentacyclic triterpene to have been isolated from a fungus, though several have been obtained from lichens,[211] where they are probably biosynthesized by the phycobiont, and from higher plants.

Antheridiol (**127**) is a sex hormone of the water mould *Achlya bisexualis*.[1] The presence of a C_2 side-chain at C-24 is rare among fungal products, though several sterols with 24-ethyl or 24-ethylidene groups (formed by two successive methylations) have been isolated from higher plants and marine organisms. Stigmasterol (**128**) has been detected (by g.l.c.) in the yeast *Debaromyces hansenii*,[212] and fucosterol (**129**) in several Phycomycetes (by combined g.l.c./mass spectroscopy).[213] Stigmast-7-enol and stigmast-7,24(28)-dienol have been isolated from the uredospores of flax rust,[214] but they are thought to be formed in the host plant.

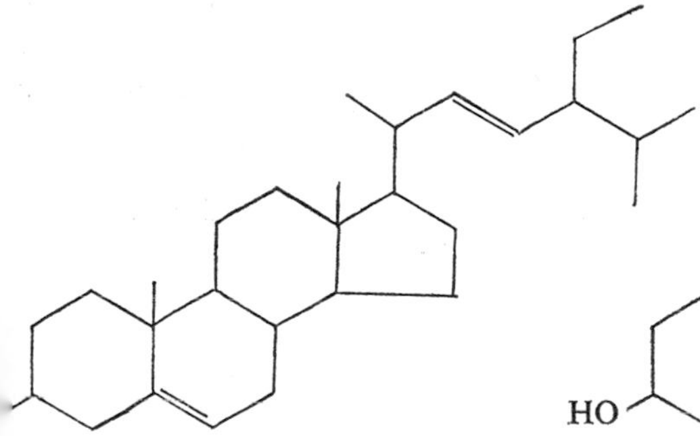

128

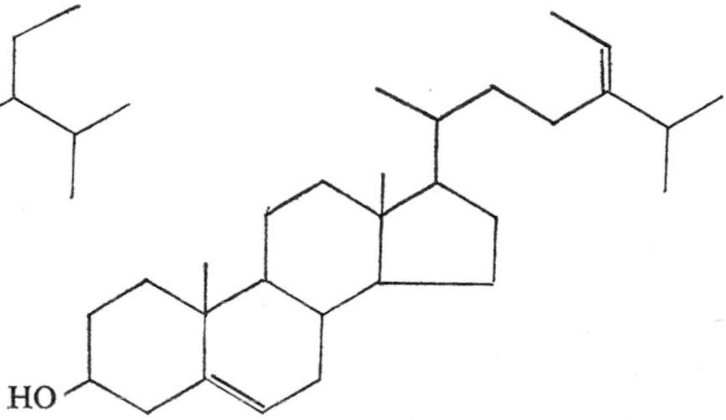

129

B.5.b.vi. *The Fusidanes and Protostanes*

130

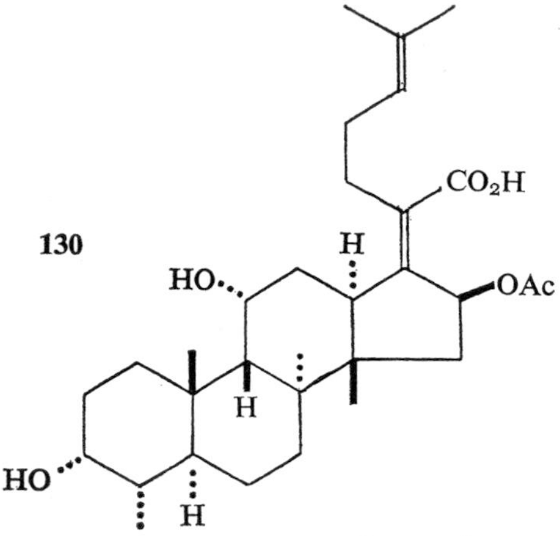

fusidic acid (ramycin)
Fusidium coccineum [215]
  Cephalosporium sp.,[216] *Mucor
  ramannianus*[217,218]

131

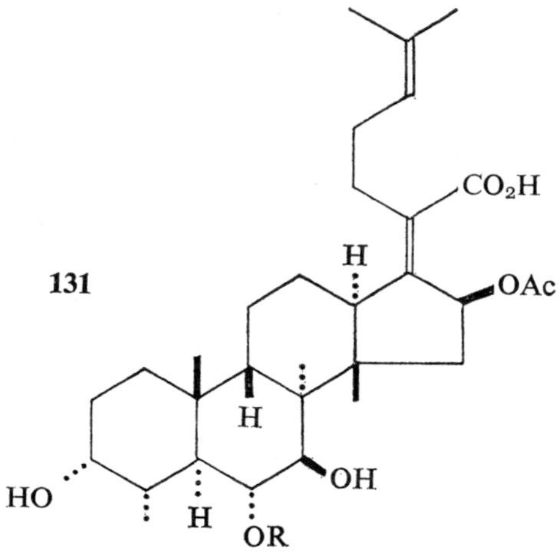

a: R = Ac, cephalosporin P$_1$
      Cephalosporium sp.[219]

b: R = H, desacetylcephalosporin P$_1$
      C. acremonium[220c]

132

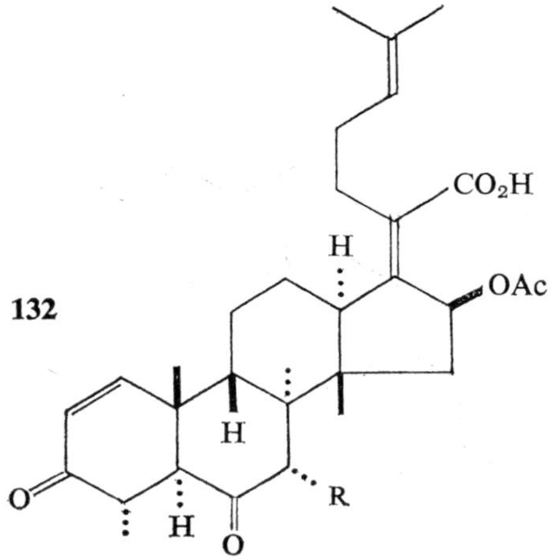

a: R = OAc, helvolic acid
      Aspergillus fumigatus mut. *helvo*
      C. caerulens,[222] *Emericellopsi.
      terricola*,[222] *Acrocylindrium ory:*

b: R = OH, helvolinic acid
      C. caerulens[224]

c: R = H, 7-desacetoxyhelvolic acid
      C. caerulens[224]

133

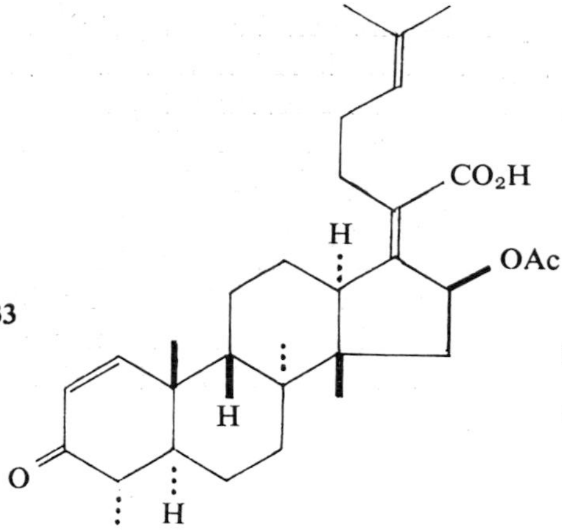

3-oxo-16β-acetoxyfusida-1,17(20)[16,21-*cis*],
24-trien-21-oic acid
C. caerulens[225]

134

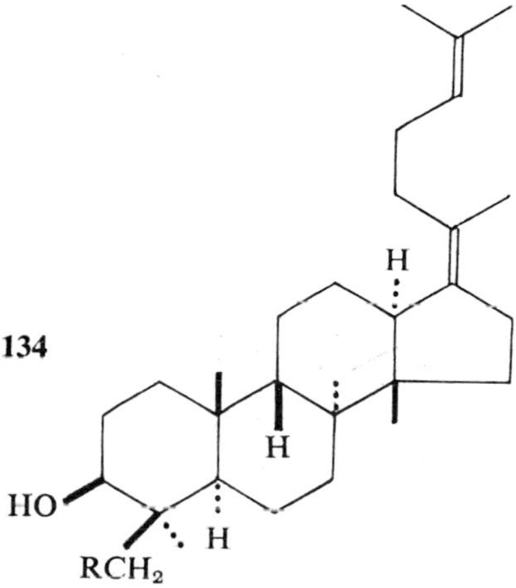

a: R = OH, 3β-hydroxy-4β-hydroxymethylfusida-
17(20)[16,21-*cis*],24-diene
C. caerulens[226]

b: R = H, 3β-hydroxy-4β-methylfusida-
17(20)[16,21-*cis*],24-diene
C. caerulens[227]

35

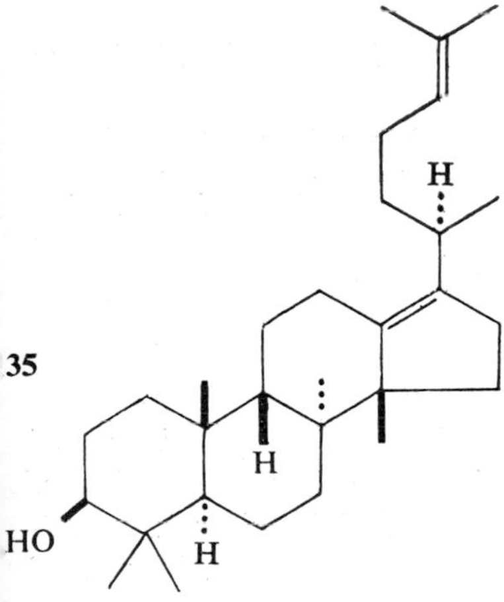

3β-hydroxyprotosta-13(17),24-diene
(fusisterol)
C. caerulens,[227] *F. coccineum*[83]

Fusidic acid (130), cephalosporin P_1 (131a) and helvolic acid (132a) are antibiotics produced by various Fungi Imperfecti. All have now been obtained from *Cephalosporium* spp., and *Fusidium coccineum*, from which fusidic acid was first isolated, is taxonomically related to moulds from the genera *Cephalosporium* and *Penicillium*.[228] Although it was the last to be isolated, fusidic acid was the first to have its structure completely defined by chemical[229] and X-ray crystallographic[230] methods. Early structural work on cephalosporin P_1[231] and helvolic acid[232] was handicapped by the omission of a methyl group subsequently detected by mass spectroscopy[233] and n.m.r. spectroscopy.[234] Structures (131a) and (132a) considered on

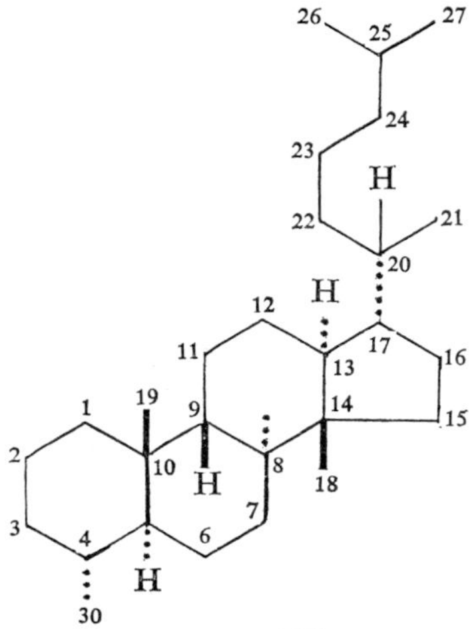

136

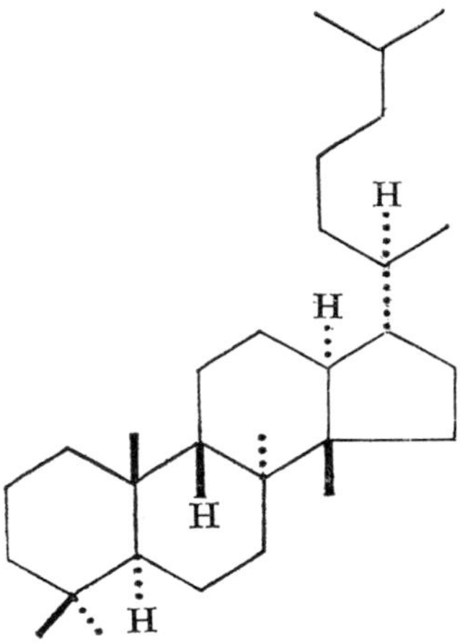

137

chemical grounds to be the most likely for cephalosporin P_1 and helvolic acid, have been confirmed by the correlation of helvolic acid with fusidic acid.[223,225]

The systematic nomenclature of the compounds is based upon the fusidane skeleton (136),[229] in which the α-configuration is assigned to the 17-side-chain on biogenetic considerations, and the S-configuration at C-20 is arbitrary. The fusidane skeleton has the stereochemistry of the intermediate (93) which has long been postulated[235] between squalene and lanosterol but which had not hitherto been exemplified. The structures of the fusidanes thus provide support for two features of lanosterol biosynthesis—the boat conformation of ring B during squalene cyclization and the 9β-hydrogen of the intermediate—for which no direct experimental evidence is available. Tritium-labelled squalene-2,3-oxide is converted to fusidic acid by *F. coccineum* without randomization of labelling,[236] a result of interest in view of the α-configuration of the 3-hydroxyl group of the fusidanes; presumably an isomerization occurs, as in the biosynthesis of the bile acids.

The structures of compounds (134) and (135), which are "prototype" steroids, have led to the suggestion[227] that their nomenclature be based on the "protostane" nucleus (137).

B.5.b.vii. *Viridin and Related Compounds*

138

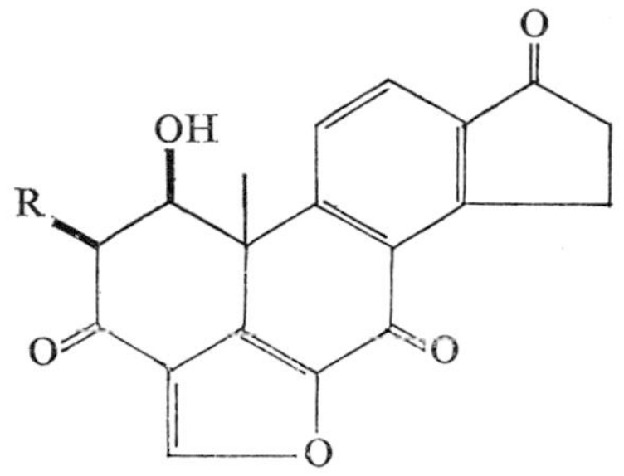

a: R = OMe, viridin   L[243]
        Gliocladium virens[237]

b: R = H, desmethoxyviridin
        Apiospora camptospora[238]

139

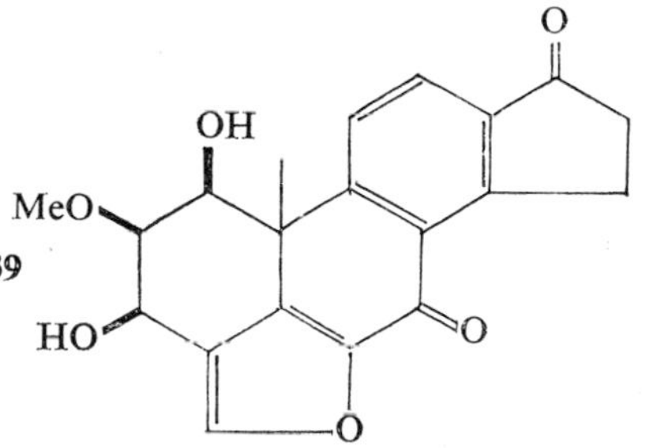

viridiol
Gliocladium virens[239]

140

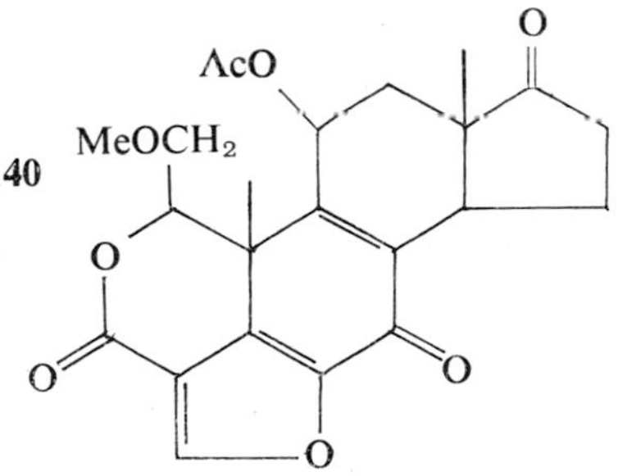

wortmannin
Penicillium wortmanni[240]

Structure (138a) for the highly antifungal metabolite viridin was deduced by a combination of chemical degradations[241] and spectroscopy.[242] The incorporation of [2-^{14}C]mevalonate into viridin gives the labelling pattern shown in (141), consistent with its biosynthesis via a steroidal, rather than a diterpenoid intermediate.[243]

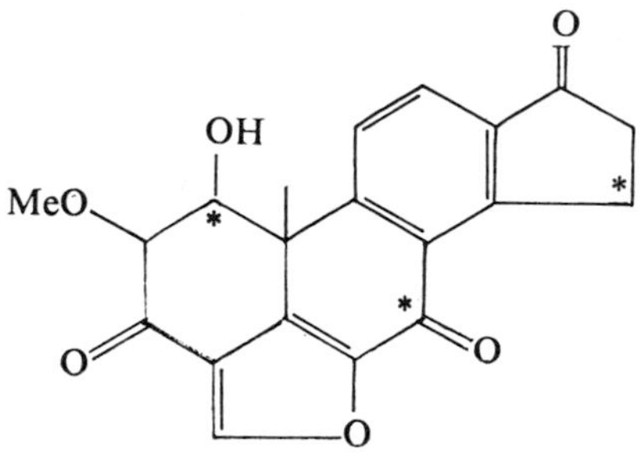

141

B.6. Carotenoids

The carotenoids are C_{40} pigments present in all green plants and in many species of algae, bacteria and fungi. They include open-chain compounds, e.g. lycopene (142), monocyclic compounds, e.g. γ-carotene (143) and compounds with two rings, e.g. β-carotene (144). β-Carotene and related compounds are important as precursors of vitamin A (145), and are produced commercially both as provitamins and as colouring matters for foodstuffs (vitamin A is a trace constituent of *Phycomyces*). Potential commercial processes for the production of β-carotene by fermentation are now available.

The carotenoids form a large and specialist subject, and here we shall comment briefly upon their biosynthesis and then discuss the role of the trisporic acids in carotenogenesis in some Phycomycetes. Lists of the carotenoids produced by fungi are given by Miller[13] and by Shibata *et al.*,[244] and a good general account of the subject is that of Davis.[245]

B.6.a. *Biosynthesis*

The biosynthesis of the carotenoids by head-to-head linkage of geranylgeranyl pyrophosphate resembles the biosynthesis of squalene from farnesyl pyrophosphate. However, whereas squalene formation is a reductive process involving NADPH, the formation of the carotenoids is not, and the first product is phytoene (146) in which the two geranylgeranyl units are linked through a *cis* double bond. In this process the protons derived from the 5S-position of mevalonate are lost from C-1 of geranylgeranyl pyrophosphate.[246] Stepwise dehydrogenation of phytoene leads to the carotenoids, but the stage at which the cyclizations occur has not been clearly established.

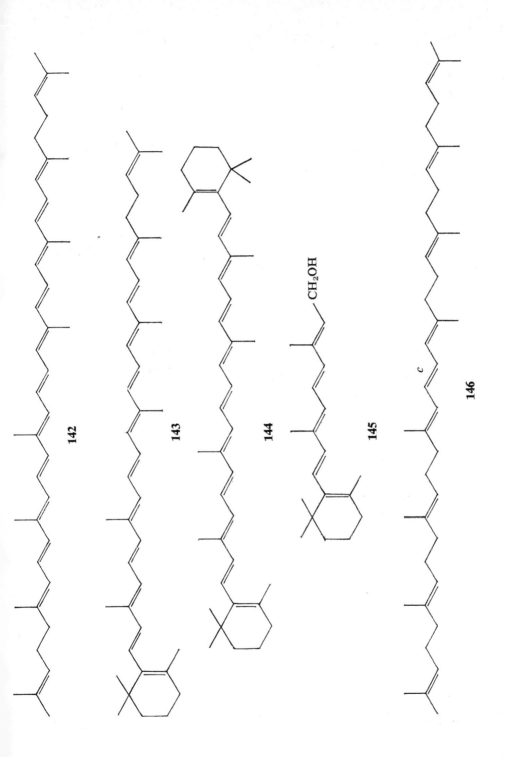

142

143

144

145

146

B.6.b. *The Trisporic Acids*

147

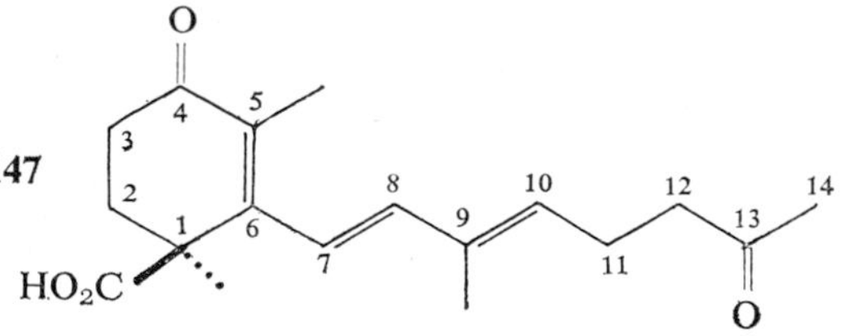

trisporic acid B†
Blakeslea trispora[247,248]
Mucor mucedo[249]

148

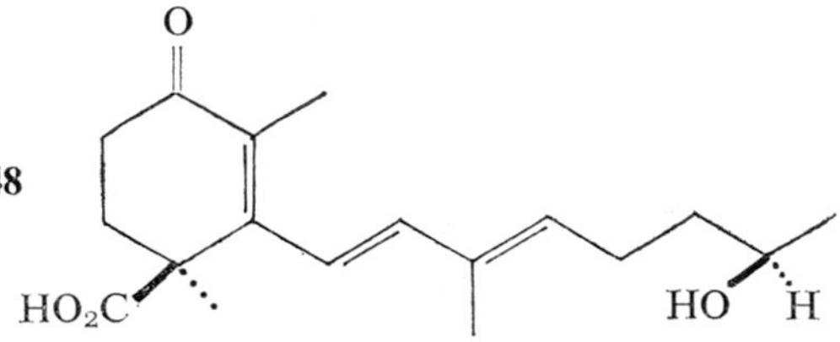

trisporic acid C†
B. trispora[247,250] *M. mucedo*[249]

149

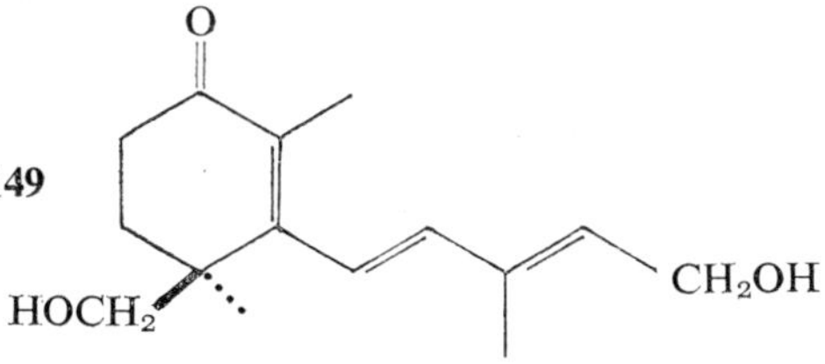

trisporone
B. trispora[251]

150

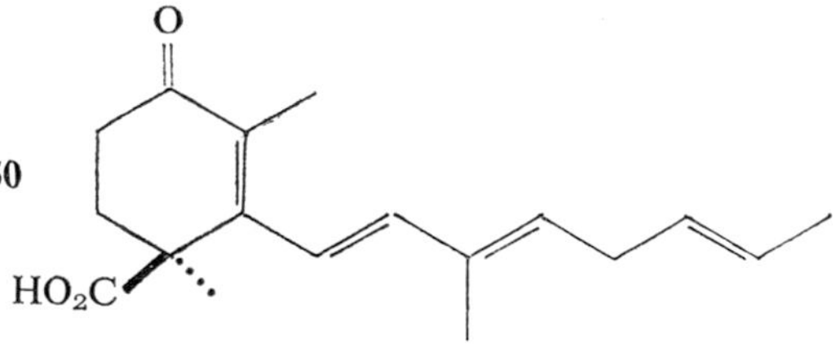

$\Delta^{12,13}$-anhydrotrisporic acid C†
B. trispora[252]

† These compounds occur as both *cis* and *trans* isomers about the 9,10-double bond.[249,252]

Some species of the Mucorales are heterothallic (p. 12), i.e. the presence of two types of mycelium (*plus* and *minus*) is required for the formation of sexual spores. Combined cultures also produce larger quantities of carotenoids than do *plus* or *minus* strains alone,[253] and this increased carotenogenesis was shown[247] to be mediated by the trisporic acids produced in combined cultures. It had previously been established that the formation of sexual spores was under hormonal control, and the hormones of *B. trispora* and *M. mucedo* have now been identified as the trisporic acids.[249,252,254] The trisporic acids are also produced when the *plus* and *minus* strains are grown in the same vessel but separated by a membrane,[254] so that the hormones themselves are produced in response to a soluble substance, and physical contact of the two strains is not required. Since the hormones are formed in combined cultures of *B. trispora plus* and *Zygorhynchus moelleri* (a homothallic species) but not in combined cultures of *B. trispora minus* and *Z. moelleri*,[254] it appears that the *plus* strain carries out the rate-limiting step in their biosynthesis.

The trisporic acids are formed from β-carotene by way of vitamin A (retinal), and the oxidation of the *gem*-dimethyl system is stereospecific.

REFERENCES

1. A. W. Barksdale, *Science*, 1969, **166**, 831.
2. C. G. Elliott and B. A. Knights, *J. Sci. Food Agric.*, 1969, **20**, 406, and refs. cited.
3. R. B. Clayton (ed.), "Methods in Enzymology. Vol. 15. Steroids and Terpenes". Academic Press, London and New York, 1969.
4. From the large literature of this subject, three reviews may be mentioned: J. H. Richards and J. B. Hendrickson, "The Biosynthesis of Steroids, Terpenes and Acetogenins", Benjamin, New York, 1964; J. D. Bu'Lock, "The Biosynthesis of Natural Products", McGraw-Hill, London, 1965; R. B. Clayton, *Quart. Rev.*, 1965, **19**, 168, 201.
5. G. Popjak and J. W. Cornforth, *Biochem. J.*, 1966, **101**, 553; J. W. Cornforth, *Angew. Chem., Intern. Ed.*, 1968, **7**, 903.
6. J. W. Cornforth, R. H. Cornforth, C. Donninger and G. Popjak, *Proc. Roy. Soc. London*, 1966, **163B**, 492.
7. J. E. Baldwin, R. E. Hackler and D. P. Kelly, *Chem. Commun.*, 1968, 537; *J. Amer. Chem. Soc.*, 1968, **90**, 4758; R. B. Bates and D. Feld, *Tetrahedron Lett.*, 1968, 417; J. E. Baldwin and D. P. Kelly, *Chem. Commun.*, 1968, 899; G. M. Blackburn and W. D. Ollis, *Chem. Commun.*, 1968, 1261.
8. H. C. Rilling, *J. Biol. Chem.*, 1966, **241**, 3233.
9. H. C. Rilling and W. W. Epstein, *J. Amer. Chem. Soc.*, 1969, **91**, 1041.
10. G. Popjak, J. Edmond, K. Clifford and V. Williams, *J. Biol. Chem.*, 1969, **244**, 1897.
11. E. Sprecher, *Planta Med.*, 1965, **13**, 418.
12. L. L. Medsker, D. Jenkins and J. F. Thomas, *Environ. Sci. Technol.*, 1969, **3**, 476.
13. M. W. Miller, "The Pfizer Handbook of Microbial Metabolites", McGraw-Hill, New York, 1961.
14. W. O. Godtfredsen and S. Vangedal (a) *Proc. Chem. Soc.*, 1964, 188; (b) *Acta Chem. Scand.*, 1965, **19**, 1088.
15. E. Härri, W. Loeffler, H. Stahelin, Ch. Stoll, Ch. Tamm and D. Wiesinger, *Helv. Chim. Acta*, 1962, **45**, 839; B. Böhner, E. Fetz, E. Härri, H. P. Sigg, Ch. Stoll and Ch. Tamm, *Helv. Chim. Acta*, 1965, **48**, 1079.
16. G. G. Freeman and R. I. Morrison, *Nature*, 1948, **162**, 30.
17. E. R. H. Jones and G. Lowe, *J. Chem. Soc.*, 1960, 3959.
18. B. Achilladelis and J. R. Hanson, *Phytochemistry*, 1968, **7**, 589.
19. J. R. Hanson, personal communication.

20. E. T. Glaz, E. Scheiber, J. Gyimesi, I. Horvath, K. Steczek, A. Szentirmai, and G. Bohus, *Nature*, 1959, **184**, 908; E. T. Glaz, E. Csanyi and J. Gyimesi, *Nature*, 1966, **212**, 617; J. Gyimesi and A. Melera, *Tetrahedron Lett.*, 1967, 1665.

21. B. Achilladelis and J. R. Hanson, *Phytochemistry*, 1969, **8**, 765.

22. P. W. Brian, A. W. Dawkins, J. F. Grove, H. G. Hemming, D. Lowe and G. L. F. Norris, *J. Exptl. Bot.*, 1961, **12**, 1.

23. (a) A. W. Dawkins, J. F. Grove and B. K. Tidd, *Chem. Commun.*, 1965, 27; (b) A. W. Dawkins, *J. Chem. Soc.*, 1966, 116; (c) B. K. Tidd, *J. Chem. Soc.*, 1967, 218; (d) E. Flury, R. Mauli and H. P. Sigg, *Chem. Commun.*, 1965, 26; (e) H. P. Sigg, R. Mauli, E. Flury and D. Hauser, *Helv. Chim. Acta*, 1965, **48**, 962; (f) J. F. Grove, *Chem. Commun.*, 1969, 1266.

24. J. R. Bamburg, N. V. Riggs and F. M. Strong, *Tetrahedron*, 1968, **24**, 3329.

25. J. R. Bamburg and F. M. Strong, *Phytochemistry*, 1969, **8**, 2405.

26. T. Tatsuno, Y. Fujimoto and Y. Morita, *Tetrahedron Lett.*, 1969, 2823; Y. Ueno, I. Ueno, T. Tatsuno, K. Ohokubo and H. Tsunoda, *Experientia*, 1969, **25**, 1062.

27. J. F. Grove, *J. Chem. Soc. C*, 1970, 378.

28. M. Okuchi, M. Itoh, Y. Kaneko and S. Doi, *Agr. Biol. Chem. (Tokyo)*, 1968, **32**, 394.

29. Ch. Tamm and J. Gutzwiller, *Helv. Chim. Acta*, 1962, **45**, 1726; 1965, **48**, 157.

30. J. Gutzwiller and Ch. Tamm, *Helv. Chim. Acta*, 1965, **48**, 177.

31. E. Fetz, B. Böhner, and Ch. Tamm, *Helv. Chim. Acta*, 1965, **48**, 1669.

32. W. Zürcher and Ch. Tamm, *Helv. Chim. Acta*, 1966, **49**, 2594.

33. B. Böhner and Ch. Tamm, *Helv. Chim. Acta*, 1966, **49**, 2527.

34. B. Böhner and Ch. Tamm, *Helv. Chim. Acta*, 1966, **49**, 2547.

35. W. O. Godtfredsen, J. F. Grove and Ch. Tamm, *Helv. Chim. Acta*, 1967, **50**, 1666.

36. J. F. Grove, *J. Chem. Soc. C*, 1968, 810.

37. S. Abrahamsson and B. Nilsson, *Proc. Chem. Soc.*, 1964, 188.

38. J. Gutzwiller, R. Mauli, H. P. Sigg and Ch. Tamm. *Helv. Chim. Acta*, 1964, **47**, 2234.

39. A. T. McPhail and G. A. Sim, *Chem. Commun.*, 1965, 350.

40. G. Snatzke and Ch. Tamm, *Helv. Chim. Acta*, 1967, **50**, 1618.

41. P. de Mayo, E. Y. Spencer and R. W. White, *Can. J. Chem.*, 1961, **39**, 1608; 1963, **41**, 2996.

42. S. Tamura, A. Sakurai, K. Kainuma and M. Takai, *Agr. Biol. Chem. (Tokyo)*, 1963, **27**, 738; 1965, **29**, 216.

43. P. de Mayo, R. E. Williams and E. Y. Spencer, *Can. J. Chem.*, 1965, **43**, 1357.

44. P. de Mayo, J. R. Robinson, E. Y. Spencer and R. W. White, *Experientia*, 1962, **18**, 359.

45. D. C. Aldridge and W. B. Turner, *J. Chem. Soc. C*, 1970, 686.

46. P. de Mayo and R. E. Williams, *J. Amer. Chem. Soc.*, 1965, **87**, 3275.

47. M. Anchel, A. Hervey and W. J. Robbins, *Proc. Nat. Acad. Sci. U.S.*, 1950, **36**, 300.

48. T. C. McMorris, M. S. R. Nair and M. Anchel, *J. Amer. Chem. Soc.*, 1967, **89**, 4562.

49. S. Takahashi, H. Iinuma, T. Takita, K. Maeda and H. Umezawa, *Tetrahedron Lett.*, 1969, 4663.

50. T. C. McMorris and M. Anchel, *J. Amer. Chem. Soc.*, 1963, **85**, 831; 1965, **87**, 1594.

51. K. Nakanishi, M. Ohashi, M. Tada and Y. Yamada, *Tetrahedron*, 1965, **21**, 1231.
52. A. Ichihara, H. Shirahama and T. Matsumoto, *Tetrahedron Lett.*, 1969, 3965.
53. M. S. R. Nair, H. Takeshita, T. C. McMorris and M. Anchel, *J. Org. Chem.*, 1969, **34**, 240.
54. C. Basset, R. T. Sherwood, J. A. Kepler and P. B. Hamilton, *Phytopathology*, 1967, **57**, 1046; J. A. Kepler, M. E. Wall, J. E. Mason, C. Basset, A. T. McPhail and G. A. Sim, *J. Amer. Chem. Soc.*, 1967, **89**, 1260.
55. F. Kavanagh, A. Hervey and W. J. Robbins, *Proc. Nat. Acad. Sci. U.S.*, 1949, **35**, 343.
56. J. J. Dugan, P. de Mayo, M. Nisbet, J. R. Robinson and M. Anchel, *J. Amer. Chem. Soc.*, 1966, **88**, 2838.
57. N. G. Heatley, M. A. Jennings and H. W. Florey, *Brit. J. Exp. Pathol.*, 1947, **28**, 35.
58. F. W. Comer, F. McCapra, I. H. Qureshi, J. Trotter and A. I. Scott, *Chem. Commun.*, 1965, 310; *Tetrahedron*, 1967, **23**, 4761.
59. J. N. Ashley, B. C. Hobbs and H. Raistrick, *Biochem. J.*, 1937, **31**, 385.
60. D. H. R. Barton and N. H. Werstiuk, *Chem. Commun.*, 1967, 30.
61. D. S. Tarbell, R. M. Carman, D. D. Chapman, K. R. Huffmann and N. J. McCorkindale, *J. Amer. Chem. Soc.*, 1960, **82**, 1005.
62. A. J. Birch and S. F. Hussain, *J. Chem. Soc. C*, 1969, 1473.
63. (a) H. P. Sigg and H. P. Weber, *Helv. Chim. Acta*, 1968, **51**, 1395; (b) T. Sassa, H. Kaise and K. Munakata, *Agr. Biol. Chem. (Tokyo)*, 1970, **34**, 649.
64. W. H. Nutting, H. Rapaport and L. Machlis, *J. Amer. Chem. Soc.*, 1968, **90**, 6434.
65. H. Nishikawa, *Agr. Biol. Chem. (Tokyo)*, 1962, **26**, 696.
66. S. Natori, H. Ogawa and H. Nishikawa, *Chem. Pharm. Bull. (Tokyo)*, 1964, **12**, 236.
67. S. Natori, Y. Inouye and H. Nishikawa, *Chem. Pharm. Bull. (Tokyo)*, 1967, **15**, 380.
68. E. Heilbronner and R. W. Schmid, *Helv. Chim. Acta*, 1954, **37**, 2018.
69. F. Sorm, V. Benesova and V. Herout, *Chem. Listy*, 1953, **47**, 1856.
70. D. J. Bertelli and J. H. Crabtree, *Tetrahedron*, 1968, **24**, 2079.
71. R. Bentley and D. Chen, *Phytochemistry*, 1969, **8**, 2171.
72. T. Goto, H. Kakisawa and Y. Hirata, *Tetrahedron*, 1963, **19**, 2079.
73. G. A. Ellestad, R. H. Evans Jr. and M. P. Kunstmann, *Tetrahedron*, 1969, **25**, 1323.
74. Y. Nawata, K. Ando, G. Tamura, K. Arima and Y. Iitaka, *J. Antibiotics (Tokyo)*, 1969, **22**, 511.
75. S. Nozoe and K. T. Suzuki, *Tetrahedron Lett.*, 1969, 2457.
76. S. Nozoe, K. T. Suzuki and S. Okuda, *Tetrahedron Lett.*, 1968, 3643.
77. K. Hirai, S. Nozoe, K. Tsuda, Y. Iitaka, K. Ishibashi and M. Shirasaka, *Tetrahedron Lett.*, 1967, 2177.
78. K. Kawashima, K. Nakanishi, M. Tada and H. Nishikawa, *Tetrahedron Lett.*, 1964, 1227.
79. B. E. Cross, R. H. B. Galt, J. R. Hanson, P. J. Curtis, J. F. Grove and A. Morrison, *J. Chem. Soc.*, 1963, 2937.
80. J. Polonsky, Z. Baskevitch, N. Cagnoli Bellavita and P. Ceccherelli, *Chem. Commun.*, 1968, 1404; *Bull. Soc. Chim. France*, 1970, 1912.
81. F. Kavanagh, A. Hervey and W. J. Robbins, *Proc. Nat. Acad. Sci. U.S.*, 1951, **37**, 570.

82. F. Kavanagh, A. Hervey and W. J. Robbins, *Proc. Nat. Acad. Sci. U.S.*, 1952, **38**, 555.
83. D. Arigoni, *Gazz. Chim. Ital.*, 1962, **92**, 884; *Pure Appl. Chem.*, 1968, **17**, 331.
84. A. J. Birch, D. W. Cameron, C. W. Holzapfel and R. W. Rickards, *Chem. and Ind.*, 1963, 374; A. J. Birch, C. W. Holzapfel and R. W. Rickards, *Tetrahedron*, 1966, **Suppl. 8**, 359.
85. G. A. Ellestad, R. H. Evans Jr. and M. P. Kunstmann, *J. Amer. Chem. Soc.*, 1969, **91**, 2134.
86. C. H. Calzadilla, G. Ferguson, S. A. Hutchinson and N. J. McCorkindale, unpublished.
87. G. G. Freeman and R. I. Morrison, *Biochem. J.*, 1948, **43**, xxiii; S. E. Michael, *Biochem. J.*, 1948, **43**, xxiv.
88. G. A. Ellestad, B. Green, A. Harris, W. B. Whalley and H. Smith, *J. Chem. Soc.*, 1965, 7246.
89. A. Robertson, W. R. Smithies and E. Tittensor, *J. Chem. Soc.*, 1949, 879.
90. A. Harris, A. Robertson and W. B. Whalley, *J. Chem. Soc.*, 1958 (a) 1799; (b) 1807.
91. A. J. Birch, R. W. Rickards, H. Smith, A. Harris and W. B. Whalley (a) *Proc. Chem. Soc.*, 1958, 223; (b) *Tetrahedron*, 1959, **7**, 241.
92. J. J. Britt and D. Arigoni, *Proc. Chem. Soc.*, 1958, 224.
93. C. W. Holzapfel and P. S. Steyn, *Tetrahedron*, 1968, **24**, 3321; A. J. Allison, J. D. Connolly and K. H. Overton, *J. Chem. Soc. C*, 1968, 2122.
94. W. B. Whalley, B. Green, D. Arigoni, J. J. Britt and C. Djerassi, *J. Amer. Chem. Soc.*, 1959, **81**, 5520; C. Djerassi, B. Green, W. B. Whalley and C. G. de Grazia, *J. Chem. Soc.*, 1966, 624.
95. A. I. Scott, D. W. Young, S. A. Hutchinson and N. S. Bhacca, *Tetrahedron Lett.*, 1964, 849.
96. B. Green, A. Harris and W. B. Whalley, *Chem. and Ind.*, 1958, 1369.
97. A. I. Scott, S. A. Sutherland, D. W. Young, L. Guglielmetti, D. Arigoni and G. A. Sim, *Proc. Chem. Soc.*, 1964, 19.
98. B. Achilladelis and J. R. Hanson, *J. Chem. Soc. C*, 1969, 2010.
99. B. Achilladelis and J. R. Hanson, *Phytochemistry*, 1968, **7**, 589.
100. C. W. Holzapfel, A. J. Birch and R. W. Rickards, *Phytochemistry*, 1969, **8**, 1009.
101. P. W. Brian, J. F. Grove and J. MacMillan, *Progr. Chem. Org. Nat. Prod.*, 1960, **18**, 350; J. F. Grove, *Quart. Rev.*, 1961, **15**, 56; J. F. Grove. *In* "Biochemistry of Industrial Microorganisms" (C. Rainbow and A. H. Rose, eds.), p. 320, Academic Press, New York and London, 1963; B. O. Phinney and C. A. West, *Ann. Rev. Plant. Physiol.*, 1960, **11**, 411.
102. B. E. Cross, R. H. B. Galt and J. R. Hanson, *Tetrahedron*, 1962, **18**, 451.
103. J. R. Hanson, *Tetrahedron*, 1966, **22**, 701.
104. R. H. B. Galt, *Tetrahedron*, 1968, **24**, 1337.
105. B. E. Cross and K. Norton, *J. Chem. Soc.*, 1965, 1570.
106. B. E. Cross, *J. Chem. Soc.*, 1966, 501.
107. R. H. B. Galt, *J. Chem. Soc.*, 1965, 3143.
108. J. R. Hanson, *Tetrahedron*, 1967, **23**, 733.
109. R. H. B. Galt, D. M. Harrison and J. MacMillan, *Tetrahedron Lett.*, 1968, 3137.
110. B. E. Cross, R. H. B. Galt and J. R. Hanson, *J. Chem. Soc.*, 1963, 2944.
111. E. P. Serebryakov, N. S. Kobrina, A. V. Simolin and V. F. Kucherov, *Chem. and Ind.*, 1968, 1770.
112. B. E. Cross, R. H. B. Galt and J. R. Hanson, *J. Chem. Soc.*, 1963, 3783.

113. B. E. Cross, R. H. B. Galt and J. R. Hanson, *J. Chem. Soc.*, 1963, 5052.
114. J. M. Lehn and S. Huneck, *Z. Naturforsch*, 1965, **20B**, 1013.
115. A. J. Birch, R. W. Rickards and H. Smith, *Proc. Chem. Soc.*, 1958, 192.
116. E. Wenkert, *Chem. and Ind.*, 1955, 282.
117. B. E. Cross, R. H. B. Galt and J. R. Hanson, *J. Chem. Soc.*, 1964, 295.
118. J. R. Hanson and A. F. White, *J. Chem. Soc. C*, 1969, 981.
119. I. Shechter and C. A. West, *J. Biol. Chem.*, 1969, **244**, 3200.
120. B. E. Cross and J. C. Stewart, *Tetrahedron Lett.*, 1968, 5195.
121. B. E. Cross and J. C. Stewart, *Tetrahedron Lett.*, 1968, 6321.
122. B. D. Cavell and J. MacMillan, *Phytochemistry*, 1967, **6**, 1151.
123. (a) R. H. B. Galt, *J. Chem. Soc.*, 1965, 3143; (b) B. E. Cross, R. H. B. Galt and K. Norton, *Tetrahedron*, 1968, **24**, 231.
124. J. E. Graebe, D. T. Dennis, C. D. Upper and C. A. West, *J. Biol. Chem.*, 1965, **240**, 1847.
125. A. J. Verbiscar, G. Cragg, T. A. Geissman and B. O. Phinney, *Phytochemistry*, 1967, **6**, 807.
126. D. T. Dennis and C. A. West, *J. Biol. Chem.*, 1967, **242**, 3293.
127. T. A. Geissman, A. J. Verbiscar, B. O. Phinney and G. Cragg, *Phytochemistry*, 1966, **5**, 933.
128. J. R. Hanson and A. F. White, *Chem. Commun.*, 1969, 410.
129. B. E. Cross, K. Norton and J. C. Stewart, *J. Chem. Soc. C*, 1968, 1054.
130. J. R. Hanson and A. F. White, *Chem. Commun.*, 1969, 1071.
131. Y. Iitaka, I. Watanabe, I. T. Harrison and S. Harrison, *J. Amer. Chem. Soc.*, 1968, **90**, 1092; T. Riós and S. Pérez C., *Chem. Commun.*, 1969, 214; T. Ríos and F. Gómez G., *Tetrahedron Lett.*, 1969, 2929.
132. K. Tsuda, S. Nozoe, M. Morisaki, K. Hirai, A. Itai, S. Okuda, L. Canonica, A. Fiecchi, M. Galli Kienle and A. Scala, *Tetrahedron Lett.*, 1967, 3369.
133. M. Orsenigo, *Phytopathol. Zeit.*, 1957, **29**, 189.
134. K. Ishibashi and M. Nakamura, *J. Agr. Chem. Soc. Japan*, 1958, **32**, 739.
135. K. Ishibashi, *J. Agr. Chem. Soc. Japan*, 1961, **35**, 323; 1962, **36**, 226; *J. Antibiotics (Tokyo)*, 1962, **15A**, 88. H. Ohkawa and T. Tamura, *Agr. Biol. Chem. (Tokyo)*, 1966, **30**, 285.
136. S. Nozoe, M. Morisaki, K. Tsuda, Y. Iitaka, N. Takahashi, S. Tamura, K. Ishibashi and M. Shirasaka, *J. Amer. Chem. Soc.*, 1965, **87**, 4968.
137. S. Nozoe, M. Morisaki, K. Fukushima and S. Okuda, *Tetrahedron Lett.*, 1968, 4457.
138. A. Ballio, E. B. Chain, P. de Leo, B. F. Erlanger, M. Mauri and A. Tonolo, *Nature*, 1964, **203**, 297.
139. L. Canonica, A. Fiecchi, M. Galli Kienle and A. Scala, *Tetrahedron Lett.*, 1966, 1211.
140. S. Nozoe, K. Hirai and K. Tsuda, *Tetrahedron Lett.*, 1966, 2211.
141. A Itai, S. Nozoe, K. Tsuda, S. Okuda, Y. Iitaka and Y. Nakagama, *Tetrahedron Lett.*, 1967, 4111.
142. S. Nozoe, A. Itai, K. Tsuda and S. Okuda, *Tetrahedron Lett.*, 1967, 4113.
143. A. Ballio, M. Brufani, C. G. Casinovi, S. Cerrini, W. Fedeli, R. Pellicciari, B. Santurbano and A. Vaciago, *Experientia*, 1968, **24**, 631.
144. K. D. Barrow, D. H. R. Barton, E. B. Chain, C. Conlay, T. V. Smale, R. Thomas and E. S. Waight, *Chem. Commun.*, 1968, 1195; E. Hough, M. B. Hursthouse, S. Neidle and D. Rogers, *Chem. Commun.*, 1968, 1197; K. D. Barrow, D. H. R. Barton, E. B. Chain, U. F. W. Ohnsorge and R. Thomas, *Chem. Commun.*, 1968, 1198.

145. L. Canonica, A. Fiecchi, M. Galli Kienle, B. M. Ranzi and A. Scala, *Tetrahedron Lett.*, 1966, 3035.
146. S. Nozoe and M. Morisaki, *Chem. Commun.*, 1969, 1319.
147. S. Nozoe, M. Morisaki, K. Tsuda and S. Okuda, *Tetrahedron Lett.*, 1967, 3365.
148. L. Canonica, A. Fiecchi, M. Galli Kienle, B. M. Ranzi, A. Scala, T. Salvatori and E. Pella, *Tetrahedron Lett.*, 1967, 3371.
149. L. Canonica, A. Fiecchi, M. Galli Kienle, B. M. Ranzi and A. Scala, *Tetrahedron Lett.*, 1967, 4657.
150. L. Canonica, A. Fiecchi, M. Galli Kienle, B. M. Ranzi and A. Scala, *Tetrahedron Lett.*, 1968, 275.
151. S. Nozoe, M. Morisaki, S. Okuda and K. Tsuda, *Tetrahedron Lett.*, 1968. 2347.
152. P. D. G. Dean, P. R. O. de Montellano, K. Bloch and E. J. Corey, *J. Biol. Chem.*, 1967, **242**, 3014; J. D. Willett, K. B. Sharpless, K. E. Lord, E. E. van Tamelen and R. B. Clayton, *J. Biol. Chem.*, 1967, **242**, 4182.
153. D. J. Hanahan and S. J. Wakil, *J. Amer. Chem. Soc.*, 1953, **75**, 273; W. G. Dauben and T. W. Hutton, *J. Amer. Chem. Soc.*, 1956, **78**, 2647; W. G. Dauben, T. W. Hutton, and G. A. Boswell, *J. Amer. Chem. Soc.*, 1959, **81**, 403.
154. W. G. Dauben and J. H. Richards, *J. Amer. Chem. Soc.*, 1956, **78**, 5329; W. G. Dauben, Y. Ban and J. H. Richards, *J. Amer. Chem. Soc.*, 1957, **79**, 968.
155. W. Lawrie, J. McLean, P. L. Pauson and J. Watson, *Chem. Commun.*, 1965, 623.
156. E. I. Mercer and M. W. Johnson, *Phytochemistry*, 1969, **8**, 2329.
157. H. Danielsson and K. Bloch, *J. Amer. Chem. Soc.*, 1957, **79**, 500.
158. W. G. Dauben, G. J. Fonken and G. A. Boswell, *J. Amer. Chem. Soc.*, 1957, **79**, 1000.
159. G. J. Alexander, A. M. Gold and E. Schwenk, *J. Amer. Chem. Soc.*, 1957, **79**, 2967; G. J. Alexander and E. Schwenk, *J. Amer. Chem. Soc.*, 1957, **79**, 4554.
160. L. W. Parks, *J. Amer. Chem. Soc.*, 1958, **80**, 2023.
161. J. T. Moore and J. L. Gaylor, *J. Biol. Chem.*, 1969, **244**, 6334.
162. E. Lederer, *Biochem. J.*, 1964, **93**, 449.
163. L. J. Goad, A. S. A. Hamman, A. Dennis and T. W. Goodwin, *Nature*, 1966, **210**, 1322.
164. M. Akhtar, P. F. Hunt and M. A. Parvez, *Chem. Commun.*, 1966, 565; *Biochem. J.*, 1967, **103**, 616.
165. D. H. R. Barton, D. M. Harrison and G. P. Moss, *Chem. Commun.*, 1966, 595.
166. M. Akhtar, M. A. Parvez and P. F. Hunt, *Biochem. J.*, 1966, **100**, 38C; 1969, **113**, 727.
167. I. Smedley-MacLean, *Biochem. J.*, 1928, **22**, 22.
168. Y. S. Chen and R. H. Haskins, *Can. J. Chem.*, 1963, **41**, 1647.
169. T. G. Halsall and G. C. Sayer, *J. Chem. Soc.*, 1959, 2031.
170. H. Wieland, F. Rath and H. Hesse, *Ann. Chem.*, 1941, **548**, 34.
171. D. H. R. Barton and J. D. Cox, *J. Chem. Soc.*, 1949, 214.
172. H. Wieland and W. Benend, *Ann. Chem.*, 1943, **554**, 1.
173. D. H. R. Barton and J. D. Cox, *J. Chem. Soc.*, 1948, 1354.
174. G. R. Pettit and J. C. Knight, *J. Org. Chem.*, 1962, **27**, 2696.

175. G. Goulston and E. I. Mercer, *Phytochemistry*, 1969, **8**, 1945.
176. D. H. R. Barton and T. Bruun, *J. Chem. Soc.*, 1951, 2728.
177. O. N. Breivak, J. L. Owades and R. F. Light, *J. Org. Chem.*, 1954, **19**, 1734; K. Petzoldt, M. Kühne, E. Blanke, K. Kieslich and E. Kaspar, *Ann. Chem.*, 1967, **709**, 203.
178. H. Wieland and G. Coutelle, *Ann. Chem.*, 1941, **548**, 270.
179. G. H. Alt and D. H. R. Barton, *J. Chem. Soc.*, 1954, 1356.
180. P. Wieland and V. Prelog, *Helv. Chim. Acta*, 1947, **30**, 1028.
181. K. E. Schulte, G. Rücker and H. Fachmann, *Tetrahedron Lett.*, 1968, 4763.
182. G. R. Pettit and J. C. Knight, *J. Org. Chem.*, 1962, **27**, 2696.
183. H. K. Adam, I. M. Campbell and N. J. McCorkindale, *Nature*, 1967, **216**, 397.
184. E. Schwenk and G. J. Alexander, *Arch. Biochem. Biophys.*, 1958, **76**, 65.
185. H. Katsuki and K. Bloch, *J. Biol. Chem.*, 1967, **242**, 222.
186. D. H. R. Barton, D. M. Harrison and D. A. Widdowson, *Chem. Commun.*, 1968, 17.
187. T. Bimpson, L. J. Goad and T. W. Goodwin, *Chem. Commun.*, 1969, 297.
188. M. Akhtar, W. A. Brooks and I. A. Watkinson, *Biochem. J.*, 1969, **115**, 135.
189. M. Akhtar and M. A. Parvez, *Biochem. J.*, 1968, **108**, 527.
190. J. J. Beereboom. H. Fazakerley and T. G. Halsall, *J. Chem. Soc.*, 1957, 3437.
191. T. G. Halsall, R. Hodges and G. C. Sayer, *J. Chem. Soc.*, 1959, 2036.
192. J. M. Guider, T. G. Halsall and E. R. H. Jones, *J. Chem. Soc.*, 1954, 4471.
193. W. Lawrie, J. McLean and J. Watson, *J. Chem. Soc. C*, 1967, 1776.
194. A. Gaudemer, J. Polonsky, R. Gmelin, H. K. Adam and N. J. McCorkindale, *Bull. Soc. Chim. France*, 1967, 1844.
195. N. Entwistle and A. D. Pratt, *Tetrahedron*, 1968, **24**, 3949; 1969, **25**, 1449.
196. F. T. Bond, D. S. Fullerton, L. A. Sciuchetti and P. Catalfomo, *J. Amer. Chem. Soc.*, 1966, **88**, 3882.
197. (a) R. M. Gascoigne, J. S. E. Holker, B. J. Ralph and A. Robertson, *J. Chem. Soc.*, 1951, 2346; (b) F. N. Lahey and P. H. A. Strasser, *J. Chem. Soc.*, 1951, 873.
198. L. A. Cort, R. M. Gascoigne, J. S. E. Holker, B. J. Ralph, A. Robertson and J. J. H. Simes, *J. Chem. Soc.*, 1954, 3713.
199. I. Yamamoto, *Agr. Biol. Chem. (Tokyo)*, 1961, **25**, 400.
200. J. Fried, P. Grabowich, E. F. Sabo and A. I. Cohen, *Tetrahedron*, 1964, **20**, 2297.
201. A. Bowers, T. G. Halsall, E. R. H. Jones and A. J. Lemin, *J. Chem. Soc.*, 1953, 2548; A. Bowers, T. G. Halsall and G. C. Sayer, *J. Chem. Soc.*, 1954, 3070.
202. J. J. H. Simes, M. Wootton, B. J. Ralph and J. T. Pinhey, *Chem. Commun.*, 1969, 1150.
203. T. G. Halsall and R. Hodges, *J. Chem. Soc.*, 1954, 2385.
204. G. Goulston, L. J. Goad and T. W. Goodwin, *Biochem. J.*, 1967, **102**, 15C.
205. W. W. Epstein and G. van Lear, *J. Org. Chem.*, 1966, **31**, 3434.
206. H. K. Adam, T. A. Bryce, I. M. Campbell, N. J. McCorkindale, A. Gaudemer, R. Gmelin and J. Polonsky, *Tetrahedron Lett.*, 1967, 1461.
207. S. Shibata, S. Natori, K. Fujita, I. Kitagawa and K. Watanabe, *Chem. Pharm. Bull. (Tokyo)*, 1958, **6**, 608.
208. T. A. Bryce I. M. Campbell and N. J. McCorkindale, *Tetrahedron*, 1967, **23**, 3427.
209. Y. Tsuda and K. Isobe, *Tetrahedron Lett.*, 1965, 3337.

210. T. C. McMorris and A. W. Barksdale, *Nature*, 1967, **215**, 320; G. P. Arsenault, K. Biemann, A. W. Barksdale and T. C. McMorris, *J. Amer. Chem. Soc.*, 1968, **90**, 5635.
211. S. Huneck. *In* "Progress in Phytochemistry" (L. Reinhold and Y. Liwschitz, eds.), Vol. I. Interscience, London, 1968.
212. E. Merdinger and E. M. Devine, *J. Bacteriol.*, 1965, **89**, 1488.
213. N. J. McCorkindale, S. A. Hutchinson, B. A. Pursey, W. T. Scott and R. Wheeler, *Phytochemistry*, 1969, **8**, 861.
214. L. L. Jackson and D. S. Frear, *Phytochemistry*, 1968, **7**, 651.
215. W. O. Godtfredsen, S. Jahnsen, H. Lorck, K. Roholt and L. Tybring, *Nature*, 1962, **193**, 987.
216. Belgian Patent 619,287.
217. P. J. van Dijck and P. de Somer, *J. Gen. Microbiol.*, 1958, **18**, 377.
218. H. Vanderhaeghe, P. van Dijck, and P. de Somer, *Nature*, 1965, **205**, 710.
219. H. S. Burton and E. P. Abraham, *Biochem. J.*, 1951, **50**, 168.
220. (a) T. G. Halsall, E. R. H. Jones, G. Lowe and C. E. Newall, *Chem. Commun.*, 1966, 685; (b) P. Oxley, *Chem. Commun.*, 1966, 729; (c) T. S. Chou, E. J. Eisenbraun and R. T. Rapala, *Tetrahedron*, 1969, **25**, 3341; (d) S. Okuda, S. Iswasaki, M. I. Sair, Y. Machida, A. Inone and K. Tsuda, *Tetrahedron Lett.*, 1967, 2295.
221. E. Chain, H. W. Florey, M. A. Jennings and T. I. Williams, *Brit. J. Exp. Pathol.*, 1943, **24**, 108.
222. S. Okuda, S. Iwasaki, K. Tsuda, Y. Sano, T. Hata, S. Udagawa, Y. Nakayama and H. Yamaguchi, *Chem. Pharm. Bull. (Tokyo)*, 1964, **12**, 121.
223. W. von Daehne, H. Lorch and W. O. Godtfredsen, *Tetrahedron Lett.*, 1968, 4843.
224. S. Okuda, Y. Nakayama and K. Tsuda, *Chem. Pharm. Bull. (Tokyo)*, 1966, **14**, 436.
225. S. Okuda, Y. Sato, T. Hattori and M. Wakabayashi, *Tetrahedron Lett.*, 1968, 4847.
226. S. Okuda, Y. Sato, T. Hattori and H. Igarashi, *Tetrahedron Lett.*, 1968, 4769.
227. T. Hattori, H. Igarashi, S. Iwasaki, and S. Okuda, *Tetrahedron Lett.*, 1969, 1023.
228. J. Nicot, *C.R. Acad. Sci., Paris, Ser. D.*, 1968, **267**, 290.
229. W. O. Godtfredsen, W. von Daehne, S. Vangedal, A. Marquet, D. Arigoni and A. Melera, *Tetrahedron*, 1965, **21**, 3505.
230. A. Cooper, *Tetrahedron*, 1966, **22**, 1379.
231. B. M. Baird, T. G. Halsall, E. R. H. Jones and G. Lowe, *Proc. Chem. Soc.*, 1961, 257.
232. N. L. Allinger and J. L. Coke, *J. Org. Chem.*, 1961, **26**, 4522, and references cited therein.
233. J. F. Lynch, J. M. Wilson, H. Budzikiewicz and C. Djerassi, *Experientia*, 1963, **19**, 211.
234. A. Melera, *Experientia*, 1963, **19**, 565.
235. A. Eschenmoser, L. Ruzicka, O. Jeger and D. Arigoni, *Helv. Chim. Acta*, 1955, **38**, 1890.
236. W. O. Godtfredsen, H. Lorck, E. E. van Tamelen, J. D. Willett and R. B. Clayton, *J. Amer. Chem. Soc.*, 1968, **90**, 208.
237. P. W. Brian and J. C. McGowan, *Nature*, 1945, **156**, 144.
238. B. F. Burrows, personal communication.
239. J. S. Moffatt, J. D. Bu'Lock and T. H. Yuen, *Chem. Commun.*, 1969, 839.

240. J. MacMillan, A. E. Vantsone and S. K. Yeboah, *Chem. Commun.*, 1968, 613.
241. J. F. Grove, J. S. Moffatt and E. B. Vischer, *J. Chem. Soc.*, 1965, 3803; P. McCloskey, *J. Chem. Soc.*, 1965, 3811; J. S. Moffatt, *J. Chem. Soc.*, 1965, 734.
242. J. F. Grove, J. S. Moffatt and P. McCloskey, *Chem. Commun.*, 1965, 343; *J. Chem. Soc.*, 1966, 743.
243. J. F. Grove, *J. Chem. Soc. C*, 1969, 549.
244. S. Shibata, S. Natori and S. Udagawa, "List of Fungal Products", p. 24. University of Tokyo Press, 1964.
245. J. B. Davis. In "Rodd's Chemistry of Carbon Compounds" (S. Coffey, ed.), 2nd edition, Vol. IIB, p. 231, Elsevier, Amsterdam, 1968.
246. M. J. Buggy, G. Britton and T. W. Goodwin, *Biochem. J.*, 1969, **114**, 641.
247. L. Caglioti, G. Cainelli, B. Camerino, R. Mondelli, A. Prieto, A. Quilico, T. Salvatori and A. Selva, *Chim. Ind. (Milan)*, 1964, **46**, 961.
248. L. Caglioti, G. Cainelli, B. Camerino, R. Mondelli, A. Prieto, A. Quilico, T. Salvatori and A. Selva, *Tetrahedron*, 1966, **Suppl. 7**, 175.
249. D. J. Austin, J. D. Bu'Lock and G. W. Gooday, *Nature*, 1969, **223**, 1178.
250. G. Cainelli, P. Grasselli, and A. Selva, *Chim. Ind. (Milan)*, 1967, **49**, 628.
251. G. Cainelli, B. Camerino, P. Grasselli, R. Mondelli, S. Morrocchi, A. Prieto, A. Quilico and A. Selva, *Chim. Ind. (Milan)*, 1967, **49**, 748.
252. T. Reschke, *Tetrahedron Lett.*, 1969, 3435.
253. C. W. Hesseltine, *U.S. Dept. Agr. Tech. Bull.*, 1961, **1245**, 33; D. M. Thomas and T. W. Goodwin, *Phytochemistry*, 1967, **6**, 355.
254. H. van den Ende, *J. Bacteriol.*, 1968, **96**, 1298.

CHAPTER 7

Secondary Metabolites Derived from Intermediates of the Tricarboxylic Acid Cycle

A. THE TRICARBOXYLIC ACID CYCLE

THE TRICARBOXYLIC acid (TCA) cycle (Scheme 1) completes the oxidation of glucose to carbon dioxide and water, and provides intermediates for amino-acid biosynthesis (Chapter 8) and for the biosynthesis of secondary metabolites which form the subject of this chapter. The withdrawal of intermediates for synthetic purposes is compensated for either by the operation of the glyoxylate cycle (Scheme 1), which provides a point of entry into the cycle for a second molecule of acetate, or by carboxylation of pyruvate to oxaloacetate (the Wood–Werkman reaction).

Because labelled acetate is used so frequently in biosynthetic studies, it is interesting to follow the fate of the carbon atoms of acetate in the TCA cycle (Scheme 2). Three points merit comment. First, although citric acid is a symmetrical molecule, the enzyme system carrying out its conversion to *cis*-aconitic acid is not, so that the carboxyl groups of citric acid are not equivalent. As a result, C-1 of acetate is incorporated mainly into the 5-carboxyl group of glutamic acid derived from α-oxoglutarate (Scheme 3). Secondly, succinic and fumaric acids behave symmetrically, so that C-1 of acetate is incorporated equally into both carboxyl groups. As a result, C-1 of acetate is incorporated into both carboxyl groups of α-oxoglutarate, though to an unequal degree, during the second turn of the cycle. Thirdly, C-2 of acetate is incorporated into both methylene groups of succinate, and since one of these becomes a carboxyl group during the second turn of the cycle C-2 of acetate is incorporated into all the carbons of the C_4 dicarboxylic acids. In contrast, no matter how many turns of the cycle are completed, C-1 of acetate is incorporated only into the carboxyl groups of the C_4 dicarboxylic acids. This accounts for the greater randomization of radioactivity often observed when [2-^{14}C]acetate is used in the study of polyketide biosynthesis.

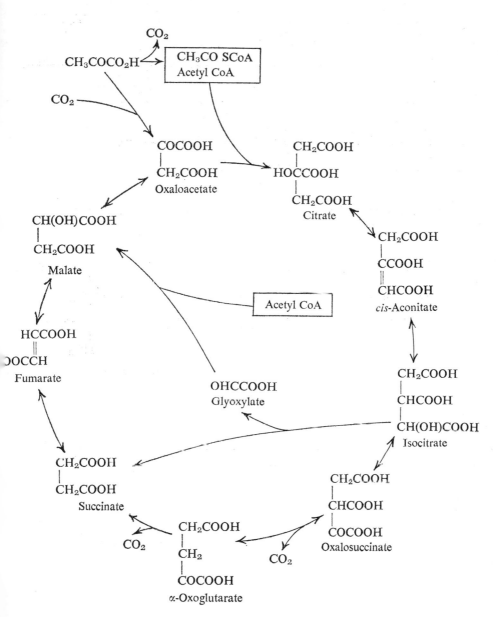

Scheme 1: The tricarboxylic acid and glyoxylate cycles

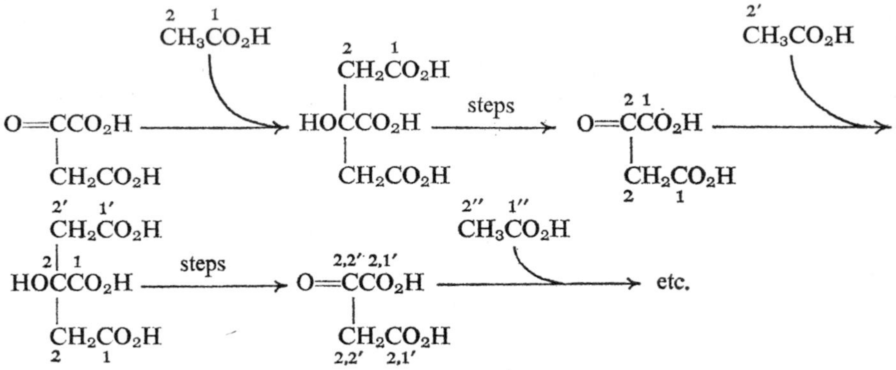

$$\overset{2}{C}H_3\overset{1}{C}O_2H$$

$$\overset{2}{C}H_2\overset{1}{C}O_2H$$

$$O{=}CCO_2H \longrightarrow HOCCO_2H \xrightarrow{\text{steps}} O{=}\overset{2}{C}\overset{1}{C}O_2H \longrightarrow$$

$$\underset{CH_2CO_2H}{|} \qquad \underset{CH_2CO_2H}{|} \qquad \underset{\overset{|}{CH_2}\overset{}{CO_2H}}{|}$$

$$\overset{2'}{C}H_2\overset{1'}{C}O_2H \qquad\qquad\qquad \overset{2''}{C}H_3\overset{1''}{C}O_2H$$

$$\overset{2|}{HO}\overset{1}{C}CO_2H \xrightarrow{\text{steps}} O{=}\overset{2,2'}{C}\overset{2,1'}{C}O_2H \longrightarrow \text{etc.}$$

$$\underset{\overset{2}{C}H_2\overset{1}{C}O_2H}{|} \qquad\qquad \underset{\overset{2,2'}{C}H_2\overset{2,1'}{C}O_2H}{|}$$

$\overset{2'}{C}H_3\overset{}{C}O_2H$ (top right)

Scheme 2: The distribution of the carbon atoms of acetate in TCA cycle intermediates. 1,1′,1″, . . . refer to C-1 of acetate entering at the first, second, third, . . . turn of the cycle

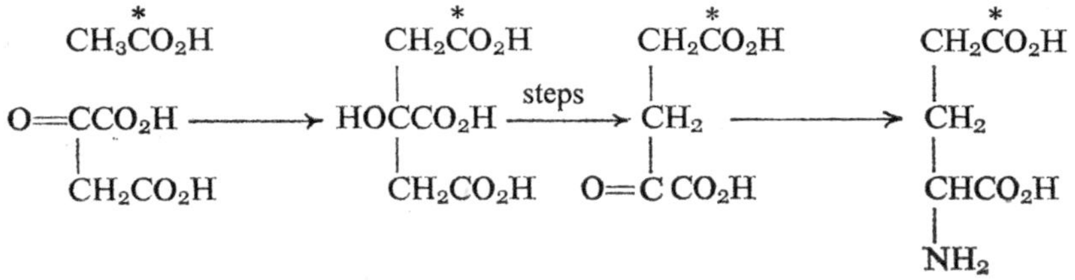

$$\overset{*}{C}H_3CO_2H \qquad \overset{*}{C}H_2CO_2H \qquad \overset{*}{C}H_2CO_2H \qquad \overset{*}{C}H_2CO_2H$$

$$O{=}CCO_2H \longrightarrow HOCCO_2H \xrightarrow{\text{steps}} CH_2 \longrightarrow CH_2$$

$$\underset{CH_2CO_2H}{|} \qquad \underset{CH_2CO_2H}{|} \quad \underset{O{=}C\ CO_2H}{|} \qquad \underset{CHCO_2H}{|}$$

$$\underset{NH_2}{|}$$

Scheme 3: The incorporation of C-1 of acetate into glutamate

B. COMPOUNDS RELATED TO TRICARBOXYLIC ACID CYCLE INTERMEDIATES

1   (structure: epoxysuccinic acid)

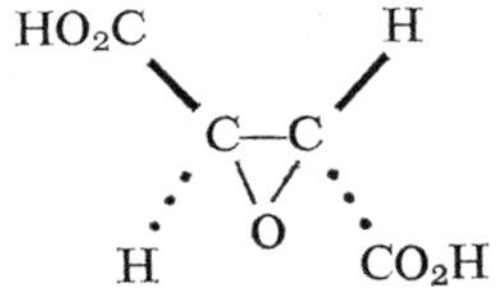

epoxysuccinic acid L^{10}
Paecilomyces varioti,[1] *Penicillium viniferum*,[2] *Aspergillus fumigatus*[3]

2      $HO_2CCH_2CCO_2H$
                $\underset{CH_2}{\|}$

itaconic acid $L^{11,12}$
Aspergillus itaconicus, A. terreus,[4] *P. charlesii*[5]

3    $HO_2CCH_2\overset{\overset{\displaystyle OH}{|}}{\underset{\underset{\displaystyle CH_2OH}{|}}{C}}CO_2H$ itatartaric acid
A. terreus,[6] *A. itaconicus*[7]

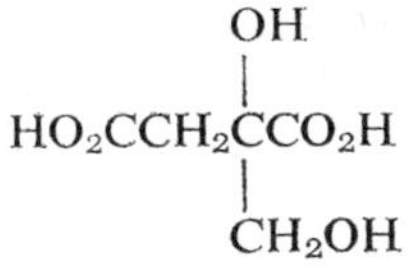

4    $HO_2CCH_2CH=CHCO_2H$ glutaconic acid
A. niger[8]

5

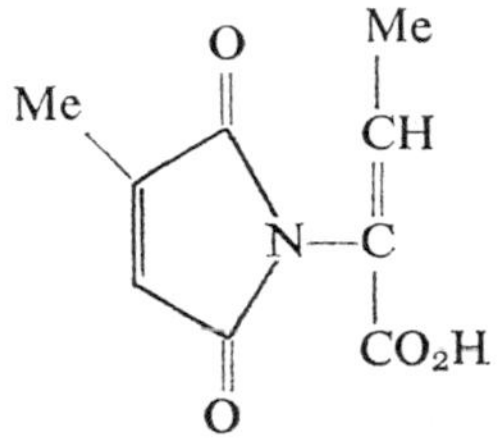

pencolide
Penicillium multicolor[9]

All the normal TCA cycle intermediates have been detected in fungi, sometimes in large quantity; for example, citric acid is produced commercially by fermentation of *Aspergillus niger*. When a primary metabolite such as citric acid is produced in non-physiological quantity it becomes a secondary metabolite, but here we are more concerned with compounds which have no known metabolic role.

Epoxysuccinic acid (**1**) has been shown by ^{14}C and [18]O labelling experiments to be derived from fumaric acid and molecular oxygen.[10]

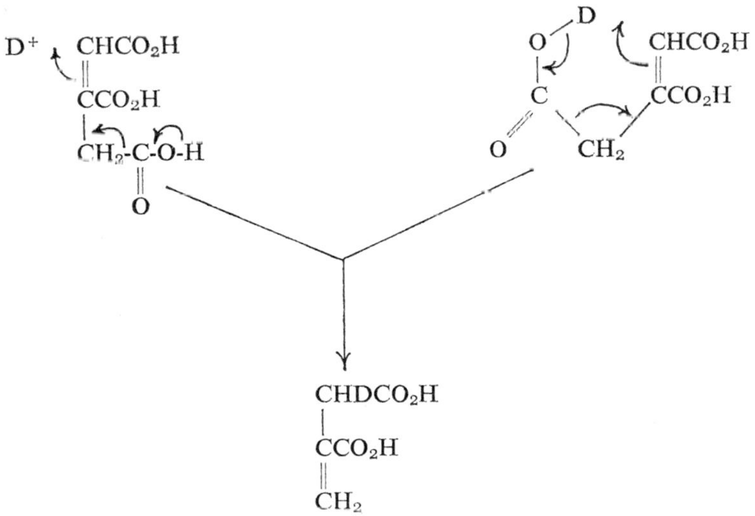

Scheme 4: Possible mechanisms for the decarboxylation of *cis*-aconitic acid

Experiments with labelled precursors,[11,12] in some cases using cell-free preparations from *A. terreus*,[12] have established that itaconic acid (**2**) is formed by decarboxylation of *cis*-aconitic acid. Experiments in the presence of D_2O show that the decarboxylation proceeds by one of the mechanisms of Scheme 4, i.e. that decarboxylation is accompanied by double-bond migration, the normal process for $\beta\gamma$-unsaturated acids.

Arpai[13] has prepared from a mutant of *A. terreus* a cell-free system which will oxidize itaconic acid to itatartaric acid (**3**).

Although glutaconic acid (**4**) could be derived from α-oxoglutaric acid (hence its inclusion here), in mammalian systems it is formed by dehydrogenation of glutaric acid during lysine catabolism. Pencolide (**5**) could arise from citraconic acid and threonine.

C. COMPOUNDS FORMED BY CONDENSATION OF AN ACETATE-DERIVED CHAIN WITH A TCA CYCLE INTERMEDIATE

A number of fungal products result, at least formally, from condensation of an acetate-derived chain, often a fatty acid, with an intermediate of the tricarboxylic acid cycle, usually a C_4 dicarboxylic acid. (Although some compounds included here could, in principle, be derived from a C_3 compound such as pyruvate, the compounds whose biosynthesis has been studied are, in fact, derived from a C_4 intermediate.) The compounds will be discussed in three groups: (1) those which arise by condensation of a carboxyl group of the C_4 dicarboxylic acid with the α-methylene group of the fatty acid, (2) those which arise by condensation of the keto-group of oxaloacetic acid with the α-methylene group of the fatty acid and (3) the nonadrides, in which compounds of type (2) have undergone a cyclic dimerization.

C.1. CARBOXYL-METHYLENE CONDENSATION PRODUCTS

6

γ-methyltetronic acid
Penicillium charlesii,[14] *P. fellutanum*[15]

7

carlosic acid   *L*[28]
P. charlesii[16,17], *P. fellutanum*,[15]
P. cinerascens[18]

8 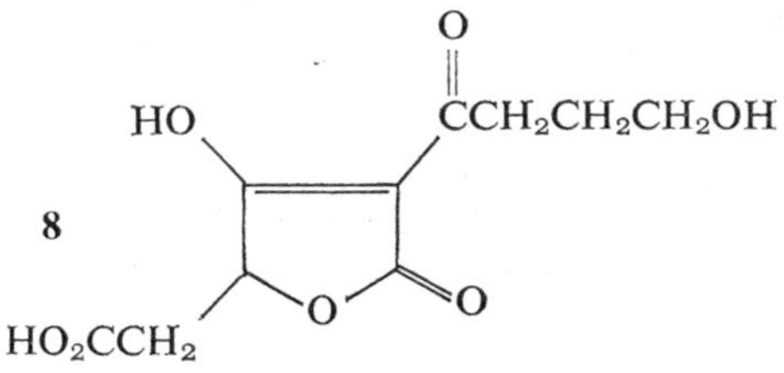   carlic acid
        P. charlesii[16,17]

9 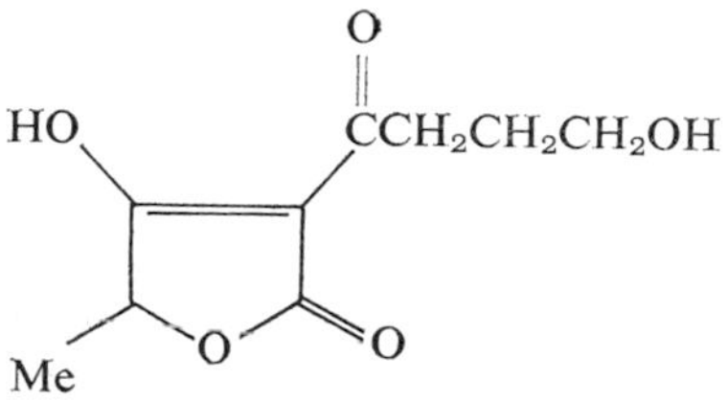   carolic acid   L[29]
        P. charlesii,[16,19] P. fellutanum[15]

10 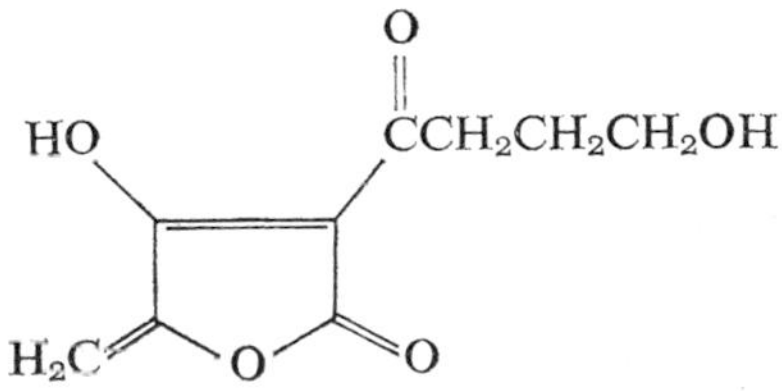   dehydrocarolic acid
        P. cinerascens[18]

11 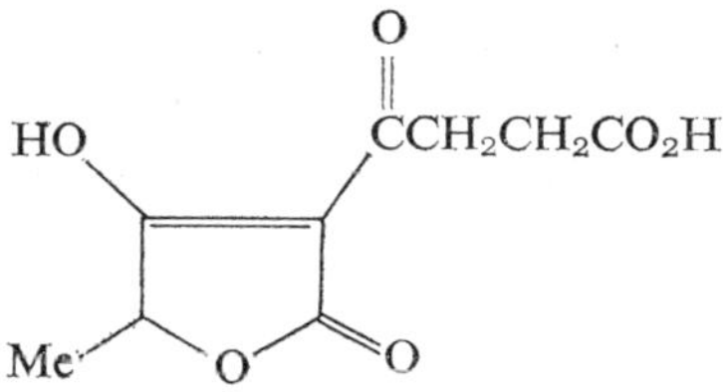   carolinic acid
        P. charlesii[16,19]

12 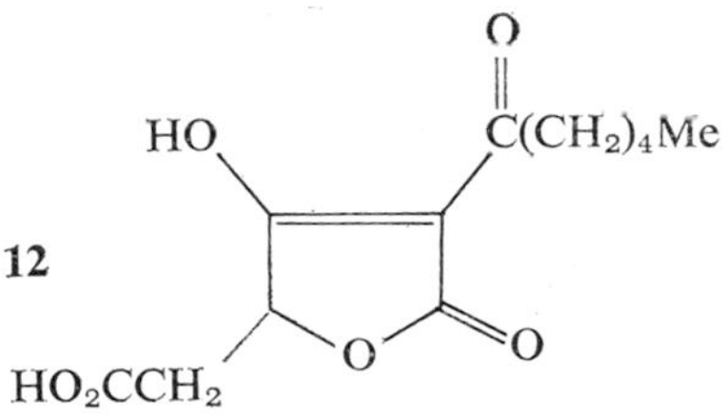   viridicatic acid
        P. viridicatum[20]

13 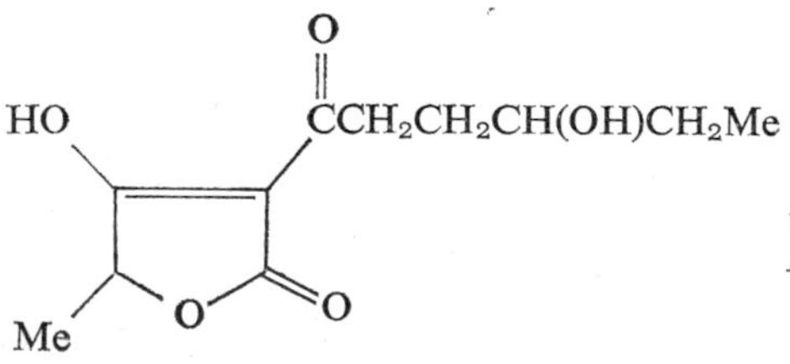   terrestric acid
        P. terrestre[21]

14

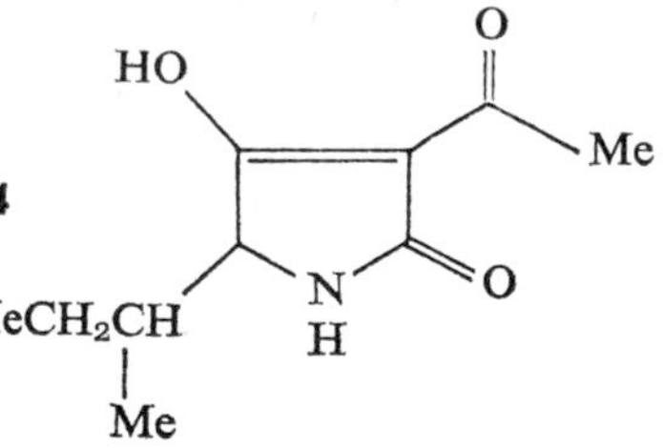

tenuazonic acid L[29]
Alternaria tenuis[22]

15

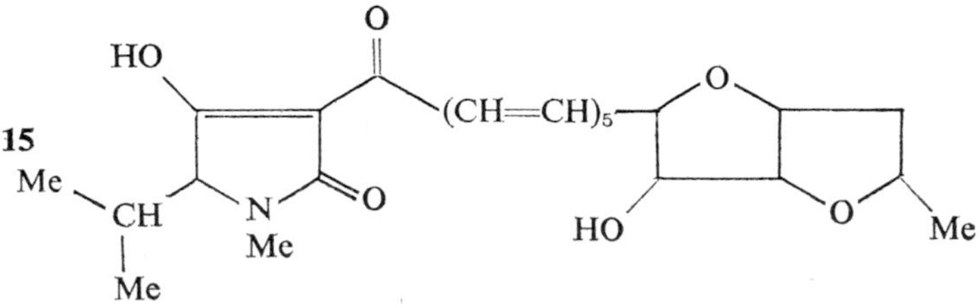

erythroskyrin L
Penicillium
islandicum[23]

This group comprises mainly the tetronic acids, though we also include two tetramic acids in which the C_4 dicarboxylic acid moiety is replaced by an amino-acid. In many of the compounds, decarboxylation has occurred so that the C_4 fragment is no longer discernible, but the relationship between carlosic acid (**7**), carlic acid (**8**), and carolic acid (**9**), and between viridicatic acid (**12**) and terrestric acid (**13**) strongly suggests that the compounds containing a C_3 moiety arise by decarboxylation of an intermediate containing a C_4 moiety (see below).

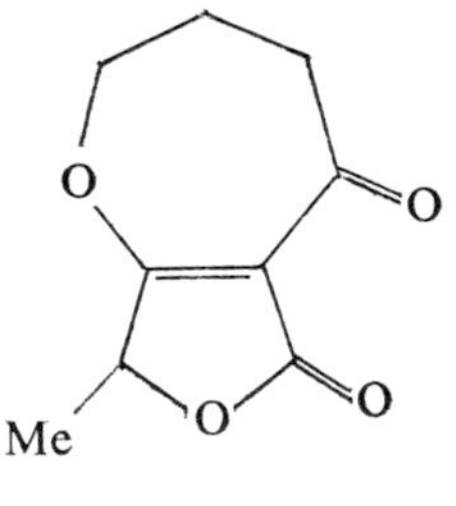

16 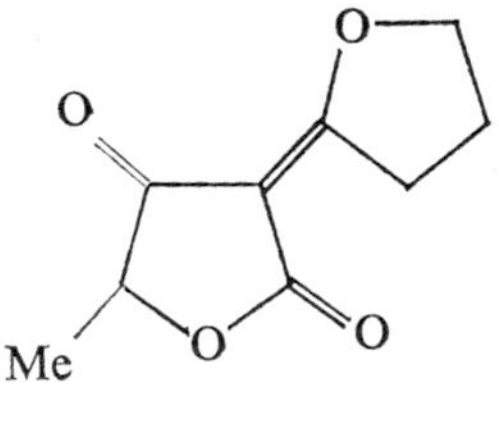   **17**

The acyltetronic acids possessing a hydroxyl group in the side-chain form cyclic ethers which are usually written in the seven-membered ring form, e.g. (**16**) for carolic acid, first proposed by Raistrick and his colleagues. Plimmer[24] has produced n.m.r. evidence in support of this structure and Sudo *et al.*[25] drew the same conclusion from the n.m.r. spectrum of their synthetic material.

However, Burrows and Turner[26] find that several of the signals in the n.m.r. spectra of carolic acid and of dehydrocarolic acid are doubled, suggesting that the five-membered ring system (17) is formed, the doubling of the signals resulting from *cis–trans* isomerism about the exocyclic double bond. The formation of a five-membered rather than a seven-membered ring seems more likely *a priori*, and was first suggested by Dean.[27]

Carbon atoms 1,2 and 5–8 of carlosic acid (7) and carolic acid (9) are derived from acetate and malonate,[28] which are inefficient precursors of carbon atoms 3,4,9 and 10. The best incorporation of radioactivity into the latter group of atoms was obtained with [2,3-^{14}C]succinic acid as substrate, when little incorporation into the side-chain occurred. These results are consistent with the condensation of a C_4 dicarboxylic acid with a β-keto-acid to give carlosic acid which can then be hydroxylated at C-8 and decarboxylated to give carolic acid. The relationship of dehydrocarolic acid (10) to the other tetronic acids is of interest—is it a precursor of, or derived from, carolic acid or are both compounds formed from a common intermediate?

Tenuazonic acid (14) is derived from acetoacetate and isoleucine,[29] and erythroskyrin (15) from one acetyl unit, nine malonyl units, and valine.[30]

The aspertetronins (p. 358) are tetronic acids in which the C_4 residue is replaced by an unusual branched-chain residue.

C.2. CARBONYL-METHYLENE CONDENSATION PRODUCTS

18

R^1O_2C $CH_2CO_2R^2$
  $\diagdown$ $\diagup$
  CH—CH
  $|$ $|$
$CH_3(CH_2)_{12}CH_2$ CO_2H

R^1 or R^2 = Me, rangiformic acid
Lichens[31]

19

HO_2C CH_2CO_2H
  $\diagdown$ $\diagup$
  CH—COH
  $|$ $|$
$CH_3(CH_2)_8CH_2$ CO_2H

decylcitric acid
Penicillium spiculisporum[32]

20

R^1O_2C $CH_2CO_2R^2$
  $\diagdown$ $\diagup$
  CH—COH
  $|$ $|$
$CH_3(CH_2)_{12}CH_2$ CO_2H

a: R^1 or R^2 = Me, caperatic acid
Lichens[31]

b: R^1 = R^2 = H, norcaperatic acid
Cantharellus floccosus,[33] *Polyporus fibrillosus*[34]

21

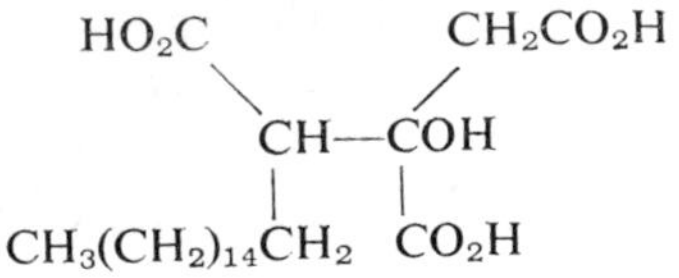

$$HO_2C$$ $$CH_2CO_2H$$

$$CH—COH$$

$$CH_3(CH_2)_{14}CH_2 \quad CO_2H$$

agaricic acid
Polyporus officinalis[35]

22

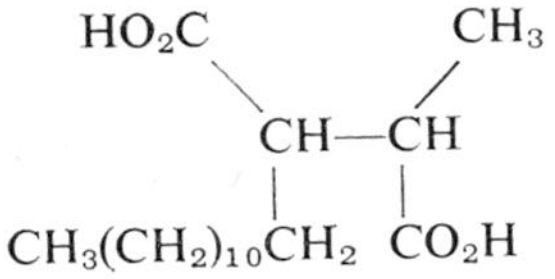

$$HO_2C$$ $$CH_3$$

$$CH—CH$$

$$CH_3(CH_2)_{10}CH_2 \quad CO_2H$$

roccellic acid
Lichens[31]

23

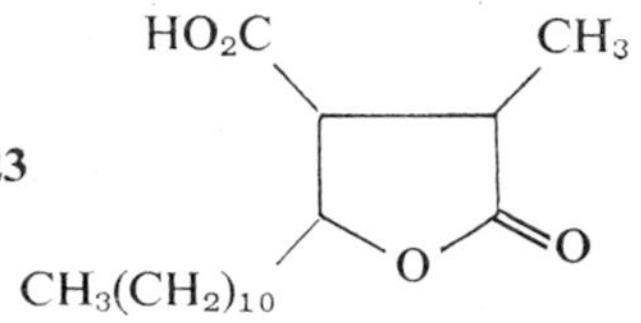

$$HO_2C$$ $$CH_3$$

$$CH_3(CH_2)_{10}$$

nephrosteranic acid
Lichens[31]

24

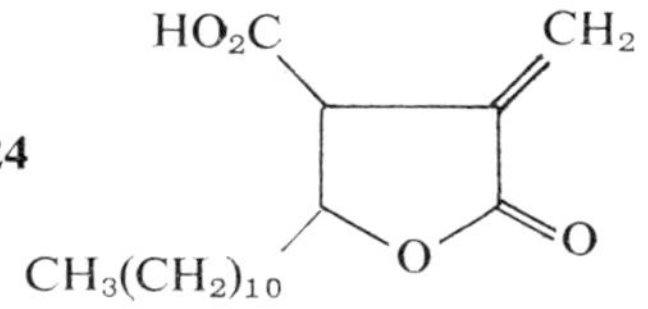

$$HO_2C$$ $$CH_2$$

$$CH_3(CH_2)_{10}$$

nephrosterinic acid
Lichens[31]

25

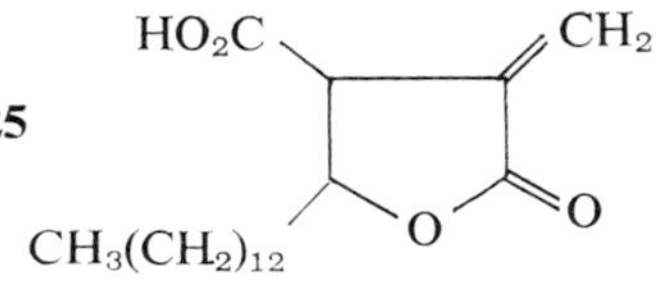

$$HO_2C$$ $$CH_2$$

$$CH_3(CH_2)_{12}$$

protolichesterinic acid L^{45}
Lichens[31]

26

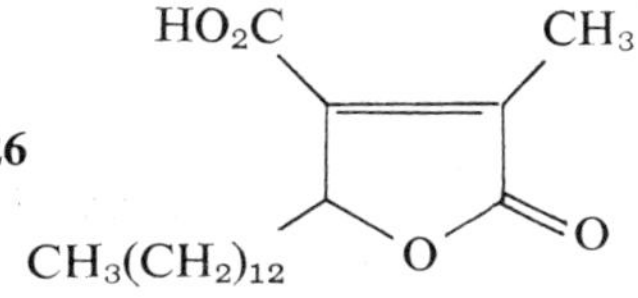

$$HO_2C$$ $$CH_3$$

$$CH_3(CH_2)_{12}$$

lichesterinic acid
Lichens[31]

27
HO$_2$C   CH$_3$
CH$_3$(CH$_2$)$_{12}$

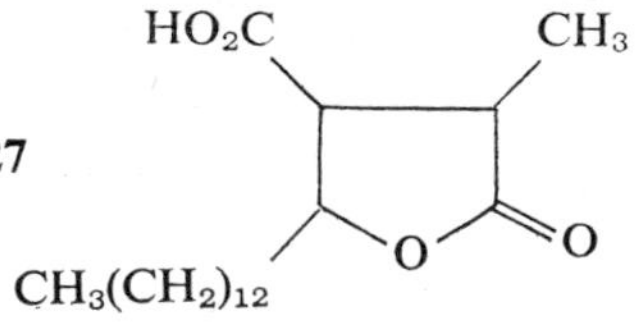

nephromopsic acid
Lichens[31]

28
CH$_2$CO$_2$H
CH$_3$(CH$_2$)$_{10}$

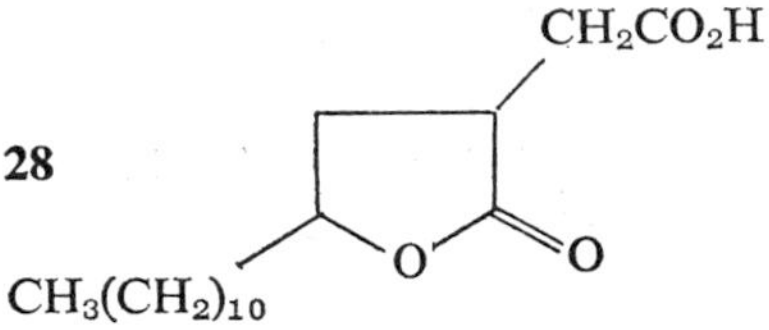

acaranoic acid
Acarospora chlorophana[36]

29
CH$_2$CO$_2$H
CH$_3$(CH$_2$)$_{10}$

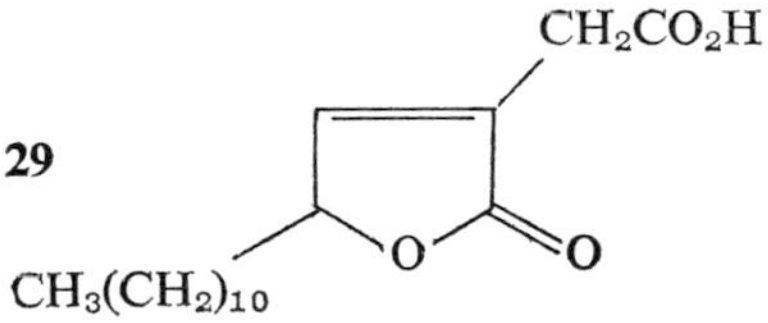

acarenoic acid
A. chlorophana[36]

30
CH$_2$
CH$_3$(CH$_2$)$_3$

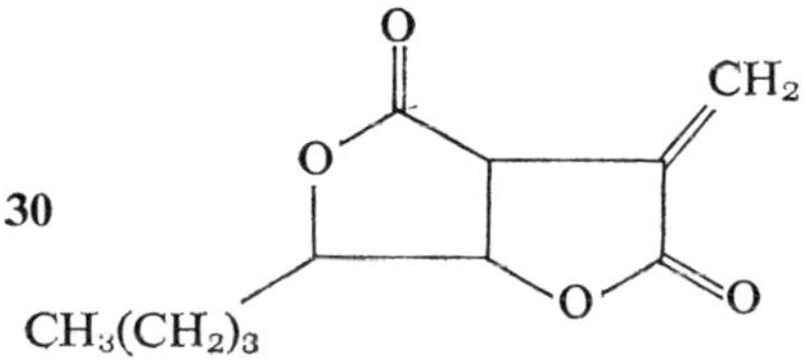

canadensolide
Penicillium canadense[37]

31
CH$_3$
CH$_3$(CH$_2$)$_3$

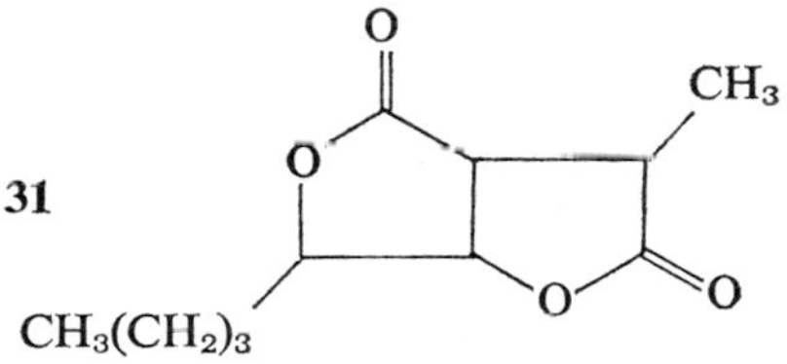

dihydrocanadensolide
P. canadense,[37] *Aspergillus indicus*[38]

32
CH$_2$
CO$_2$H
CH$_3$(CH$_2$)$_3$

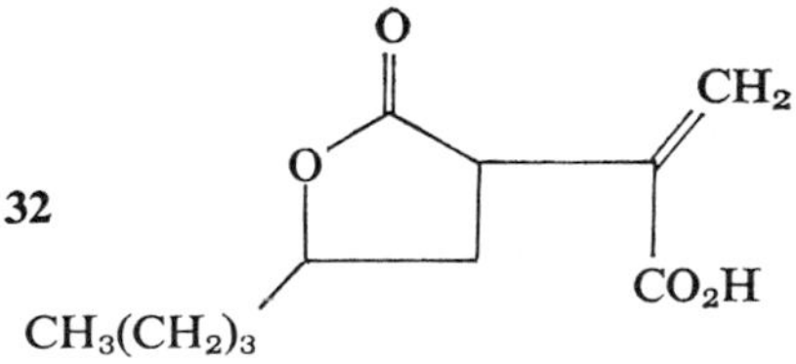

canadensic acid
P. canadense[39]

33

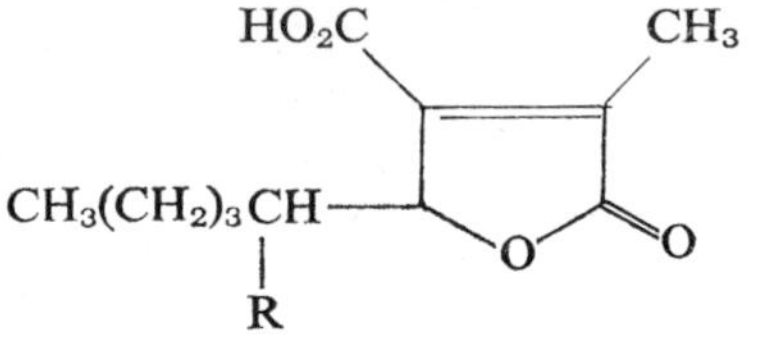

$$HO_2C \quad\quad CH_3$$

$$CH_3(CH_2)_3CH \underset{R}{-}$$

a: R = H, isocanadensic acid
b: R = OH, hydroxyisocanadensic acid
  P. canadense[39]

34

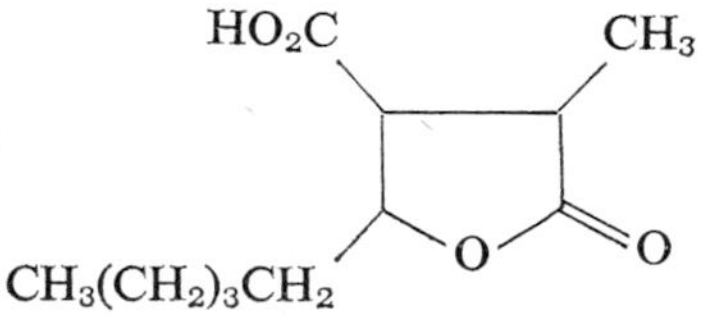

$$HO_2C \quad\quad CH_3$$

$$CH_3(CH_2)_3CH_2$$

dihydroisocanadensic acid
P. canadense[39]

35

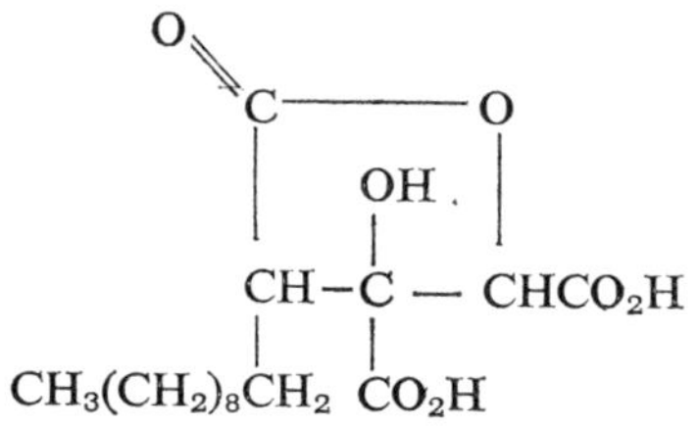

$$CH-\underset{OH}{\overset{.}{C}} - CHCO_2H$$

$$CH_3(CH_2)_8CH_2 \quad CO_2H$$

minioluteic acid
P. minio-luteum[40]

36

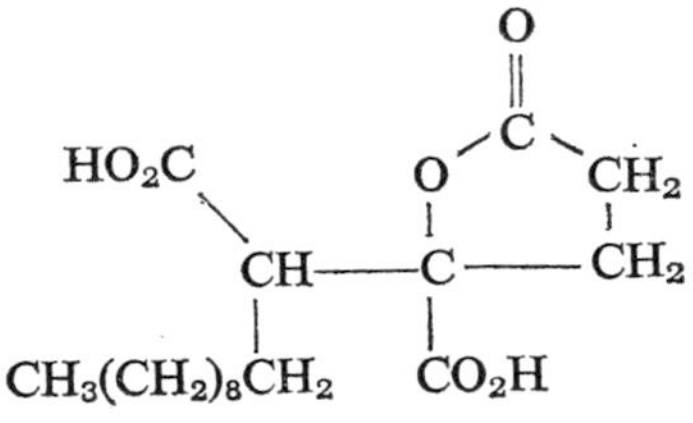

$$HO_2C \quad\quad O \quad CH_2$$

$$CH \underset{CH_3(CH_2)_8CH_2}{-} C \underset{CO_2H}{-} CH_2$$

spiculisporic acid   *L*[43]
Penicillium spp.[41]

37

$$\underset{1\ 2}{HO_2CCH_2}\underset{3}{\overset{\overset{\displaystyle 14}{\overset{\displaystyle Me}{|}}}{C}}=\underset{4}{CH}\underset{5}{CH}=\underset{6}{CH}\underset{7}{CH}=\underset{8}{CH}$$

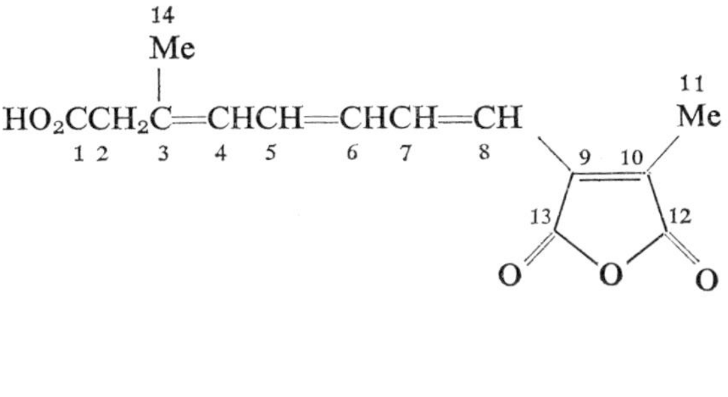

itaconitin
Aspergillus itaconicus[42]

This group can be derived formally by condensation of the carbonyl group of oxaloacetate with the α-methylene group of a fatty acid (when the acid is acetic, the product is citric acid and the process is one of the reactions of the TCA cycle). The products of such condensations are exemplified by decylcitric acid (19), caperatic acid (20) and agaricic acid (21), while in spiculisporic acid (36) oxaloacetate has been replaced by α-oxoglutarate. Removal of the hydroxyl group from caperatic acid (20), perhaps by elimination followed by reduction, would give rise to rangiformic acid (18). In many compounds, especially among the lichen products, a hydroxyl group in the β-position of the fatty acid chain is involved in lactone formation; the hydroxyl group may result from hydroxylation of a compound such as caperatic acid (20) or may represent a residual β-oxygen from the biosynthesis of the chain. Decarboxylation has occurred in several of the compounds, possibly by the mechanism established for the biosynthesis of itaconic acid (see p. 284), to give methylene compounds such as nephrosterinic acid (24) which might be reduced to the methyl compounds such as nephrosteranic acid (23).

The enzymatic synthesis of spiculisporic acid (36) from lauryl coenzyme A and α-oxoglutarate has been reported.[43] In the course of these experiments it was found[32] that *P. spiculisporum* slowly lost its ability to produce spiculisporic acid and produced instead (−)-decylcitric acid (19). The identity of the organism was confirmed at the Centraalbureau voor Schimmelcultures, Baarn, and a fresh culture from that source again produced spiculisporic acid. This is an interesting example of an organism losing the capacity to produce a secondary metabolite (a common occurrence!) while acquiring the ability to produce a closely related metabolite. The enzyme responsible for the synthesis of (−)-decylcitric acid has been isolated.[44] The "degenerate" strain of *P. spiculisporum* contains an isomerase, so that although (−)-decylcitric acid is the product of the synthetase, the major metabolite of *P. spiculisporum* is (+)-decylcitric acid.

In the only study of the biosynthesis of an α-methylene lactone, Bloomer et al.[45] found that radioactivity from [1,4-^{14}C]succinate was incorporated mainly into the lactone carbonyl group of (+)-protolichesterinic acid (25) by *Cetraria islandica*.

Acetate and malonate are efficiently incorporated into C-1 to C-9 and C-13 of itaconitin (37)[42b] (C-1 is derived from C-2 of acetate), and C-14 is derived from the C$_1$-pool. The origin of C-10 to C-12 is less clear. The low incorporation of radioactivity into these positions from [1,5-^{14}C]citric acid, [2,3-^{14}C]succinic acid or [1-^{14}C]pyruvic acid suggests that the TCA cycle is a poor source of the carbon atoms. [6-^{14}C]Glucose was incorporated more efficiently and gave unequal labelling of C-10 and C-11, again suggesting that these carbon atoms do not originate in the TCA cycle. It is suggested[42b] that C-10 to C-12 of itaconitin are derived from phosphoenolpyruvate, possibly after carboxylation to oxaloacetate.

In canadensolide (**30**) and related compounds hydroxylation has occurred in the γ-position of the fatty acid moiety resulting in the formation of a second lactone ring. Avenaciolide (**38**), a metabolite of *Aspergillus avenaceus*,[46]

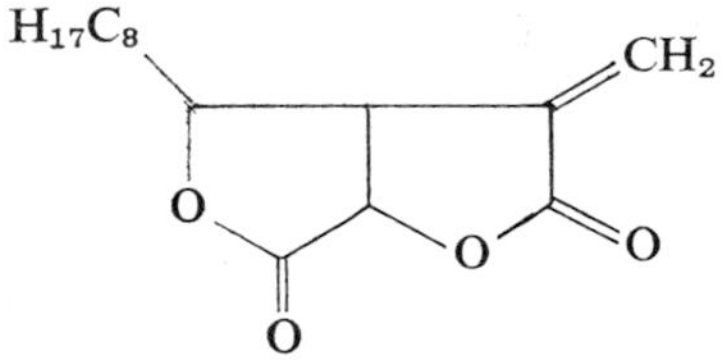

38

is structurally similar to canadensolide but is biogenetically "wrong" in the present context. It could result from condensation of succinic acid with a β-keto-acid as indicated in Scheme 5. A similar problem is encountered in the biosynthesis of canescin (p. 125).

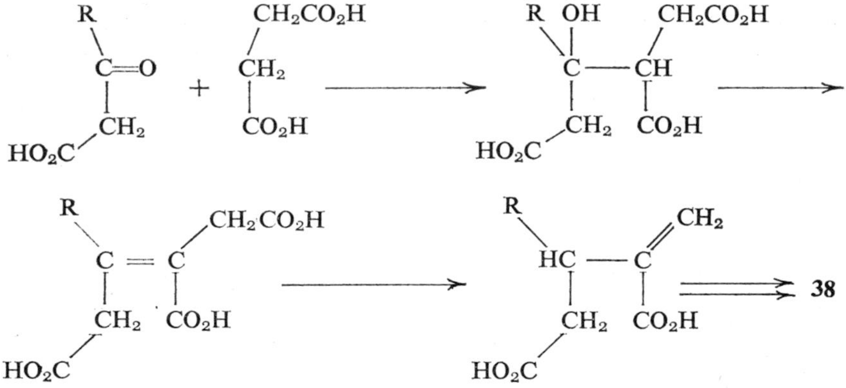

Scheme 5: Possible biosynthesis of avenaciolide (**38**)

C.3. THE NONADRIDES

39

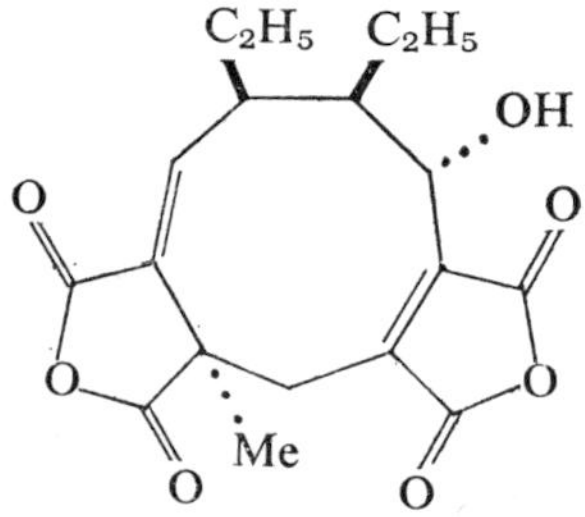

glauconic acid   *L*[54]
Penicillium glaucum[47]

40 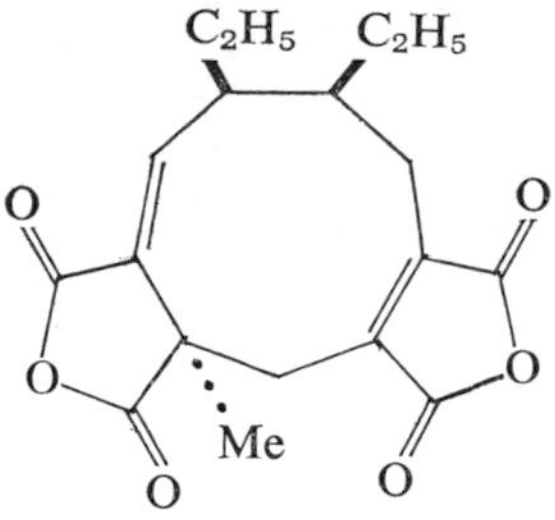 glaucanic acid (glauconic acid II)
P. glaucum[50]

41 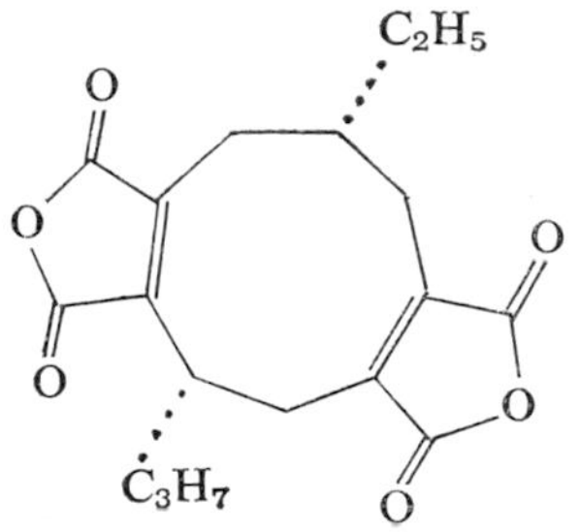 byssochlamic acid
Byssochlamys fulva[48]

42 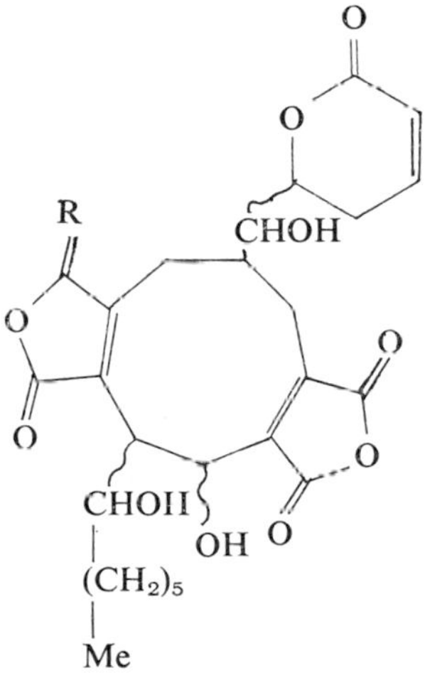 a: R = H, OH, rubratoxin A
b: R = O, rubratoxin B
Penicillium rubrum[49]

The structures of glauconic, glaucanic and byssochlamic acids have recently been elucidated by chemical[50] and X-ray crystallographic[51] methods, and their chemistry and biosynthesis have been reviewed.[52] With the isolation of the rubratoxins, and of an isomer of byssochlamic acid from *Helminthosporium heveae*,[53] the class is a growing one.

Inspection of the structures of the nonadrides suggested[52] that they might be formed by dimerization of C_9 precursors (Scheme 6) of the type discussed

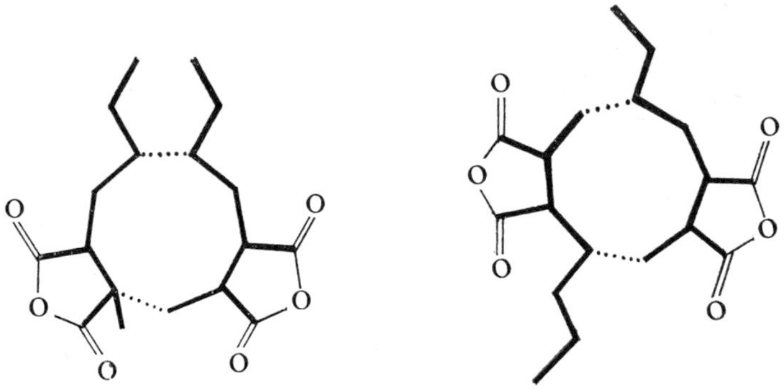

Scheme 6: Alternative cyclizations of C_9 precursors to give either glauconic or byssochlamic acid

in Section C.2, i.e. derived by condensation of a fatty acid with oxaloacetate. This biosynthetic pathway has been confirmed[54] for glauconic acid using labelled acetate, pyruvate, succinate, and glucose. The tritiated anhydride (43) and the ^{14}C-labelled anhydride (44) were incorporated into glauconic

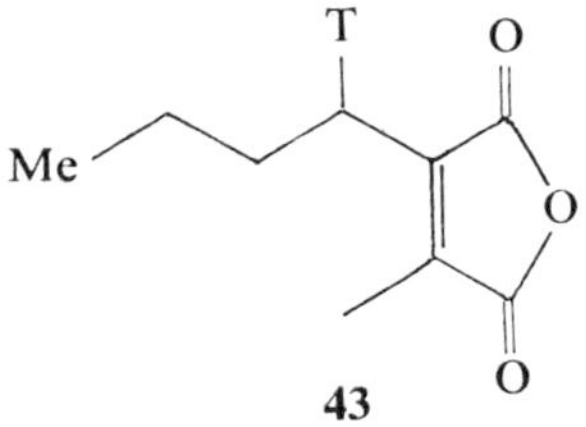

acid to the extent of 0·25 and 51·5% respectively so that the sequence is probably (43)→(44)→glauconic acid.

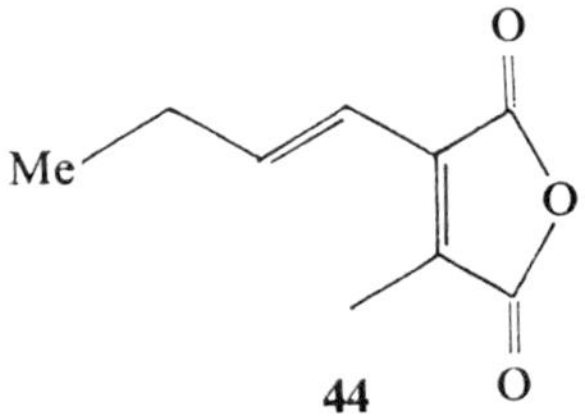

REFERENCES

1. K. Sakaguchi and T. Inoue, *J. Agr. Chem. Soc., Japan*, 1940, **16**, 1015.
2. K. Sakaguchi and T. Inoue, *J. Agr. Chem. Soc., Japan*, 1938, **14**, 1517.
3. J. H. Birkinshaw, A. Bracken and H. Raistrick, *Biochem. J.*, 1945, **39**, 70.
4. C. T. Calam, A. E. Oxford and H. Raistrick *Biochem. J.*, 1939, **33**, 1488.
5. S. Lybing and L. Reio, *Acta Chem. Scand.*, 1958, **12**, 1575.
6. F. H. Stodola, M. Friedkin, A. J. Moyer and R. D. Coghill, *J. Biol. Chem.*, 1945, **161**, 739.
7. T. Kobayashi, T. Tabuchi and K. Kitahara, *J. Agr. Chem. Soc. Japan*, 1961, **35**, 541.

8. S. Baba and K. Sakaguchi, *Bull. Agr. Chem. Soc., Japan*, 1942, **18**, 93.
9. J. H. Birkinshaw, M. G. Kalyanpur and C. E. Stickings, *Biochem. J.*, 1963, **86**, 237.
10. K. Aida and J. W. Foster, *Nature*, 1962, **196**, 672; L J. Wilkoff and W. R. Martin, *J. Biol. Chem.*, 1963, **238**, 843.
11. R. H. Corzo and E. L. Tatum, *Fed. Proc.*, 1953, **12**, 470.
12. R. Bentley and C. P. Thiessen, *J. Biol. Chem.*, 1957, **226**, 673, 689, 703.
13. J. Arpai, *Nature*, 1958, **182**, 661; *J. Bacteriol.*, 1959, **78**, 153.
14. P. W. Clutterbuck, H. Raistrick and F. Reuter, *Biochem. J.*, 1935, **29**, 1300.
15. V. C. Vora, *J. Sci. Ind. Res. (India)*, 1954, **13B**, 504.
16. P. W. Clutterbuck, W. N. Haworth, H. Raistrick, G. Smith and M. Stacey, *Biochem. J.*, 1934, **28**, 94.
17. P. W. Clutterbuck, H. Raistrick and F. Reuter, *Biochem. J.*, 1935, **29**, 871.
18. A. Bracken and H. Raistrick, *Biochem. J.*, 1947, **41**, 569.
19. P. W. Clutterbuck, H. Raistrick and F. Reuter, *Biochem. J.*, 1935, **29**, 300.
20. J. H. Birkinshaw and M. S. Samant, *Biochem. J.*, 1960, **74**, 369.
21. J. H. Birkinshaw and H. Raistrick, *Biochem. J.*, 1936, **30**, 2194.
22. T. Rosett, R. H. Sankhala, C. E. Stickings, M. E. U. Taylor and R. Thomas *Biochem. J.*, 1957, **67**, 390; C. E. Stickings, *Biochem. J.*, 1959, **72**, 332.
23. B. H. Howard and H. Raistrick, *Biochem. J.*, 1954, **57**, 212; J. Shoji, S. Shibata, U. Sankawa, H. Taguchi and Y. Shibanuma, *Chem. Pharm. Bull. (Tokyo)*, 1965, **13**, 1240.
24. J. R. Plimmer, *J. Org. Chem.*, 1964, **29**, 511.
25. R. Sudo, A. Kaneda and N. Itoh, *J. Org. Chem.*, 1967, **32**, 1844.
26. B. F. Burrows and R. W. Turner, unpublished.
27. F. M. Dean. "Naturally-occurring Oxygen Ring Compounds", p. 74. Butterworths, London, 1963.
28. R. Bentley, D. S. Bhate and J. G. Keil, *J. Biol. Chem.*, 1962, **237**, 859.
29. C. E. Stickings and R. J. Townsend, *Biochem. J.*, 1961, **78**, 412.
30. S. Shibata, U. Sankawa, H. Taguchi and K. Yamasaki, *Chem. Pharm. Bull. (Tokyo)*, 1966, **14**, 474.
31. For references see M. W. Miller. "The Pfizer Handbook of Microbial Metabolites". McGraw-Hill, New York, 1961.
32. S. Gatenbeck and A. Måhlen, *Acta Chem. Scand.*, 1968, **22**, 2613.
33. J. T. Miyata, V. E. Tyler and L. R. Brady, *Lloydia*, 1966, **29**, 43.
34. G. Sullivan, *J. Pharm. Sci.*, 1968, **57**, 1804.
35. H. Thoms and J. Vogelsang, *Ann. Chem.*, 1907, **357**, 145.
36. G. Bendz, J. Santesson and L. Tibell, *Acta Chem. Scand.*, 1966, **20**, 1181; J. Santesson, *Acta Chem. Scand.*, 1967, **21**, 1993.
37. N. J. McCorkindale, J. L. C. Wright, P. W. Brian, S. M. Clarke and S. A. Hutchinson, *Tetrahedron Lett.*, 1968, 727.
38. A. J. Birch, A. A. Qureshi and R. W. Richards, *Aust. J. Chem.*, 1968, **21**, 2775.
39. N. J. McCorkindale, personal communication.
40. J. H. Birkinshaw and H. Raistrick, *Biochem. J.*, 1934, **28**, 828.
41. P. W. Clutterbuck, H. Raistrick and M. L. Pintoul, *Trans. Roy. Soc. London, Ser. B*, 1931, **220**, 301.
42. (a) S. Nakajima, *Chem. Pharm. Bull. (Tokyo)*, 1965, **13**, 73; (b) V. Sankawa and S. Shibata, *Chem. Pharm. Bull. (Tokyo)*, 1969, **17**, 2025.
43. S. Gatenbeck and A. Måhlen. *In* "Proc. Congress Antibiotics, Prague, 1964". p. 540, Butterworths, London.
44. S. Gatenbeck and A. Måhlen, *Acta Chem. Scand.*, 1968, **22**, 2617.

296   7. SECONDARY METABOLITES FROM TCA CYCLE

45. J. L. Bloomer, W. R. Eder and W. F. Hoffman, *Chem. Commun.*, 1968, 354; J. L. Bloomer and W. F. Hoffman, *Tetrahedron Lett.*, 1969, 4339; J. L. Bloomer and W. F. Hoffman, *Tetrahedron Lett.*, 1969, 4339; J. L. Bloomer, W. R. Eder and W. F. Hoffmann, *J. Chem. Soc. C*, 1970, 1848.
46. D. Brookes, B. K. Tidd and W. B. Turner, *J. Chem. Soc.*, 1963, 5385.
47. N. Wijkman, *Ann. Chem.*, 1931, **485**, 61.
48. H. Raistrick and G. Smith, *Biochem. J.*, 1933, **27**, 1814.
49. M. O. Moss, F. V. Robinson, A. B. Wood, H. M. Paisley and J. Feeney, *Nature*, 1968, **220**, 767; M. O. Moss, A. B. Wood and F. V. Robinson, *Tetrahedron Lett.*, 1969, 367.
50. D. H. R. Barton and J. K. Sutherland, *J. Chem. Soc.*, 1965, 1769; D. H. R. Barton, L. M. Jackman, L. Rodriguez-Hahn and J. K. Sutherland, *J. Chem. Soc.*, 1965, 1772; D. H. R. Barton, L. D. S. Godinho and J. K. Sutherland, *J. Chem. Soc.*, 1965, 1779; J. E. Baldwin, D. H. R. Barton and J. K. Sutherland, *J. Chem. Soc.*, 1965, 1787.
51. G. Ferguson, J. M. Robertson and G. A. Sim, *Proc. Chem. Soc.*, 1962, 385; T. A. Hamor, I. C. Paul, J. M. Robertson and G. A. Sim, *J. Chem. Soc.*, 1963, 5502.
52. J. K. Sutherland, *Prog. Chem. Org. Nat. Prod.*, 1967, **25**, 131.
53. W. B. Turner, unpublished.
54. J. L. Bloomer, C. E. Moppett and J. K. Sutherland, *J. Chem. Soc. C*, 1968, 588.

CHAPTER 8

Secondary Metabolites Derived from Amino-acids

FUNGI PRODUCE a variety of nitrogen-containing secondary metabolites most of which are derived from the common amino-acids. We have already mentioned a few compounds which are derived primarily by other routes with the incorporation of an amino-acid, usually through an amide linkage, and in the next chapter we shall discuss the cytochalasins, in which a phenyl-alanine residue is linked through carbon-bonds to an acetate-derived chain. In this chapter we shall be concerned with compounds whose carbon skeletons are derived mainly, or wholly, from amino-acids with retention of the nitrogen atom(s), but will ignore the amines which result from decarboxyla-tion of the common amino-acids. Three of the classes of compound—the ergot alkaloids, the *Amanita* toxins and the siderochromes—form specialist topics which have been extensively reviewed elsewhere. These are treated rather briefly here and complete lists are not given, though the classes are exemplified.

A. AMINO-ACID BIOSYNTHESIS

The carbon skeletons of the essential amino-acids are derived from inter-mediates of glucose catabolism and of the tricarboxylic acid (TCA) cycle. The amino-group is introduced by transamination, the usual nitrogen donor

$$RCOCO_2H + R'CH(NH_2)CO_2H \rightleftharpoons RCH(NH_2)CO_2H + R'COCO_2H$$

being glutamic acid formed by amination of α-oxoglutaric acid of the TCA cycle. Nitrogen also becomes available by amination of pyruvate to alanine and by addition of ammonia to fumarate to give aspartate.

Several amino-acids are derived from each precursor, and we shall briefly consider the biosynthesis of some of these groups of amino-acids emphasising the products and processes which are most involved in fungal secondary metabolism.

A.1. THE SERINE GROUP

Serine is derived from 3-phosphoglycerate, an intermediate of the Embden–Meyerhof pathway of glycolysis, and is the precursor of glycine (Scheme 1). In the formation of glycine, the hydroxymethylene group of serine is transferred to tetrahydrofolic acid (FH_4) and enters the C_1-pool; thus C-1 and C-6 of glucose become available for biological methylations. Transamination of glycine yields glyoxylic acid which is hydrolysed to formate and carbon dioxide; C-2 of glycine and of serine enter the C_1-pool in this way. Serine is also the precursor of cysteine.

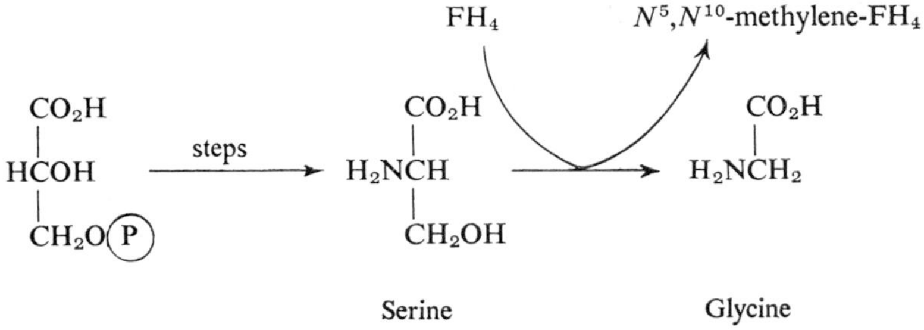

Scheme 1: The biosynthesis of serine and glycine

A.2. THE ASPARTIC ACID GROUP

Aspartic acid is formed from oxaloacetate by transamination and is a precursor of asparagine, threonine, isoleucine, lysine (in bacteria but not fungi, see below) and methionine. Methionine is of particular interest to us

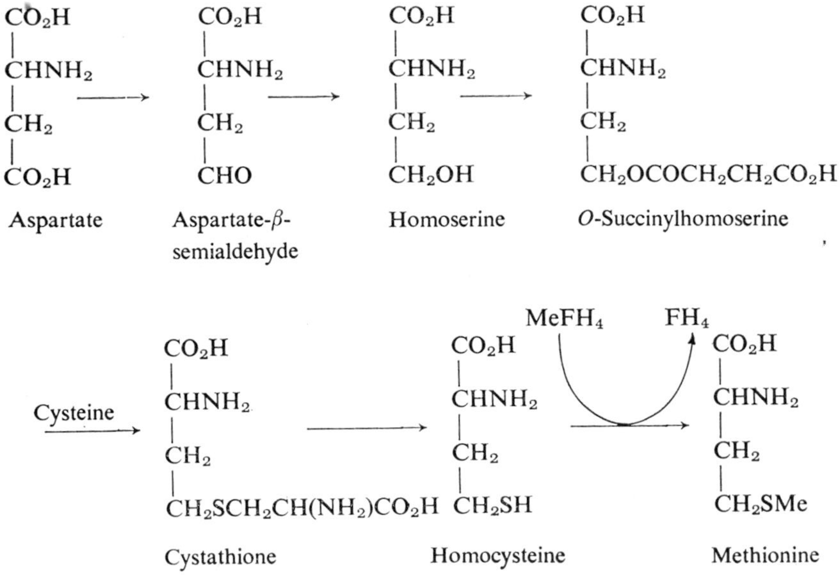

Scheme 2: The biosynthesis of methionine

as an intermediate in biological methylations, and its biosynthesis is outlined in Scheme 2, from which it will be seen that its sulphur atom is derived from cysteine and its S-methyl group from the C_1-pool.

A.3. LYSINE

There are two pathways to lysine. In bacteria it is formed by condensation of aspartate semialdehyde with pyruvate to give 2,3-dihydrodipicolinic acid which is converted to lysine via diaminopimelic acid (Scheme 3).

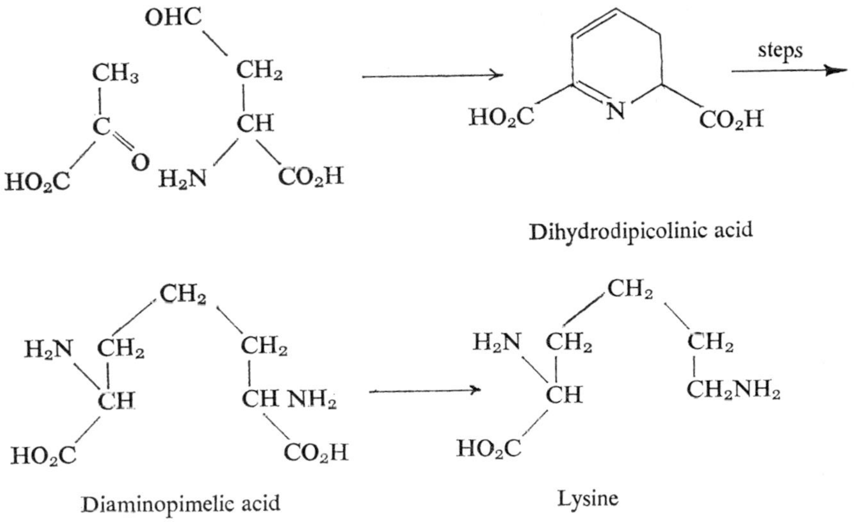

Dihydrodipicolinic acid

Diaminopimelic acid   Lysine

Scheme 3: The biosynthesis of lysine from aspartate semialdehyde

$CH_3COSCoA$ CH_2CO_2H $HOCHCO_2H$

Homocitrate   Homoisocitrate

α-Oxoadipate   α-Aminoadipate   Lysine

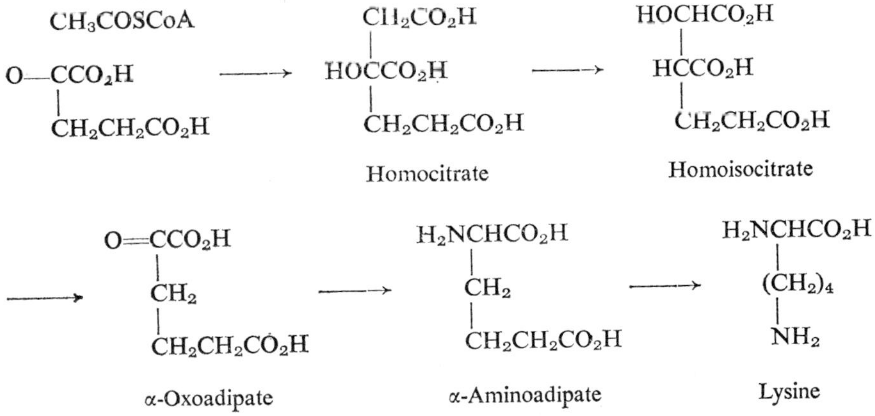

Scheme 4: The biosynthesis of lysine via α-aminoadipic acid

In fungi, the more usual pathway to lysine is via α-aminoadipate (Scheme 4). We shall see when we discuss dipicolinic acid, which is of widespread occurrence in bacterial spores and is also produced by *Penicillium citreo-viride*, that there is controversy as to whether its biosynthesis is related to the bacterial (Scheme 3) or fungal (Scheme 4) route to lysine. α-Aminoadipic acid is involved in penicillin biosynthesis.

A.4. THE AROMATIC AMINO-ACIDS

We saw in Chapter 3 that phenylalanine and tyrosine are derived from prephenic acid. The third essential aromatic amino-acid, tryptophan, is derived from chorismic acid, another intermediate of the shikimic acid pathway, by the reactions of Scheme 5.

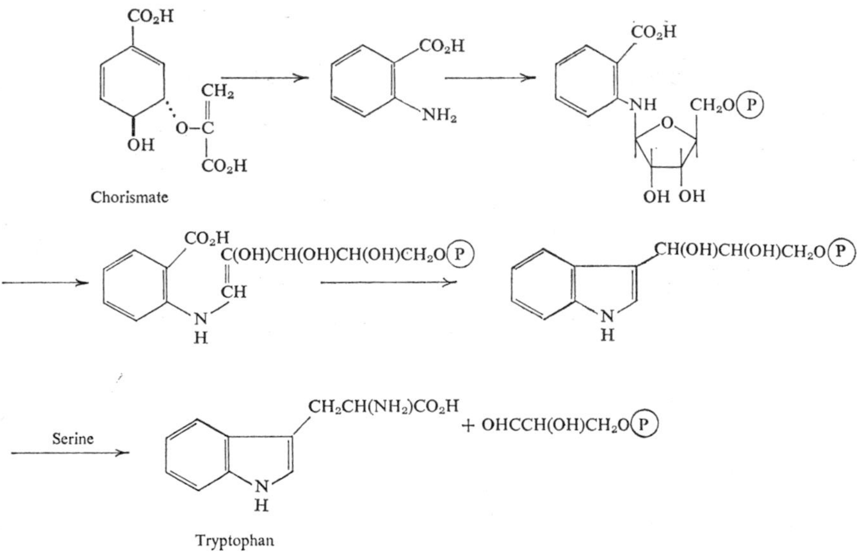

Scheme 5: The biosynthesis of tryptophan

B. SECONDARY METABOLITES DERIVED FROM A SINGLE AMINO-ACID

B.1. MISCELLANEOUS COMPOUNDS

1 OHCN(OH)CH$_2$CO$_2$H

hadacidin L^{12}
Penicillium frequentans,[1] *P. purpurescens*[2]

2 CH$_2$=CHCH$_2$CH$_2$N⁺Me$_3$Cl⁻

3-butenyltrimethylammonium chloride
Amanita muscaria[3]

3

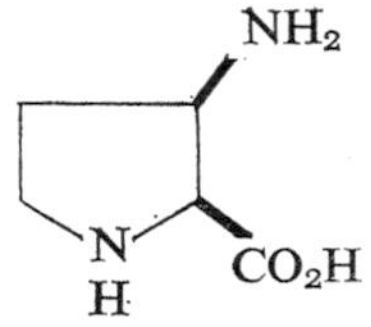

3-amino-L-proline
Morchella esculenta[4]

4 MeCH$_2$CCH(NH$_2$)CO$_2$H
‖
CH$_2$

β-methylene-L-(+)-norvaline
Lactarius helvus[5]

5

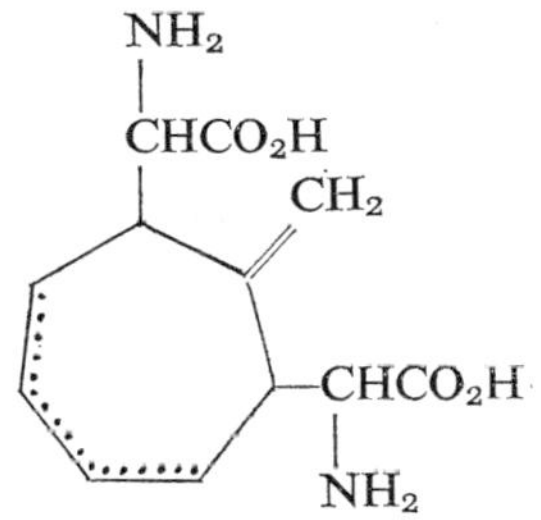

2-methylenecycloheptene-1,3-diglycine
L. helvus[6]

6

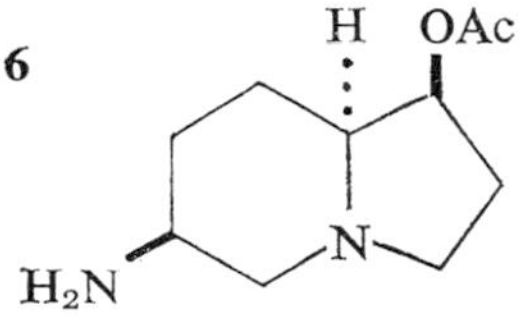

slaframine $L^{10,11}$
Rhizoctonia leguminicola[7]

7

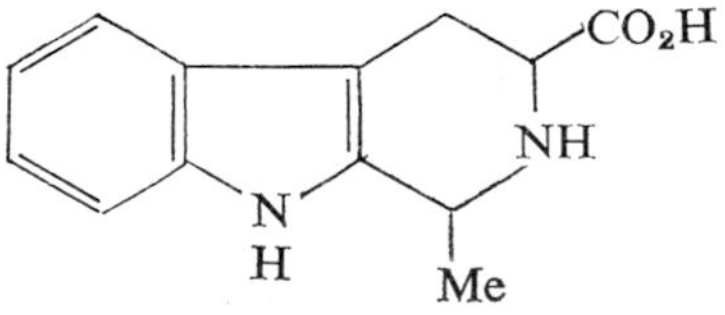

tetrahydro-1-methyl-β-carboline-3-carboxylic acid
Amanita muscaria[8]

Though these compounds bear no relationship to one another, they are discussed together because they appear to be derived from amino-acids (though this has been established only for hadacidin and slaframine) and do not fit conveniently into any other section. Hadacidin (1) is an inhibitor of tumours and of plant-growth. Slaframine (6) was isolated as a result of the physiological effects it produces in cattle whose fodder is contaminated by *R. leguminicola*.[9,10]

The biosynthesis of hadacidin in *P. aurantio-violaceum* has been studied by Stevens and Emery.[12] Radioactivity from [1-^{14}C]glycine was incorporated almost exclusively into the glycyl moiety of hadacidin but formate, C-3 of serine, and C-2 of glycine were incorporated into both the formyl and glycyl moieties; these results are as expected for the derivation of the formyl group via the C_1-pool (see p. 297). *N*-Hydroxyglycine was efficiently incorporated into hadacidin but nitroacetic acid, glyoxylic acid oxime, and *N*-formylglycine were not incorporated. The oxygen of the *N*-hydroxyl group is derived from atmospheric oxygen. These results are accommodated by the sequence of Scheme 6.

$$H_2NCH_2CO_2H \quad \rightarrow \quad HN(OH)CH_2CO_2H \quad \rightarrow \quad OHCN(OH)CH_2CO_2H$$

Scheme 6: The biosynthesis of hadacidin

Lysine is incorporated into slaframine via pipecolic acid,[10] the α-nitrogen of lysine becoming the ring-nitrogen of slaframine. A sequence such as that of Scheme 7 has been suggested[11] for the biosynthesis of slaframine.

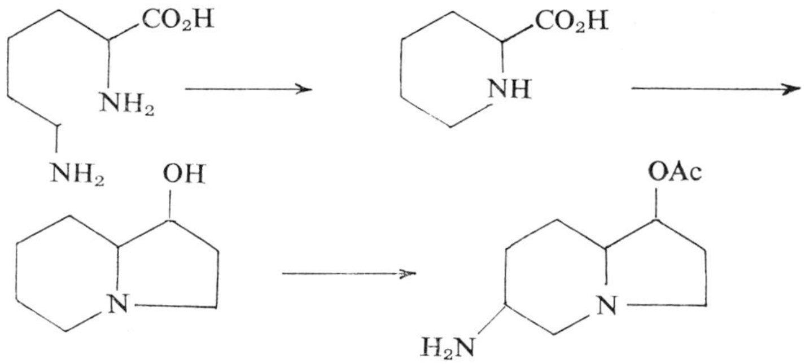

Scheme 7: The incorporation of pipecolic acid into slaframine

B.2. NITRO-COMPOUNDS

8        $O_2NCH_2CH_2CO_2H$ β-nitropropionic acid L[20,23–25]
                                    Aspergillus flavus,[13] *A. oryzae*[14]
                                    A. wentii,[15] *A. avenaceus*,[17]
                                    Penicillium atrovenetum[18]

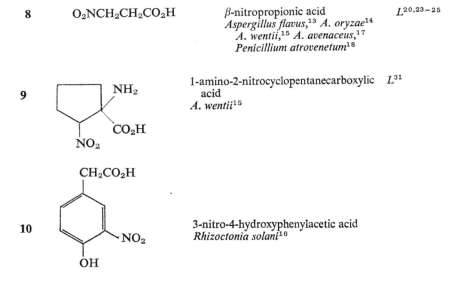

9   1-amino-2-nitrocyclopentanecarboxylic L[31]
          acid
          A. wentii[15]

10   3-nitro-4-hydroxyphenylacetic acid
          Rhizoctonia solani[16]

β-Nitropropionic acid (**8**) was first obtained from higher plants (for references see ref. 24), where it occurs as a glycoside and is responsible for the toxicity of some species. The first isolation from a fungus was from *A. flavus* and it is produced in high yield by *P. atrovenetum*. 1-Amino-2-nitro-cyclopentanecarboxylic acid (**9**) was detected as a result of its plant-growth regulating effects.[19]

Birch *et al.*[20] found that radioactivity from sodium [^{14}C]bicarbonate and from [4-^{14}C]aspartic acid was specifically incorporated into the carboxyl group of β-nitropropionic acid but that radioactivity from labelled β-alanine (derived from biologically labelled β-nitropropionic acid) was not incorporated into β-nitropropionic acid. These results suggest the sequence of Scheme 8 and explain the fact[18,21,22] that production of β-nitropropionic

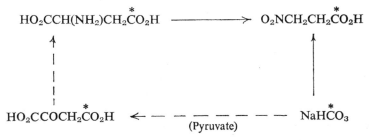

Scheme 8: Incorporation of aspartic acid and carbon dioxide into β-nitropropionic acid

acid is greater on media containing ammonia, rather than nitrate, as nitrogen source. In experiments with [^{15}N,U-^{14}C]aspartic acid, Gatenbeck and Forsgren[23] found that although some transamination had occurred, the excess of ^{15}N in β-nitropropionic acid was nearly twice that in the medium, i.e. that the amino-group of aspartic acid was used more efficiently than ammonia from the medium. Birkinshaw and Dryland[24] found that the distribution of radioactivity in β-nitropropionic acid formed in the presence of [1-^{14}C]acetate and [2-^{14}C]pyruvate is consistent with incorporation of these precursors via oxaloacetate of the tricarboxylic acid cycle with loss of C-1; some of the pyruvate is carboxylated directly to oxaloacetate (see p. 280) and the unequal radioactivity of C-2 and C-3 of pyruvate-derived β-nitropropionic acid excludes a symmetrical intermediate such as fumarate, contrary to an earlier suggestion.[21] In the pyruvate experiment, aspartic acid from the mycelium had almost the same specific activity (after subtraction of the activity of C-1) as β-nitropropionic acid, and the two compounds had the same distribution of radioactivity. Finally, Shaw and McCloskey[25] used ^{15}N- and ^{18}O-labelled precursors to study the source of the nitro-group of β-nitropropionic acid and found (1) that NH$_4$$^+$ is used in preference to NO$_3$$^-$, (2) that ^{18}O from nitrate is not incorporated into the nitro-group of β-nitropropionic acid and (3) that the amino-group of aspartate is used in preference to ammonia. They also found that L-aspartic acid (but not the D-isomer) contributes to the carbon skeleton of β-nitropropionic acid.

It is thus firmly established that the carbon skeleton of β-nitropropionic acid is derived from aspartic acid and it seems likely that the nitro-group arises by oxidation of the amino-group *in situ*. During a search for intermediates between aspartic acid and β-nitropropionic acid Shaw[26] found that extracts of *P. atrovenetum* catalyse the reduction of β-nitroacrylic acid to β-nitropropionic acid in the presence of NADPH (but not NADH). The specificity of the enzyme (of several possible substrates, only β-nitroacrylic acid was reduced) and the fact that β-nitroacrylic acid inhibits the incorporation of radioactivity from aspartate into β-nitropropionic acid by whole mycelium suggest that β-nitroacrylic acid is a natural intermediate for β-nitropropionic acid (though β-nitroacrylic acid has not been detected in *P. atrovenetum*).

The biosynthesis of β-nitropropionic acid in the higher plant *Indigofera spicata* appears to follow a different course from that in fungi.[27] Radioactivity from [^{14}C]aspartate was not incorporated into β-nitropropionic acid in either whole plants or cuttings, whereas radioactivity from [2-^{14}C]malonate and [2-^{14}C]malonylhydroxamate was incorporated. The radioactivity was incorporated specifically in cuttings, but in whole plants some radioactivity appeared at C-3 of β-nitropropionic acid.

The presence in *A. flavus* of a peroxidase which will oxidize β-nitropropionic acid (but not nitromethane, nitroethane or nitropropane) to nitrite led Marshall and Alexander[28] to suggest that β-nitropropionic acid might be involved in the oxidation of ammonia to nitrite. In support of this they cite the work of Eylar and Schmidt[29] who found that only fifteen of nearly a thousand organisms were able to form nitrite. All fifteen were fungi and thirteen were identified as strains of *A. flavus*. In view of the recent isolation of β-nitropropionic acid from *A. wentii* it is interesting that this organism, too, has been shown to produce nitrite and nitrate.[30]

Radioactivity from [1-^{14}C]lysine is incorporated specifically into the carboxyl group of 1-amino-2-nitrocyclopentanecarboxylic acid (9).[31] The amino-acid (9) is a particularly convenient substrate for studying the incorporation of ^{15}N from doubly-labelled lysine since the nitro-group and the amino-group can be obtained separately as ammonia and can serve as a control for each other. The results of experiments with [α-^{15}N,1-^{14}C]lysine and with [ε-^{15}N,1-^{14}C]lysine show that the nitrogen atoms of lysine are retained as the nitrogen atoms of 1-amino-2-nitrocyclopentanecarboxylic acid,[32] suggesting a sequence such as that of Scheme 9.

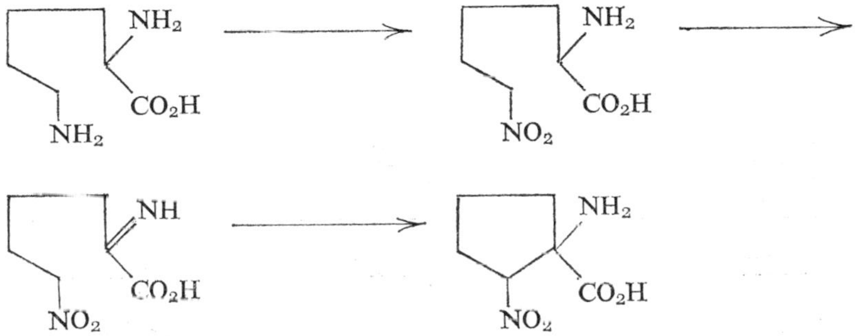

Scheme 9: Possible route for the incorporation of lysine into
1-amino-2-nitrocyclopentanecarboxylic acid

A phytotoxin from *Rhizoctonia solani* has been tentatively formulated as (11).[33] There is little evidence for structure (11) but if it is correct the

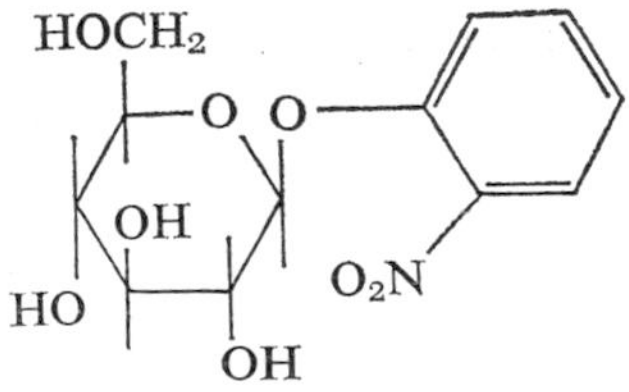

11

compound is of interest for its structural relationship to compound **(10)** and because of the rare occurrence of glycosides among fungal products.

B.2.a. *Nitro-compounds from Other Sources*

Several nitro-compounds have been isolated from bacteria, the most important being the antibiotic chloramphenicol **(12,** Scheme 10). The

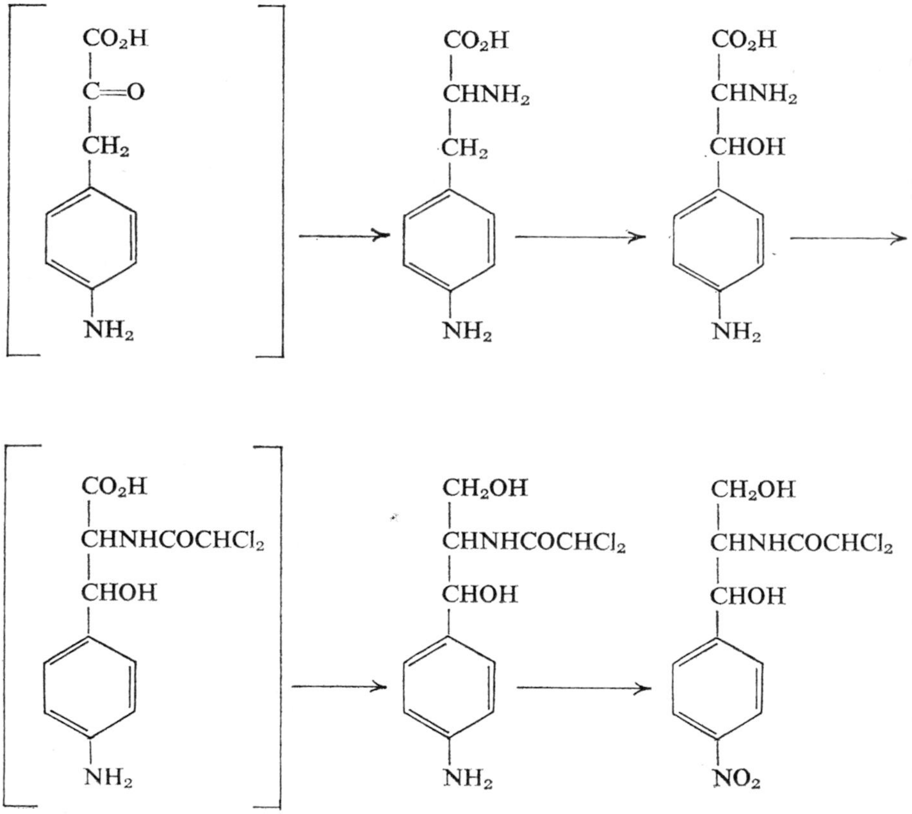

Scheme 10: The biosynthesis of chloramphenicol

biosynthesis of chloramphenicol has been studied in detail[34] and the most recent results[35] are consistent with the route of Scheme 10, the nitro-group arising by direct oxidation of an amino-group as in the case of β-nitropropionic acid and 1-amino-2-nitrocyclopentanecarboxylic acid. The biosynthesis of aureothin (13), an antibiotic from *Streptomyces thioluteus*[36] also seems likely to involve direct oxidation of an amino-group since *S. thioluteus* will oxidize *p*-aminobenzoate and related compounds to nitro-compounds.[37]

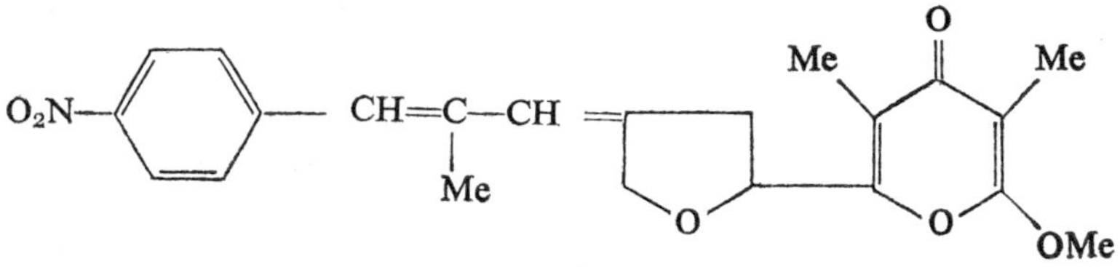

13

Apart from β-nitropropionic acid, aristolochic acid-I (14a),[38] aristolochic acid-II (14b) and miserotoxin (15),[39] are the only nitro-compounds to have been isolated from higher plants. The nitro-group of aristolochic acid-I is derived from the amino-group of tyrosine.[40]

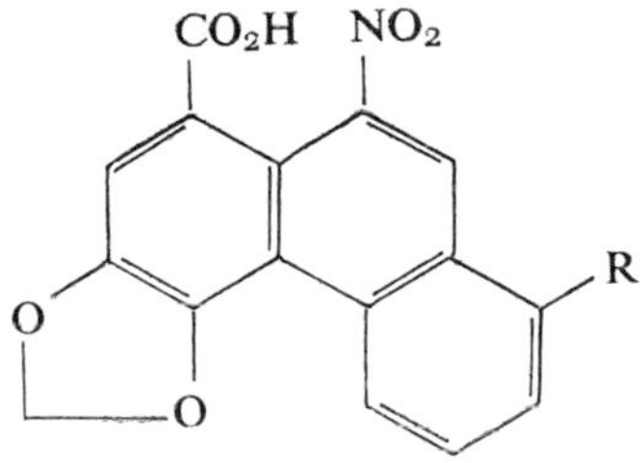

14a   R = OMe
14b   R = H

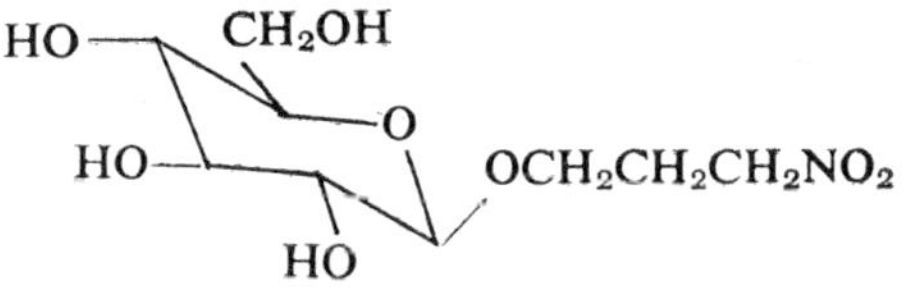

15

B.3. AGARITINE AND RELATED COMPOUNDS

16

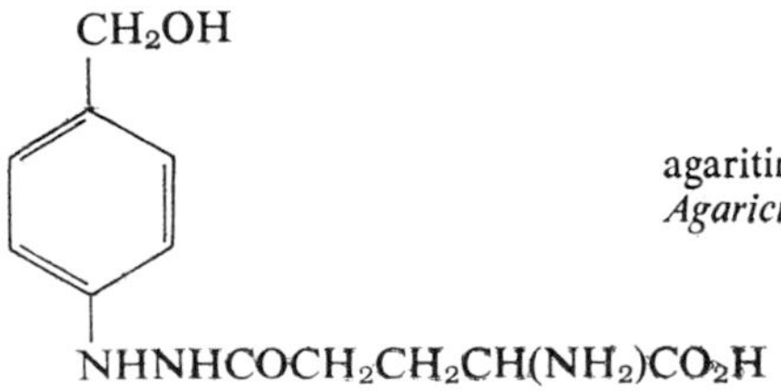

agaritine
Agaricus bisporus[41]

17

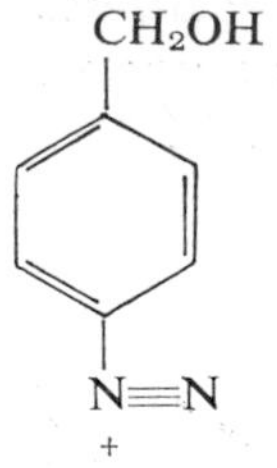

CH$_2$OH

N≡N
+

p-hydroxymethylbenzenediazonium ion
A. bisporus[42]

18

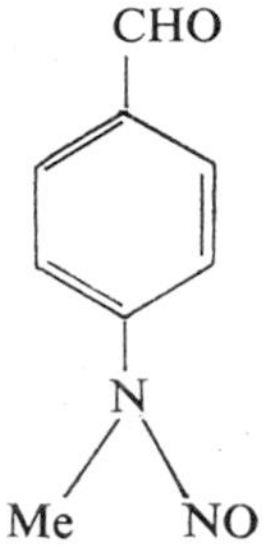

CHO

N
Me   NO

N-methylnitrosoaminobenzaldehyde
Clitocybe suaveolens[43]

Agaritine (**16**) was obtained from the press-juice of *A. bisporus*, the commercial mushroom of the United States, and has been detected in several other species of *Agaricus*. The diazonium compound (**17**) was isolated (by coupling reaction) from the basal stalk of *A. bisporus* during a study of the distribution of agaritine in the mushroom. An enzyme which catalyzes the hydrolysis of agaritine to *p*-hydroxymethylphenylhydrazine and glutamic acid has been partially purified.[44] The same enzyme also catalyzes the transfer of glutamate from agaritine to *p*-hydroxyaniline to give glutamic acid *p*-hydroxyanilide which had previously been isolated from *A. hortensis*.[45]

Several antibiotics from *Streptomyces* spp. contain N–N bonds including diazo-, hydrazido- and *N*-nitroso-groups. Gyromitrin (**19**), a toxin from the Ascomycete *Gyromitra esculenta*,[46] is an *N*-formylhydrazone.

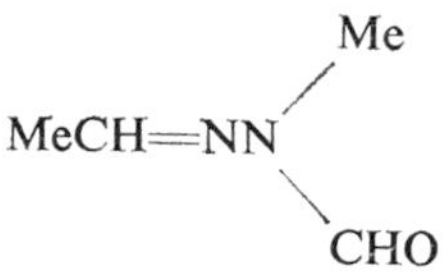

Me
/
MeCH=NN
\
CHO

19

B.4. INSECTICIDAL METABOLITES OF AGARICACEAE

Amanita muscaria, the common toadstool with the distinctive bright red cap speckled with white, has long been known to be toxic to flies (*Musca* spp.), a property from which both its systematic and its common name (Fly Agaric) have been derived. Five compounds with insecticidal properties have been isolated from *A. muscaria* or related Agaricaceae.

20

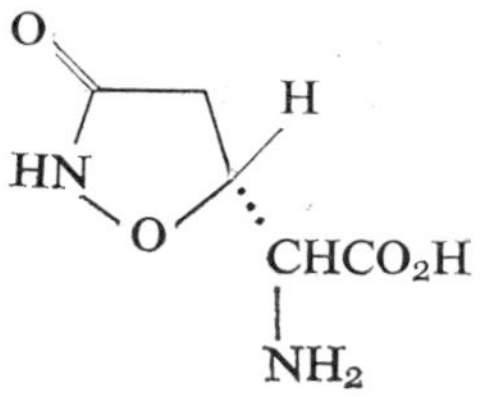

tricholomic acid
Tricholoma muscarium[47]

21

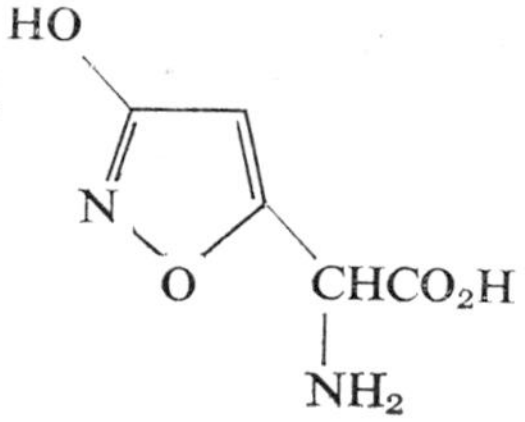

ibotenic acid (Amanita factor C)
Amanita muscaria,[48,49] *A. pantherina*,[48]
A. strobiliformis[50]

22

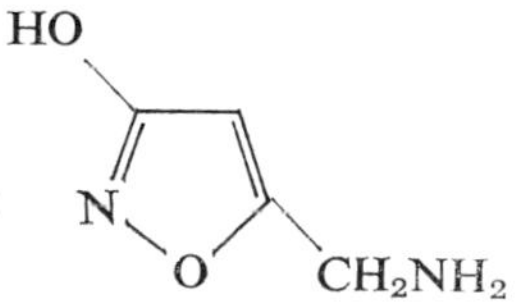

pantherine (agarin, Amanita factor B, muscimol)
A. pantherina,[51] *A. muscaria*[52]

23

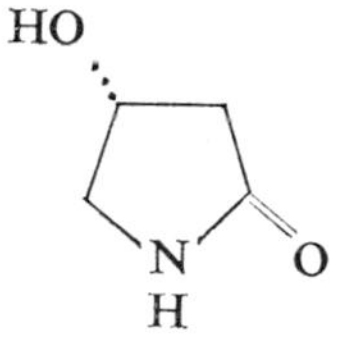

(−)-4-hydroxypyrrolidone
A. muscaria[8]

24

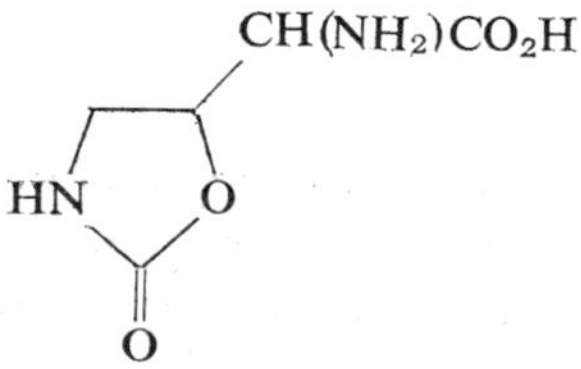

muscazone
A. muscaria[53]

Compounds (20) to (23) are formally derived from the amino-acid (25) which has not been obtained from a natural source. They are related structurally to the antibiotic cycloserine (26) produced by *Streptomyces* spp.

$HO_2CCH(NH_2)CH(OH)CH_2CO_2H$

25

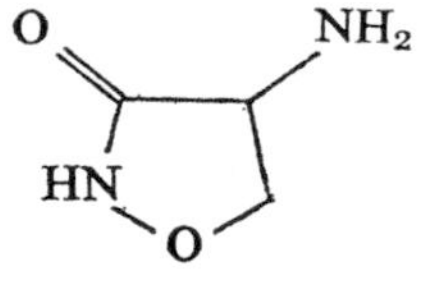

26

B.5. COMPOUNDS DERIVED FROM ANTHRANILIC ACID

27   2-pyruvoylaminobenzamide
Penicillium chrysogenum P. notatum[54]

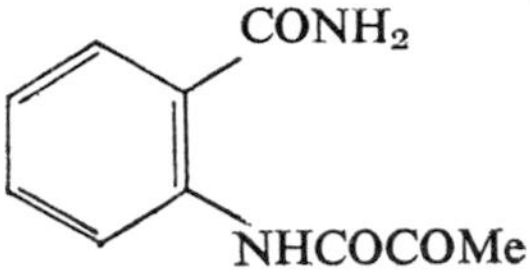

28   cinnabarin (polystictin)
Polyporus cinnabarinus[55]

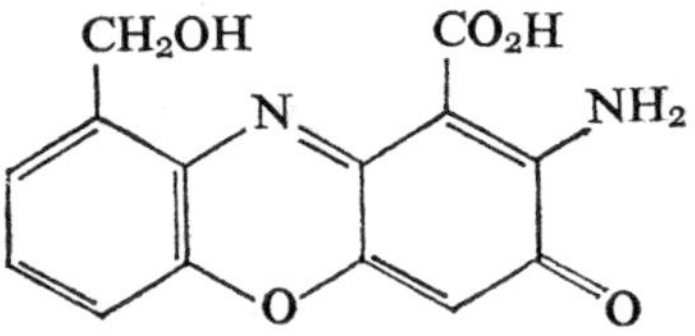

29   cinnabarinic acid
P. cinnabarinus[56]

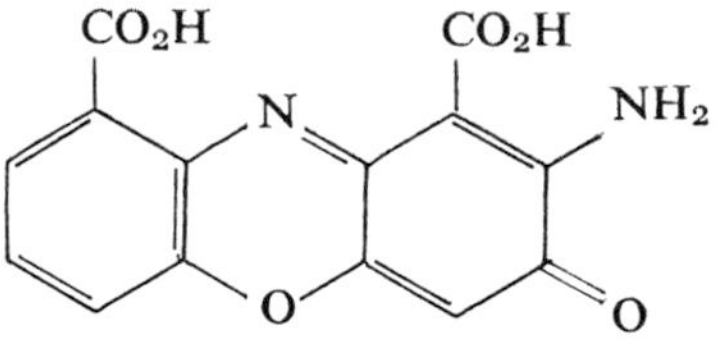

30   tramesanguin
P. cinnabarinus[57]

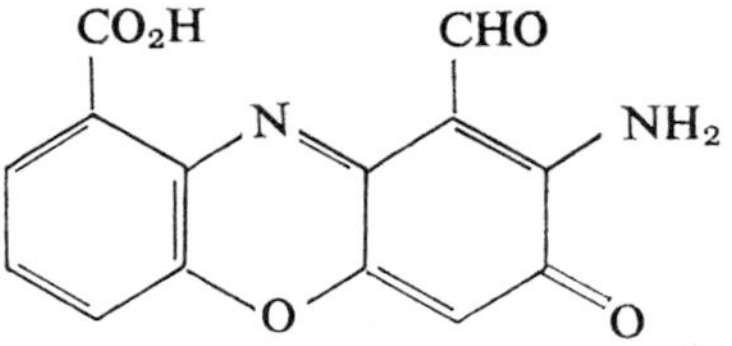

The phenoxazones are red pigments isolated from the fruiting bodies of *P. cinnabarinus* (for which there are a number of synonyms[55]). They presumably arise by oxidative coupling of 3-hydroxyanthranilic acid or related compounds, a process which is well-known *in vitro*. 3-Hydroxyanthranilic acid is a degradation product of tryptophan (p. 43) and a dihydroderivative, 6-amino-5-hydroxy-1,3-cyclohexadiene-1-carboxylic acid (**31**), has been obtained in yields of 10 g/litre from *Streptomyces aureofaciens*.[58] Circumstantial evidence for the production of phenoxazones by coupling of aminophenols is provided by the co-occurrence of *o*-aminophenol (questiomycin B)

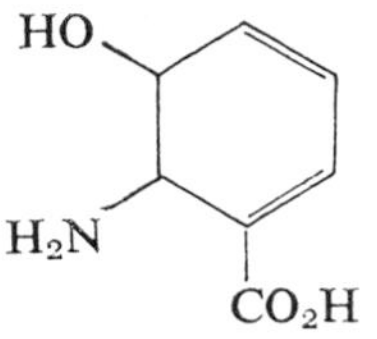

31 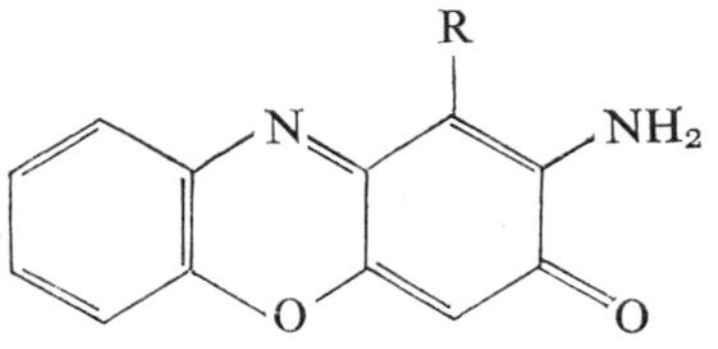   32

and questiomycin A (**32**, *R* = H) in a *Streptomyces* sp.[59] 2-Amino-1-carboxy-3*H*-phenoxazin-3-one (**32**, *R* = CO$_2$H) has been isolated from a *Nocardia* sp.[60]

B.6. PYRIDINE DERIVATIVES

33

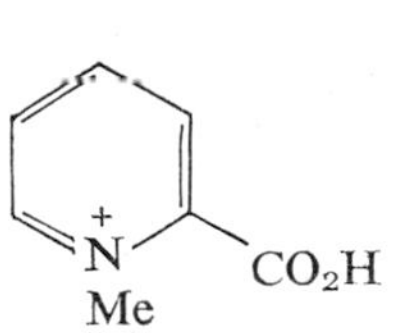

trigonelline
Polyporus sulphureus[61]

34

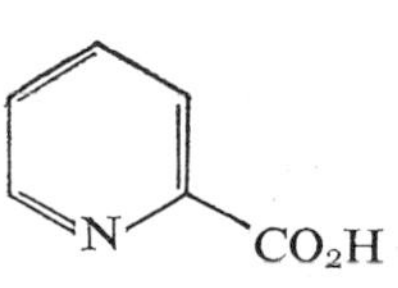

homarine
P. sulphureus[61]

35

α-pycolic acid
Piricularia oryzae[62]

36

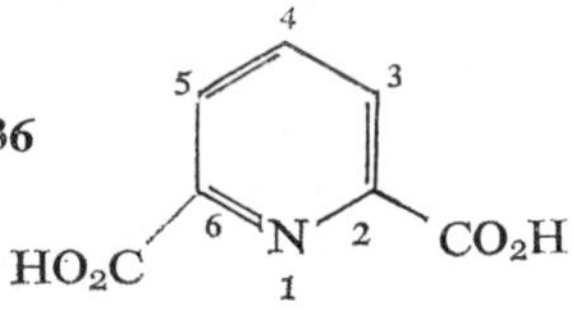

dipicolinic acid L^{68-70}
Penicillium citreo-viride[63]

37

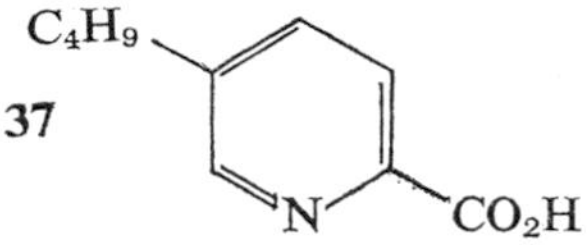

fusaric acid L^{74-76}
Fusarium spp.[64,65]

38

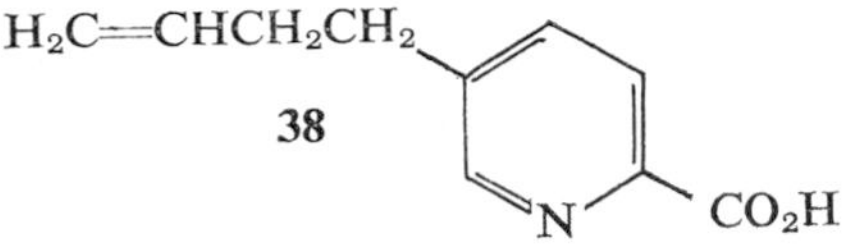

dehydrofusaric acid
Fusarium spp.[65]

Although the pyridine nucleus is present in a variety of natural products, ranging from alkaloids to essential cofactors, it has rarely been encountered among fungal secondary metabolites. Trigonelline and homarine were detected in small quantity in *P. sulphureus* but occur widely in higher plants and marine crustaceans. Dipicolinic acid was first obtained as a natural product from bacterial spores, where it occurs widely if not universally. Fusaric acid was isolated as a plant-growth inhibitor from *Fusarium moniliforme* (*Gibberella fujikuroi*) by Yabuta *et al.*[64] during a study of the plant-growth promoters (gibberellins) produced by the organism. It has since been isolated from a number of *Fusaria* including *F. oxysporium-lycopersica* which also produces dehydrofusaric acid.[65] *F. oxysporium-lycopersica* causes a wilt disease of plants and it has been suggested[66] that fusaric acid is responsible for the symptoms, though more recent evidence[67] does not support this view.

The pyridine ring arises by a variety of routes in nature; indeed nicotinic acid (39), one of the most important natural derivatives of pyridine, is formed by two independent routes. Nicotinic acid is involved in the biosynthesis of diphosphopyridinenucleotide (NAD) and triphosphopyridinenucleotide (NADP), the cofactors for most biological dehydrogenations, and is made available for this purpose in animals and in some micro-organisms by degradation of tryptophan via 3-hydroxyanthranilic acid. Nicotinic acid is also involved in the biosynthesis of the alkaloid nicotine, and for this purpose it is derived by condensation of a C_3 and a C_4 unit, possibly glyceraldehyde and aspartic acid respectively; this route to nicotinic acid is also used by some fungi and bacteria.

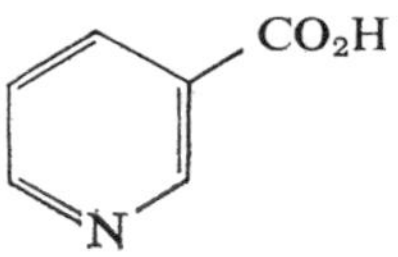

39

The biosynthesis of dipicolinic acid has been studied in bacterial spores and, more recently, in *P. citreo-viride*. It is firmly established[68,70] that the immediate precursor of the pyridine ring is $\alpha\varepsilon$-diaminopimelic acid or the corresponding diketo- or aminoketoacids (ambiguity arises because diketopimelic acid reacts non-enzymatically with ammonium acetate to give dipicolinic acid[71] and because the amino- and keto-acids are probably in equilibrium in micro-organisms). Diketopimelic acid has been detected in *P. citreo-viride* grown on a minimal medium,[70] and a fraction catalysing the synthesis of dipicolinic acid from the diketopimelate-ammonia complex has been obtained from sporangia of *Penicillium*.[72]

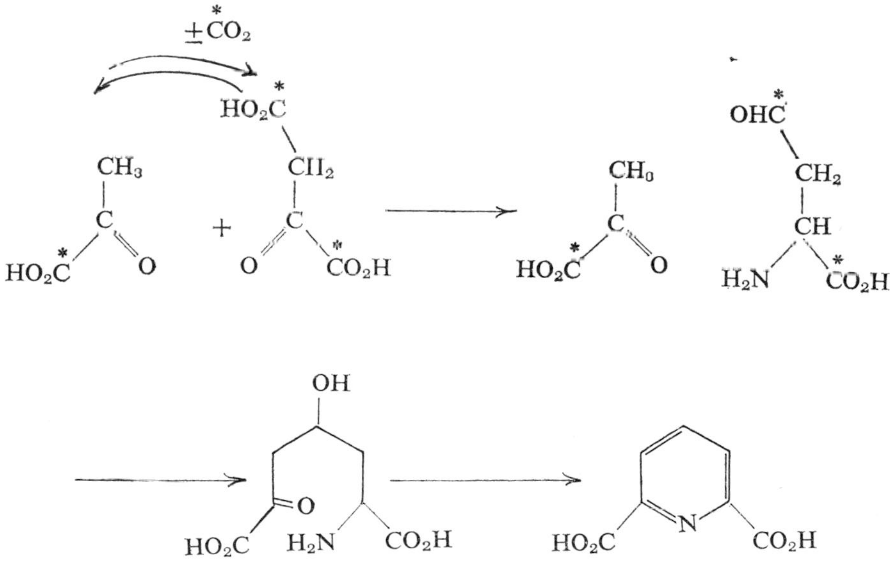

Scheme 11: Biosynthesis of dipicolinic acid from $C_3 + C_4$ intermediates[69]

The origin of the C_7 intermediate is less certain. Kanie *et al.*[69] studied the incorporation of $^{14}CO_2$,[1-^{14}C]-, [2-^{14}C]- and [6-^{14}C]glucose, and [1-^{14}C]- and [2-^{14}C]glycerol into dipicolinic acid, degrading the radioactive product so as to obtain, separately, the carboxyl carbons, C-2 and C-6, C-3 and C-5, and C-4. Their results are consistent with the sequence shown in Scheme 11, though the intermediates have not been rigorously defined. The same route had previously been proposed[73] for the biosynthesis of dipicolinic acid in bacterial spores.

On the other hand, the experiments of Tanenbaum and Kaneko,[70] which did not involve degradation of the pyridine ring of the labelled dipicolinic acid, gave results which are more consistent with the sequence of Scheme 12, which links the biosynthesis of dipicolinic acid with that of lysine which is also formed from α-ketoadipic acid.

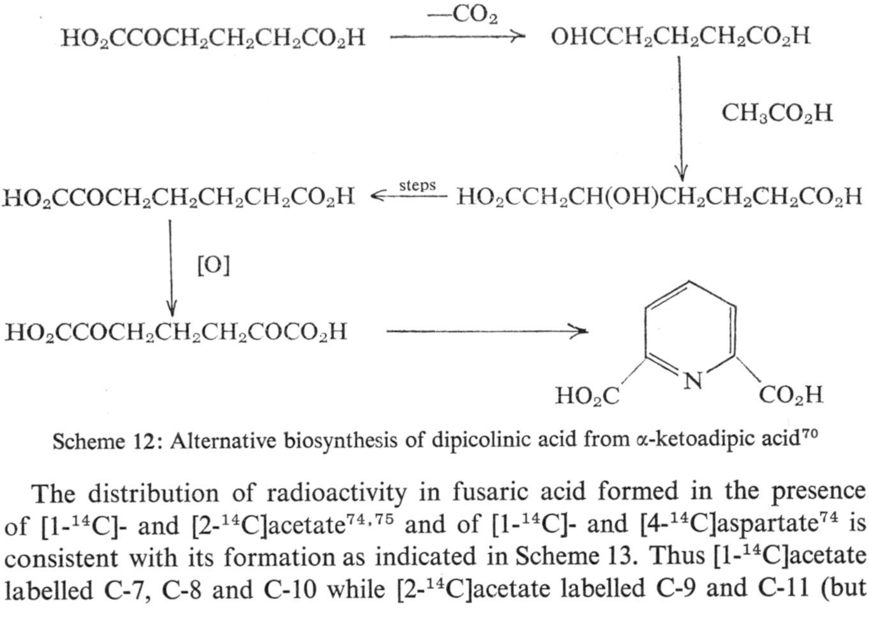
—CO_2
$HO_2CCOCH_2CH_2CH_2CO_2H$
$OHCCH_2CH_2CH_2CO_2H$
CH_3CO_2H
$HO_2CCOCH_2CH_2CH_2CH_2CO_2H$
steps
$HO_2CCH_2CH(OH)CH_2CH_2CH_2CO_2H$
[O]
$HO_2CCOCH_2CH_2CH_2COCO_2H$
N
HO_2C
CO_2H

Scheme 12: Alternative biosynthesis of dipicolinic acid from α-ketoadipic acid[70]

The distribution of radioactivity in fusaric acid formed in the presence of [1-^{14}C]- and [2-^{14}C]acetate[74,75] and of [1-^{14}C]- and [4-^{14}C]aspartate[74] is consistent with its formation as indicated in Scheme 13. Thus [1-^{14}C]acetate labelled C-7, C-8 and C-10 while [2-^{14}C]acetate labelled C-9 and C-11 (but

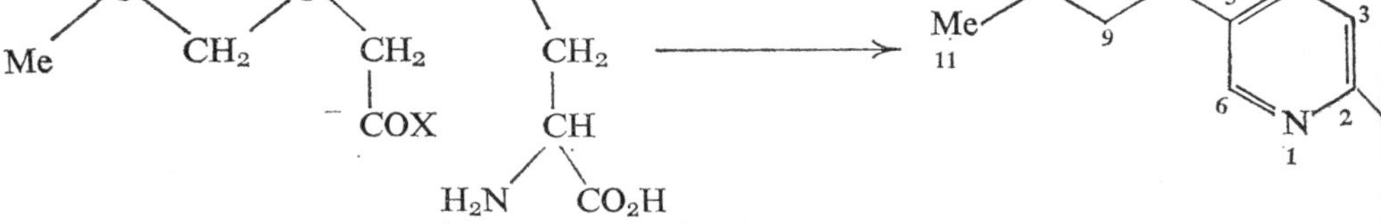
O
O
C
C
Me
CH_2
CH_2
COX
HO_2C
CH_2
CH
H_2N
CO_2H
Me
11
10
9
8
5
4
3
6
N
2
1

Scheme 13: Incorporation of aspartate and acetate into fusaric acid not C-10); both labelled the pyridine ring. Fusaric acid produced in the presence of [1-^{14}C]aspartate had two-thirds of its radioactivity at C-7 and the remainder in the ring; [4-^{14}C]aspartate gave the opposite distribution. In the latter experiment, the distribution of radioactivity between C-7 and the pyridine ring was similar to that between C-1 and C-4 of protein aspartate, so that the radioactivity of the ring is probably located at position 4.

Experiments with [1-^{13}C]- and [2-^{13}C]acetate and with [4-^{13}C]aspartate, using n.m.r. spectroscopy to analyse the product, support the above results.[76] An experiment with [^{15}N,1-^{14}C]aspartate gave the rather surprising result that ^{15}N was more efficiently incorporated into fusaric acid than was ^{14}C. Possibly aspartic acid provides the nitrogen of fusaric acid by transamination to some intermediate further along the biosynthetic sequence.

B.7. INDOLE DERIVATIVES

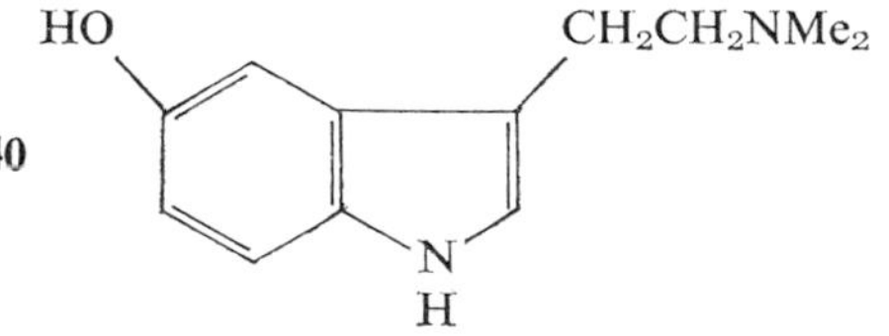

bufotenin
Amanita mappa[77]

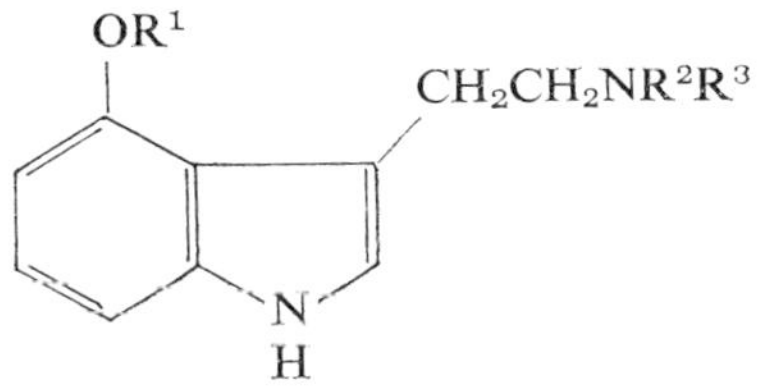

	R^1	R^2	R^3		
a	H$_2$PO$_3$	Me	Me	psilocybin	L[85]
b	H	Me	Me	psilocin	
				Psilocybe spp.[78]	
c	H$_2$PO$_3$	H	Me	baeocystin	
				P. baeocystis[79]	
d	H$_2$PO$_3$	H	H	norbaeocystin	
				P. baeocystis[71]	

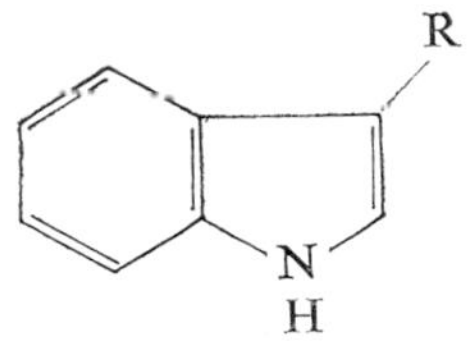

a: R = CHO, indole-3-aldehyde
b: R = CO$_2$H, indole-3-carboxylic acid
        Lasiodiplodia theobromae[80]
c: R = CHMeCO$_2$H, indole-3-isopropionic   L[86]
        acid
        Claviceps sp.[81]

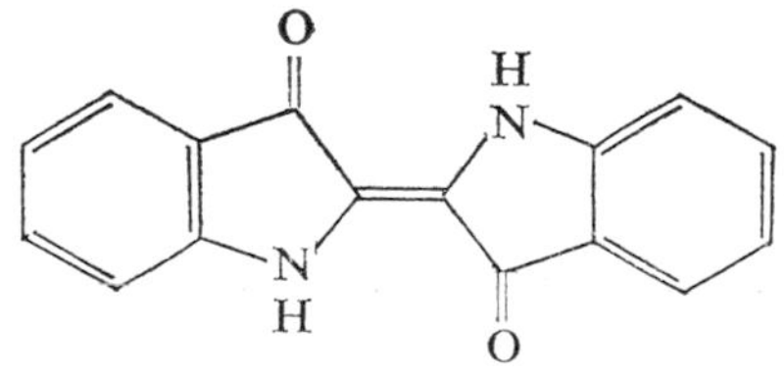

indigo
Schizophyllum commune,[82]
Agaricus campestris[83]

The indole nucleus, derived from tryptophan, is present in many fungal products but here we are concerned with simple substituted indole derivatives. Psilocybin (**41a**) and psilocin are the halucinogenic principles of the Mexican magic-fungus "Teonanacatl" used in religious ceremony; an interesting account of the history of the drug and of the isolation of the compounds is available.[84] The biosynthesis of psilocybin has been studied in submerged cultures of *Psilocybe cubensis*,[85] the results suggesting the sequence of Scheme 14 though this may not be a unique pathway (see also ref. 79).

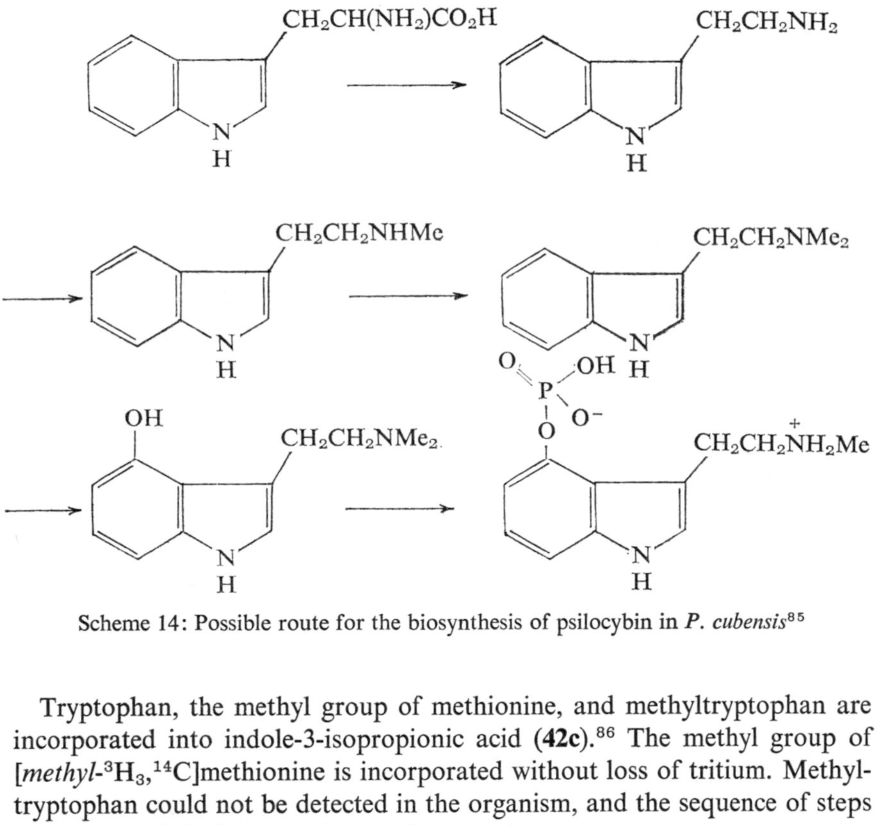

Scheme 14: Possible route for the biosynthesis of psilocybin in *P. cubensis*[85]

Tryptophan, the methyl group of methionine, and methyltryptophan are incorporated into indole-3-isopropionic acid (**42c**).[86] The methyl group of [*methyl*-^{3}H$_3$,^{14}C]methionine is incorporated without loss of tritium. Methyltryptophan could not be detected in the organism, and the sequence of steps leading to indole-3-isopropionic acid is not known.

B.8. THE ERGOT ALKALOIDS

This is a specialist subject which has been adequately reviewed elsewhere and only a general account of the field will be given, with references to review articles except for one or two recent results. A complete list of the compounds is not presented.

Ergot, the rye-borne sclerotia of the fungus *Claviceps purpurea* (see also the ergochromes), has a long history both as a cause of social disaster and as a drug. Bread made from contaminated rye was the cause of the poisoning (now referred to as ergotism but known earlier as Holy Fire or St. Anthony's Fire) of the populations of large areas, while small doses of ergot were used in obstetrics to induce uterine contractions. The physiological effects are due to a complex mixture of alkaloids[87] some of which are derivatives of lysergic acid (see below). While the obstetric use of the compounds has declined, being now restricted to the control of haemorrhage after childbirth, interest in lysergic acid has been maintained by Hofmann's observation in 1959 of the hallucinogenic effects of lysergic acid diethylamide (LSD) which has since been widely used as a research tool in psychiatry, and even more widely by those intent on "taking a trip". Lysergic acid amide has recently been recognised as the hallucinogenic component of the plants *Ipomea tricolor* and *Rivea corymbosa* which have long been used by Mexican Indians in religious ceremonies.

The ergot alkaloids proper are amides of lysergic acid (**44a**) which occur in interconvertible pairs epimeric at C-8. The amine part of the amide ranges

in complexity from ammonia to complicated cyclic tripeptides such as that present in ergotamine (**44b**). A second class of alkaloids, the so-called clavine type, were subsequently isolated from ergot of wild grasses. These have the general formula (**45**) where R^1, R^2 and R^3 are H, OH or OAc; as in the case of the ergot alkaloids pairs of C-8 epimers occur.

The ergot alkaloids can now be produced by cultivation of *Claviceps* spp. on aqueous media.[88] The *Claviceps* spp. are classified according to the grass from which they were isolated, the classes being *Elymus*, *Agropyron*, *Pennisetum*, *Paspalum* and *Secale*. The *Paspalum* strain (*Claviceps paspali*) is particularly useful in that it produces in stirred culture high yields of lysergic acid methylcarbinolamide [**44**, R = NHCH(OH)Me] which is readily converted to lysergic acid amide. In surface culture, good yields of the alkaloids

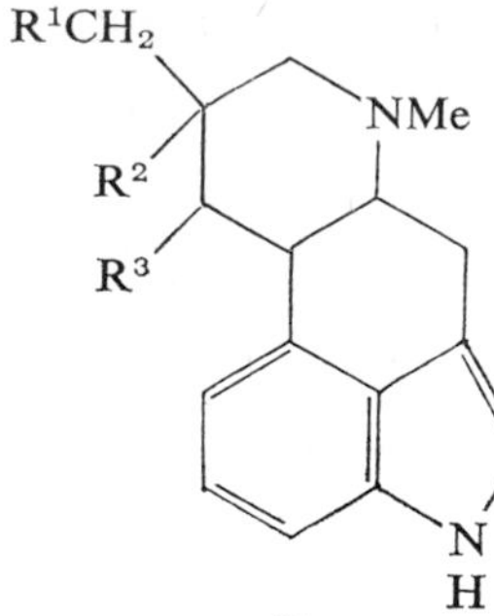

45 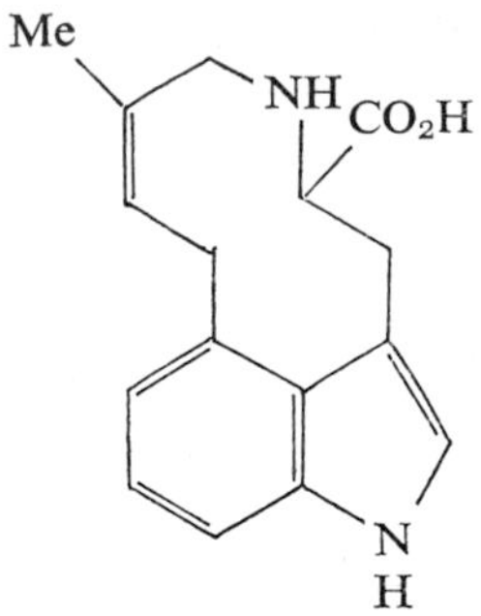   **46**

are obtained by replacement cultures. Compounds of this type have also been obtained from species of *Aspergillus* and *Penicillium*.

A comprehensive review of the biosynthesis of the ergot alkaloids has recently appeared[89] and the present state of knowledge is summarized in Scheme 15. It has recently been shown[90] that 4-dimethylallyltryptophan (**47**)

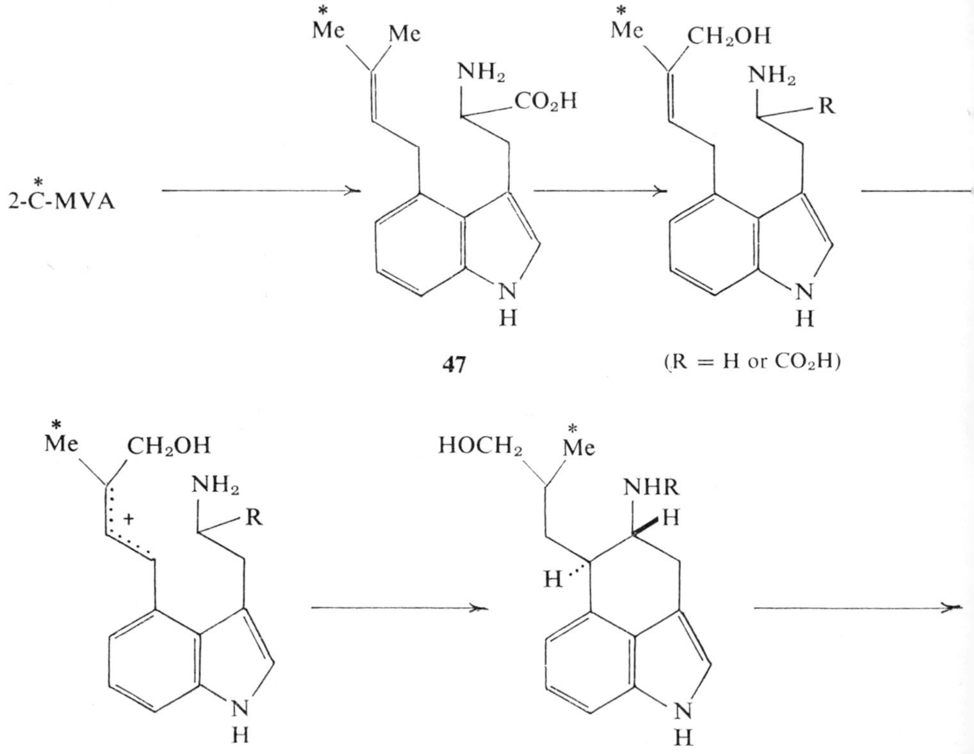

47   (R = H or CO$_2$H)

(R = H or CO$_2$H)   Chanoclavine-I (R = Me)

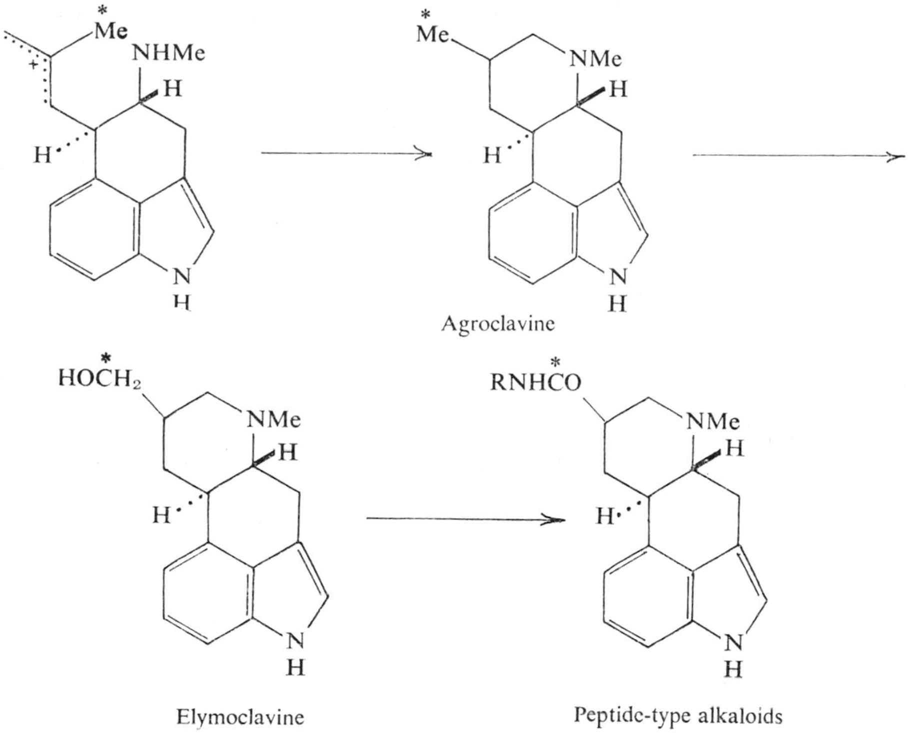

Agroclavine

Elymoclavine   Peptide-type alkaloids

Scheme 15: The biosynthesis of the ergot alkaloids showing the incorporation of radioactivity from [2-^{14}C]mevalonate

accumulates under anaerobic conditions, the first demonstration that *Claviceps* is capable of synthesizing this presumed intermediate, and a new metabolite, clavicipitic acid (**46**), has been isolated from *Claviceps* cultures to which ethionine had been added.[91] Clavicipitic acid is also present in small amounts in uninhibited cultures so that it is a normal metabolite.

B.9. THE CYCLOPIAZONIC ACIDS

48

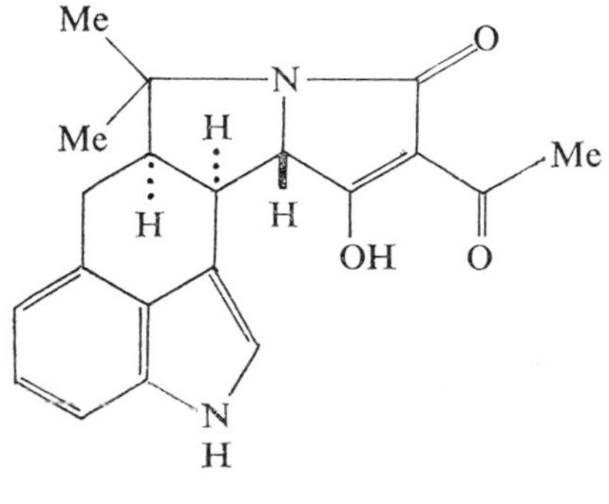

cyclopiazonic acid
Penicillium cyclopium[92]

49

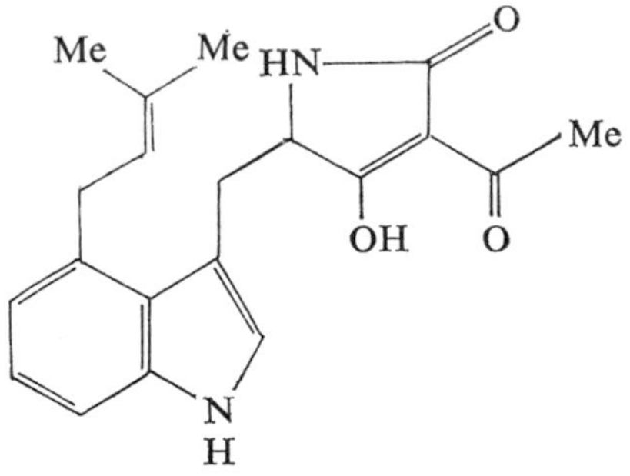

β-cyclopiazonic acid
P. cyclopium[93]

The 4-dimethylallyltryptophan moiety of the ergot alkaloids is also present in cyclopiazonic acid (**48**), where it has undergone condensation with a (formal) acetoacetyl unit to form a tetramic acid. β-Cyclopiazonic acid is produced by *P. cyclopium* in the absence of trace metals and is a precursor of cyclopiazonic acid on full medium. Acetate, mevalonate, and tryptophan are incorporated as expected into cyclopiazonic acid.[93]

C. SECONDARY METABOLITES DERIVED FROM TWO AMINO-ACIDS

C.1. DIKETOPIPERAZINES AND RELATED COMPOUNDS

A variety of fungal metabolites possess a diketopiperazine ring or are derived from intermediates of this type. We can divide the compounds into (i) simple diketopiperazines, in which the ring system is unchanged, (ii) aspergillic acid and related compounds, in which the ring system has been oxidized at the nitrogen atoms or has lost one of the carbonyl oxygens, (iii) gliotoxin and related compounds, in which the diketopiperazine system is bridged by sulphur and (iv) viridicatin and related compounds, in which the ring has undergone contraction.

C.1.a. *Simple Diketopiperazines*

50

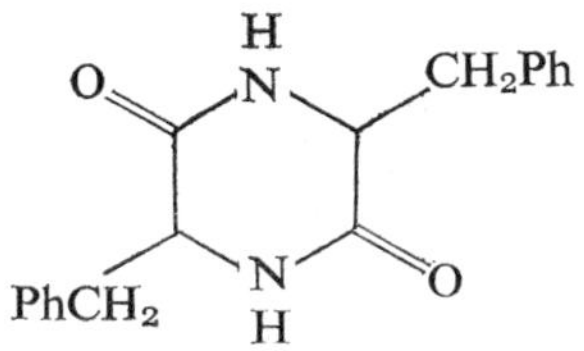

L-phenylalanine anhydride
Penicillium nigricans[94]
Streptomyces noursei[95]

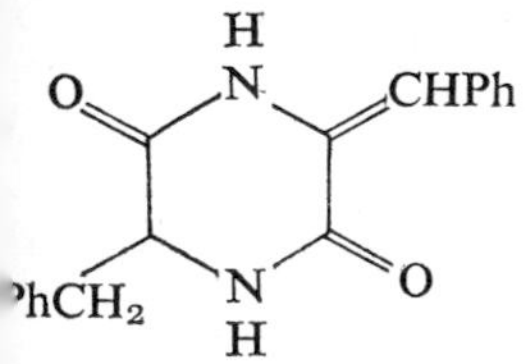

3-benzyl-6-benzylidene diketopiperazine
S. noursei[95,96]

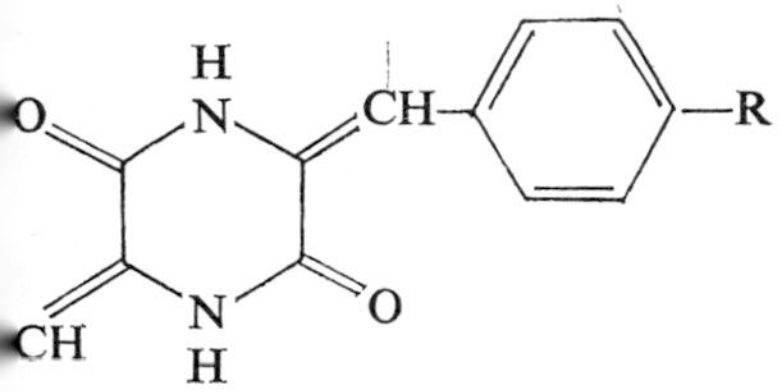

a: R = H, 3,6-dibenzylidene diketopiperazine
    S. noursei[95]
b: R = OMe, 3-anisylidene-6-benzylidene-
    diketopiperazine
    S. thioluteus[97]

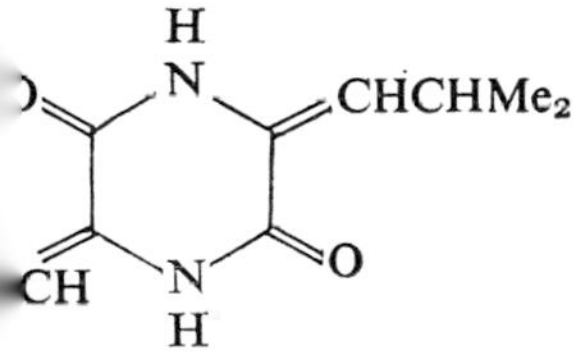

albonoursin
S. noursei[95,96,98]

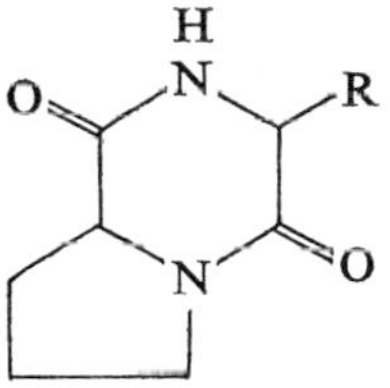

a: R = (CH₃)₂CHCH₂, prolylleucyl anhydride
b: R = (CH₃)₂CH, prolylvalyl anhydride
    Aspergillus ochraceus,[99] *Metarrhizium anisopliae,*[00]
    Rosellinia necatrix[100]

c: R — PhCH₂, prolylphenylalanyl anhydride
    R. necatrix[100]

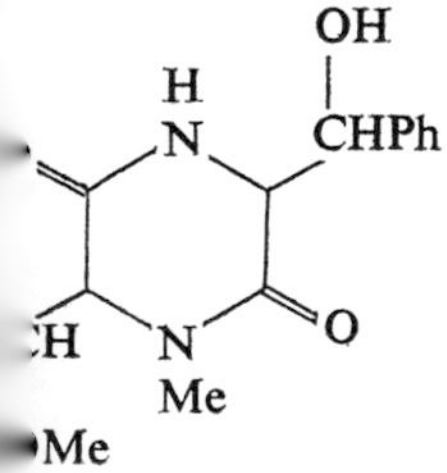

picroroccellin
Roccella fuciformis[101]

56

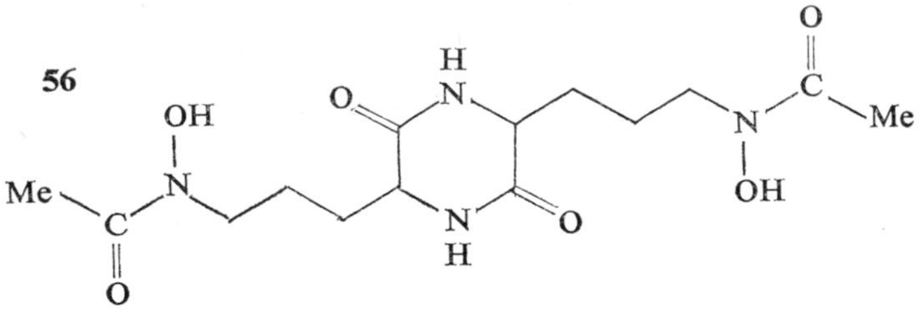

rhodotorulic acid
Rhodotorula pilimanae[1c

57

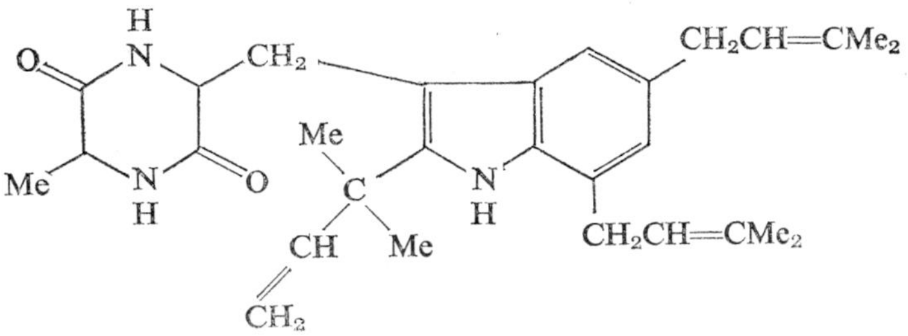

echinulin   L[109
Aspergillus echinulatus[103]
A. repens, A ruber,
A. amstelodami[104]
A. chevalieri[105]

58

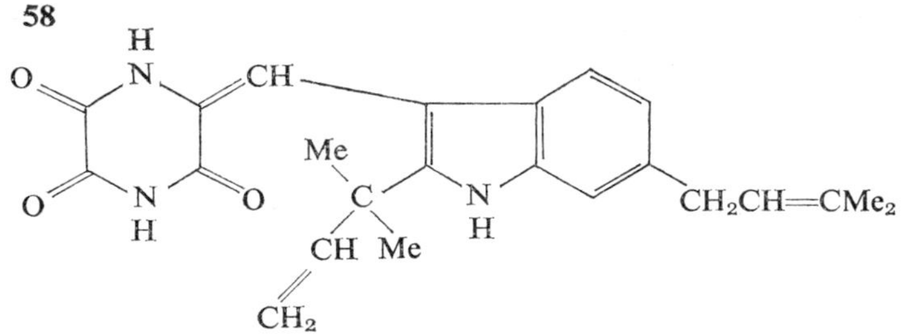

neoechinulin
A. amstelodami[106]

59

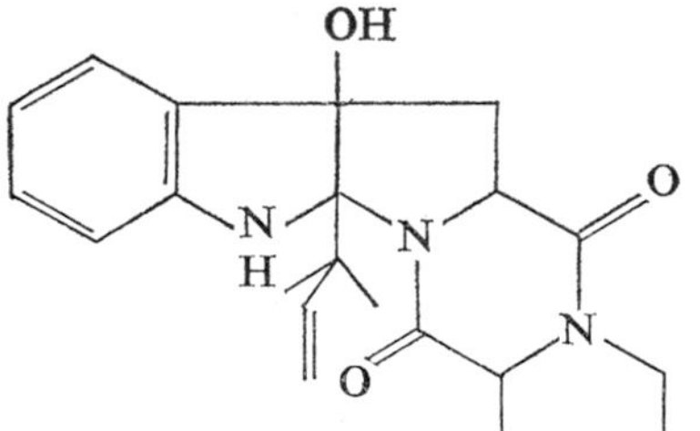

brevianamide E
Penicillium brevi-compactum[107]

60

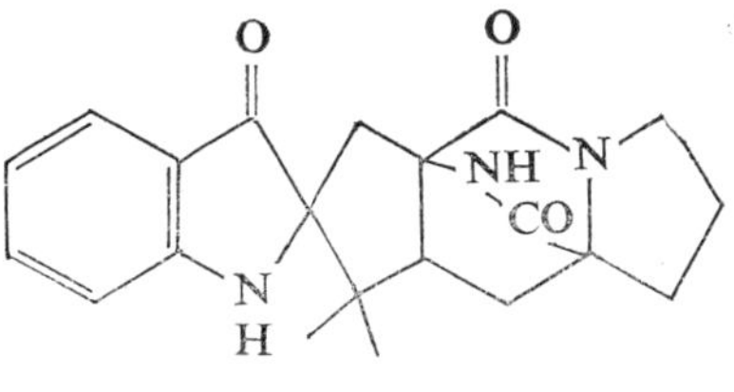

brevianamide A
P. brevi-compactum[107]

Diketopiperazines derived from the common amino-acids are frequently produced in fermentations and doubt has been expressed[108] as to whether they are true metabolites; in particular, prolylleucyl anhydride (54a) has been isolated from peptone and from corn-steep liquor, both commonly used in fermentation media. However, Brown et al.[95] could not detect diketopiperazines in the yeast extract used in their medium (though whether before or after autoclaving is not clear), so that the dibenzyl (50) and dehydrodibenzyl diketopiperazines (51) to (53) seem genuine products of the fermentation.

Picroroccellin (55), a lichen product, is derived from two oxygenated phenylalanine moieties of the type which might be involved in the biosynthesis of cyclopenin (p. 333). Rhodotorulic acid (56), isolated from a red yeast is formed from two molecules of N-acetyl-N-hydroxyornithine, an amino-acid which is also present in the siderochromes (p. 343). Like the siderochromes, rhodotorulic acid is a growth factor for *Bacillus* and *Arthrobacter*; it does not, however, antagonize the antibacterial action of albomycin.

The structure of echinulin (57) was derived largely as a result of the degradative studies of Quilico and his colleagues, but Birch et al.[109] made an important contribution by applying biosynthetic considerations to the problem. Using [2-^{14}C]mevalonate or [1-^{14}C]acetate they were able to show that three isoprene units are present in echinulin. They also showed that [1-^{14}C]alanine is specifically incorporated into the diketopiperazine ring, and subsequently Birch and Farrar[110] showed that the [*methylene*-^{14}C]tryptophan is incorporated into echinulin. MacDonald and Slater[111] found that [*carboxyl*-^{14}C]tryptophan is incorporated into echinulin without randomization of the label and that L-tryptophan is incorporated more efficiently than the D-isomer (a result consistent with the current opinion of the absolute stereochemistry of echinulin[112]). The substitution of tryptophan by isoprene units also occurs in the formation of the ergot alkaloids (p. 318) and cyclopiazonic acid (p. 319).

The isoprenoid side-chain of brevianamide E (59) corresponds to one of those in echinulin, and the formation of a second tetrahydropyrrole ring also occurs in the biosynthesis of the sporidesmins (74)–(78). Brevianamide A (60) could arise by rearrangement of brevianamide E.

C.1.b. *Aspergillic Acid and Related Compounds*

61

flavacol
Aspergillus flavus[113]

62

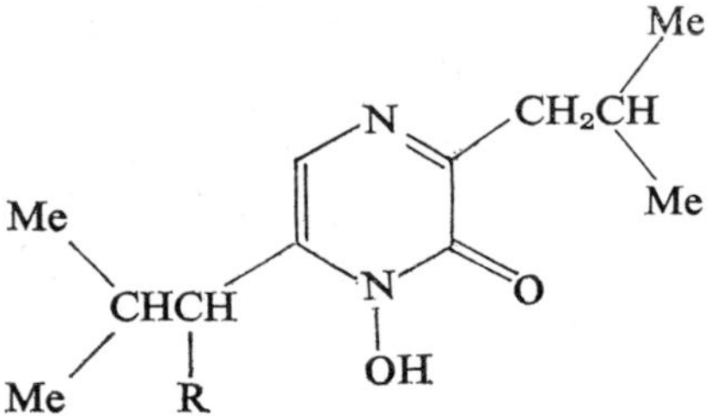

a : R = H, neoaspergillic acid L^{123}
    A. sclerotiorum[114]

b : R = OH, neohydroxyaspergillic acid L^{123}
    A. sclerotiorum[114,115]

63

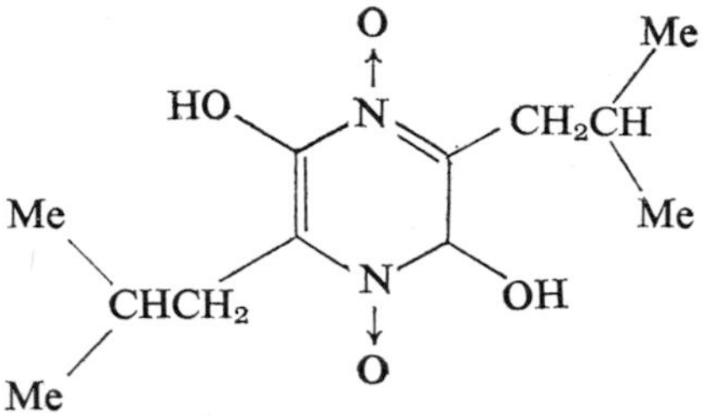

pulcherrimic acid L^{123}
Candida pulcherrima[116]

64

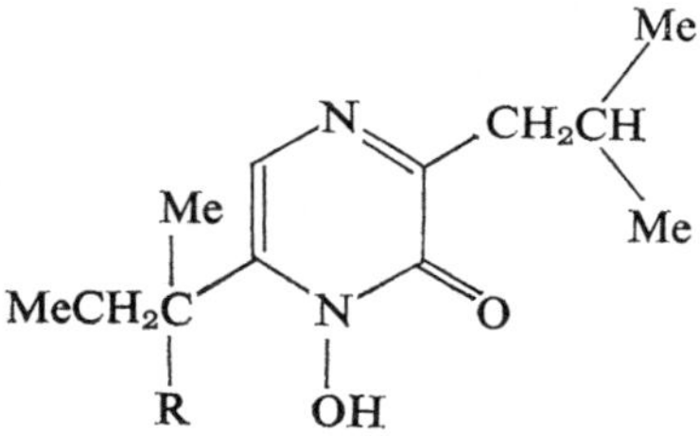

a : R = H, aspergillic acid L^{123}
    A. flavus[117]

b : R = OH, hydroxyaspergillic acid L^{123}
    A. flavus[118]

65

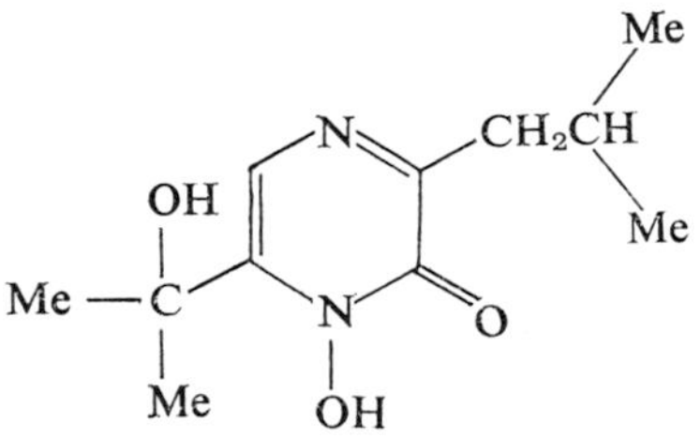

muta-aspergillic acid
A. oryzae[119]

66

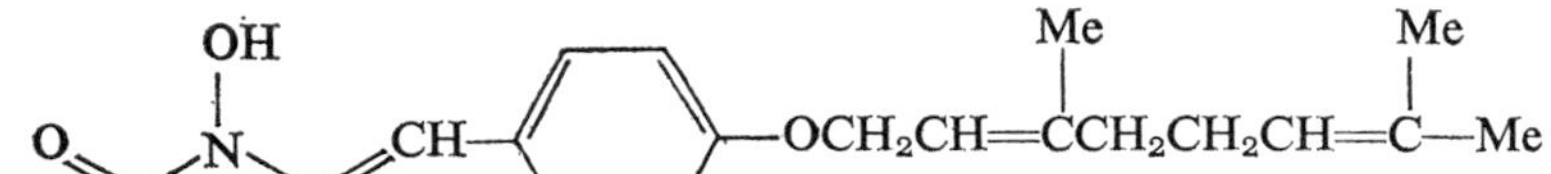

mycelianamide L^{127}
Penicillium griseofulvum[120]

In these compounds the diketopiperazine system has been modified by loss of one of the oxygen atoms or by oxygenation of one or both of the nitrogen atoms. The compounds are tautomeric, with the hydroxamic acid form (a) predominating in solution. Several of the compounds were isolated

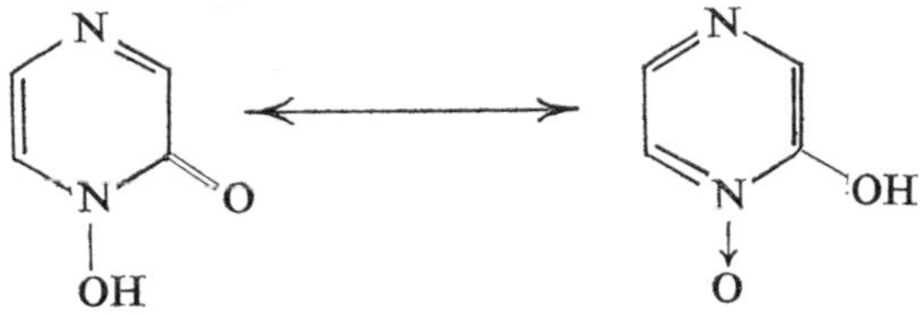

(a)   (b)

as a result of their antibacterial activity. Pulcherrimic acid (63) was isolated as a red iron-complex called pulcherrimin and seems to occur in two isomeric forms.[121] The ultraviolet spectrum of synthetic muta-aspergillic acid (65) differs slightly from that of the natural material so that there is some doubt as to its structure.[122]

The biosynthesis of several of the compounds has been studied by MacDonald and his co-workers (for a review see ref. 123). Using labelled

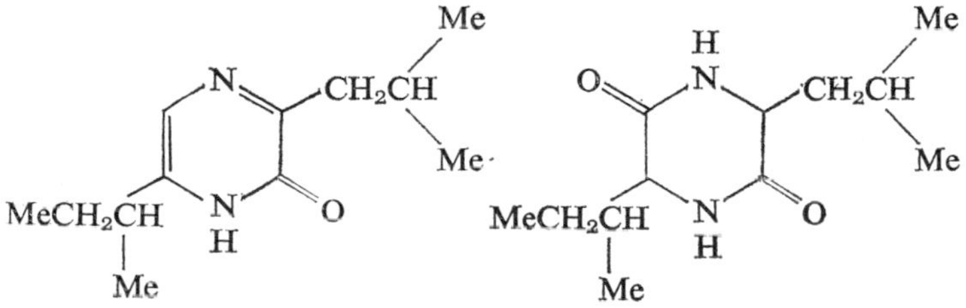

67   68

precursors, MacDonald[124] showed that aspergillic acid (64a) and hydroxy-aspergillic acid (64b) are formed from leucine and isoleucine in A. flavus and that aspergillic acid was a precursor of hydroxyaspergillic acid; radio-activity from aspergillic acid and hydroxyaspergillic acid is not incorporated into mycelial protein. Micetich and MacDonald[114] found that washed mycelium of A. sclerotiorum converts deoxyaspergillic acid (67) to aspergillic acid; on the other hand isoleucylleucyl anhydride (68) gives higher activity in mycelial protein of A. flavus than in aspergillic acid.

A similar picture was obtained[125] for the biosynthesis of neoaspergillic acid (62a) and neohydroxyaspergillic acid (62b) from leucine in A. sclero-tiorum. Flavacol (61) was converted to neoaspergillic acid and neohydroxyaspergillic acid but radioactivity from leucine anhydride was incorporated to a greater extent into mycelial leucine than into the hydroxamic acids. The results may be summarized as in Scheme 16.

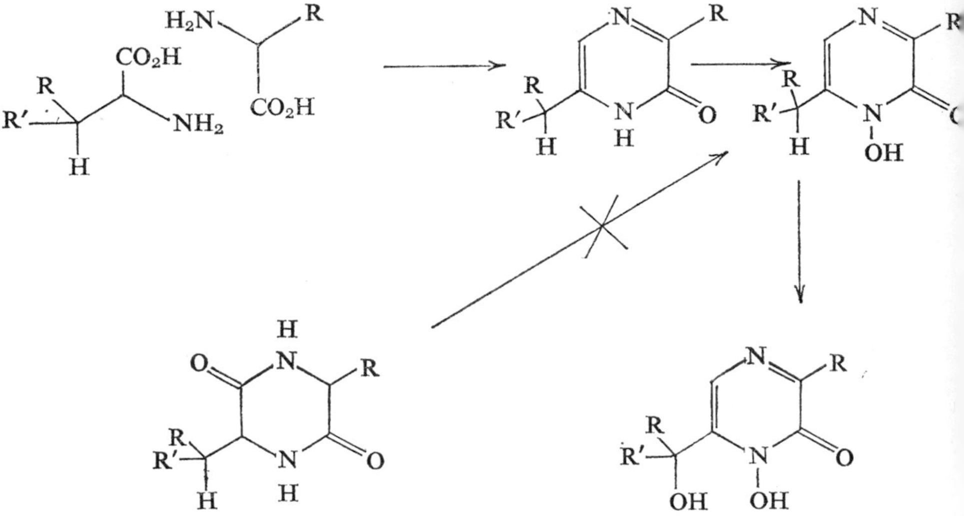

Scheme 16: Biosynthesis of the aspergillic acids

MacDonald has shown[126] that L-leucine anhydride is derived from L-leucine and is incorporated into pulcherrimic acid without labelling mycelial leucine. The biosynthesis of pulcherrimic acid thus follows the route of Scheme 17, differing from that of Scheme 16 for the aspergillic acids.

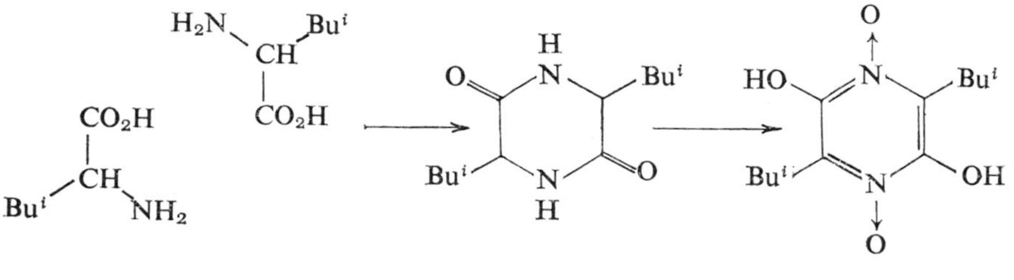

Scheme 17: Biosynthesis of pulcherrimic acid

The heterocyclic ring of mycelianamide (66) is derived formally from alanine and a dehydrotyrosine which is etherified with geraniol. Birch and his co-workers[127] have shown that the side-chain is derived from mevalonate without randomization of label in the *gem*-dimethyl group (at the time of these experiments a wrong structure had been assigned to the side-chain but the results are unaffected). The biosynthesis of the rest of the molecule has not been studied.

C.1.c. *Gliotoxin and Related Compounds*

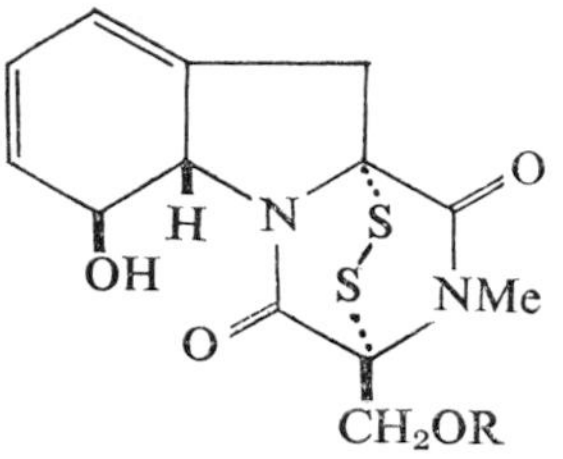

I[149,152]

a: R = H, gliotoxin
  Widespread among Fungi Imperfecti[128]
b: R = Ac, gliotoxin acetate
  Penicillium terlikowski[129]

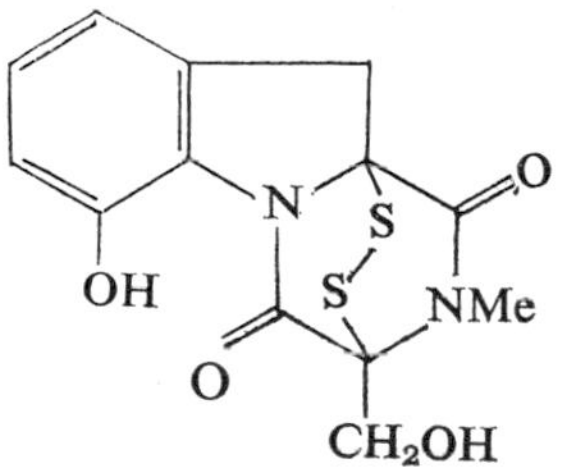

dehydrogliotoxin
P. terlikowski[130]

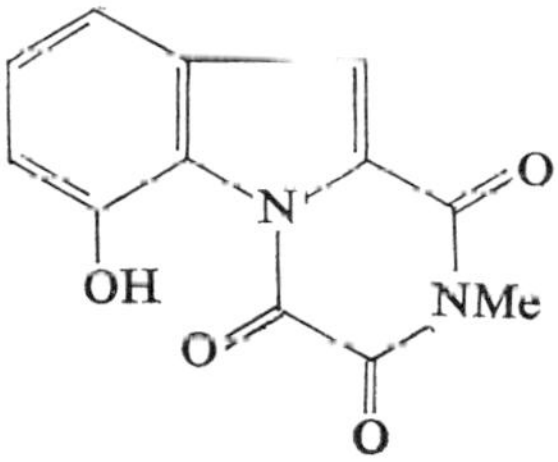

dioxopyrazinoindole A†
P. terlikowski[131]

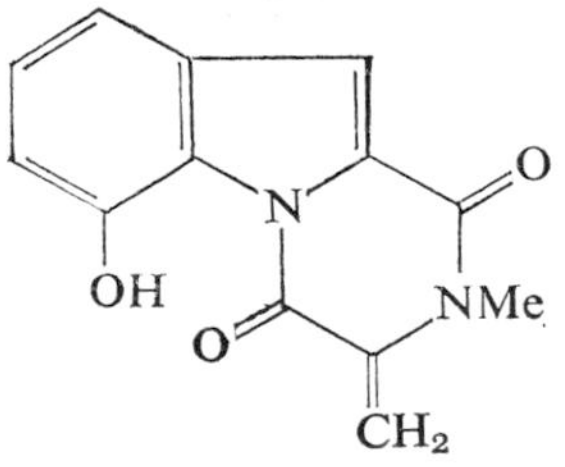

dioxopyrazinoindole B†
P. terlikowski[132]

73

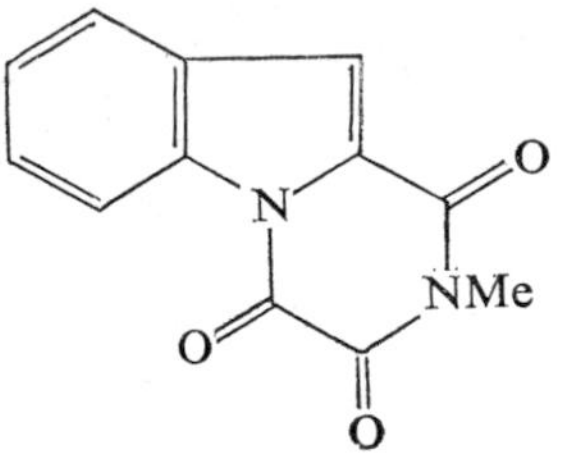

dioxopyrazinoindole C†
P. terlikowski[131]

74

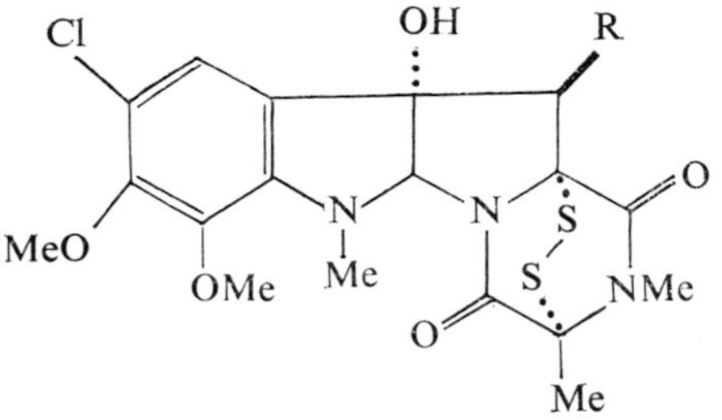

a: R = OH, sporidesmin   *L*[153]
b: R = H, sporidesmin B
Sporidesmium bakeri (= Pithomyces chartarum[132])

75

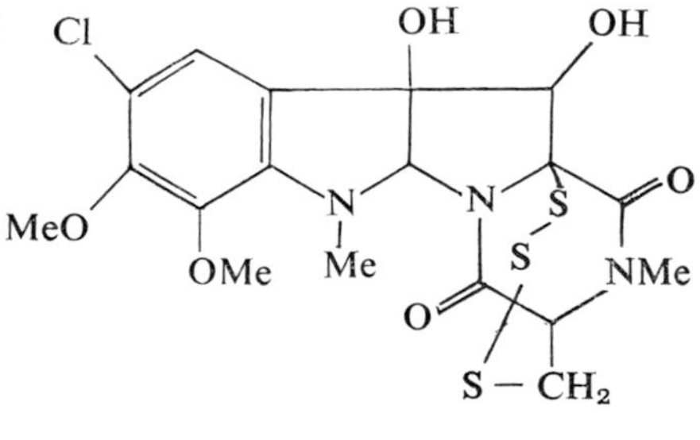

sporidesmin C
Pithomyces chartarum[133]

76

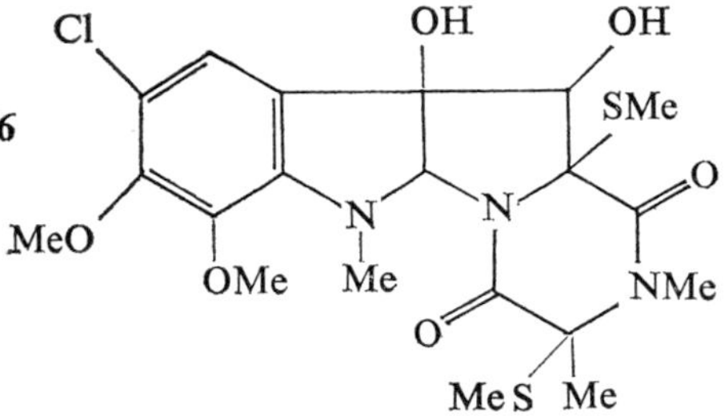

sporidesmin D
Pithomyces chartarum[134,136]

† These trivial names have been coined for ease of indexing.

77

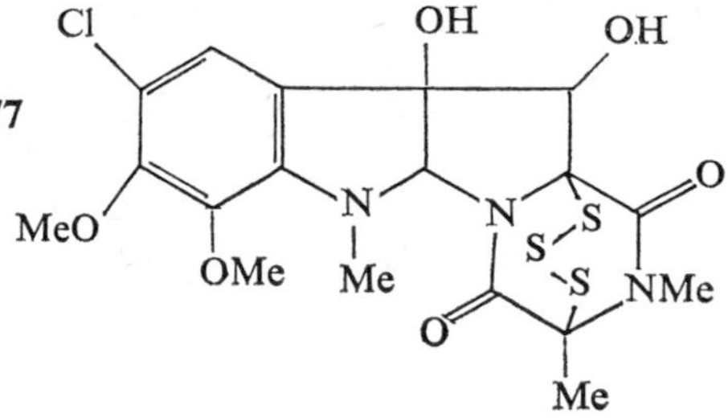

sporidesmin E
P. chartarum[135]

78

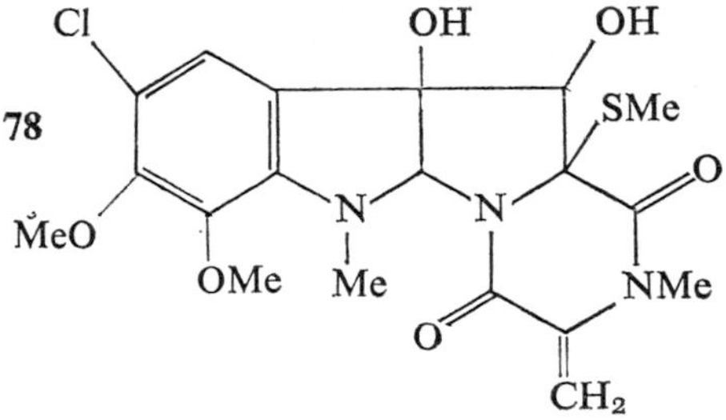

sporidesmin F
P. chartarum[136]

79

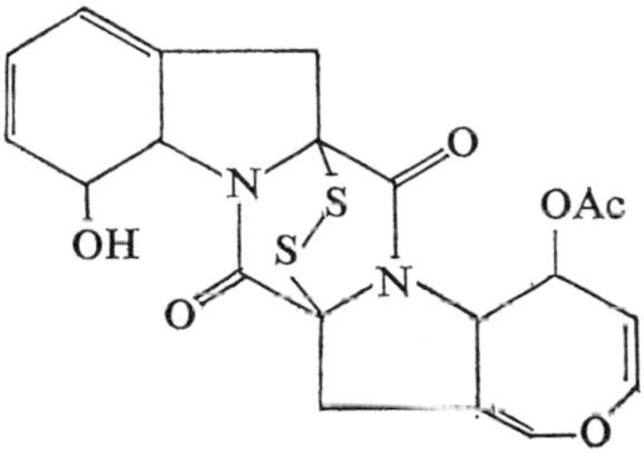

apoaranotin
Arachniotus aureus[137]

80

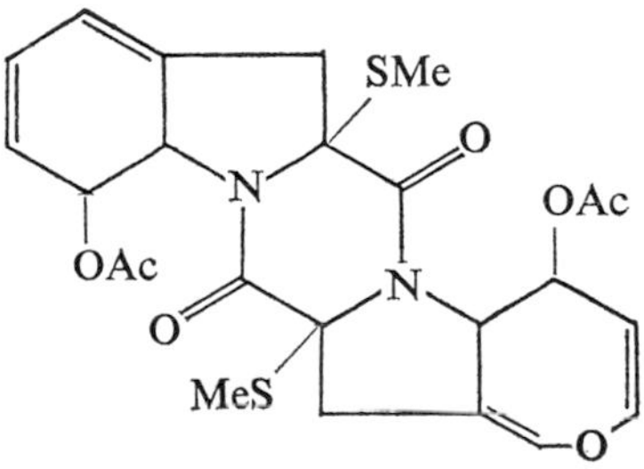

bisdethiodi(methylthio)acetylapo-
aranotin (BDAA)
Arachniotus aureus[137]

81

a: R = H, aranotin
b: R = Ac, acetylaranotin
    (antibiotic LL-S88α)
    Arachniotus aureus,[138] *Aspergillus*
    terreus[139]

82

bisdethiodi(methylthio)acetyl-
    aranotin (BDA) (antibiotic LL-S88β)
    Arachniotus aureus,[138]
    Aspergillus terreus[139]

Gliotoxin (**69a**), a highly antifungal and antibacterial metabolite of several Fungi Imperfecti, was first isolated in 1936 by Weindling and Emerson[140] from a *Trichoderma* sp. (for a discussion of the genus of the producing organism, see Brian[141]). The chemistry of the compound was extensively studied by Johnson and his co-workers[142] and the correct structure (**69a**) was finally proposed in 1958[143] and confirmed by X-ray analysis[144] which also revealed the absolute configuration as in (**69a**).

The sporidesmins were detected as a result of a correlation of facial eczema of cattle and sheep with fungal infestation of pastures (for references, see ref. 145). When *Pithomyces chartarum* (= *Sporidesmium bakeri*), one of the micro-organisms involved, was grown as surface culture on potato–carrot medium and fed to sheep, the animals showed symptoms similar to those of facial eczema. Isolation of sporidesmin and the determination of its structure by chemical[132] and X-ray[146] methods followed. The absolute configuration of sporidesmin was deduced[144] by comparison of its C.D. curve with that of gliotoxin.

Aranotin (**81a**), apoaranotin (**79**), and the reduced compounds (**80**) and (**82**) were isolated as a result of their antiviral activity. The structure and absolute configuration of BDA (**82**) was determined by X-ray analysis,[147] and the configuration of acetylaranotin was deduced.[148] Acetylaranotin and BDA have been isolated independently as antibiotics LL-S88α and LL-S88β respectively and the structure and stereochemistry of the former was deduced by X-ray analysis.

Using ^{14}C-labelled precursors[149,150] or [15]N- and ^{13}C-labelled precursors
with mass spectroscopic analysis,[151] it has been shown that phenylalanine
and m-tyrosine, but not acetate or tryptophan, are incorporated into the
indolecarboxylic acid moiety of gliotoxin, that the three-carbon moiety
arises from serine or from glycine (probably via serine), and that the N-methyl
group is derived from methionine, from C-3 of serine, or from C-2 of glycine,
(i.e. from the C_1-pool). Using [15]N- and doubly-labelled precursors,[151,152]
it was shown that ^{15}N from glycine was incorporated into both nitrogen
atoms of gliotoxin but that ^{15}N from phenylalanine was incorporated only
into N-5 of gliotoxin with considerable dilution by transamination. Glutamic
and aspartic acids contribute towards both nitrogen atoms but not to the
carbon skeletons. It seems that though both phenylalanine and serine are in
equilibrum with a nitrogen pool, there is little exchange of nitrogen between
them.

Many details of the biosynthesis of gliotoxin (and of the related compounds) remain obscure—is a phenylalanylserine or an indole-2-carboxyl-
serine involved, or is phenylalanine converted to m-tyrosine before being used
for gliotoxin biosynthesis; at what stage is the aromatic ring reduced; at
what stage does N-methylation occur; at what stage is sulphur introduced
and how is the disulphide bridge formed? Theories have been advanced
concerning two of these points. It has been suggested[131] that the disulphide
bridge might arise by addition of sulphur to a diene such as (83) and a search

83

was made for such compounds among metabolites of T. viride. This resulted
in the isolation of compounds (71)–(73) in very low (3–25 μg/litre) yield, but
whether these compounds are biosynthetic intermediates, biodegradation
products or merely artefacts remains unresolved.

Neuss et al.[137] have drawn attention to the fact that a benzene oxide
intermediate could give rise to the gliotoxin-type ring system or, via benzene
oxide–oxepin equilibrium, to the aranotin-type ring system (Scheme 18).

Tryptophan, alanine, and the methyl group of methionine are incorporated
into sporidesmin.[153]

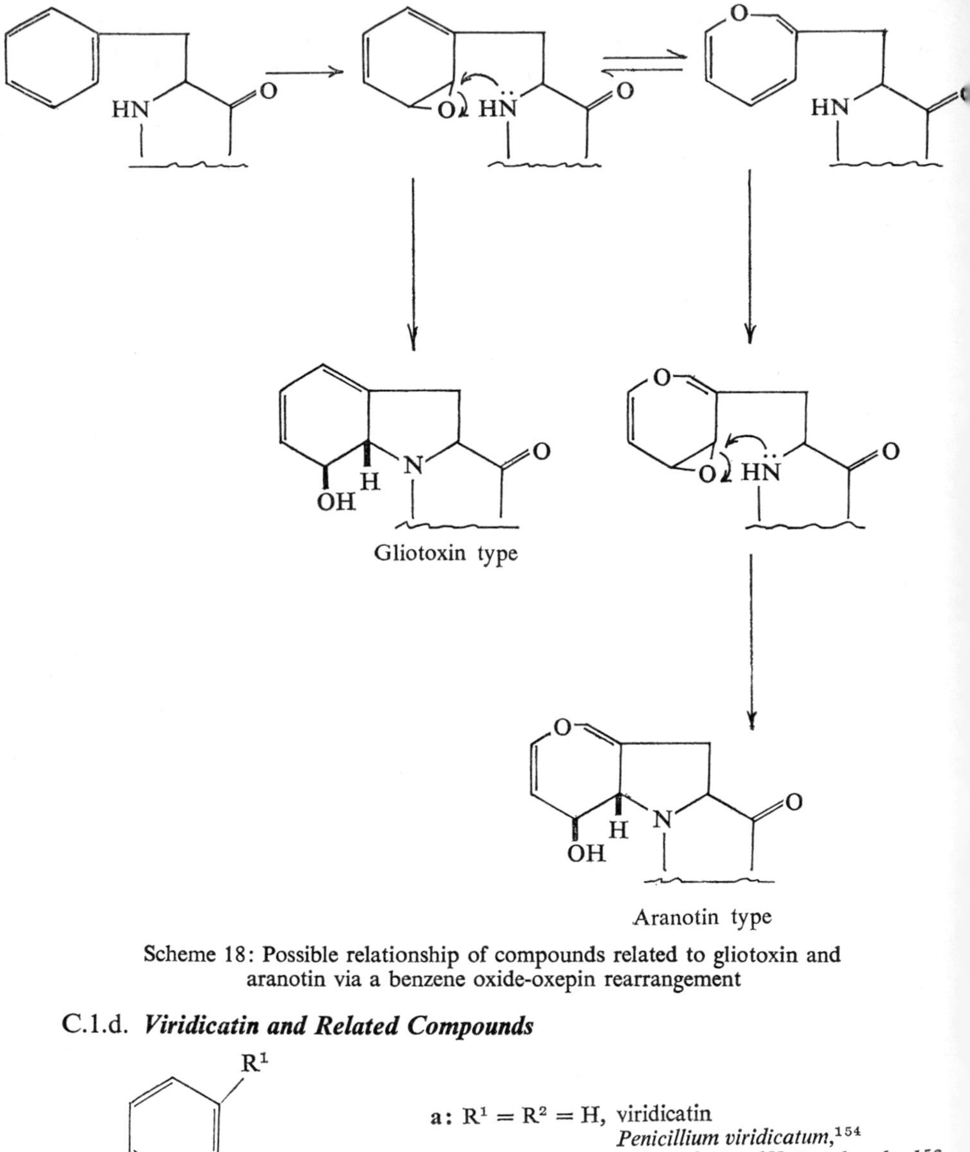

Gliotoxin type

Aranotin type

Scheme 18: Possible relationship of compounds related to gliotoxin and
aranotin via a benzene oxide-oxepin rearrangement

C.1.d. *Viridicatin and Related Compounds*

84

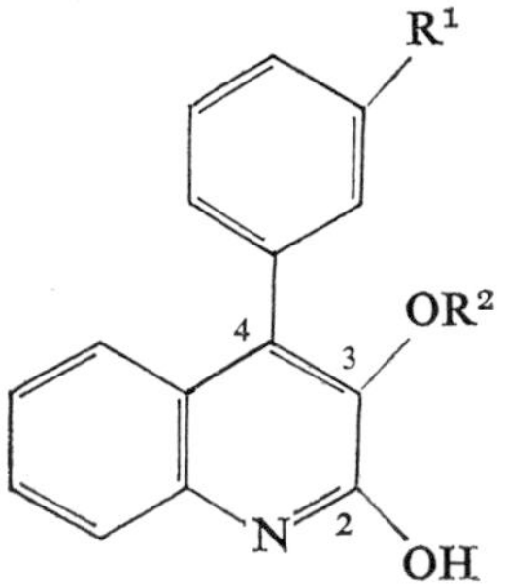

a: $R^1 = R^2 = H$, viridicatin
   Penicillium viridicatum,[154]
   P. cyclopium,[155] *P. puberulum*[156]

b: $R^1 = H$, $R^2 = Me$, 3-*O*-methylviridicatin
   P. puberulum[156]

c: $R^1 = OH$, $R^2 = H$, viridicatol
   P. viridicatum, P. cyclopium[157]

85

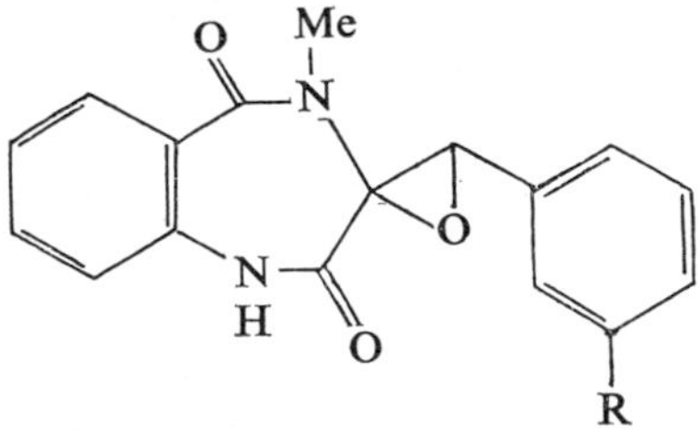

a: R = H, cyclopenin
   P. cycopium[155,157]

b: R = OH, cyclopenol
   P. viridicatum, P. cyclopium[157]

These compounds are obtained from a closely related group of *Penicillia*.[155] The structure of viridicatin was established by Bracken, Pocker and Raistrick[155] who also showed that cyclopenin (actually a mixture of cyclopenin and cyclopenol[157]) gave viridicatin on acid hydrolysis, suggesting that cyclopenin might be a biosynthetic precursor of viridicatin. Meanwhile, Luckner and Mothes[158] had shown that the phenyl ring and C-2, C-3 and C-4 of viridicatin are derived from phenylalanine, and that the benzene ring of the quinoline system is derived from anthranilic acid with loss of its carboxyl group. With the elucidation[157] of the correct structures of cyclopenin and cyclopenol, which are formally derivable from anthranilic acid and phenylalanine or *m*-tyrosine, the biological formation of viridicatin and viridicatol from cyclopenin and cyclopenol became highly likely and was confirmed by the isolation[159] of an enzyme capable of carrying out the transformation (Scheme 19).

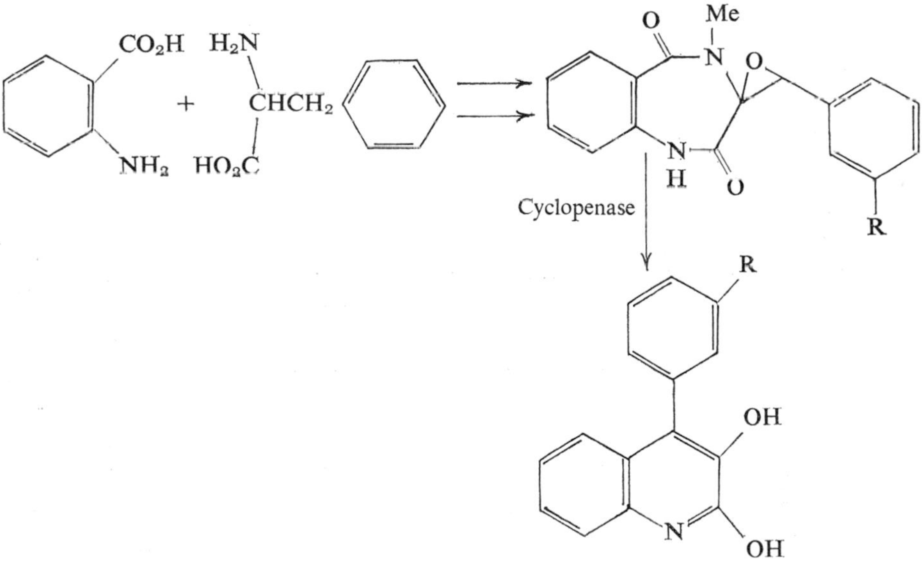

Scheme 19: Biosynthesis of cyclopenin and viridicatin

The oxygen atoms of the epoxide rings of cyclopenin and cyclopenol and of the *m*-hydroxyl group of cyclopenol are derived from molecular oxygen.[160]

C.2. Lycomarasmine and Related Compounds

86 $HO_2CCH_2CH(CO_2H)NHCH_2CH(CO_2H)NHCH_2CONH_2$ lycomarasmine
Fusarium lycopersici[161]

87 $HO_2CCH_2CH(CO_2H)NHCH_2CH(CO_2H)NHCH_2CH(NH_2)CO_2H$ aspergillomarasmine A
Aspergillus flavus oryzae[16]

88 $HO_2CCH_2CH(CO_2H)NHCH_2CH(CO_2H)NHCH_2CO_2H$ aspergillomarasmine B
(lycomarasmic acid)
A. flavus oryzae[162]

89 $HO_2CCH_2CH(CO_2H)\!-\!N$ anhydroaspergillomarasmir
A. flavus oryzae[162]

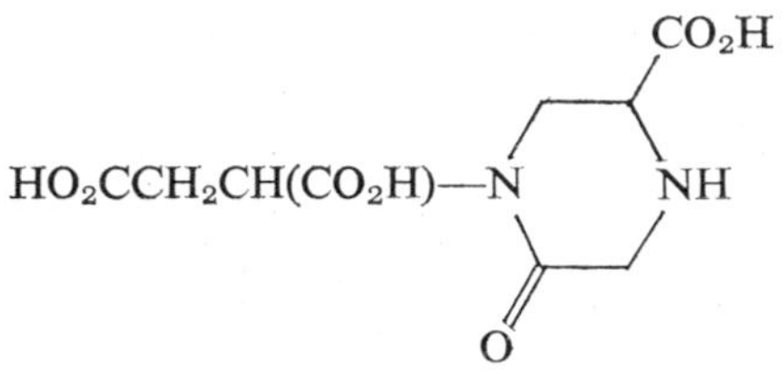

These compounds are formed by linkage of two or more amino-acid residues through amine rather than amide bonds. Lycomarasmine and aspergillomarasmines A and B are phytotoxins. Both aspergillomarasmine B (lycomarasmic acid) and anhydroaspergillomarasmine B (Substance J) had been obtained as degradation products of lycomarasmine prior to their isolation from *A. flavus oryzae*.

C.3. Penicillins and Cephalosporins

The penicillins are the most important compounds to have been isolated from fungi. Their discovery marked the beginning of a new era in medicine and they have been of enormous commercial importance, an importance which has been maintained by the discovery of 6-aminopenicillanic acid and its conversion to semi-synthetic penicillins with improved therapeutic properties.

The literature of the penicillins has been extensively reviewed,[163,164] and the history of their development, from Fleming's first observation of the antagonism ("antibiosis") between a strain of *Penicillium* and a *Staphyllococcus* to the large-scale production of penicillin G (**90d**) during World War II, makes fascinating reading.[163a] Here we shall present a brief account of the

historical background and deal in somewhat greater detail with the biosynthesis of the compounds.

The original contaminant isolated by Fleming was a strain of *P. notatum*
and the early work was carried out with this organism. Subsequently[163a,163c]
penicillins were isolated from a number of *Penicillia* and *Aspergilli*, and
from a *Cephalosporium*. The organisms now used for the commercial production of penicillin are mutants of *P. chrysogenum*.

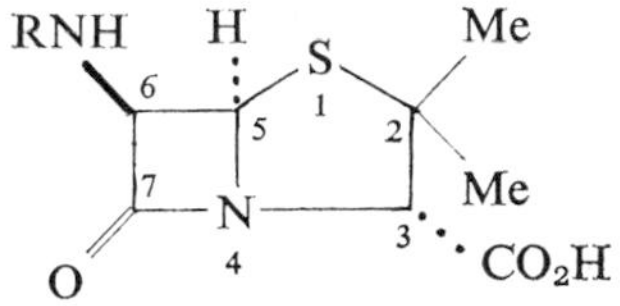

90

a R = H	6-aminopenicillanic acid (6-APA)
b R = Me CH$_2$CH=CHCH$_2$CO	penicillin F
c R = Me(CH$_2$)$_4$CO	dihydropenicillin F
d R = PhCH$_2$CO	penicillin G
e R = Me(CH$_2$)$_6$CO	penicillin K
f R = D-α-aminoadipoyl	penicillin N (cephalosporin N, synnematin B)
g R = L-α-aminoadipoyl	isopenicillin N
h R = *p*-HOC$_6$H$_4$CH$_2$CO	penicillin X

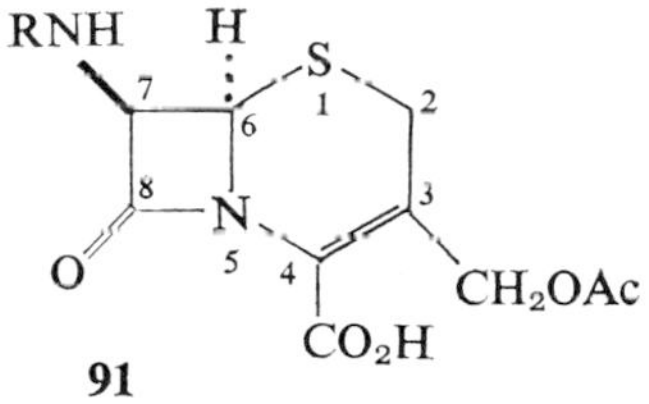

91

a R = H	7-aminocephalosporanic acid
b R = D-α-aminoadipoyl	cephalosporin C

It soon became apparent that "penicillin" is not a single compound. The
first two penicillins to be obtained in a pure state were F **(90b)** (from surface
fermentations in England) and G **(90d)** (from stirred fermentations in America),

and the isolation of penicillins K (90e) and X (90h) quickly followed. The realization that the limiting factor in penicillin production is availability of side-chain precursors led to the addition of carboxylic acids to the fermentation media to give increased yields of "natural" penicillins and also to give new pencillins with a variety of side-chains, the choice being limited by the antifungal activity of the precursor and by the ability of the fungus to degrade some side-chains. The isolation[165] of 6-aminopenicillanic acid (6-APA) (90a) from fermentations on media containing no side-chain intermediates removed these limitations and made possible the partial synthesis of a wide variety of new penicillins some of which possess improved therapeutic properties (notably penicillinase-resistance and oral administration). 6-APA can also be obtained by enzymatic hydrolysis of penicillin G,[166] and this is the basis of its commercial production.

During a study of the antibiotics produced by a *Cephalosporium* sp., two water-soluble compounds—cephalosporin N (90f) and cephalosporin C (91b)—were isolated. Cephalosporin N was shown[167] to be a new penicillin and was renamed penicillin N; it is identical with synnematin B[168,169] isolated from *Cephalosporium salmosynnematum* (perfect stage = *Emericellopis*[169]). Penicillin N has also been obtained from a *Streptomyces* sp. Cephalosporin C (91b) has the same side-chain as penicillin N but a new heterocyclic ring-system.[170] Isopenicillin N (90g), with the L-configuration of the α-aminoadipoyl side-chain, has been isolated from *P. chrysogenum*.[171]

The biosynthesis of the penicillins[164] poses two problems—the formation of the heterocyclic system of 6-APA and the introduction of the various acyl groups. 6-APA is formally derived (92) from cysteine and valine and both are incorporated intact into the penicillin molecule[172] As expected, L-cysteine is a

92

more efficient precursor than D-cysteine, but although C-3 of the penicillin nucleus has the D-configuration, L-valine is incorporated more efficiently than D-valine.[173b,174] The nitrogen atom of valine is retained though some transamination occurs.[175] The retention of tritium of [2-^{3}H]cysteine excludes the intervention of an αβ-unsaturated derivative of cysteine in penicillin biosynthesis.[176]

The above are the only firmly established facts relating to penicillin biosynthesis. The isolation from *P. chrysogenum* of the tripeptide δ-(α-amino-adipoyl)cysteinylvaline (93),[176] which is an open-chain analogue of isopenicillin N, has led to the suggestion that this compound is a precursor of isopenicillin N which might be transformed, via penicillin N and transacylation, into the other penicillins. Support for this hypothesis was provided by the (subsequent) isolation of isopenicillin N from *P. chrysogenum*.[172] A pathway of this type would also account for the inefficient incorporation of cysteinylvaline into penicillins since δ-(α-aminoadipoyl)cysteine might be formed first and then condensed with valine to give the tripeptide (93).

The role of 6-APA in penicillin biosynthesis is not clear. According to the above theory, 6-APA is not a precursor of the penicillins but is a product of their hydrolysis. On the other hand, three groups[177] have recently shown the existence of an enzyme which catalyses the acylation of 6-APA by the

$$\text{HO}_2\text{CCH(CH}_2)_3\text{CONHCH}\text{——}\overset{\displaystyle\text{SH}}{\underset{\displaystyle\text{O}}{\text{CH}_2}}\quad \underset{\displaystyle\text{CHMe}_2}{\text{CHMe}_2}$$

93

coenzyme A derivatives of acids, suggesting that 6-APA is the precursor of the penicillins (though this would not, of course, preclude the initial formation of 6-APA from penicillin N or isopenicillin N).

The biosynthesis of cephalosporin C is clearly related to that of the penicillins and might involve a common intermediate—perhaps the tripeptide (93) or a compound with the penicillanic acid ring system. The penicillin ring system can be converted chemically to that of cephalosporin C,[178] but this conversion has not been achieved biologically and there is some evidence that the pathways diverge at an earlier stage. Radioactivity from α-amino-adipic acid,[180,181] valine,[179,180,182] and a mixture of DL- and *meso*-cystine[180,181] is incorporated into cephalosporin C. Acetate is incorporated into both the *O*-acetyl group and the α-aminoadipoyl moiety, which arises from acetate and α-oxoglutarate via homocitrate.

Thus the steps between the precursor amino-acids and the antibiotics remain obscure (a hypothetical pathway for the biosynthesis of the penicillins and cephalosporins has been discussed[164a]). A major difficulty has been the impermeability of the fungi to many of the likely intermediates; it may be that experiments with cell-free systems will overcome this problem.

D. SECONDARY METABOLITES DERIVED FROM MORE THAN TWO AMINO-ACIDS

D.1. Miscellaneous Peptides

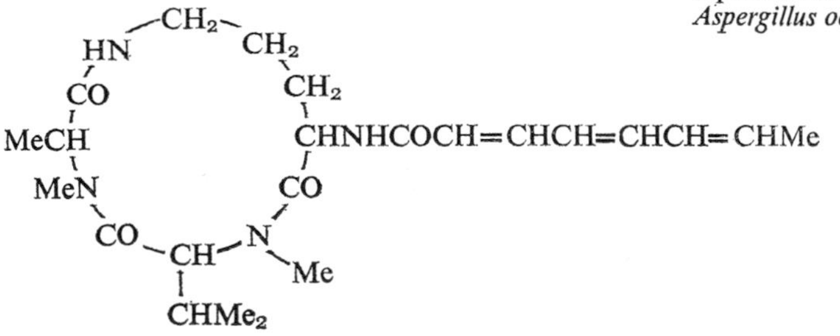

aspochracein
Aspergillus ochraceus[183]

94

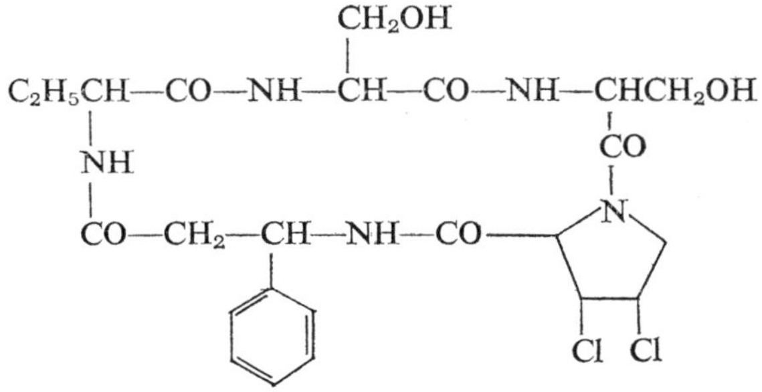

islanditoxin
Penicillium islandicum[184]

95

96

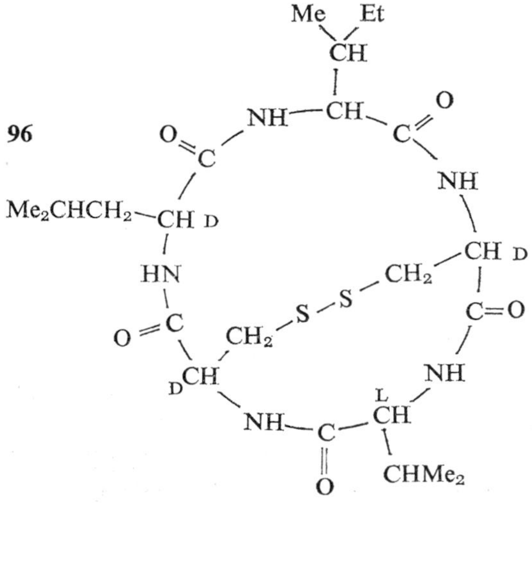

malformin A   *L*[190.19]
Aspergillus niger,[185]
A. ficuum, A. awamori,
A. phoenicis[186]

97

$CHMe_2$

CH (D) — CO — NH

NH

CO

CH (L)

$PhCH_2$ NH — CO — CH (D)

CH_2Ph

NH

CH (L) — $CHMe_2$

CO

NH

fungisporin
Penicillium spp.,
Aspergillus spp.[187]

Although many antibiotics produced by bacteria, especially *Streptomyces* and *Bacillus*, are cyclic peptides, none of the fungal compounds has been reported as antibacterial. Aspochracein (**94**) is insecticidal, islanditoxin (**95**) is responsible for the toxicity of foodstuffs contaminated by *P. islandicum*, malformin A (**96**) causes curvatures and malformations of bean plants[188] and curvatures of cornshoots,[189] and fungisporin (**97**) is a sublimable component of the spores of some *Penicillia* and *Aspergilli*.

The biosynthesis of malformin A by washed cells of *A. niger*[190] and by an enzyme preparation[191] has been studied by Yukioka and Winnick. They found that the component amino-acids are incorporated into malformin and that L- as well as D-leucine and cysteine are incorporated, the isomerization probably occurring at an intermediate peptide level. The biosynthesis of malformin by the cell-free system was not affected by inhibitors of protein biosynthesis so that peptide and protein biosynthesis seem unconnected; the same conclusion has been reached for several of the antibacterial peptides.[192]

D.2. THE AMANITA TOXINS

Mushroom poisoning is most commonly caused by ingestion of *Amanita* species, notably *A. phalloides*, which resemble the edible mushroom in appearance and flavour. The toxic principles are peptides whose structures have been determined largely by Wieland and his co-workers.[193] About ten toxins are known and they can be divided into the phalloidin and amanitin groups, which differ in their biological activity and in some chemical properties, notably the presence of tryptophan and a sulphide bridge in the former and 6-hydroxytryptophan and a sulphoxide bridge in the latter. The two groups are exemplified by phalloidin (**98**) and α-amanitin (**99**). An interesting recent development is the isolation of an antitoxin, antamanide (**100**).[194]

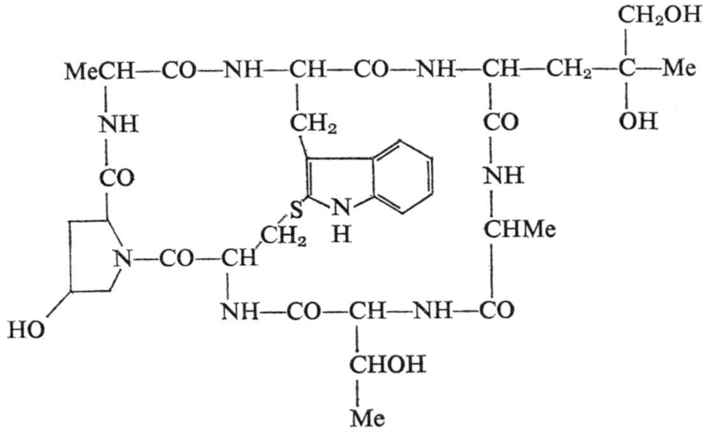

$$\begin{array}{c} \text{CH}_2\text{OH} \\ | \\ \text{MeCH—CO—NH—CH—CO—NH—CH—CH}_2\text{—C—Me} \end{array}$$

$$\begin{array}{ccc} \text{NH} & \text{CH}_2 & \text{CO} \quad \text{OH} \\ | & & | \\ \text{CO} & & \text{NH} \end{array}$$

N—CO—CH ... CH₂ H ... CHMe

HO ... NH—CO—CH—NH—CO

CHOH

Me

98   Phalloidin

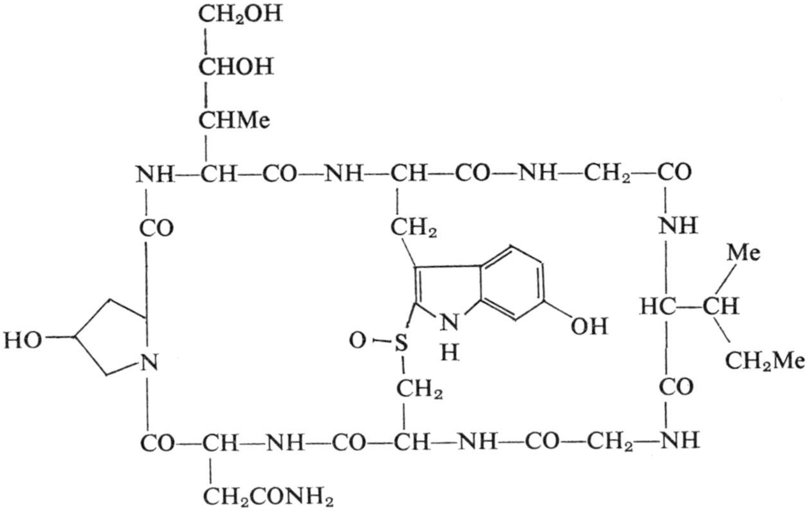

CH₂OH
|
CHOH
|
CHMe
|
NH—CH—CO—NH—CH—CO—NH—CH₂—CO
|
CO ... CH₂ ... NH

Me
|
HC—CH
|
HO ... N ... N H OH ... CO ... CH₂Me

O—S
|
CH₂

CO—CH—NH—CO—CH—NH—CO—CH₂—NH
|
CH₂CONH₂

99   α-Amanitin

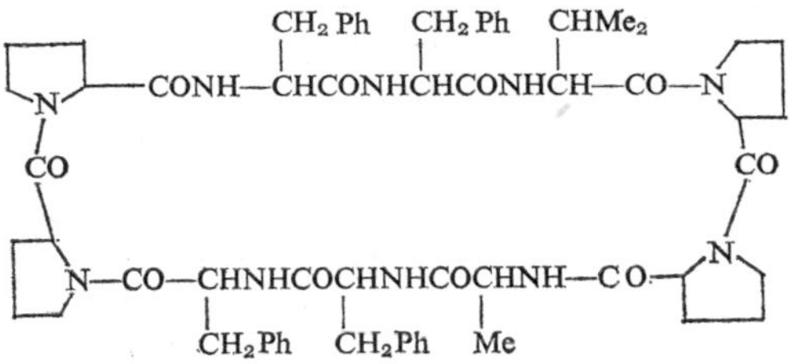

$$
\begin{array}{c}
\text{CH}_2\text{Ph} \quad \text{CH}_2\text{Ph} \quad \text{CHMe}_2 \\
| \qquad\quad | \qquad\quad | \\
\text{CONH—CHCONHCHCONHCH—CO—N}
\end{array}
$$

100 Antamanide

D.3. DEPSIPEPTIDES

101

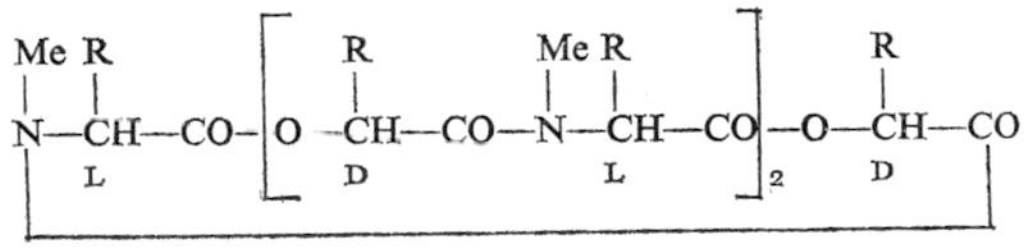

$$
\begin{array}{c}
\text{Me R} \qquad\quad \text{R} \qquad \text{Me R} \qquad\quad \text{R} \qquad\qquad \text{Me} \\
|\;\; | \qquad\quad | \qquad\quad |\;\; | \qquad\quad | \qquad\qquad | \\
\text{N—CH—CO—}\!\left[\text{O—CH—CO—N—CH—CO}\right]_2\!\text{—O—CH—CO} \\
\quad\text{L} \qquad\qquad \text{D} \qquad\qquad \text{L} \qquad\qquad \text{D}
\end{array}
$$

a: R = Et-CH, enniatin A
b: R = Me₂CH, enniatin B
Fusarium spp.[195]

102

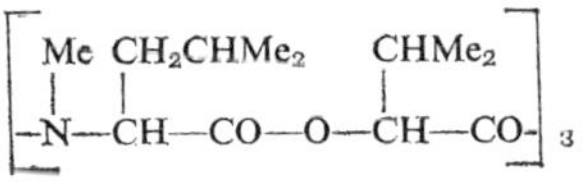

$$
\left[\text{—N—CH—CO—O—CH—CO—}\right]_3
$$

with Me, CH₂CHMe₂, CHMe₂ substituents

enniatin C
Fusarium spp.[196]

103

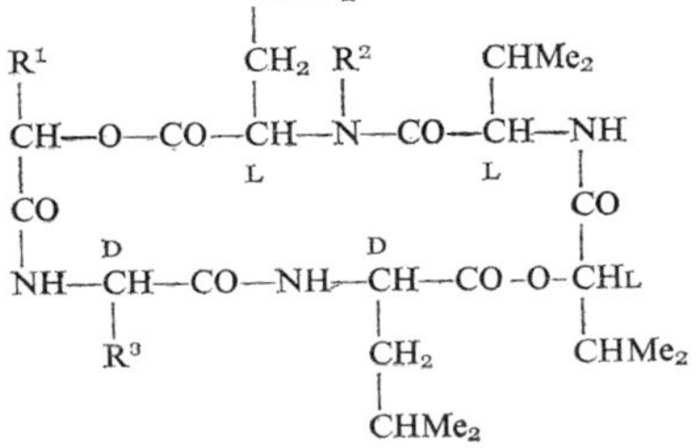

(structure with R¹, R², R³, CHMe₂, CH₂, CO, NH groups; L and D configurations)

	R^1	R^2	R^3	
a	Me$_2$CH	Me	Me$_2$CH	sporidesmolide I[197]
b	Me$_2$CH	Me	Me$_2$CHCH$_2$	sporidesmolide II[197]
c	Me$_2$CH	H	Me$_2$CH	sporidesmolide III[197,198]
d	Me$_2$CHCH$_2$	Me	Me$_2$CH	sporidesmolide IV[199]

Sporidesmium bakeri
(= *Pithomyces chartarum*)

104

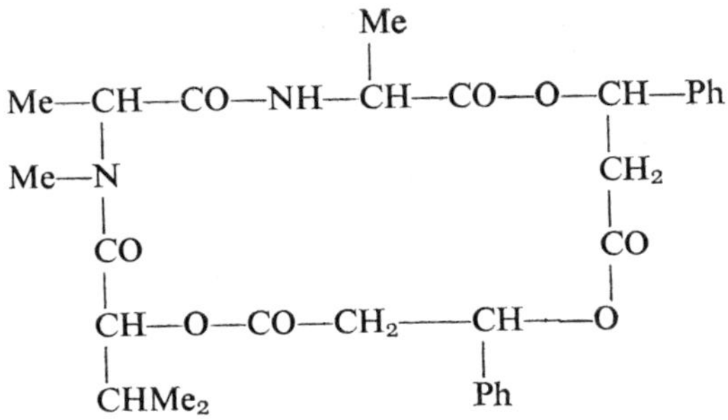

Me—CH—CO—NH—CH(^{Me})—CO—O—CH—Ph

pithomycolide
Sporidesmium bakeri
(= *Pithomyces chartarum*)[200]

105

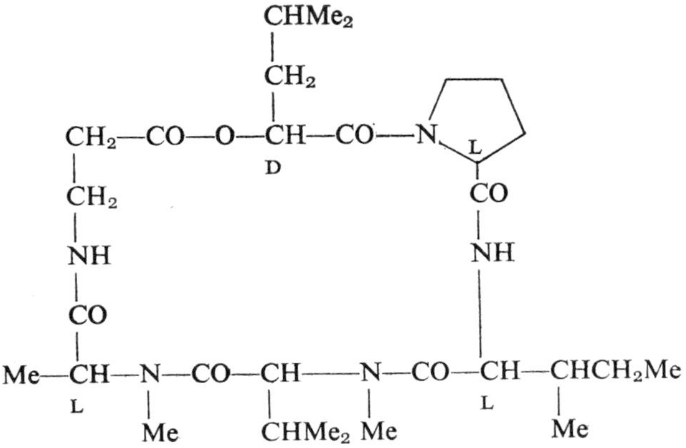

destruxin B
Oospora destructor,
Aspergillus ochrace

106

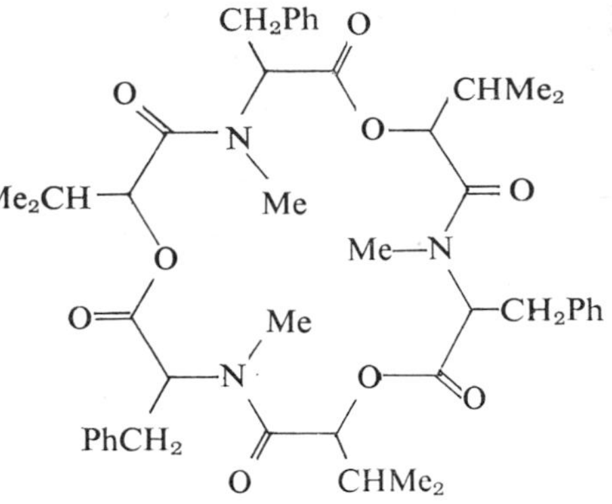

beauvericin
Beauvaria bassiana[202]

Depsipeptides are compounds containing both amide and ester linkages. The enniatins (**101**) and (**102**) are active against *Mycobacterium tuberculosis*; their structures have been proved by synthesis.[203] The sporidesmolides (**103**) and pithomycolide (**104**) were isolated during the search for the toxins (sporidesmins) of *S. bakeri*; they have no reported biological activity. Destruxin B (**105**) is an insecticide, and beauvericin (**106**) is toxic to brine shrimps.

D.4. THE SIDEROCHROMES

The siderochromes contain three hydroxamic acid residues complexed with a ferric ion. The class can be divided into the sideromycins, which are antibiotics from Actinomycetes, and the sideramines, which are bacterial growth-factors isolated mainly from fungi (ferrichrome A, which has neither biological activity, is an exception to this classification). The sideramines can be readily detected by their ability to reverse the antibacterial activity of albomycin, the first sideromycin to be isolated. They are produced by several *Aspergilli* and best yields are obtained when iron is excluded from the medium.

The subject has been reviewed[204] and here we need only note the general structural features. The hydroxamic acid groups of all the fungal sidero-chromes are derived from three molecules of acylated *N*-hydroxyornithine which are linked together with three glycines (as in ferrichrome), two glycines and one serine (as in ferricrocin), or one glycine and two serines (as in ferrichrysin, ferrirubin and ferrirhodin) to form a cyclic peptide. The acyl group can be acetyl (as in ferrichrome, ferricrocin and ferrichrysin) or *trans*-5-hydroxy-3-methylpent-2-enoyl (as in ferrirubin or ferrirhodin). Thus ferricrocin has structure (**107**).

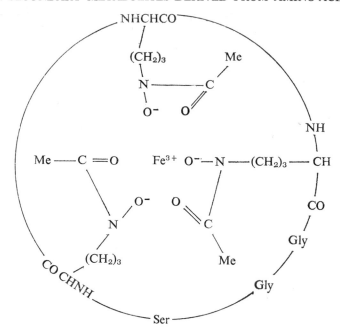

107   Ferricrocin

REFERENCES

1. E. A. Kaczka, C. O. Gitterman, E. L. Dulaney and K. Folkers, *Biochemistry* 1962, **1**, 340.
2. R. A. Gray, G. W. Gauger, E. L. Dulaney, E. A. Kaczka and H. B. Woodruff, *Plant Physiol.*, 1964, **39**, 204.
3. P. L. Schuller and C. A. Salemink, *Planta Med.*, 1962, **10**, 327.
4. S. Hatanaka, *Phytochemistry*, 1969, **8**, 1305.
5. B. Levenberg, *J. Biol. Chem.*, 1968, **243**, 6009.
6. E. Honkanen, T. Moisio, A. I. Virtanen and A. Melera, *Acta Chem. Scand.*, 1964, **18**, 1319.
7. D. P. Rainey, E. B. Smalley, M. H. Crump and F. M. Strong, *Nature*, 1965, **205**, 203; R. A. Gardiner, K. L. Rinehart, J. J. Snyder and H. P. Broquist, *J. Amer. Chem. Soc.*, 1968, **90**, 5639.
8. T. Matsumoto, W. Trueb, R. Gwinner and C. H. Eugster, *Helv. Chim. Acta*, 1969, **52**, 716.
9. E. B. Smalley, R. E. Nichols, M. H. Crump and J. N. Henning, *Phytopathology*, 1962, **52**, 753.
10. S. D. Aust, H. P. Broquist and K. L. Rinehart, *Biotechnol. Bioeng.*, 1968, **10**, 403.
11. J. J. Snyder and H. P. Broquist, 158th *A.C.S. Meeting*, 1969, *Abstr. BIOL.* 210.
12. R. L. Stevens and T. F. Emery, *Biochemistry*, 1966, **5**, 74.

13. M. T. Bush, A. Goth and H. L. Dickison, *J. Pharmacol.*, 1945, **84**, 262; M. T. Bush, O. Touster and J. E. Brockman, *J. Biol. Chem.*, 1951, **188**, 685.
14. S. Nakamura and C. Shimoda, *J. Agr. Chem. Soc. Japan*, 1954, **28**, 909.
15. B. F. Burrows and W. B. Turner, *J. Chem. Soc. C*, 1966, 255.
16. H. Aoki, T. Sassa and T. Tamura, *Nature*, 1963, **200**, 575.
17. D. Brookes, B. K. Tidd and W. B. Turner, *J. Chem. Soc.*, 1963, 5385.
18. H. Raistrick and A. Stössl, *Biochem. J.*, 1958, **68**, 647.
19. P. W. Brian, G. W. Elson, H. G. Hemming and M. Radley, *Nature*, 1965, **207**, 998.
20. A. J. Birch, B. J. McLoughlin, H. Smith and J. Winter, *Chem. and Ind.*, 1960, 840.
21. J. W. Hylin and H. Matsumoto, *Arch. Biochem. Biophys.*, 1960, **93**, 542.
22. P. D. Shaw and N. Wang, *J. Bacteriol.*, 1964, **88**, 1629.
23. S. Gatenbeck and B. Forsgren, *Acta Chem. Scand.*, 1964, **18**, 1750.
24. J. H. Birkinshaw and A. M. L. Dryland, *Biochem. J.*, 1964, **93**, 478.
25. P. D. Shaw and J. A. McCloskey, *Biochemistry*, 1967, **6**, 2247.
26. P. D. Shaw, *Biochemistry*, 1967, **6**, 2253.
27. E. Candlish, L. J. La Croix, and A. M. Unrae, *Biochemistry*, 1969, **8**, 182.
28. K. C. Marshall and M. Alexander, *J. Bacteriol.*, 1962, **83**, 572.
29. O. R. Eylar and E. L. Schmidt, *J. Gen. Microbiol.*, 1959, **20**, 473.
30. E. Malavolta, R. de Camargo and H. P. Haag, *Bol. Inst. Zimotéc* (São Paulo), 1955, No. 13.
31. B. F. Burrows, S. D. Mills and W. B. Turner, *Chem. Commun.*, 1965, 75.
32. S. D. Mills, unpublished results.
33. U.S. Patent 3,179,653 (*Chem. Abs.*, 1965, **63**, 1170).
34. Cf. D. Gottlieb. *In* "Antibiotics. Volume II. Biosynthesis" (D. Gottlieb and P. D. Shaw, eds.), p. 37. Springer-Verlag, New York, 1967.
35. R. McGrath, L. C. Vining, F. Sala, and D. W. S. Westlake, *Can. J. Biochem.*, 1968, **46**, 587.
36. K. Yamada, H. Nakata and Y. Hirata, *J. Chem. Soc. Japan*, 1960, **81**, 340.
37. S. Kawai, K. Kobayashi, T. Oshima and F. Egami, *Arch. Biochem. Biophys.*, 1965, **112**, 537.
38. M. Pailer, *Prog. Chem. Org. Nat. Prod.*, 1960, **18**, 66.
39. F. Stermitz, F. A. Norris and M. C. Williams, *J. Amer. Chem. Soc.*, 1969, **91**, 4599.
40. F. Comer, H. P. Tiwari and I. D. Spenser, *Can. J. Chem.*, 1969, **47**, 481.
41. B. Levenberg, *J. Biol. Chem.*, 1964, **239**, 2267; E. G. Daniels, R. B. Kelly and J. W. Hinman, *J. Org. Chem.*, 1962, **27**, 3229.
42. B. Levenberg, *Biochim. Biophys. Acta*, 1962, **63**, 212.
43. H. Herrmann, *Z. Physiol. Chem.*, 1961, **326**, 13.
44. H. J. Gigliotti and B. Levenberg, *J. Biol. Chem.*, 1964, **239**, 2274.
45. J. Jadot, J. Casimir and M. Renard, *Biochim. Biophys. Acta*, 1960, **43**, 322.
46. P. H. List and P. Luft, *Arch. Pharm.*, 1968, **301**, 294.
47. T. Takemoto and T. Nakajima, *J. Pharm. Soc. Japan*, 1964, **84**, 1183, 1230.
48. T. Takemoto, T. Nakajima and R. Sakuma, *J. Pharm. Soc. Japan*, 1964, 84, 1233.
49. K. Bowden, A. C. Drysdale and G. A. Mogey, *Nature*, 1965, **206**, 1359.
50. T. Takemoto, T. Nakajima and T. Yokobe, *J. Pharm. Soc. Japan*, 1964, **84**, 1232.
51. M. Onda, H. Fukushima and M. Akagawa, *Chem. Pharm. Bull (Tokyo)*, 1964, **12**, 751.

52. K. Bowden and A. C. Drysdale, *Tetrahedron Lett.*, 1965, 727; G. F. R. Müller and C. H. Eugster, *Helv. Chim. Acta*, 1965, **48**, 910.
53. R. Good, G. R. F. Müller and C. H. Eugster, *Helv. Chim. Acta*, 1965, **48**, 927; R. Reiner and C. H. Eugster, *Helv. Chim. Acta*, 1967, **50**, 128.
54. P. J. Suter and W. B. Turner, *J. Chem. Soc. C*, 1967, 2240.
55. J. Gripenberg, *Acta Chem. Scand.*, 1951, **5**, 590; G. W. K. Cavill, B. J. Ralph, J. R. Tetaz and R. L. Werner, *J. Chem. Soc.*, 1953, 525; J. Gripenberg, *Acta Chem. Scand.*, 1958, **12**, 603.
56. J. Gripenberg, E. Honkanen and O. Patoharju, *Acta Chem. Scand.*, 1957, **11**, 1485.
57. J. Gripenberg, *Acta Chem. Scand.*, 1963, **17**, 703.
58. J. R. D. McCormick, J. Reichenthal, U. Hirsch and N. O. Sjolander, *J. Amer. Chem. Soc.*, 1962, **84**, 3711.
59. K. Anzai, K. Isono, K. Okuma and S. Suzuki, *J. Antibiotics (Tokyo)*, 1960, **13**, 125.
60. N. N. Gerber, *Biochemistry*, 1966, **5**, 3824.
61. P. H. List, *Planta Med.*, 1958, **6**, 424.
62. K. Tamari and J. Kaji, *J. Agr. Chem. Soc. Japan*, 1954, **28**, 254.
63. J. Ooyama, N. Nakamura and O. Tanabe, *Bull. Agr. Chem. Soc. Japan*, 1960, **24**, 743.
64. T. Yabuta, K. Kambe and T. Hayashi, *Bull. Agr. Chem. Soc. (Japan)*, 1934, **10**, 1059.
65. Ch. Stoll, J. Renz and E. Gäumann, *Phytopathol. Z.*, 1957, **29**, 388.
66. E. Gäumann, *Phytopathology*, 1957, **47**, 342.
67. M. S. Kuo and R. P. Scheffer, *Phytopathology*, 1964, **54**, 1041; I. S. Mansour and R. P. Scheffer, *Phytopathology*, 1966, **56**, 888.
68. P. H. Hodson and J. W. Foster, *J. Bacteriol.*, 1966, **91**, 562.
69. M. Kanie, S. Fujimoto and J. W. Foster, *J. Bacteriol.*, 1966, **91**, 570.
70. S. W. Tanenbaum and K. Kaneko, *Biochemistry*, 1964, **3**, 1314.
71. J. F. Powell and R. E. Strange, *Nature*, 1959, **184**, 878.
72. C. Kawasaki, J. Sukurai and M. Kondo, *Nippon Saikingaku Zasshi*, 1968, **23**, 772.
73. H. H. Martin and J. W. Foster, *J. Bacteriol.*, 1958, **76**, 167.
74. T. A. Dobson, D. Desaty, D. Brewer and L. C. Vining, *Can. J. Biochem.*, 1967, **45**, 809.
75. R. D. Hill, A. M. Unrau and D. T. Canvin, *Can. J. Chem.*, 1966, **44**, 2077.
76. D. Desaty, A. G. McInnse, D. G. Smith and L. C. Vining, *Can. J. Biochem.*, 1968, **46**, 1293.
77. T. Wieland, W. Motzel and H. Merz, *Ann. Chem.*, 1953, **581**, 10.
78. A. Hofmann, R. Heim, A. Brack, H. Kobel, A. Frey, H. Ott, Th. Petrzilka and F. Troxler, *Helv. Chim. Acta*, 1959, **42**, 1557.
79. A. Y. Leung and A. G. Paul, *J. Pharm. Sci.*, 1967, **56**, 146; 1968, **57**, 1667.
80. D. C. Aldridge, S. Galt, D. Giles and W. B. Turner, *J. Chem. Soc. C*, in the press.
81. T. Yamano, *J. Agr. Chem. Soc. Japan*, 1961, **35**, 1284.
82. P. G. Miles, H. Lund and J. R. Raper, *Arch. Biochem. Biophys.*, 1956, **62**, 1.
83. H. Falanghe and P. A. Bobbio, *Arch. Biochem. Biophys.*, 1962, **96**, 430.
84. R. Heim, *Actualites Pharmacol.*, p. 171. Masson, Paris, 12th series, 1959.
85. S. Agurell and J. L. G. Nilsson, *Acta Chem. Scand.*, 1968, **22**, 1210.
86. U. Hornemann, M. K. Speedie, K. M. Kelley, L. H. Hurley and H. G. Floss, *Arch. Biochem. Biophys.*, 1969, **131**, 430.

87. A. Stoll and A. Hofmann. *In* "The Alkaloids. Chemistry and Physiology" (R. H. F. Manske, ed.), Vol. VIII. p. 725. Academic Press, London and New York, 1965.

88. M. Abe and S. Yamatodani. *In* "Progress in Industrial Microbiology" (D. J. D. Hockenhull, ed.), Vol. V. Heywood, London, 1964.

89. R. Voigt, *Die Pharmazie*, 1968, **23**, 285, 353, 419.

90. J. E. Robbers and H. G. Floss, *Arch. Biochem. Biophys.*, 1968, **126**, 967.

91. J. E. Robbers and H. G. Floss, *Tetrahedron Lett.*, 1969, 1857.

92. C. W. Holzapfel, *Tetrahedron*, 1968, **24**, 2101.

93. C. W. Holzapfel and D. C. Wilkins, 5th *Int. Symp. Chem. Nat. Prod.*, London, 1968, *Abstr.* C65.

94. J. H. Birkinshaw and Y. S. Mohammed, *Biochem. J.*, 1962, **85**, 523.

95. R. Brown, C. Kelley and S. E. Wiberley, *J. Org. Chem.*, 1965, **30**, 277.

96. M. Vondráček and Z. Vaněk, *Chem. and Ind.*, 1964, 1686.

97. N. N. Gerber, *J. Org. Chem.*, 1967, **32**, 4055.

98. G. S. Rosenfeld, L. I. Rostovtseva, V. M. Baikina, D. M. Trakhtenberg and A. S. Khokhlov, *Antibiotiki*, 1963, **8**, 201; A. S. Khokhlov and G. B. Lokshin, *Tetrahedron Lett.*, 1963, 1881; A. S. Khokhlov, G. B. Lokshin, N. S. Vul'fson and V. I. Zaretskii, *Izv. Akad. Nauk SSSR, Ser. Khim.*, 1966, 1191.

99. Y. Kodaira, *Agr. Biol. Chem. (Tokyo)*, 1961, **25**, 261.

100. Y.-S. Chen, *Bull. Agr. Chem. Soc. Japan*, 1960, **24**, 372.

101. M. O. Forster and W. B. Saville, *J. Chem. Soc.*, 1922, 816.

102. C. L. Atkin and J. B. Neilands, *Biochemistry*, 1968, **7**, 3734.

103. A. Quilico and L. Panizzi, *Chem. Ber.*, 1943, **76**, 348.

104. A. Quilico and C. Cardani, *Atti Accad. Nazl. Lincei, Rend. Classe Sci. Fis. Mat. Nat.*, 1950, **9**, 220, (*Chem. Abs.*, 1951, **45**, 3909).

105. Z. Kitamura, U. Kurimoto and M. Yokoyama, *J. Pharm. Soc. Japan*, 1956, **76**, 972.

106. M. Barbetta, G. Casnati, A. Pochini and A. Selva, *Tetrahedron Lett.*, 1969, 4457.

107. A. J. Birch and J. J. Wright, *Chem. Commun.*, 1969, 644.

108. L. A. Mitscher, M. P. Kunstmann, J. H. Martin, W. W. Andres, R. H. Evans, K. J. Sax and E. L. Patterson, *Experientia*, 1967, **23**, 796.

109. A. J. Birch, G. E. Blance, S. David and H. Smith, *J. Chem. Soc.*, 1961, 3128.

110. A. J. Birch and K. R. Farrar, *J. Chem. Soc.*, 1963, 4277.

111. J. C. MacDonald and G. P. Slater, *Can. J. Microbiol.*, 1966, **12**, 455.

112. E. Houghton and J. E. Saxton, *Tetrahedron Lett.*, 1968, 5475.

113. G. Dunn, G. T. Newbold and F. S. Spring, *J. Chem. Soc.*, 1949, 2586.

114. R. G. Micetich and J. C. MacDonald, *J. Chem. Soc.*, 1964, 1507.

115. U. Weiss, F. Strelitz, H. Flon and I. N. Asheshov, *Arch. Biochem. Biophys.*, 1958, **74**, 150.

116. A. J. Kluyver, J. P. van der Walt and A. J. can Triet, *Proc. Nat. Acad. Sci. U.S.*, 1953, **39**, 583; A. H. Cook and C. A. Slater, *J. Chem. Soc.*, 1956, 4133.

117. E. C. White and J. H. Hill, *J. Bacteriol.*, 1943, **45**, 433; J. D. Dutcher, *J. Biol. Chem.*, 1947, **171**, 321.

118. A. E. O. Menzel, O. Wintersteiner and G. Rake, *J. Bacteriol.*, 1943, **46**, 109; J. D. Dutcher, *J. Biol. Chem.*, 1958, **232**, 785.

119. S. Nakamura, *Bull. Agr. Chem. Soc. Japan*, 1960, **24**, 629; 1961, **25**, 74.

120. A. E. Oxford and H. Raistrick, *Biochem. J.*, 1948, **42**, 323; A. J. Birch, R. A. Massy-Westropp and R. W. Rickards, *J. Chem. Soc.*, 1956, 3717; R. B. Bates, J. H. Schauble and M. Soucek, *Tetrahedron Lett.*, 1963, 1683.

121. J. C. MacDonald, *Can. J. Chem.*, 1963, **41**, 165.
122. M. Sugiyama, M. Masaki, and M. Ohta, *Tetrahedron Lett.*, 1967, 845.
123. J. C. MacDonald. *In* "Antibiotics. Vol. II. Biosynthesis" (D. Gottlieb and P. D. Shaw, eds.), p. 43. Springer-Verlag, New York, 1967.
124. J. C. MacDonald, *J. Biol. Chem.*, 1961, **236**, 512; 1962, **237**, 1977.
125. J. C. MacDonald, *J. Biol. Chem.*, 1965, **240**, 1692.
126. J. C. MacDonald, *Biochem. J.*, 1965, **96**, 533.
127. A. J. Birch, R. J. English, R. A. Massy-Westropp and H. Smith, *J. Chem. Soc.*, 1958, 369; A. J. Birch, M. Kocor, N. Sheppard and J. Winter, *J. Chem. Soc.*, 1962, 1502
128. For references see S. Wilkinson and J. F. Spilsbury, *Nature*, 1965, **206**, 619.
129. J. R. Johnson, A. R. Kidwai and J. S. Warner, *J. Amer. Chem. Soc.*, 1953, **75**, 2110.
130. G. Lowe, A. Taylor and L. C. Vining, *J. Chem. Soc. C*, 1966, 1799.
131. M. S. Ali, J. S. Shannon and A. Taylor, *J. Chem. Soc. C*, 1968, 2044.
132. J. W. Ronaldson, A. Taylor, E. P. White and R. J. Abraham, *J. Chem. Soc.*, 1963, 3172; R. Hodges, J. W. Ronaldson, A. Taylor and E. P. White, *Chem. and Ind.*, 1963, 42.
133. R. Hodges and J. S. Shannon, *Aust. J. Chem.*, 1966, **19**, 1059.
134. R. Rahman and A. Taylor, *Chem. Commun.*, 1967, 1032.
135. D. Brewer, R. Rahman, S. Safe and A. Taylor, *Chem. Commun.*, 1968, 1571; R. Rahman, S. Safe and A. Taylor, *J. Chem. Soc. C*, 1969, 1665.
136. W. D. Jamieson, R. Rahman and A. Taylor, *J. Chem. Soc. C*, 1969, 1564.
137. N. Neuss, R. Nagarajan, B. B. Molloy and L. L. Huckstep, *Tetrahedron Lett.*, 1968, 4467.
138. R. Nagarajan, L. L. Huckstep, D. H. Lively, D. C. DeLong, M. M. Marsh and N. Neuss, *J. Amer. Chem. Soc.*, 1968, **90**, 2980.
139. D. B. Cosulich, N. R. Nelson and J. H. van der Hende, *J. Amer. Chem. Soc.*, 1968, **90**, 6519; P. A. Miller, P. W. Trown, W. Fulmor, G. O. Morton and J. Karliner, *Biochem. Biophys. Res. Commun.*, 1968, **33**, 219.
140. R. Weindling and O. H. Emerson, *Phytopathology*, 1936, **26**, 1068.
141. P. W. Brian, *Nature*, 1944, **154**, 667.
142. Cf. J. R. Johnson. *In* "The Roger Adams Symposium", p. 60. Wiley, New York, 1955.
143. M. R. Bell, J. R. Johnson, B. S. Wildi and R. B. Woodward, *J. Amer. Chem. Soc.*, 1958, **80**, 1001.
144. A. F. Beecham, J. Fridrichsons and A. M. Mathieson, *Tetrahedron Lett.*, 1966, 3131.
145. J. Done, P. H. Mortimer, A. Taylor and D. W. Russell, *J. Gen. Microbiol.*, 1961, **26**, 207.
146. J. Fridrichsons and A. M. Mathieson, *Acta Cryst.*, 1965, **18**, 1043.
147. J. W. Moncrief, *J. Amer. Chem. Soc.*, 1968, **90**, 6516.
148. R. Nagarajan, N. Neuss and M. M. Marsh, *J. Amer. Chem. Soc.*, 1968, **90**, 6518.
149. R. J. Suhadolnik and R. G. Chenoweth, *J. Amer. Chem. Soc.*, 1958, **80**, 4391.
150. J. A. Winstead and R. J. Suhadolnik, *J. Amer. Chem. Soc.*, 1960, **82**, 1644.
151. A. K. Bose, K. G. Das, P. T. Funke, I. Kugajevsky, O. P. Shukla, K. S. Khanchandani and R. J. Suhadolnik, *J. Amer. Chem. Soc.*, 1968, **90**, 1038.
152. A. K. Bose, K. S. Khanchandani, R. Tavares and P. T. Funke, *J. Amer. Chem. Soc.*, 1968, **90**, 3593.

153. N. R. Towers and D. E. Wright, *New Zealand J. Agr. Res.*, 1969, **12**, 275.
154. K. G. Cunningham and G. G. Freeman, *Biochem. J.*, 1953, **53**, 328.
155. A. Bracken, A. Pocker and H. Raistrick, *Biochem. J.*, 1954, **57**, 587.
156. D. J. Austin and M. B. Meyers, *J. Chem. Soc.*, 1964, 1197.
157. J. H. Birkinshaw, M. Luckner, Y. S. Mohammed, K. Mothes and C. E. Stickings, *Biochem. J.*, 1963, **89**, 196; Y. S. Mohammed and M. Luckner, *Tetrahedron Lett.*, 1963, 1953.
158. M. Luckner and K. Mothes, *Tetrahedron Lett.*, 1962, 1035; *Arch. Pharm.*, 1963, **296**, 18.
159. M. Luckner, *Verh. Ges. Expt. Med. DDR*, 1964, **6**, 395; *Eur. J. Biochem.*, 1967, **2**, 74.
160. L. Nover and M. Luckner, *FEBS Letters*, 1969, **3**, 292.
161. Pl. A. Plattner and N. Clauson-Kaas, *Helv. Chim. Acta*, 1945, **28**, 188; E. Hardegger, P. Liecht, L. M. Jackman, A. Boller and Pl. A. Plattner, *Helv. Chim. Acta*, 1963, **46**, 60.
162. A. L. Haenni, M. Barbier and E. Lederer, *C. R. Hebd. Séance Acad. Sci.*, 1962, **255**, 1476; M. Robert, M. Barbier, E. Lederer, L. Roux, K. Biemann and W. Vetter, *Bull. Soc. Chim. Fr.*, 1962, 187.
163. (a) H. W. Florey, E. Chain, N. G. Heatley, M. A. Jennings, A. G. Sanders, E. P. Abraham and M. E. Florey. "Antibiotics", Vol. II, p. 631 ff. O.U.P., London, 1949. (b) H. T. Clarke, J. R. Johnson and R. Robinson. "The Chemistry of Penicillin". Princeton University Press, 1949. (c) T. Korzybski, Z. Kowzyk-Gindifer and W. Kurylowicz. "Antibiotics", Vol. II, p. 1147 ff. Pergamon Press, Oxford, 1967.
164. (a) A. L. Demain. *In* "Biosynthesis of Antibiotics" (J. F. Snell, Ed.), Vol. I, p. 30. Academic Press, London and New York, 1966. (b) E. P. Abraham and G. G. F. Newton. *In* "Antibiotics. Volume II. Biosynthesis" (D. Gottlieb and P. D. Shaw, eds.), Springer-Verlag, New York, 1967.
165. F. R. Batchelor, F. P. Doyle, J. H. C. Nayler and G. N. Rolinson, *Nature*, 1959, **183**, 257.
166. For example, see Y. Kameda, Y. Kimura, E. Toyoura and T. Omori, *Nature*, 1961, **191**, 1122; W. Kaufmann and K. Bauer, *Naturwissenschaften*, 1960, **47**, 474; G. N. Rolinson, F. R. Batchelor, D. Butterworth, J. Cameron-Wood, M. Cole, G. C. Eustace, M. V. Hart, M. Richards and E. B. Chain, *Nature*, 1960, **187**, 236.
167. E. P. Abraham, G. G. F. Newton, K. Crawford, H. S. Burton and C. W. Hale, *Nature*, 1953, **171**, 343.
168. E. P. Abraham, G. G. F. Newton, J. R. Schenk, M. P. Hargie, B. H. Olson, D. M. Schuurmans, M. W. Fisher and S. A. Fusari, *Nature*, 1955, **176**, 551.
169. J. H. Grosklags and M. E. Swift, *Mycologia*, 1957, **49**, 305.
170. I. M. Miller, E. O. Stapley and L. Chaiet, *Bacteriol. Proc.*, 1968, 32.
171. E. P. Abraham and G. G. F. Newton, *Biochem. J.*, 1961, **79**, 377.
172. E. H. Flynn, M. H. McCormick, M. C. Stamper, H. de Valeria and C. W. Godzeski, *J. Amer. Chem. Soc.*, 1962, **84**, 4594.
173. (a) H. R. V. Arnstein and P. T. Grant, *Biochem. J.*, 1954, **57**, 360; C. M. Stevens, P. Vohra and C. W. deLong, *J. Biol. Chem.*, 1954, **211**, 297; (b) H. R. V. Arnstein and M. E. Clubb, *Biochem. J.*, 1957, **65**, 618.
174. C. M. Stevens, E. Inamine and C. W. deLong, *J. Biol. Chem.*, 1956, **219**, 405.
175. C. M. Stevens and C. W. deLong, *J. Biol. Chem.*, 1958, **230**, 991.
176. H. R. V. Arnstein and D. Morris, *Biochem. J.*, 1960, **76**, 357.

177. S. Gatenbeck and U. Brunsberg, *Acta Chem. Scand.*, 1968, **22**, 1059; R. Brunner, M. Röhr and M. Zinner, *Z. Physiol. Chem.*, 1968, **349**, 95; B. Spencer, *Biochem. Biophys. Res. Commun.*, 1968, **31**, 170.
178. R. B. Morin, B. G. Jackson, R. A. Müller, E. R. Lavagnino, W. B. Scanlon and S. L. Andrews, *J. Amer. Chem. Soc.*, 1963, **85**, 1896.
179. S. C. Warren, G. G. F. Newton and E. P. Abraham, *Biochem. J.*, 1967, **103**, 902.
180. P. W. Trown, B. Smith and E. P. Abraham, *Biochem. J.*, 1963, **86**, 284.
181. S. C. Warren, G. G. F. Newton and E. P. Abraham, *Biochem. J.*, 1967, **103**, 891.
182. E. P. Abraham, G. G. F. Newton and S. C. Warren. *In* "Chemistry of Microbial Products. I. Symposium on Microbiology", 1964, No. 6, 79.
183. R. Myokei, A. Sakurai, C.-F. Chang, Y. Kodaira, N. Takahashi and S. Tamura, *Tetrahedron Lett.*, 1969, 695; *Agr. Biol. Chem. (Tokyo)*, 1969, **33**, 1491, 1501.
184. S. Marumo and Y. Sumiki, *J. Agr. Chem. Soc. Japan*, 1955, **29**, 305; S. Marumo, *Bull. Agr. Chem. Soc. Japan*, 1959, **23**, 428.
185. N. Takahashi and R. W. Curtis, *Plant Physiol.*, 1961, **36**, 30; S. Marumo and R. W. Curtis, *Phytochemistry*, 1961, **1**, 245; K. Anzai and R. W. Curtis, *Phytochemistry*, 1965, **4**, 263.
186. S. Iriuchijima and R. W. Curtis, *Phytochemistry*, 1969, **8**, 1397.
187. Y. Sumiki and K. Miyao, *J. Agr. Chem. Soc. Japan*, 1952, **26**, 27; K. Miyao, *Bull. Agr. Chem. Soc. Japan*, 1960, **24**, 23; R. O. Studer, *Experientia*, 1969, **25**, 899.
188. R. W. Curtis, *Plant Physiol.*, 1958, **33**, 17.
189. R. W. Curtis, *Science*, 1958, **128**, 661.
190. M. Yukioka and T. Winnick, *Biochim. Biophys. Acta*, 1966, **119**, 614.
191. M. Yukioka and T. Winnick, *J. Bacteriol.*, 1966, **91**, 2237.
192. E. D. Weinberg. *In* "Antibiotics. II. Biosynthesis" (D. Gottlieb and P. D. Shaw, eds.). pp. 240–275. Springer-Verglag, New York, 1967.
193. T. Wieland, *Prog. Chem. Org. Nat. Prod.*, 1967, **25**, 214.
194. T. Wieland, G. Lüben, H. Ottenheym and H. Schiefer, *Ann. Chem.*, 1969, **722**, 173; A. Prox, J. Schmid and H. Ottenheym, *Ann. Chem.*, 1969, **722**, 179; T. Wieland, J. Faesel and H. Ottenheym, *Ann. Chem.*, 1969, **722**, 197.
195. Pl. A. Plattner, U. Nager and A. Boller, *Helv. Chim. Acta*, 1948, **31**, 594.
196. Pl. A. Plattner and U. Nager, *Helv. Chim. Acta*, 1948, **31**, 2203.
197. D. W. Russell, *J. Chem. Soc.*, 1962, 753; M. M. Shemyakin. Y. A. Ovchinnikov, V. T. Ivanov and A. A. Kiryushkin, *Tetrahedron Lett.*, 1963, 1927.
198. Y. A. Ovchinnikov, A. A. Kiryushkin and M. M. Shemyakin, *Tetrahedron Lett.*, 1965, 1111.
199. E. Bishop and D. W. Russell, *Biochem. J.*, 1964, **92**, 19P; A. A. Kiryushkin, Y. A. Ovchinnikov and M. M. Shemyakin, *Tetrahedron Lett.*, 1965, 143; *Zh. Obshch. Khim.*, 1966, **36**, 620.
200. L. H. Briggs, L. D. Colebrook, B. R. Davis and P. W. LeQuesne, *J. Chem. Soc.*, 1964, 5626.
201. Y. Kodaira, *Agr. Biol. Chem. (Tokyo)*, 1962, **26**, 36; S. Tamura, S. Kuyama, Y. Kodaira and S. Higashikawa, *Agr. Biol. Chem. (Tokyo)*, 1964, **28**, 137.
202. R. L. Hamill, C. E. Higgens, H. E. Boaz and M. Gorman, *Tetrahedron Lett.*, 1969, 4255.
203. M. M. Shemyakin, Y. A. Ovchinnikov, A. A. Kiryushkin and V. T. Ivanov, *Tetrahedron Lett.*, 1963, 885; *Izv. Akad. Nauk SSSR, Ser. Khim.*, 1965, 1623;

Y. A. Ovchinnikov, V. T. Ivanov, I. I. Mikhaleva and M. M. Shemyakin, *Izv. Akad. Nauk SSSR, Ser. Khim.*, 1964, 1912; Pl. A. Plattner, K. Vogler, R. O. Studer, P. Quitt and W. Keller-Schierlein, *Helv. Chim. Acta*, 1963, **46,** 927; P. Quitt, R. O. Studer and K. Vogler, *Helv. Chim. Acta*, 1963, **46,** 1715.

204. W. Keller-Schierlein, V. Prelog and H. Zähner, *Prog. Chem. Org. Nat. Prod.*, 1964, **22,** 279.

CHAPTER 9

Miscellaneous Secondary Metabolites

FINALLY WE come to some fungal secondary metabolites which cannot be allocated with confidence to any of the previous chapters. The small size of this chapter emphasizes the point, made in Chapter 1, that most secondary metabolites arise from a few basic pathways. It also emphasizes the advanced state of our knowledge of biosynthetic processes (at least so far as the origin of the atoms is concerned). Several of the compounds listed here will undoubtedly find a place in other chapters once their mode of biosynthesis has been established experimentally.

A. THE CYTOCHALASINS

The name cytochalasin (*cytos* = cell, *chalasis* = relaxation) was coined to describe the unique effects which the compounds produce on mammalian cells in tissue culture.[6] As well as possessing novel biological properties, the cytochalasins form a new class of fungal metabolite whose biosynthesis poses an interesting problem (currently under investigation by Professor Ch. Tamm). The structures of the cytochalasins suggest that they might be formed by condensation of a polyketide chain (with introduced C_1-units) with phenylalanine, and also raise the possibility that the macrolide ring present in (**1a**) and (**1b**) arises by a Baeyer–Villiger type oxidation of a carbocyclic system of the type present in (**2**) and (**3a**). The disadvantage of

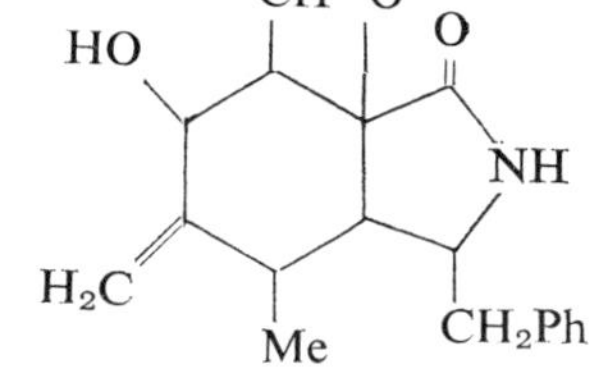

a: R = O, cytochalasin A
Helminthosporium dematioideum[1]

b: R = H, OH, cytochalasin B, phomin
H. dematioideum,[1] *Phoma* sp.[2]

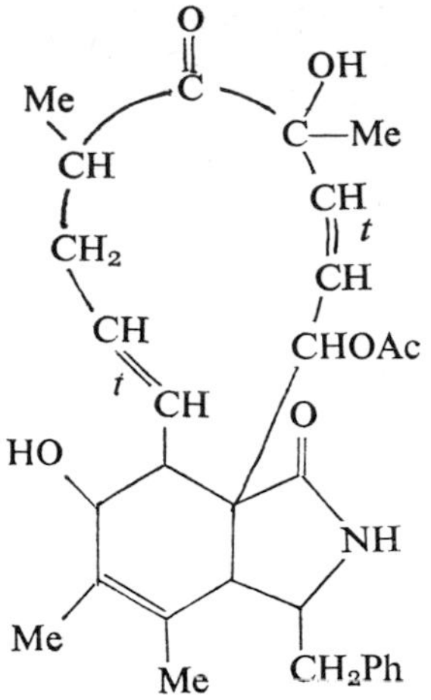

cytochalasin C
Metarrhizium anisopliae[1a,3]

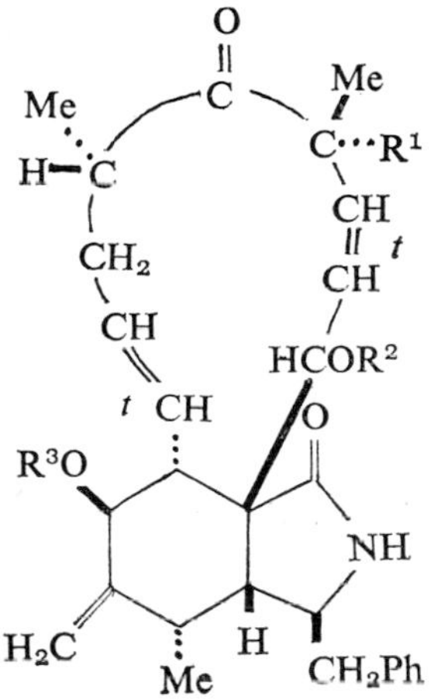

	R¹	R²	R³	
a	OH	Ac	H	cytochalasin D, zygosporin A *M. anisopliae*,[1a,3] *Zygosporium masonii*[4]
b	OH	Ac	Ac	acetylcytochalasin D, zygosporin F
c	OH	H	H	desacetylcytochalasin D, zygosporin D
d	H	H	Ac	zygosporin E *Z. masonii*[5]

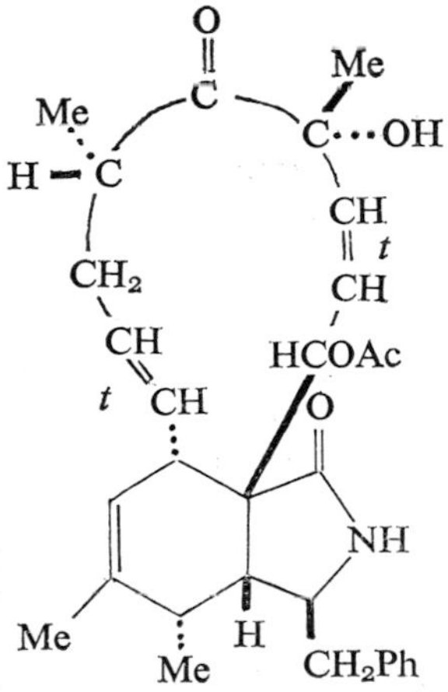

R^1 R^2 R^3

zygosporin G
Z. masonii[5]

the theoretical biosynthetic schemes which one can write for the cytochalasins is that they involve unfavourable carbonyl–carbonyl or methylene–methylene condensations. Professor Tamm has shown that phenylalanine is incorporated intact into cytochalasin B, and that acetate and formate are incorporated though detailed degradation of the resulting cytochalasin B is not yet complete.

B. UNCLASSIFIABLE METABOLITES

5

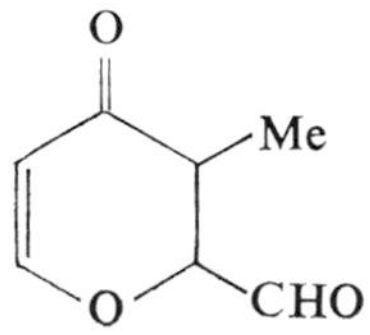

versicolin
Aspergillus versicolor[7]

6

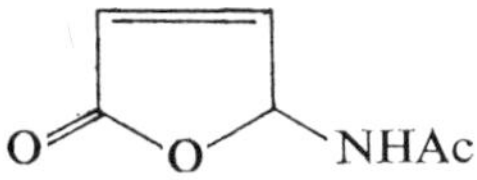

2-acetamido-2,5-dihydro-5-oxofuran
Fusarium equiseti,[8] *F. nivale*[9]

7

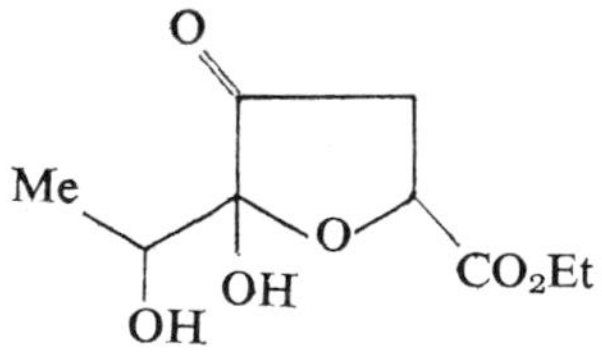

altenin
Alternaria kikuchiana[10]

8

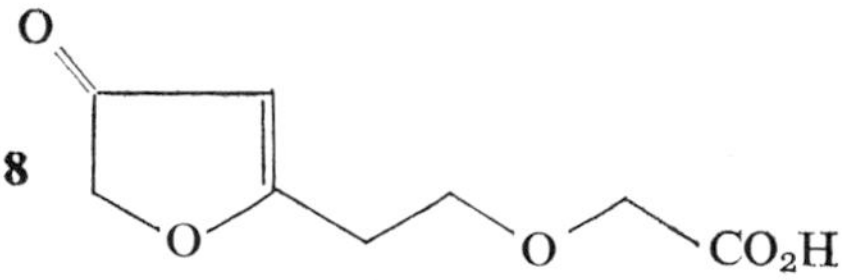

oospolide
Oospora astringens[11]

9

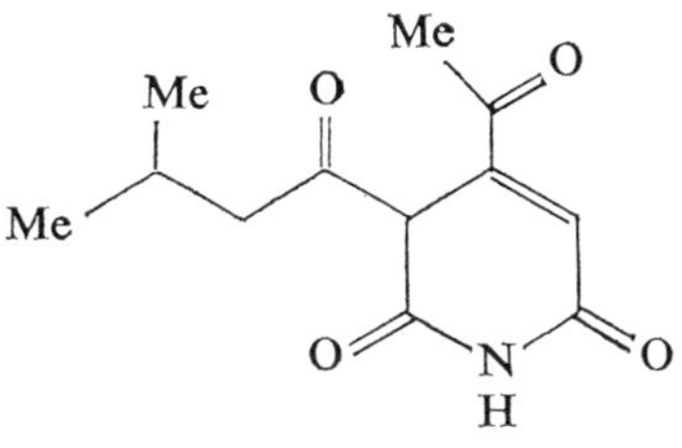

A. flavipes antibiotic
Aspergillus flavipes[12]

10

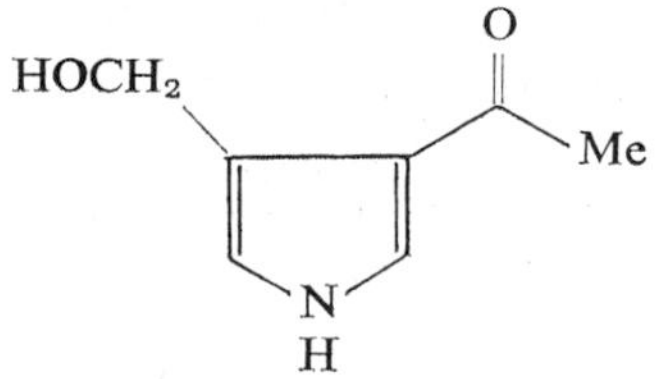

verrucarin E
Myrothecium verrucaria[13]

11

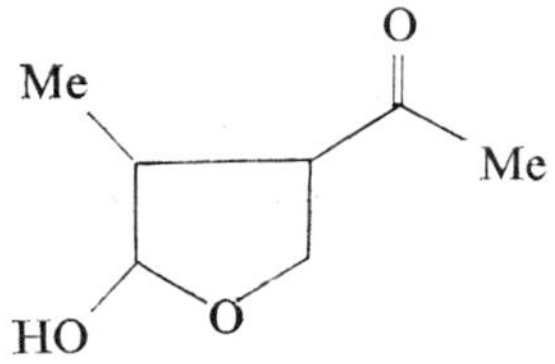

botryodiplodin
Botryodiplodia theobromae[14]

12

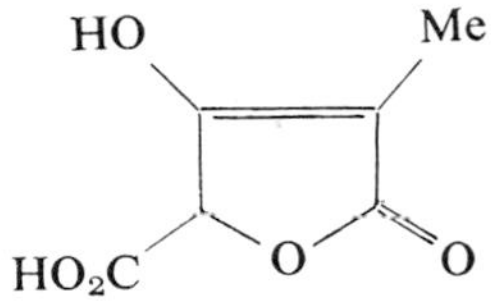

zymonic acid
Yeasts[15] L[16]

13

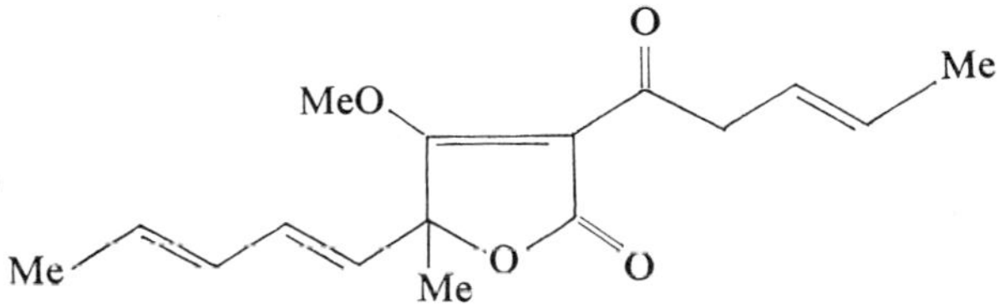

aspertetronin A
A. rugulosus[17]

14

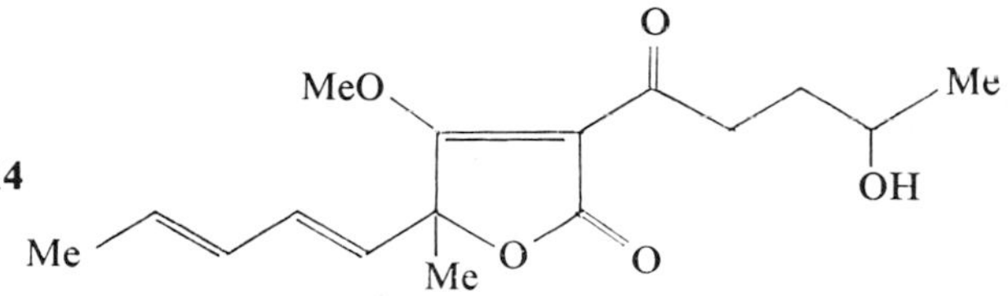

aspertetronin B
A. rugulosus[17]

15 $C_{18}H_{37}CCO_2H$
                 |
         $C_{18}H_{37}$
                 |
         $MeCHCO_2H$

fomentaric acid
Fomes fomentarius[17]

16

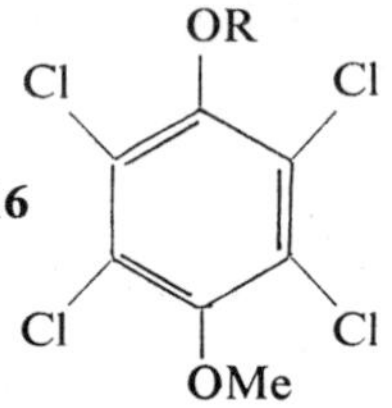

a: R = H, drosophilin A
    Drosophila subatrata[18]

b: R = Me, drosophilin A methyl ether
    Fomes fastuosus[19]

17

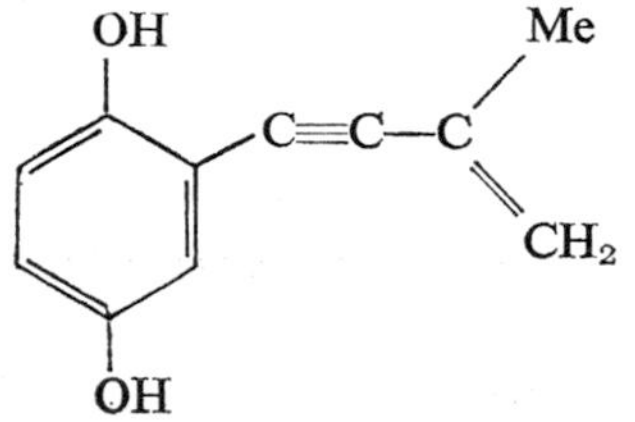

siccayne
Helminthosporium siccans[20]

18

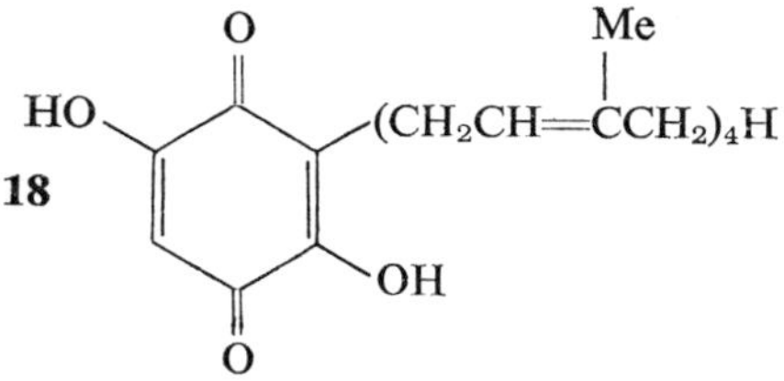

bovinone
Boletus bovinus[21]

19

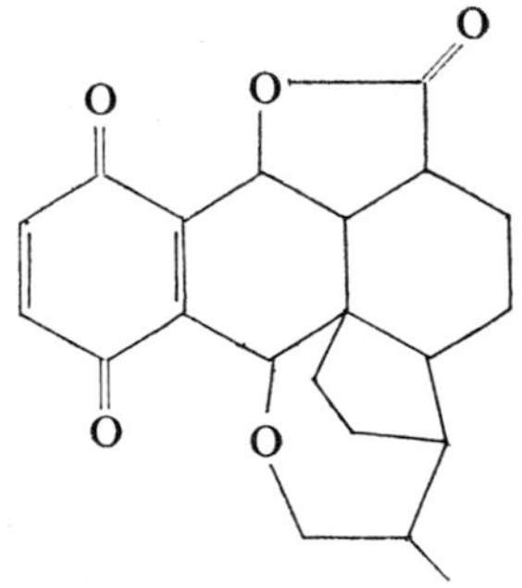

pleurotin
Pleurotus griseus[22]

20 $HO_2C(CH{=}CH)_6CO_2H$

corticrocin
Corticium croceum[23]

21

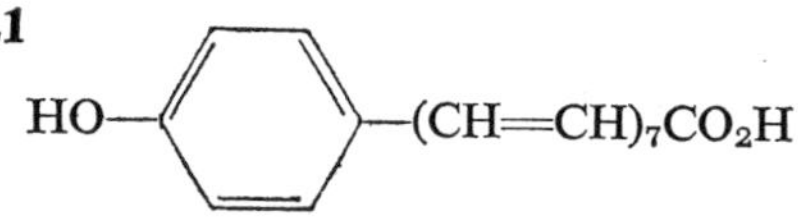

$HO-\langle\rangle-(CH{=}CH)_7CO_2H$

cortisalin
Corticium salicinum[24]

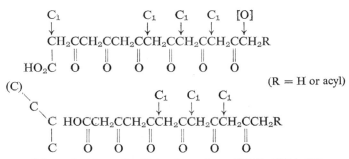

22

asperyellone, asperenone
Aspergillus awamori, A. niger[25]

23 $HOCH_2CH-CH(CH_2)_4CHCH_2C=CHC=CHCO_2H$

with Me groups and $OC-O$

antibiotic 1233A
Cephalosporium(?)sp.[26]

In the absence of labelling experiments, the biosynthetic origin of several of these compounds is ambiguous. Thus versicolin (**5**) could be a polyketide or a sugar derivative (cf. kojic acid), the aromatic rings of compounds (**16**) to (**19**) are probably derived by the shikimic acid route but could be polyketide; corticrocin (**20**) is obviously derived from acetate, but whether by the polyketide route or via a fatty acid remains to be seen; and cortisalin could be derived solely from acetate or, more likely, by condensation of a C_6–C_3 or C_6–C_1 with an acetate-derived chain. Antibiotic 1233A could be formed from acetate plus C_1-units or, as in the case of itaconitin, by incorporation of a C_3 or C_4 unit (Scheme 1).

$$C_1 \qquad\qquad C_1 \quad C_1 \quad C_1 \quad [O]$$
$$CH_2CCH_2CCH_2CCH_2CCH_2CCH_2CCH_2R$$
$$HO_2C \quad O \quad O \quad O \quad O \quad O \quad O$$

(C) ... (R = H or acyl)

$$C_1 \quad C_1 \quad C_1$$
$$C \quad HOCCH_2CCH_2CCH_2CCH_2CCH_2CCH_2R$$
$$C \quad O \quad O \quad O \quad O \quad O \quad O$$

Scheme 1: Alternative biosyntheses for antibiotic 1233A (**23**)

The *A. flavipes* antibiotic (**9**) is related structurally to the glutarimide antibiotics, e.g. cycloheximide (**24**), the glutarimide ring of which is known to be derived in part from malonic acid (* in **24**).[27]

Zymonic acid (**12**) is an artefact and the true metabolite is believed to be protozymonic acid (**25**).[16] The incorporation of [1-14C]-, [3-14C]-, and [6-14C]-glucose, [1-14C]- and [2-14C]acetate, and [*methyl*-14C] methylmalonic acid gives distributions of radioactivity in the methyl ester methyl ether of

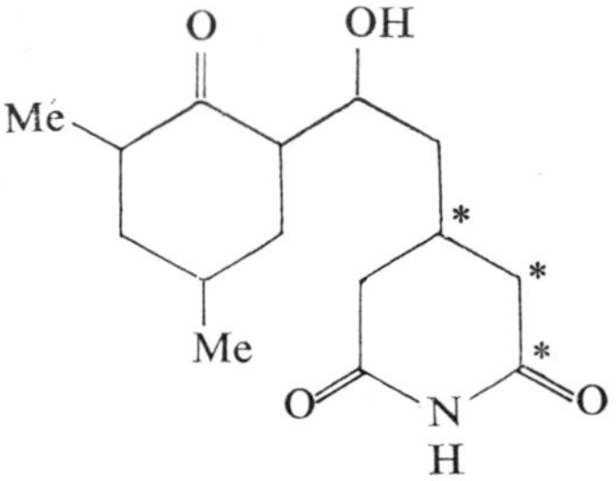

24

zymonic acid consistent with its derivation from tartronic acid and methyl-malonic acid (Scheme 2).[16]

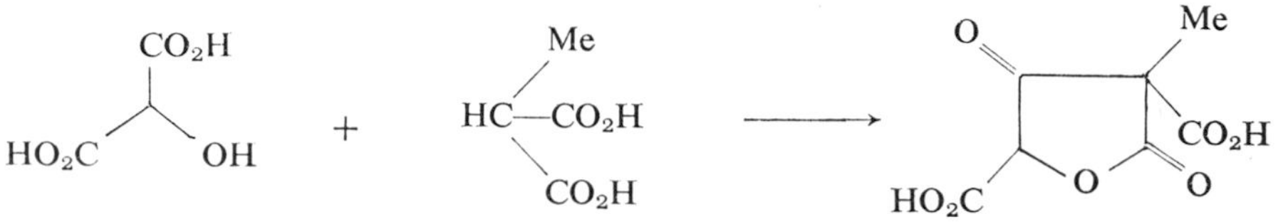

25

Scheme 2: Possible biosynthesis of protozymonic acid (**25**)

Aspertetronins A (**13**) and B (**14**) are related to the other tetronic acids (Chapter 7), but the origin of the branched-chain moiety is obscure. Condensations of the type leading to the tetronic acids may also be involved in the biosynthesis of altenin (**7**), verrucarin E (**10**) and botryodiplodin (**11**).

REFERENCES

1. D. C. Aldridge, J. J. Armstrong, R. N. Speake and W. B. Turner (a) *Chem. Commun.*, 1967, 26; (b) *J. Chem. Soc. C*, 1967, 1667.
2. W. Rothweiler and Ch. Tamm, *Experientia*, 1966, **22**, 750; *Helv. Chim. Acta*, 1970, **53**, 696.
3. D. C. Aldridge and W. B. Turner, *J. Chem. Soc. C*, 1969, 923.
4. S. Hayakawa, T. Matsushima, T. Kimura, H. Minato and K. Katagiri, *J. Antibiotics (Tokyo)*, 1968, **21**, 523; D. C. Aldridge and W. B. Turner, *J. Antibiotics (Tokyo)*, 1969, **22**, 170.
5. H. Minato and T. Katayama, *J. Chem. Soc. C*, 1970, 45.
6. S. B. Carter, *Nature*, 1967, **213**, 261.
7. A. K. Dhar and S. K. Bose, *J. Antibiotics (Tokyo)*, 1968, **21**, 156; *Tetrahedron Lett.*, 1969, 4871.
8. E. P. White, *J. Chem. Soc. C*, 1967, 346.
9. S. G. Yates, H. L. Tookey, J. J. Ellis and H. K. Burkhardt, *Tetrahedron Lett.*, 1967, 621.

10. N. Sugiyama, C. Kashima, M. Yamamoto, T. Sugaya and R. Mohri, *Bull. Chem. Soc., Japan*, 1966, **39**, 1573; N. Sugiyama, C. Kashima, Y. Hosoi, T. Ikeda and R. Mohri, *Bull. Chem. Soc., Japan*, 1966, **39**, 2470; N. Sugiyama, C. Kashima, M. Yamamoto and R. Mohri, *Bull. Chem. Soc., Japan*, 1967, **40**, 345.

11. K. Nitta, Y. Yamamoto and Y. Tsuda, *Tetrahedron Lett.*, 1967, 3013; 1968, 4231.

12. C. G. Casinovi, G. Grandolini, R. Mercantini, N. Oddo, R. Olivieri and A. Tonolo, *Tetrahedron Lett.*, 1968, 3175.

13. E. Fetz and Ch. Tamm, *Helv. Chim. Acta*, 1966, **49**, 349; P. Pfäffli and Ch. Tamm, *Helv. Chim. Acta*, 1969, **52**, 1911.

14. G. P. Arsenault, J. R. Althaus and P. V. Divekar, *Chem. Commun.*, 1969, 1414.

15. F. H. Stodola, O. L. Shotwell and L. B. Lockwood, *J. Amer. Chem. Soc.*, 1952, **74**, 5415; L. J. Haynes and J. R. Plimmer, *Quart. Rev.*, 1960, **14**, 292.

16. J. L. Bloomer and M. A. Gross, *Chem. Commun.*, 1970, 73.

17. P. Singh and S. Rangaswami, *Tetrahedron Lett.*, 1967, 149.

18. F. Kavanagh, A. Hervey and W. J. Robbins, *Proc. Nat. Acad. Sci. U.S.*, 1952, **38**, 555.

19. P. Singh and S. Rangaswami, *Tetrahedron Lett.*, 1966, 1229.

20. K. Ishibashi, K. Nose, T. Shindo, M. Arai and H. Mishima, *Ann. Rep. Sankyo Res. Lab.*, 1968, **20**, 76.

21. P. C. Beaumont and R. L. Edwards, *J. Chem. Soc. C*, 1969, 2398.

22. W. J. Robbins, F. Kavanagh and A. Hervey, *Proc. Nat. Acad. Sci. U.S.*, 1947, **33**, 171; D. Arigoni, *Pure Appl. Chem.*, 1969, **17**, 331.

23. H. Erdtman, *Acta Chem. Scand.*, 1948, **2**, 209.

24. J. Gripenberg, *Acta Chem. Scand.*, 1952, **6**, 580.

25. J. Yu, G. Tamura, N. Takahashi and K. Arima, *Agr. Biol. Chem. (Tokyo)*, 1967, **31**, 831; W. E. Jefferson, *Biochemistry*, 1967, **6**, 3479; G. Pattenden, *Tetrahedron Lett.*, 1969, 4049.

26. D. C. Aldridge, D. Giles and W. B. Turner, *Chem. Commun.*, 1970, 639.

27. Z. Vaněk, J. Cudlin, and M. Vondráček. *In* "Antibiotics II. Bisosynthesis" (D. Gottlieb and P. D. Shaw, eds.). Springer-Verlag, New York, 1967.

ADDENDUM

THIS INCLUDES references to literature which has become available to me between the preparation of the main text and the end of August, 1970, and also one or two references which were omitted from the main text. The material is presented under the appropriate section number from the main text, the first figure being the chapter number.

3.A

Allo-, *epi-*, and *epiallo*muscarine have been detected in muscarine mother-liquors,[1] and (−)-*allo*muscarine has been isolated from *Amanita muscaria*.[2] The distribution of muscarine and its epimers in *Inocybe* spp. has been studied;[3] in two species, *epi*muscarine was the major isomer.

3.D.1

The biosynthesis of cinnamamide in *Streptomyces verticillatus* involves phenylalanine ammonia-lyase.[4]

4.B.1

*Erythro-*8,9-Dihydroxy-13-oxodocosanoic acid (**1**) has been isolated from yeast.[5] 8,9,13-Triacetoxydocosanoic acid had previously been isolated.[6]

$$Me(CH_2)_8 \overset{\displaystyle O}{\overset{\|}{C}} (CH_2)_3 CH(OH)CH(OH)(CH_2)_6 CO_2 H$$

1

5.A.2(c)

On p. 81 we described six possible routes to 6-methylsalicylic acid, depending upon the timing of the various steps. Lynen and his coworkers[7] have now produced evidence for the timing of the reduction step and discuss a likely sequence for the other steps. They have isolated from *Penicillium patulum* a purified 6-methylsalicylic acid synthetase (the purified enzyme is more stable than crude extracts) which, in the absence of NADPH, converts acetyl-CoA and malonyl-CoA into triacetic acid lactone. This suggests that the reduction step occurs at the triketide level and the authors further suggest, by analogy with fatty acid biosynthesis, that dehydration follows the reduction before the next condensation reaction. The cyclization step requires that the resulting double bond be *cis*, the configuration which results from

$\beta\gamma$-elimination in the biosynthesis of unsaturated fatty acids (see p. 65); during 6-methylsalicylic acid biosynthesis, $\beta\gamma$-elimination is favoured by the formation of an $\alpha\beta$-unsaturated ketone. The sequence of steps in 6-methylsalicylic acid biosynthesis is thus envisaged as in Scheme 1.

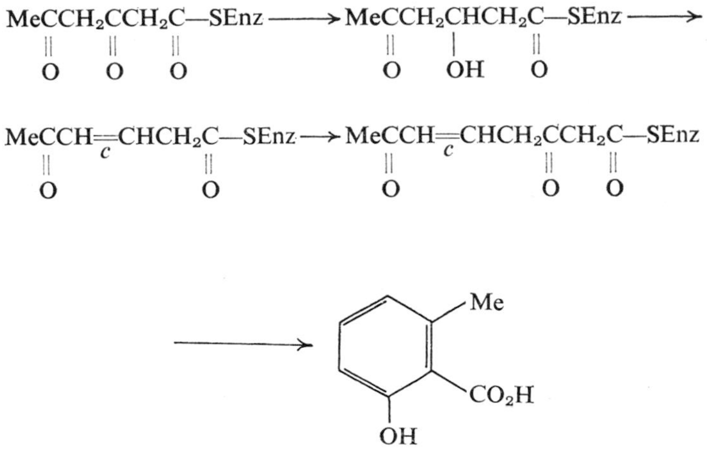

Scheme 1: A likely sequence of steps in 6-methylsalicylic acid biosynthesis

5.B.2(a)

The acetate origin of radicinin has been confirmed by experiments with ^{13}C-acetate.[8]

5.B.2(a)ii

A study of the transformation of diphenols by *Aspergillus fumigatus*[9] has produced evidence which supports a biosynthetic pathway proposed (p. 95) for fumigatin.

The following new metabolites have been isolated from *A. fumigatus*:[10] spinulosin hydrate (2), spinulosin quinol-hydrate (3), dihydrospinulosin quinol (4), and fumigatin chlorohydrin (5), which is probably an artefact derived from 3-hydroxy-4-methoxytoluquinone-1,6-epoxide (fumigatin oxide) (6), previously isolated from *A. fumigatus*. Compound (6) is incorporated into fumigatin whereas fumigatin is not incorporated into compound (6), suggesting that 6 is an intermediate between orsellinic acid and fumigatin in accord with an earlier suggestion by the same group (see p. 97) but at variance with other results, e.g. those discussed above.

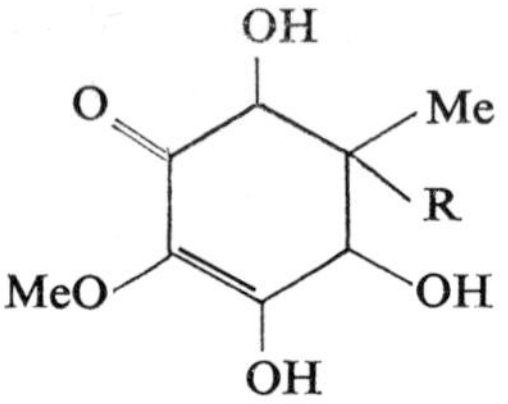

3, R=OH
4, R=H

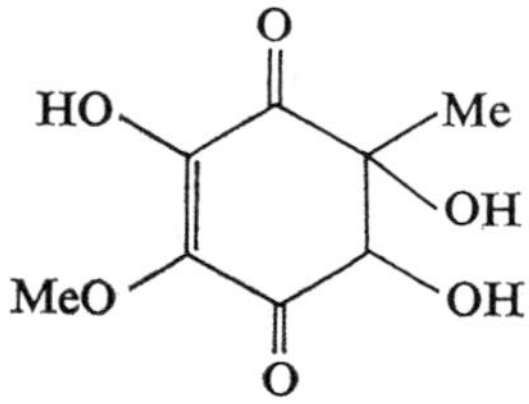

2

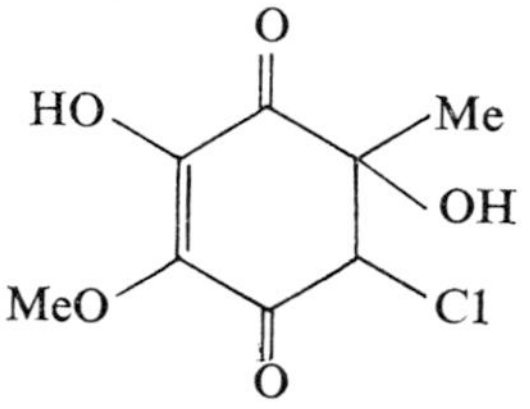

5

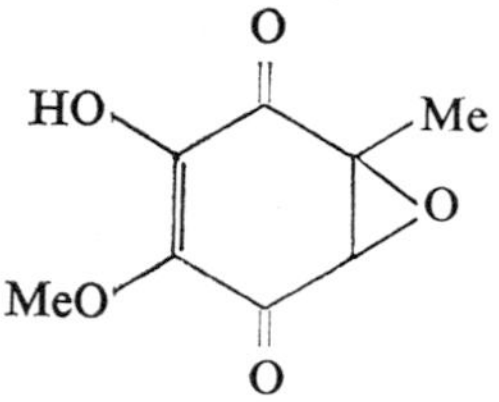

6

Phyllostine (**7**), a metabolite of a *Phyllosticta* sp.[11], is related to epoxydon.

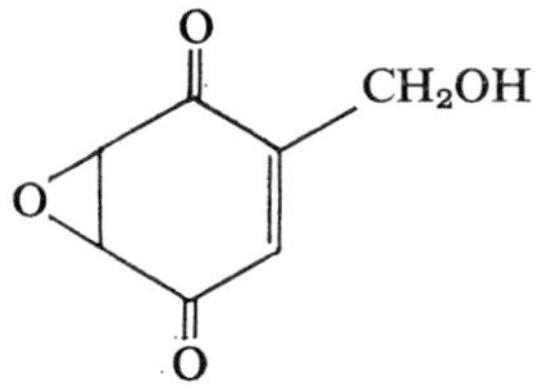

7

5.B.2(c)

Usnic acid has been isolated from an Actinomycete.[12]

5.B.3(a)

Little radioactivity from [2-^{14}C]acetate is incorporated into C-10 of

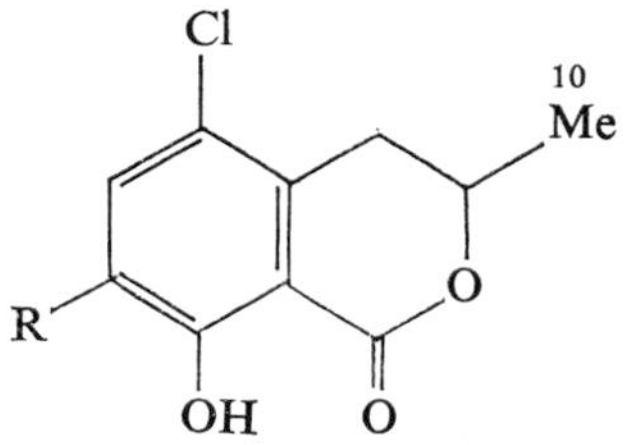

8

ochratoxin A (8),[13] a surprising result in view of its likely polyketide derivation.

5.B.3(c)

O-Methylcurvulinic acid (9) is a metabolite of *Curvularia siddiqui* and *C. ellisii*.[14]

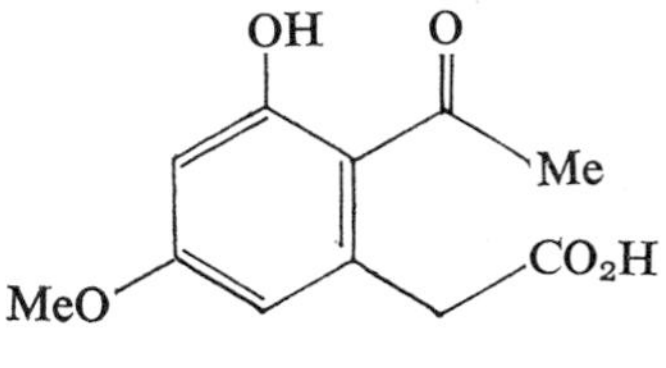

9

5.B.4(a)

The double bonds of cerulenin have the *trans* configuration.[15]

5.B.4(c)

The asymmetric centre of ascochytine has the *S*-configuration.[16] Portentol (10) is a metabolite of the lichen *Roccella fuciformis*[17] whose structure suggests

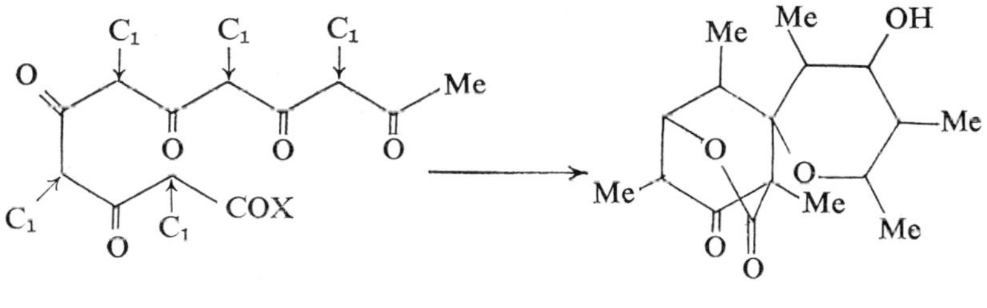

10

a propionate derivation. It is, however, formed from acetate and the C_1-pool, so that the statement (p. 76) that no fungal product incorporates propionate within the chain is still true.

5.B.5(a)

Aspergin, a metabolite of an *Aspergillus* sp., has structure (11), i.e. tetra-hydroauroglaucin.[18]

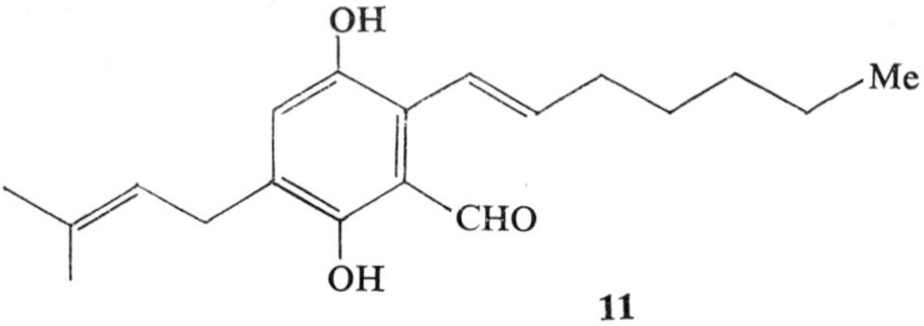

11

Fusamarin (**12**), from a *Fusarium* sp., has structure (**12**);[19] the introduction of a butyl residue onto a polyketide is a novel feature.

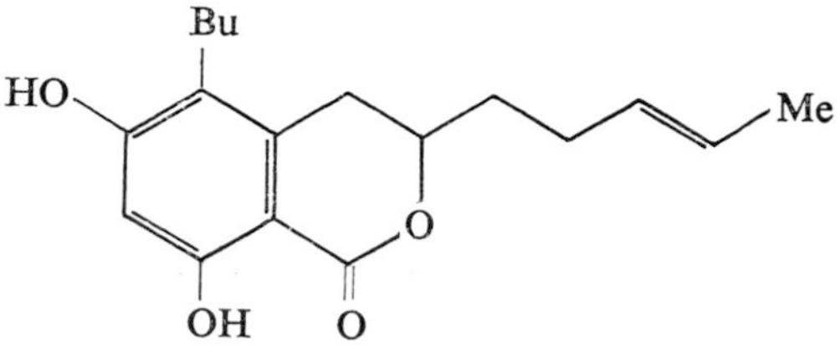

12

5.B.5(d)

[14]C-Acetate is incorporated into rubrofusarin by *Fusarium graminearum* in the expected manner.[20] In the presence of unlabelled malonate, ^{14}C-acetate gave preferential labelling of the "starter" group. *O*-Demethylanhydrofusarubin (**13**) has been isolated from *Gibberella fujikuroi*.[21]

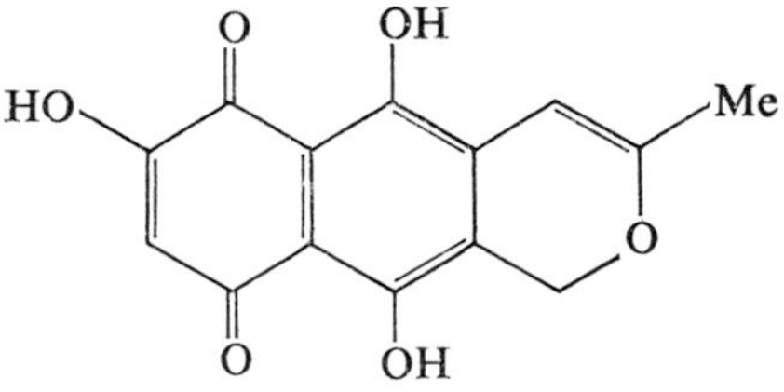

13

5.B.5(g)

The position of the *O*-methyl group of herqueinone has been defined as in (**14**) by two groups.[22,23] Both groups found that herqueinone is invariably contaminated by isoherqueinone, the product of mild base treatment of herqueinone. One group[22] suggests that herqueinone and isoherqueinone are

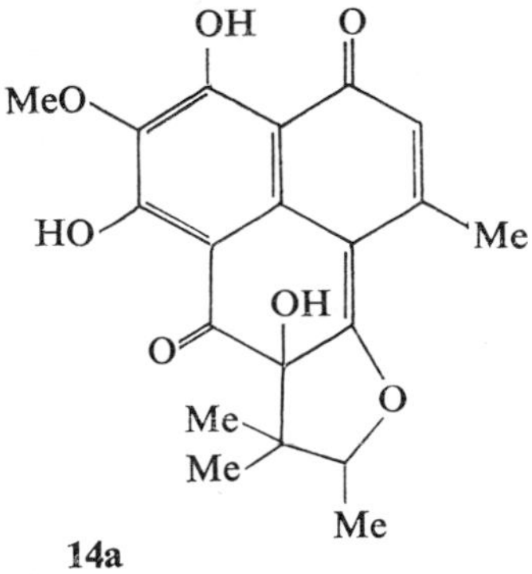

14a 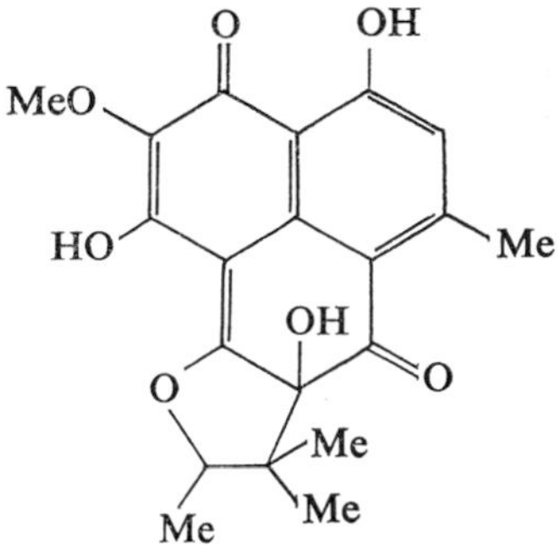   14b

epimeric at the carbon carrying the tertiary hydroxyl group, while the other[23] suggests that they are epimeric at the asymmetric centre of the side-chain. Mass spectroscopic evidence suggests that the fusion of the furan ring of herqueinone might be as in (14b) rather than in (14a) as previously assigned.

5.B.6(a)

2,5,7-Trihydroxyemodin (15) has been isolated from the lichen *Mycoblastus sanguineus*.[24] As well as dermoglaucin (16) and dermocybin (17) (see p. 157),

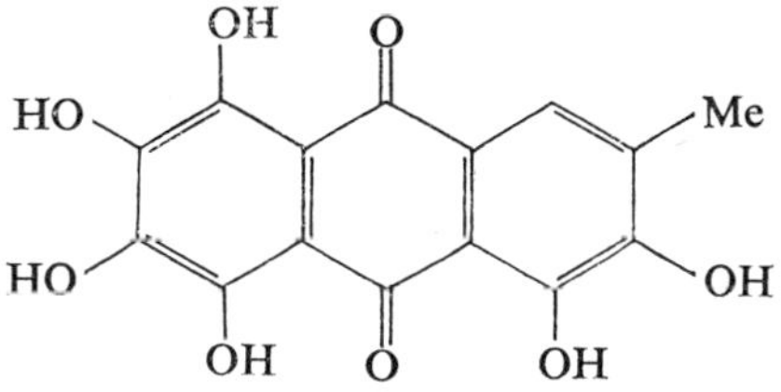

15

Dermocybe sanguinea (= *Cortinarius sanguineus*) and *D. semisanguinea* produce physcion (18), emodin (19), endocrocin (20), and four new anthraquinone carboxylic acids (21) to (24).[25]

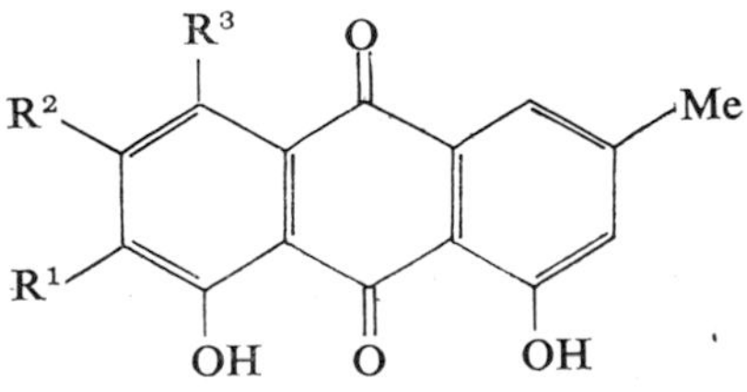

	R¹	R²	R³	
16	OH	OMe	H	dermoglaucin
17	OH	OMe	OH	dermocybin
18	H	OMe	H	physcion
19	H	OH	H	emodin

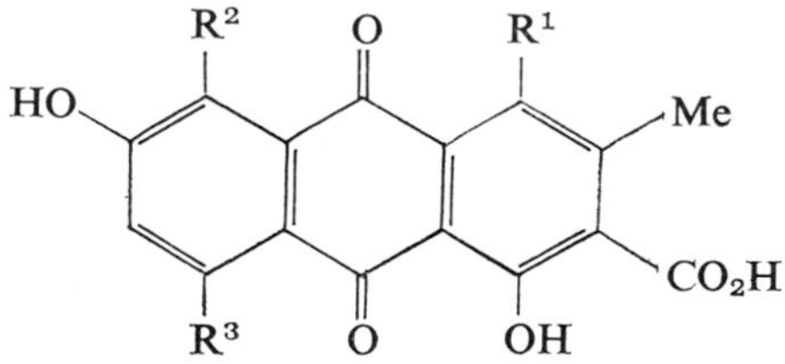

	R¹	R²	R³	
20	H	H	OH	endocrocin
21	H	H	OMe	dermolutein
22	OH	H	OMe	dermorubin
23	H	Cl	OMe	5-chlorodermolutein
24	OH	Cl	OMe	5-chlorodermorubin

5.B.6(a)iv

Secalonic acid D, an ergochrome produced by *Penicillium oxalicum*,[26] is the antipode of ergochrome AA(4,4′) (secalonic acid A). Arugosin, a metabolite

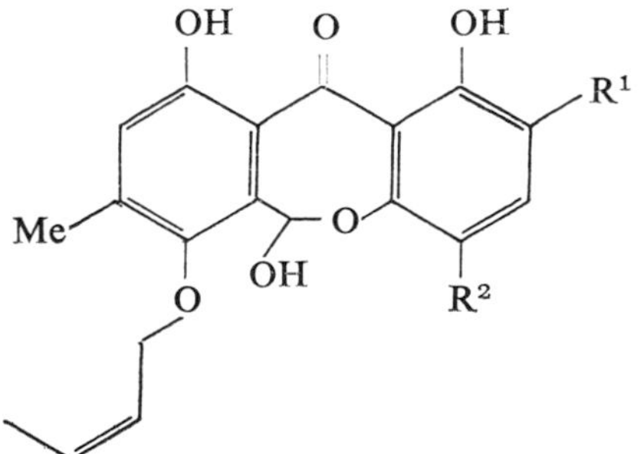

25 ; R¹=CH₂CH=CMe₂, R²=H
26 ; R¹=H, R²=CH₂CH=CMe₂

25 ; $R^1 = CH_2CH = CMe_2$, $R^2 = H$
26 ; $R^1 = H$, $R^2 = CH_2CH = CMe_2$

of *Aspergillus rugulosus*,[27] is a mixture of compounds (25) and (26), which have the carbon skeleton and oxygenation pattern of the ergochromes with introduced *O*- and *C*-isopentenyl groups.

5.B.6(c)

A new metabolite of *Penicillium hirayamae* has been shown to be 7-*epi*-5-chloroisorotiorin (27).[28]

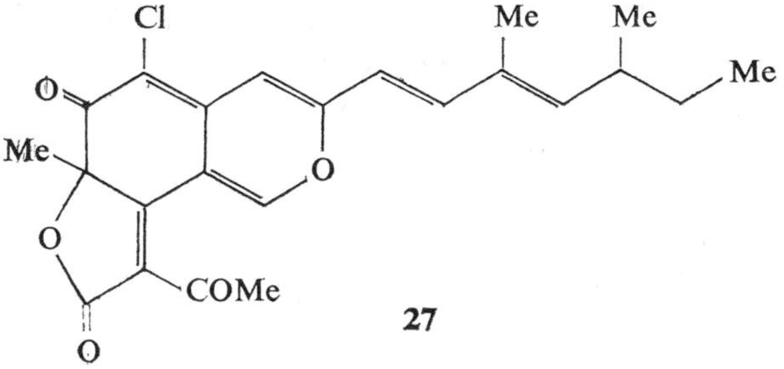

27

5.B.6(e)

Funicone, a co-metabolite of mitorubrin, mitorubrinol, and mitorubrinic acid (see p. 167) in *Penicillium funiculosum*, has structure (**28**).[29] It is suggested that funicone might be derived from an octaketide (Scheme 2).

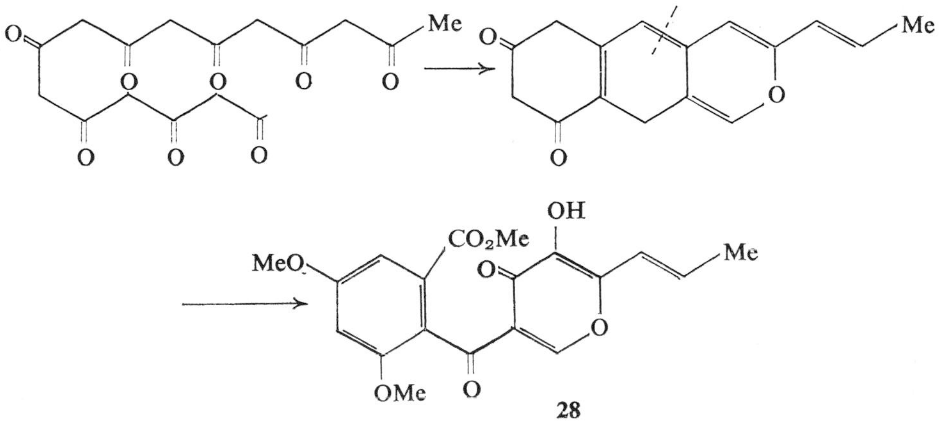

28

Scheme 2: Possible derivation of funicone from an octaketide[29]

5.B.7(c)

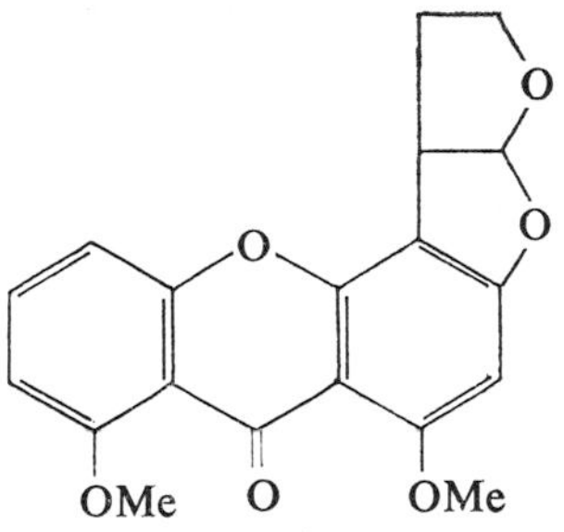

29

Dihydro-*O*-methylsterigmatocystin (**29**) has been isolated from *Aspergillus flavus*.[30] A cell-free preparation capable of aflatoxin biosynthesis has been isolated;[31] surprisingly (see p. 183) the best incorporation of radio-activity into aflatoxin was from mevalonate.

5.B.9

Nystatin aglycone has structure (**30**).[32]

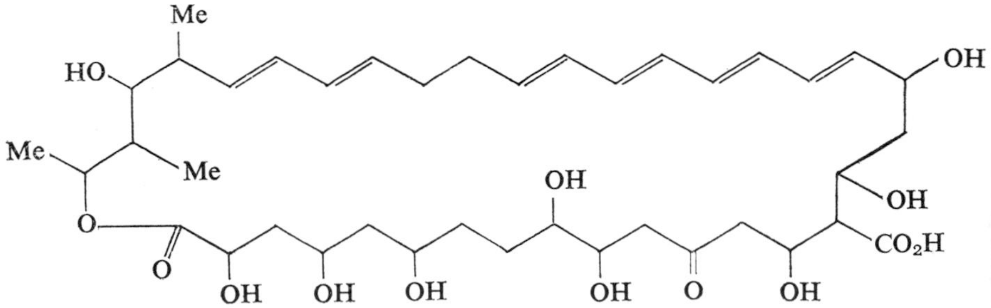

30

6.B.2(a)

4β,15-Diacetoxy-3α, 7α-dihydroxy-12,13-epoxytrichothec-9-en-8-one (niva-lenol diacetate), previously obtained from *Fusarium scirpi*, has now been isolated from *F. nivale*.[33] Experiments with tritiated mevalonate[34] support

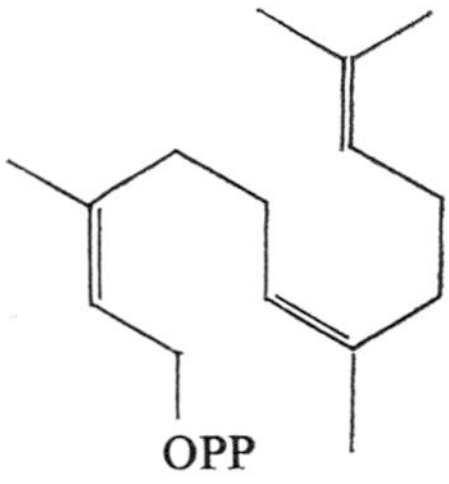

OPP

31

the intermediacy of 6,7-*trans*-farnesyl pyrophosphate, rather than the alternative folding (**31**) of a 6,7-*cis*-intermediate, in the biosynthesis of the trichothecanes. Trichodiene (**32**)[35] and trichodiol (**33**),[36] metabolites of *Trichothecium roseum*, have the trichothane skeletons.

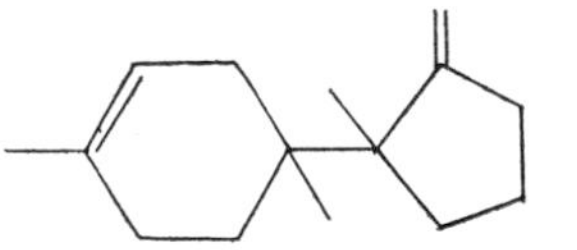

32

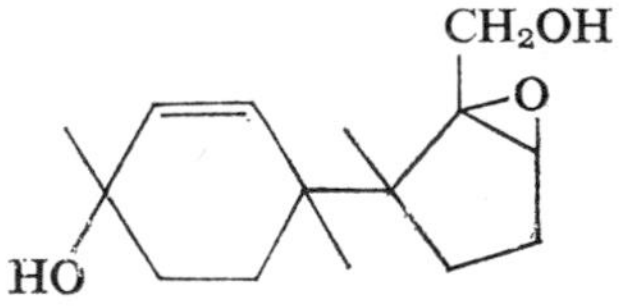

33

6.B.2(c)

Two new protoilludanes—coriolin B (**34**) and coriolin C (**35**)—have been isolated from *Coriolus consors*.[37]

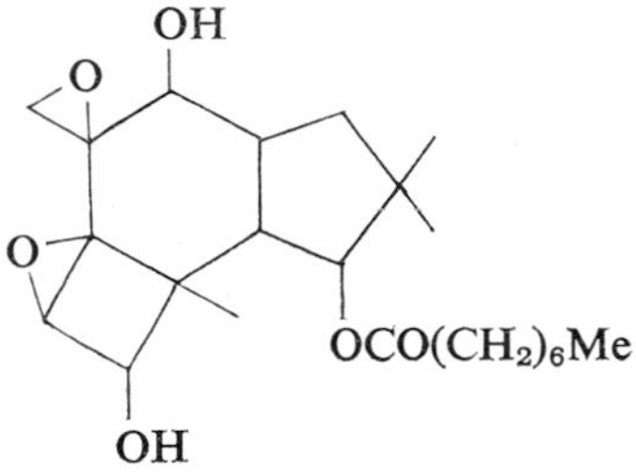

34

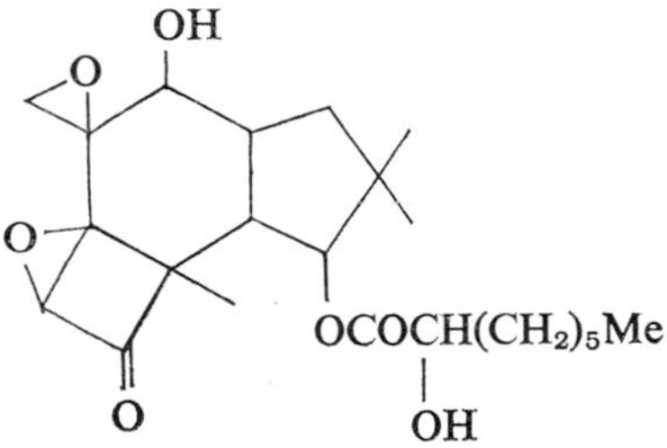

35

6.B.2(d)

Cyclonerolidol (**36**) and a co-metabolite (**37**) have been isolated from a *Trichothecium* sp. A co-metabolite (**38**) of helicobasidin (see p. 230) has been

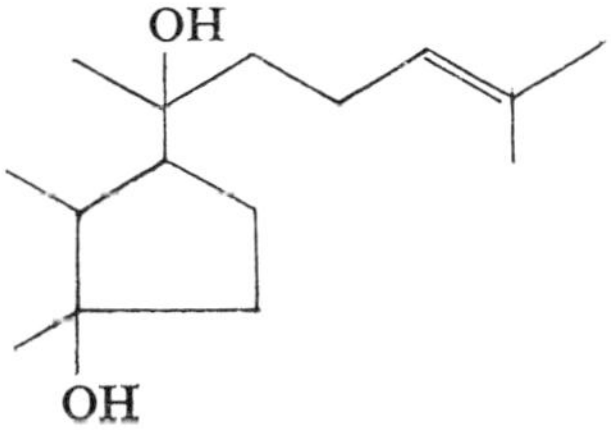

36

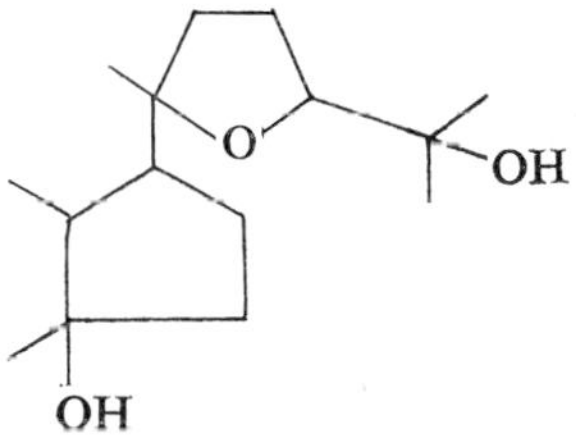

37

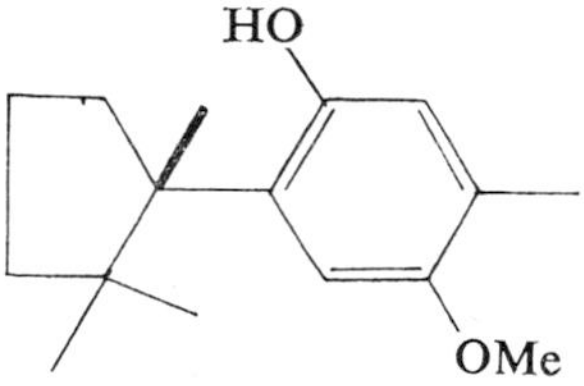

38

isolated from *Helicobasidium mompa*.[39] $4R$-[4-^{3}H,2-^{14}C]Mevalonate is incorporated into compound (**38**) with retention of all the tritium.

6.B.2(e)

Cylindrochlorin, a metabolite of a *Cylindrocladium* sp.[40], has the part-structure (39) and is obviously related to antibiotics LL-Z 1272 (p. 232).

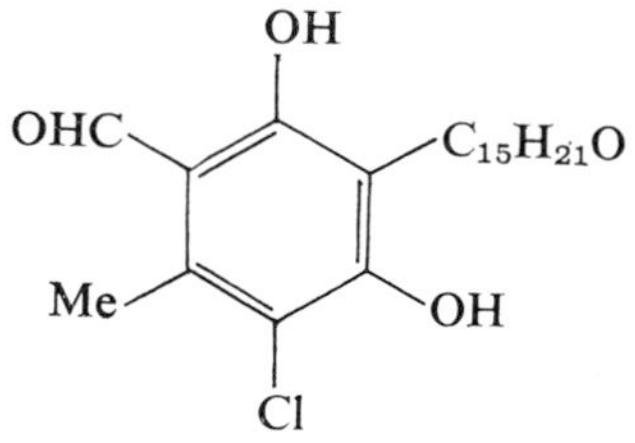

39

6.B.3(a)

Virescenosides A (40) and B (41) are β-D-altropyranosides of virescenols A and B (see p. 236), the first altrosides to be found in nature.[41] Virescenosides C, F, G, and H have structures (42), (43), (44), and (45) respectively.[42]

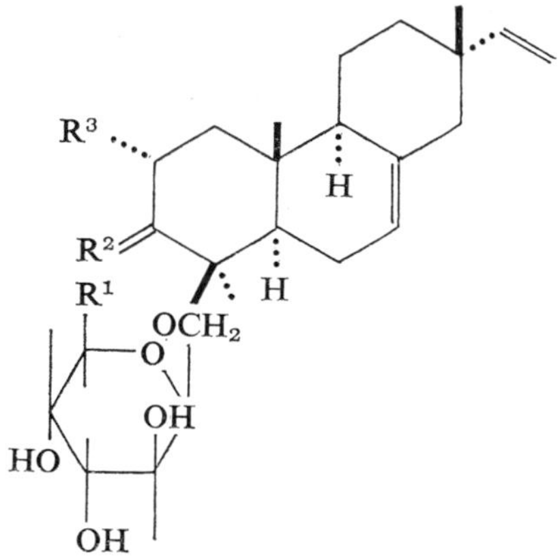

	R^1	R^2	R^3
40	CH_2OH	OH H	OH
41	CH_2OH	OH H	H
42	CH_2OH	O	H
43	CO_2H	OH H	OH
44	CO_2H	OH H	H

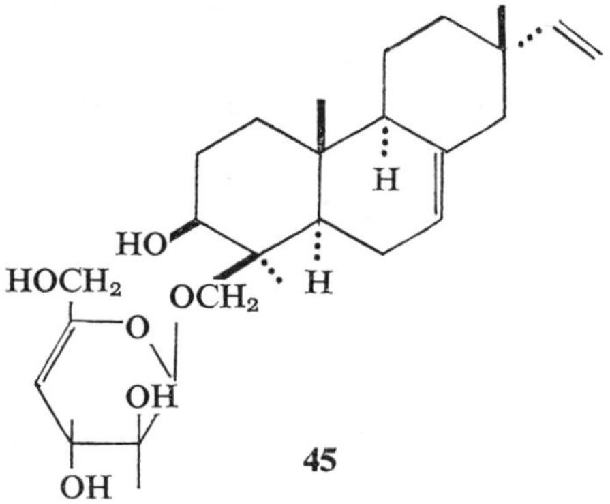

45

6.B.3(b)

Rosein III has been shown by X-ray analysis to be 11β-hydroxyrosenono-lactone (**46**).[43]

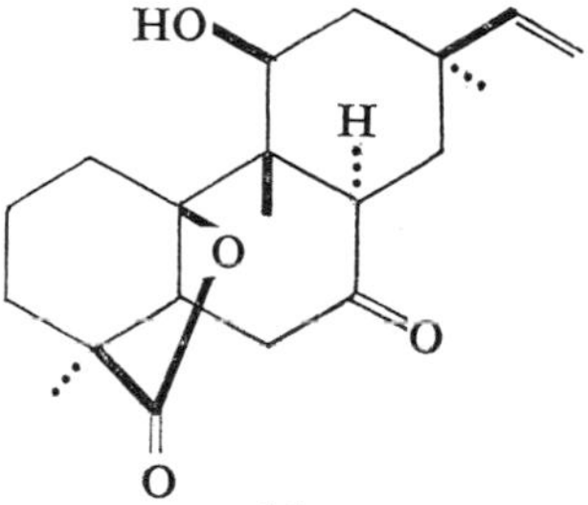

46

6.B.3(c)

The ring-contraction of a kaurane intermediate to give the gibbane skeleton requires the presence of 6β,7β-substituents (see p. 245). 6β,7β-Dihydroxykaurenoic acid (**47**) has been synthesised, labelled at the exocyclic

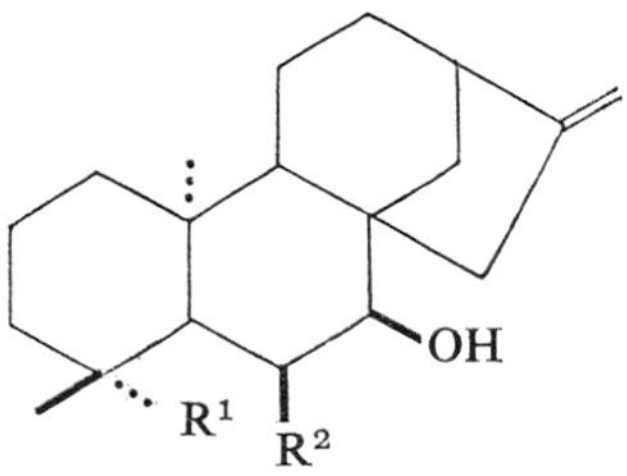

47, R¹=CO₂H, R²=OH
48, R¹=CH₂OH, R²=OH
49, R¹=CH₂OH, R²=H

47, R^1=CO$_2$H, R^2=OH
48, R^1=CH$_2$OH, R^2=OH
49, R^1=CH$_2$OH, R^2=H

methylene group, and fed to *Gibberella fujikuroi.*[44] High incorporation into fujenal was observed, but gibberellic acid and 7-hydroxy-kaurenolide were inactive. After feeding labelled kaurene to *G. fujikuroi* the diol (47) could be detected by dilution analysis so that it is a natural metabolite. It appears that once the diol (47) is present in a free state it is no longer able to participate in gibberellin biosynthesis. In accord with this, the triol (48), prepared by feeding kaurenol hemi-succinate to *G. fujikuroi*, is not incorporated into gibberellic acid though the diol (49) is incorporated to the extent of 1.8 %.[45]

6.B.4

Three co-metabolites of fusicoccin are formed from fusicoccin under mild conditions and are probably artefacts.[46]

6.B.5(a)ii

Barton and his collaborators have now published a more detailed account[47] of their investigation of ergosterol and eburicoic acid biosynthesis (see references 165 and 186 of Chapter 6). In this they report the isolation of a new yeast sterol, 3β-hydroxy-4α-methylcholesta-8(14),24-diene (50). Re-examination of their earlier results, together with new data from a cell-free

50

yeast system, leads them to conclude that 24-methylenedihydrolanosterol is not a precursor of ergosterol in *Saccharomyces cerevisiae* (contrary to earlier conclusions, see p. 254). There appear to be several possible routes to ergosterol but certain structural features appear to be necessary for a compound to serve as a precursor of ergosterol–(1) a 24,25-double bond, (2) a 8(9)-double bond, and (3) a 3β-hydroxyl group.

Barton *et al.* had shown that 24-methylenedihydrolanosterol is incorporated into eburicoic acid, and since 3β-hydroxylanosta-8,24-dien-21-oic acid has now been isolated from *Polyporus sulphureus* and is incorporated into eburicoic acid,[48] more than one route seems available for eburicoic acid biosynthesis. Ergosta-5,7,22,24(28)-tetraen-3β-ol (51) is efficiently incorporated into ergosterol by growing yeast cultures under either anaerobic or aerobic conditions.[49] The biosynthesis of $\triangle^7$-ergosterol in the alga *Chlorella vulgaris* proceeds with the retention of all the deuterium atoms of [*methyl*-$^2H_3$]methionine,[50] in contrast with the situation encountered in ergosterol

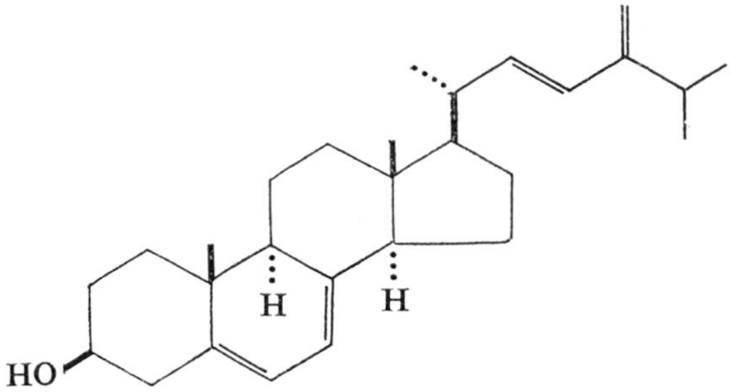

51

biosynthesis in yeast (see p. 254). 24-Ethyl compounds in the same organism retain five deuterium atoms, in contrast with the situation in higher plants.

6.B.5(b)iii

3β-Hydroxylanosta-7,9(11),24-trien-21-oic acid (**52**) has been isolated from sclerotia of *Poria cocos*.[51]

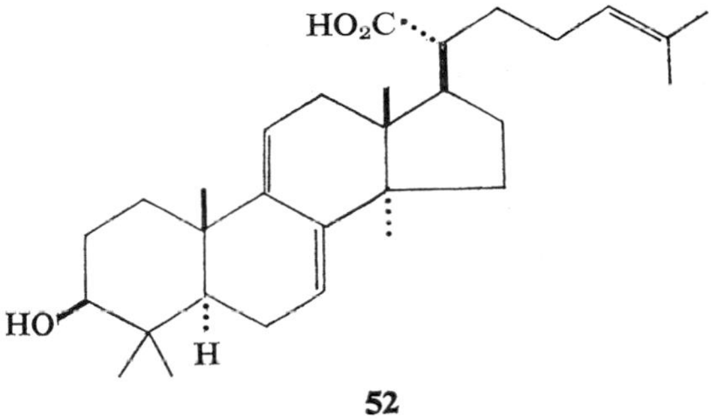

52

6.B.5(b)iv

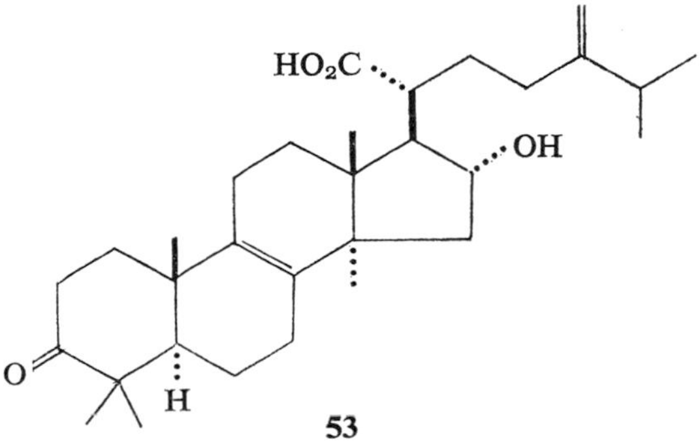

53

An examination[52] of the triterpenes of *Trametes dickinsii* has yielded the following: a mixture of polyporenic acid C and its dihydro-derivative (53); a mixture of C-20 epimers of polyporenic acid C and their dihydro-derivatives;

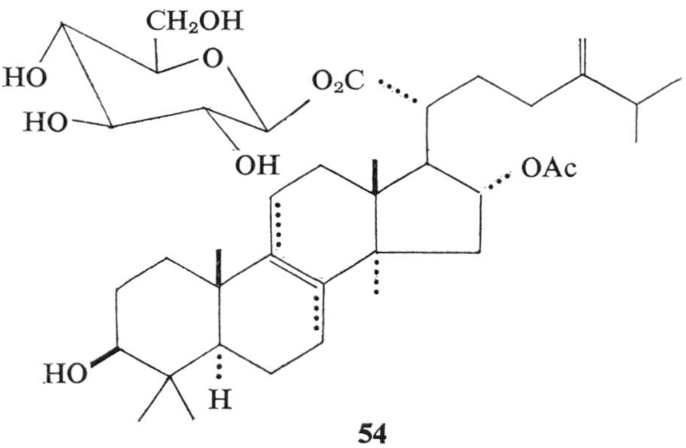

54

a mixture (54) of acetylglucosyltumulosic acid and the corresponding dehydro-derivative.

6.B.5(b)vi

The incorporation of $4R$-[4-^{3}H,2-^{14}C]mevalonate into fusidic acid confirms that the protosterols are derived *via* the intermediate (93, p. 253) with retention of tritium at C-9 and C-13.[53] Tritium is lost from C-3 in accord with the epimerization of the 3-hydroxyl group *via* a ketone. Cell-free extracts of *Emericellopsis* spp. convert mevalonate to 3β-hydroxyprotosta-17(20),24-diene.[54]

6.B.6ii

Trisporol-C, a neutral co-metabolite of the trisporic acids, has structure (55).[55] The absolute stereochemistry of the trisporic acids has been deduced,[56] and its implications for β-carotene biosynthesis discussed.

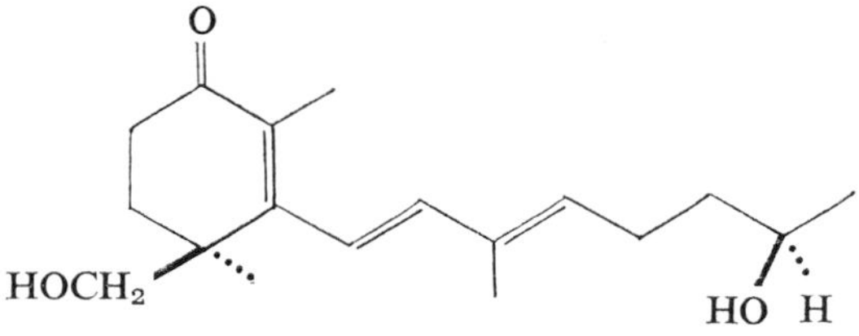

55

8.B.6

Fusaric acid and dehydrofusaric acid are interconverted by *Gibberella fujikuroi*;[57] 10-hydroxyfusaric acid (56) was also isolated. Fusaric acid and nicotinic acid are biosynthesized by different routes in *Fusarium oxysporium*.[58]

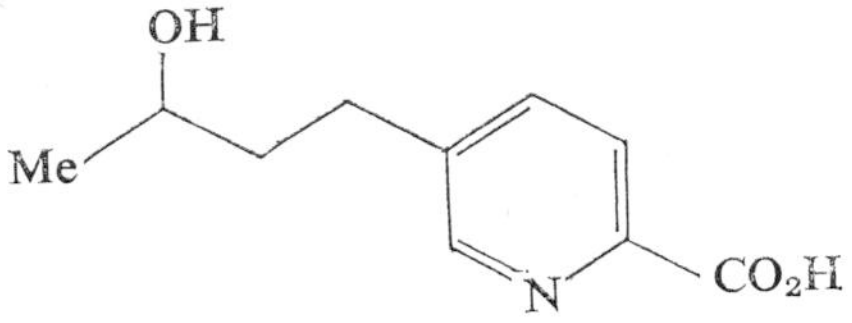

56

8.B.7

Psilocybin has been isolated from *Psilocybe semilanceata* collected in Britain[59] and from *P. subaeruginosa* collected in Australia.[60]

8.B.8

Cycloclavine (57), isolated from the higher plant *Ipomoea hildebrandtii*,[61] and rugulovasins A and B, which are stereoisomers of structure (58) from *Penicillium concavorugulosum*,[62] are new compounds related to the ergot alkaloids.

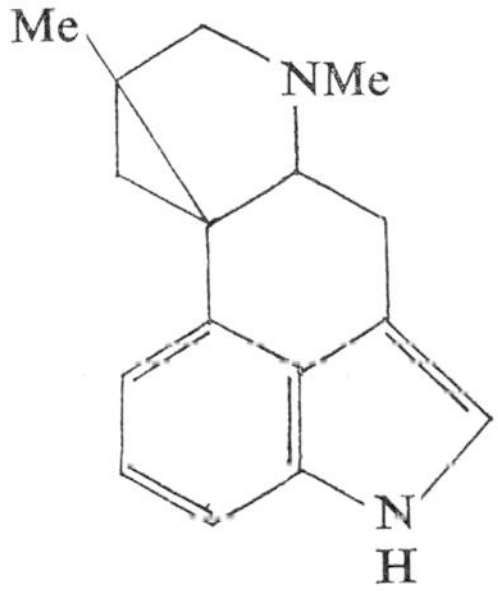

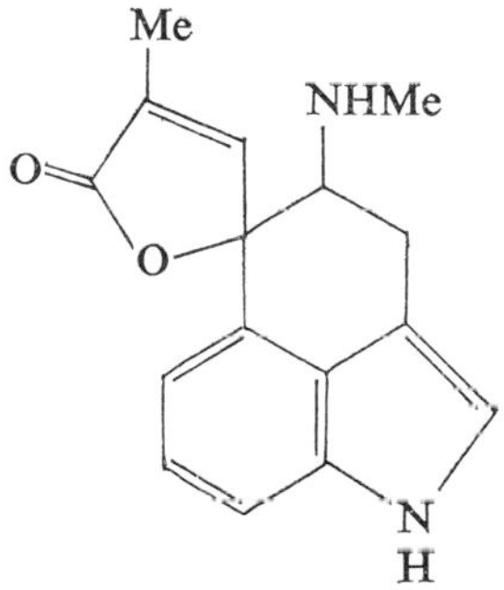

57   58

The conversion of chanoclavine-I into ergot alkaloids probably proceeds *via* the corresponding aldehyde.[63] A cell-free system for the conversion of chanoclavine-I into elymoclavine has been isolated from a strain of *Claviceps*.[64] Alanine, but not alaninol or α-methylserine, is incorporated into the alaninol residue of ergometrine, suggesting the initial formation of lysergylalanine.[65] The side chain of *N*-(α-hydroxyethyl)lysergamide is derived from the nitrogen atom and C-2 and C-3 of alanine, but results with doubly-labelled precursor suggest that alanine is not a direct precursor.[66] L-Proline is a precursor of the proline part of the peptide chain of ergotoxinine.[67]

8.C.1(c)

Chaetocin, a metabolite of *Chaetomium minutum*,[68] has structure (59).

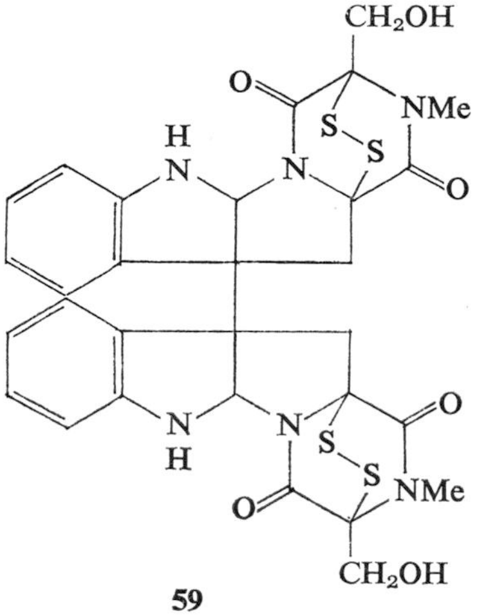

59

8.C.1(d)

Cyclopenin and cyclopenol are formed from phenylalanine, anthranilic acid, and methionine.[69] *m*-Tyrosine and dihydroxyphenylalanine are not incorporated into cyclopenol, whose *m*-hydroxyl group is probably introduced at a late stage in the biosynthesis.

8.D.1

A tetrapeptide from the lichen *Roccella cannariensis* has structure (60).[70]

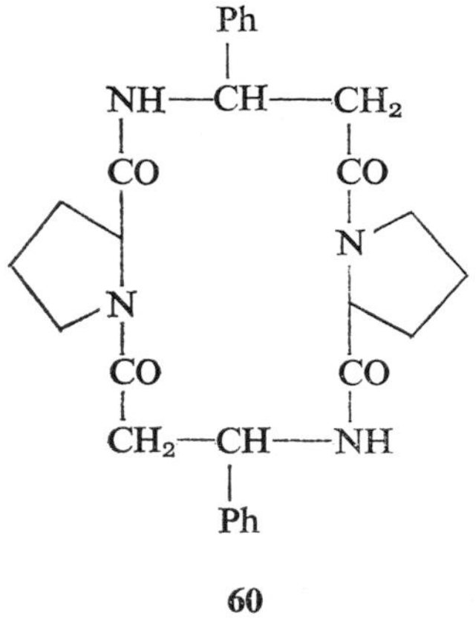

60

8.D.3

Four more insecticidal depsipeptides (*cf.* destruxin B, p. 342) have been shown to have structures (61) to (64).[71]

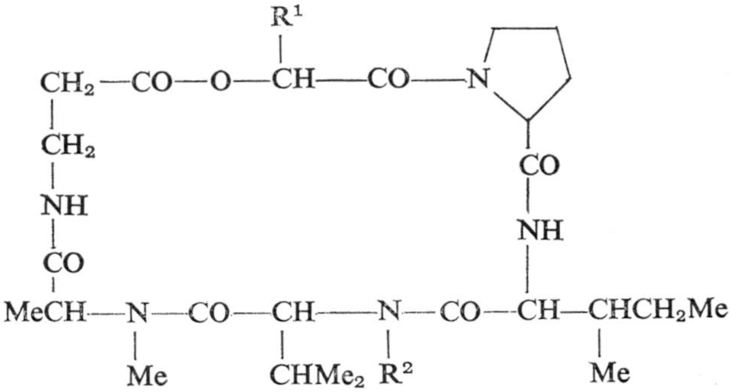

	R^1	R^2	
61	$CH_2=CHCH_2$	Me	destruxin A
62	Me_2CHCH_2	H	desmethyldestruxin B
63	$HOCH_2\diagdown_{Me}\diagup CHCH_2$	Me	destruxin C
64	$HO_2C\diagdown_{Me}\diagup CHCH_2$	Me	destruxin D

9.2

The biosynthesis of verrucarin E involves the condensation of four acetate units with loss of one carboxyl group.[72] Proline, glutamic acid, and δ-aminolaevulinic acid are not incorporated. A possible sequence is that of Scheme 3. Panepoxydon, a metabolite of *Panus* spp.[73], has structure (65). Although formally related to terreic acid, *etc.* (p. 91), it seems possible that panepoxydon is derived from shikimate rather than from acetate

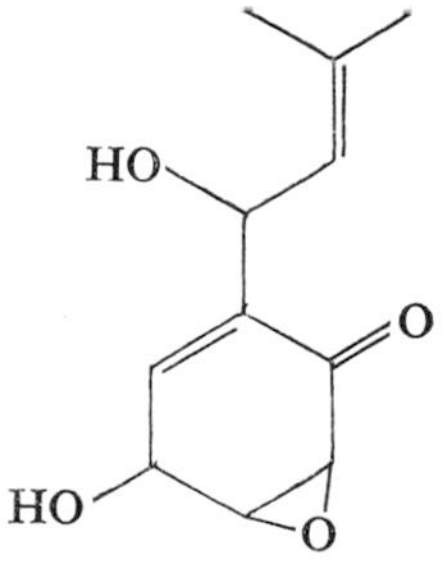

65

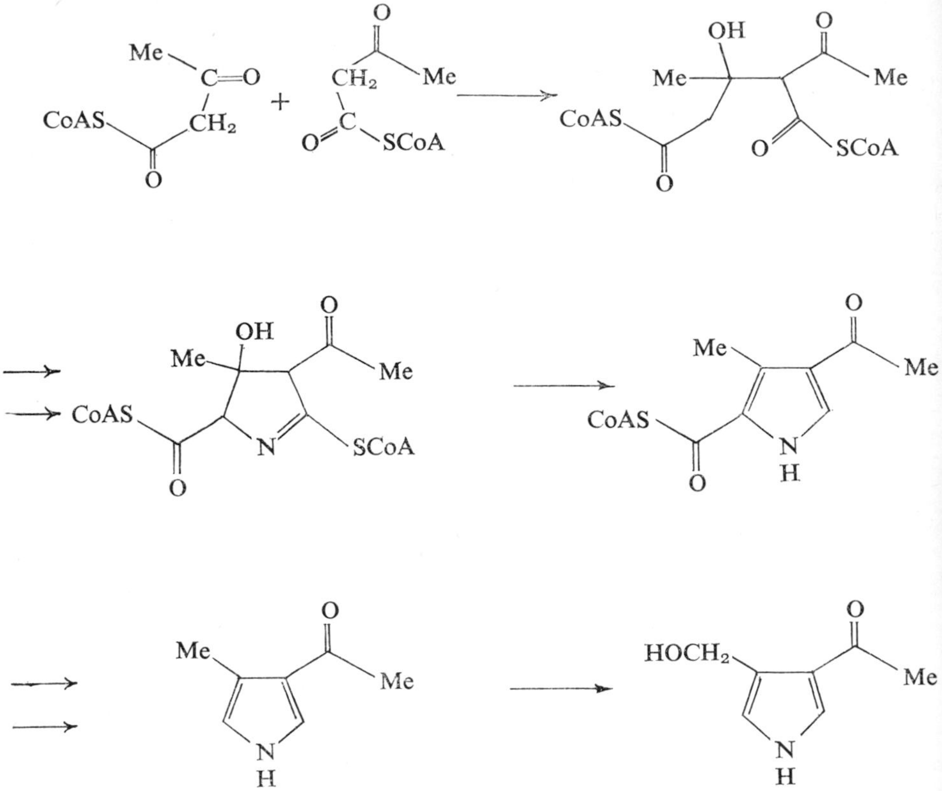

Scheme 3: A possible biosynthetic sequence for verrucarin E.[72]

Oospolide (**66**) is derived from acetate as shown.[74]

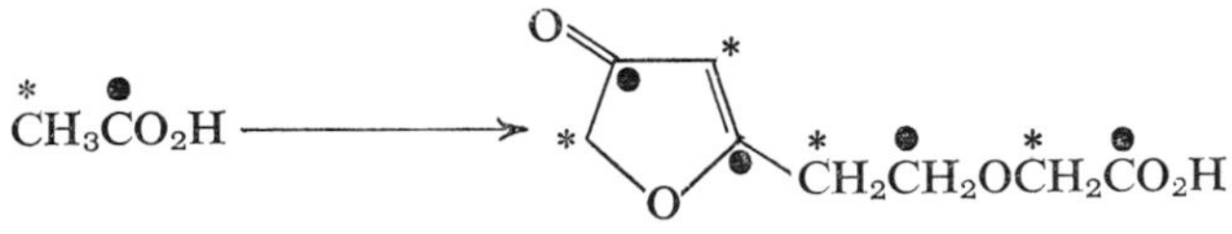

66

The structure of zymonic acid has been revised to (**67**)[75] and the compound is now thought to be an artefact derived from pyruvic acid present in the medium.

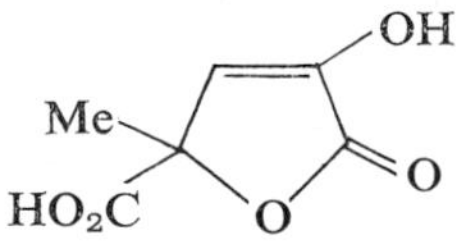

67

Lentinus edodes produces the sulphur compounds (**68**) to (**70**).

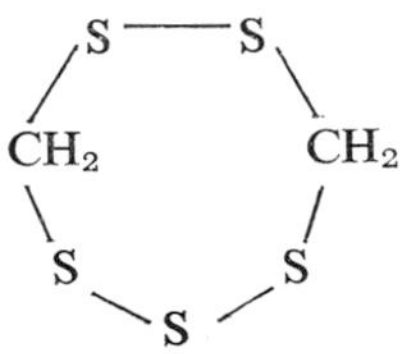

68
lenthionine
1,2,3,5,6-pentathiapane

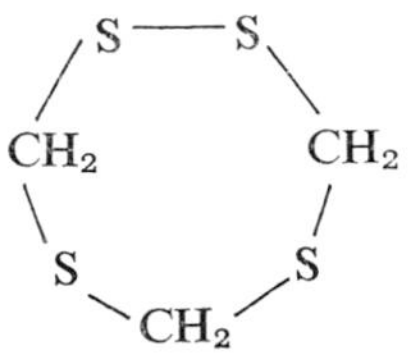

69
1,2,4,6-tetrathiapane

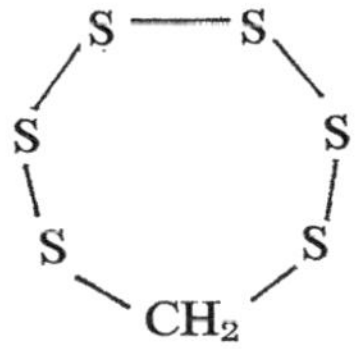

70
1,2,3,4,5,6-hexathiapane

REFERENCES

1. C. H. Eugster and E. Schleusener, *Helv. Chim. Acta*, 1969, **52**, 708.
2. *Idem, ibid.*, 1970, **53**, 130.
3. P. Catalfomo and C. H. Eugster, *ibid.*, p. 848.
4. G. S. Bezanson, D. Desaty, A. V. Emes and L. C. Vining, *Can. J. Microbiol.*, 1970, **16**, 147.
5. R. F. Vesonder and F. H. Stodola, *Can. J. Chem.*, 1969, **47**, 1247.
6. F. H. Stodola, R. F. Vesonder and L. J. Wickerham, *Biochemistry*, 1965, **4**, 1390.
7. P. Dimroth, H. Walter and F. Lynen, *Eur. J. Biochem.*, 1970, **13**, 98.
8. M. Tanabe, H. Seto and L. Johnson, *J. Amer. Chem. Soc.*, 1970, **92**, 2157.
9. P. Simonart and H. Verachtert, *Bull. Soc. Chim. Biol.*, 1969, **51**, 919.
10. Y. Yamamoto, M. Shinya and Y. Oohata, *Chem. Pharm. Bull. (Tokyo)*, 1970, **18**, 561.
11. S. Sakamura, J. Ito and R. Sakai, *Agr. Biol. Chem. (Tokyo)*, 1970, **34**, 153.
12. B. N. Bondarenko, Z. A. Lysenko, A. P. Rogozhina, D. E. Dykhovichnaya and R. P. Illarionova, *Mikrobiologiya*, 1969, **38**, 620.
13. J. W. Searcy, N. D. Davis and U. L. Diener, *Appl. Microbiol.*, 1969, **18**, 622.
14. R. G. Coombe, J. J. Jacobs and T. R. Watson, *Aust. J. Chem.*, 1968, **21**, 783.
15. S. Omura, A. Nakagawa, K. Sekikawa, M. Otani and T. Hata, *Chem. Pharm. Bull. (Tokyo)*, 1969, **17**, 2361.
16. H. Mishima and M. Kurabayashi, *7th Int. Symp. Chem. Nat. Prod., Riga*, 1970, Abstr. E 166.
17. D. J. Aberhart, K. H. Overton and S. Huneck, *J. Chem. Soc. (C)*, 1970, 1612.
18. L. B. Sokolov, L. E. Alekseeva and V. O. Kul'bakh, *7th Int. Symp. Chem. Nat. Prod., Riga*, 1970, Abstr. E 154.
19. Y. Suzuki, *Agr. Biol. Chem. (Tokyo)*, 1970, **34**, 760.
20. B. H. Mock and J. E. Robbers, *J. Pharm. Sci.*, 1969, **58**, 1560.
21. B. E. Cross, P. L. Myers and G. R. B. Webster, *J. Chem. Soc. (C)*, 1970, 930.
22. J. Cason, C. W. Koch and J. S. Correia, *J. Org. Chem.*, 1970, **35**, 179.
23. J. S. Brooks and G. A. Morrison, *Tetrahedron Lett.*, 1970, 963.
24. G. Bohman, *ibid.*, 1970, 445.
25. W. Steglich, W. Lösel and V. Austel, *Chem. Ber.*, 1969, **102**, 4104.
26. P. S. Steyn, *Tetrahedron*, 1970, **26**, 51.
27. J. A. Ballantine, D. J. Francis, C. H. Hassall and J. L. C. Wright, *J. Chem. Soc. (C)*, 1970, 1175.
28. R. W. Gray and W. B. Whalley, *Chem. Commun.*, 1970, 762.
29. L. Merlini, G. Nasini and A. Selva, *Tetrahedron*, 1970, **26**, 2739.
30. R. J. Cole, J. W. Kirksey and H. W. Schroeder, *Tetrahedron Lett.*, 1970, 3109.
31. H. G. Raj, L. Viswanathan, H. S. R. Murthy and T. A. Venkitasubramanian, *Experientia*, 1969, **25**, 1141.
32. D. G. Manwaring, R. W. Rickards and B. T. Golding, *Tetrahedron Lett.*, 1969, 5319.
33. T. Tatsuno, Y. Morita, H. Tsunoda and M. Umeda, *Chem. Pharm. Bull. (Tokyo)*, 1970, **18**, 1485.
34. B. Achilladelis, P. M. Adams and J. R. Hanson, *Chem. Commun.*, 1970, 511.
35. S. Nozoe and Y. Machida, *Tetrahedron Lett.*, 1970, 2671.

36. *Idem, ibid.*, p. 1177.
37. S. Takahashi, H. Iinuma, T. Takati, K. Maeda and H. Umezawa, *ibid.*, p. 1637.
38. S. Nozoe, M. Goi and N. Morisaki, *ibid.*, p. 1293.
39. S. Nozoe, M. Morisaki and H. Matsumoto, *Chem. Commun.*, 1970, 926.
40. A Kato, K. Ando, G. Tamura and K. Arima, *J. Antibiotics (Tokyo)*, 1970, **23**, 169.
41. N. Cagnoli Bellavita, P. Ceccherelli, M. Ribaldi, J. Polonsky and Z. Baskevitch, *Gazz. Chim. Ital.*, 1969, **99**, 1354.
42. N. Cagnoli Bellavita, P. Ceccherelli, R. Mariani, J. Polonsky and Z. Baskevitch, *Eur. J. Biochem.*, 1970, **15**, 356; N. Cagnoli Bellavita, P. Ceccherelli, M. Ribaldi, J. Polonsky and Z. Baskevitch, *7th Int. Symp. Chem. Nat. Prod.*, *Riga*, 1970, Abstr. D 37.
43. R. Guttormson, P. Main, A. J. Allison and K. H. Overton, *Chem. Commun.*, 1970, 719.
44. B. E. Cross, J. C. Stewart and J. L. Stoddart, *Phytochemistry*, 1970, **9**, 1065.
45. P. R. Jefferies, J. R. Knox and T. Ratajczak, *Tetrahedron Lett.*, 1970, 3229.
46. A. Ballio, C. G. Casinovi, G. Randazzo and C. Rossi, *Experientia*, 1970, **26**, 349.
47. D. H. R. Barton, D. M. Harrison, G. P. Moss and D. A. Widdowson, *J. Chem. Soc. (C)*, 1970, 775.
48. M. Devys and M. Barbier, *Bull. Soc. Chim. Biol.*, 1969, **51**, 925.
49. D. H. R. Barton, T. Shiori and D. A. Widdowson, *Chem. Commun.*, 1970, 939.
50. Y. Tomita, A. Uomori and H. Minato, *Phytochemistry*, 1970, **9**, 555.
51. A. Kanematsu and S. Natori, *Chem. Pharm. Bull. (Tokyo)*, 1970, **18**, 779.
52. H. Inouye, K. Tokura and T. Hayashi, *Tetrahedron Lett.*, 1970, 2811.
53. E. Caspi and L. J. Mulheirn, *J. Amer. Chem. Soc.*, 1970, **92**, 404.
54. A. Kawaguchi and S. Okuda, *Chem. Commun.*, 1970, 1012.
55. D. J. Austin, J. D. Bu'Lock and D. Drake, *Experientia*, 1970, **26**, 348.
56. J. D. Bu'Lock, D. J. Austin, G. Snatzke and L. Hruban, *Chem. Commun.*, 1970, 255.
57. D. W. Pitel and L. C. Vining, *Can. J. Biochem.*, 1970, **48**, 623.
58. D. Desaty and L. C. Vining, *ibid.*, 1967, **45**, 1953.
59. P. G. Mantle and E. S. Waight, *Trans. Brit. Mycol. Soc.*, 1969, **53**, 302.
60. J. Picker and R. W. Rickards, *Aust. J. Chem.*, 1970, **23**, 853.
61. D. Stauffacher, P. Niklaus, H. Tscherter, H. P. Weber and A. Hofman, *Tetrahedron*, 1969, **25**, 5879.
62. S. Yamatodani, Y. Asahi, A. Matsukura, S. Ohmomo and M. Abe, *Agr. Biol. Chem. (Tokyo)*, 1970, **34**, 485.
63. B. Naidoo, J. M. Cassady, G. E. Blair and H. G. Floss, *Chem. Commun.*, 1970, 471.
64. E. O. Ogunlana, B. J. Wilson, V. E. Tyler and E. Ramstad, *ibid.*, p. 775.
65. U. Nelson and S. Agurell, *Acta Chem. Scand.*, 1969, **23**, 3393.
66. N. Castagnoli, K. Corbett, E. B. Chain and R. Thomas, *Biochem. J.*, 1970, **117**, 451.
67. D. Gröger and D. Erge, *Z. Naturforsch.*, 1970, **25B**, 196.
68. D. Hauser, H. P. Weber and H. P. Sigg, *7th Int. Symp. Chem. Nat. Prod.*, *Riga*, 1970, Abstr. E 137.
69. L. Nover and M. Luckner, *Eur. J. Biochem.*, 1969, **10**, 268.
70. G. Bohman, *Tetrahedron Lett.*, 1970, 3065.
71. A. Suzuki, S. Kuyama, Y. Kodaira and S. Tamura, *Agr. Biol. Chem. (Tokyo)*, 1966, **30**, 517; A. Suzuki, H. Taguchi and S. Tamura, *ibid.*, 1970, **34**, 813.

Formula Index

References are to chapter and formula number, e.g. **3.36** refers to formula **36** in Chapter 3. Formulae in the Addendum are prefixed by the letter A.

Organism Index

Bold-face references are to the formulae of compounds isolated from each organism, those prefixed by **A** referring to formulae in the Addendum.

Author Index

Loeffler, W., 70 (19), *73*, 219 (15), 221 (15), *271*
Löfgren, N., 56 (88), *61*
Loh, L., 118 (165b), 125 (165b), *203*
Lokshin, G. B., 321 (98), *347*
Long, C. W., de 336 (173a, 174, 175), *349*
Lönnroth, I., 90 (44), 92 (44), *200*
Lorch, H., 264 (223), 266 (223), *278*
Lorck, H., 266 (236), *278*
Lord, K. E., 253 (152), *276*
Lösel, W., 157 (328), 180 (328), *208*
Lousberg, R. J. J. Ch., 146 (280), *207*
Lowe, D., 220 (22), *272*
Lowe, G., 67 (16, 17), *73*, 219 (17), 223 (17), 264 (220a), 266 (231), *271, 278*, 327 (130), *348*
Lüben, G., 339 (194), *350*
Lucas, W., 72 (31), *73*
Luckner, M., 332 (157, 158), 333 (157, 158, 159), 334 (160), *349*
Luders, W., 158 (365), *209*
Luft, P., 308 (46), *345*
Lund, H., 315 (82), *346*
Lund, N. A., 147 (285), *207*
Lundin, R. E., 88 (426), 179 (426), *211*
Lünine, B., 56 (88), *61*
Lustig, E., 88 (426), 179 (426), *211*
Lüttringhaus, A., 52 (71), *60*
Lybing, S., 282 (5), *294*
Lynch, J. F., 266 (233), *278*
Lynen, F., 62 (2), *72*, 79 (8), 80 (10), *199*
Lythgoe, B., 140 (240), *206*

M

Maass, W. S. G., 49 (63), *60*, 158 (333), *208*
McCapra, F., 176 (414), *211*, 227 (58), 229 (58), *273*
McCloskey, J. A., 9 (2), *10*, 303 (25), 304 (25), *345*
McCloskey, P., 267 (241, 242), *278, 279*
McCorkindale, N. J., 66 (8), *73*, 115 (151), *203*, 229 (61), 230 (86), 257 (183), 259 (194), 261 (206), 262 (208), 263 (213), *273, 274, 277, 278*, 289 (37, 39), 290 (39), *295*
McCormick, J. R. D., 183 (439), 186 (440, 441), *212*, 311 (58), *346*
McCormick, M. H., 336 (172), 337 (172), *349*

McCrae, W., 119 (169), *203*
MacDonald, J. C., 322 (111), 323 (111), 324 (114, 123), 325 (114, 121, 123, 124, 125), 326 (126), *347, 348*
McGahren, W. J., 119 (171), 125 (192), 126 (192), *203, 204*
McGonagle, M. P., 142 (254), 144 (250), 171 (396), *206, 210*
McGowan, J. C., 267 (237), *278*
McGrath, R., 306 (35), *345*
Machida, Y., 264 (220d), *278*
Machlis, L., 230 (64), *273*
McInnes, A. G., 134 (220), *205*, 312 (76), 315 (76), *346*
McLean, J., 253 (155), 258 (193), *276, 277*
McLoughlin, B. J., 303 (20), *345*
McMaster, W. J., 142 (253), 143 (253), *206*
MacMillan, J., 86 (22), 142 (251, 252), 144 (266), *199, 206*, 240 (101), 241 (101), 242 (109, 122), 267 (240), *274, 275, 278*
McMorris, T. C., *59*, 170 (391), *210*, 226 (48, 50), 227 (53), 229 (48, 53), 263 (210), *272, 273, 277*
McOmie, J. F. W., 141 (244), *206*
McPhail, A. T., 167 (379b), *210*, 223 (39), 227 (54), 229 (54), *272, 273*
Maeda, K., 226 (49), *272*
Måhlen, A., 287 (32), 290 (43), 291 (32, 43, 44), *295*
Mahlén, K., 3 (4), *10*
Mahmoodian, A., 157 (326), 159 (326), 169 (326), 172 (326), *208*
Majer, J., 76 (4), *199*
Majerus, P. W., 62 (2), *72*
Malavolta, E., 305 (30), *345*
Malström, L., 169 (401), 173 (401), *210*
Manchanda, A. H., 118 (165b), 125 (165b), *203*
Manson, W., 56 (84), *61*
Mansour, I. S., 312 (67), *346*
Marlow, W., 167 (381), *210*
Marquet, A., 266 (299), *278*
Marsden, D. J. S., 140 (240), *206*
Marsh, M. M., 330 (138, 148), *348*
Marshall, A. C., 51 (68), *60*
Marshall, J. R., 167 (377), *209*
Marshall, K. C., 305 (28), *345*
Martikkala, J., 45 (46), *59*

Subject Index

Bold-face entries refer to the formulae of fungal metabolites by chapter and formula number; the prefix **A** refers to formulae in the Addendum.

P

Pachybasin, **5.222a**
Pachymic acid, 262
Palitantin, **5.174a**
  biosynthesis, 72, 141, 155
Palmitic acid,
  in fungi, 62
  and brefeldin biosynthesis, 71
Panepoxydon, **A.65**
Pannaric acid, **5.70a**
Pantherine, **8.22,** 310
Papulosin, **5.222l**
Parietinic acid, **5.225b**
Patulin, **5.75**
  biosynthesis, 78, 80, 107–108
Pebrolide, **6.36a,** 231
Pencolide, **7.5**
  biosynthesis, 284
Penicillic acid, **5.74**
  biosynthesis, 106–107
Penicillins, 334–337
  biosynthesis, 336–337
  history, 334–335
  semi-synthetic, 336
  penicillin F, **8.90b**
  penicillin G, **8.90d**
  penicillin K, **8.90e**
  penicillin N, **8.90f**
  penicillin X, **8.90h**
Penicilliopsin, **5.256**
Penicillium herquei naphthalic anhy-
  dride, **5.212**
Pentaketides, 116–134
1,2,3,5,6-Pentathiapane, *see* lenthionine
Pentose phosphate cycle, 29, 33
  fate of glucose in, 34
Pentose phosphate pathway, 31
  fate of glucose in, 32
Peptides, cyclic, 338–341
Perlatolic acid, **5.172**
Phalloidin, **8.98,** 339
Phenol A, **5.124**
Phenoxazones, 310
Phenylacetamide, **3.11b**
Phenylacetic acid, **3.11a**
Phenylalanine,
  biosynthesis, 35
  and cyclopenin biosynthesis, 333
  and cytochalasin biosynthesis, 354
  and gliotoxin biosynthesis, 330

Phenylalanine—(*contd.*)
  and 5-methoxybenzofuran biosyn-
    thesis, 56
  and volucrisporin biosynthesis, 48
  and vulpinic acid biosynthesis, 50
L-Phenylalanine anhydride, **8.50**
Phenol coupling, *see* oxidative coupling
Phenyllactic acid,
  and volucrisporin biosynthesis, 48
Phlebiarubrone, **3.32**
  biosynthesis, 49
Phloroacetophenone,
  and usnic acid biosynthesis, 111
Phoenicin, **5.69**
  biosynthesis, 106
Phomazarin, **5.336**
  biosynthesis, 196
Phomin, *see* cytochalasin B
Phosphoenolpyruvate,
  and itaconitin biosynthesis, 292
Phycaron, *see* elsinochrome A
Phycobionts, 13
Phycomycetes, 12
  hormones in, 214, 270–271
Phyllostine, **A.7**
Physcion, **5.222e,** 365
Physcion anthrones, **5.229, 5.230,** 163
Picroroccellin, **8.55,** 323
Pimarane skeleton, 234, 238
Pimaricin, 191
Pinastric acid, **3.35d**
Pinicolic acid A, **6.113c**
Pinselic acid, **5.277b**
Pinselin, **5.277a**
Pipecolic acid,
  and slaframine biosynthesis, 302
Pithomycolide, **8.104**
  isolation, 343
Pleuromutilin, **6.53**
  biosynthesis, 238
Pleurotin, **9.19**
  biosynthesis, 357
Polyacetylenes, 66–70
  biosynthesis, 67–70
  structural types, 66–67
Polyene antibiotics, 191
Polyketides,
  biosynthesis, 74–84
    "starter effect", 75
    propionate in, 75–76
    mechanism of, 79–83

Polyketides,
  biosynthesis—(contd.)
    relationship to fatty acid biosynthesis, 79–80
    in cell-free systems, 79–80
    chain assembly, 80
    timing of steps, 80–83
    alkylation in, 77, 81
    two-chain theories, 83, 156, 162, 172, 195
  distribution in nature, 74
  distribution of chain-lengths, 198
  oxygenation patterns, 198
  types of cyclization, 198
Polyporenic acid A, **6.120**
  isolation of esters, 262
Polyporenic acid B, 262
Polyporenic acid C, **6.119a**
  C-20 epimers, 374
Polyporic acid, **3.26**
  and pulvinic acid biosynthesis, 49
Polystictin, *see* cinnabarin
Porphyrilic acid, **5.71**
Portentol, **A.10**
  biosynthesis, 363
Prehelminthosporal, 225
Prehelminthosporol, **6.16a**
  isolation, 225
  epimers, 225
Presiccanochromenic acid, **6.43**
Presqualene, 217–218
Primary metabolism, 1–3
Prolylleucyl anhydride, **8.54a**
Prolylphenylalanyl anhydride, **8.54c**
Prolylvalyl anhydride, **8.54b**
Prostaglandin E₂,
  biosynthesis, 71
Protocatechuic acid, **3.17**
  biosynthesis, 41
Protoilludanes, 226–229, 369
  biosynthesis, 227–229
  skeleton of, 227
Protolichesterinic acid, **7.25**
  biosynthesis, 291
Protostanes, 264–267
  skeleton, 267
Protozymonic acid, **9.25**
  biosynthesis, 357
Psilocin, **8.41b**
  biological activity, 316

Psilocybin, **8.41a**
  biological activity, 316
  biosynthesis, 316
  occurrence in Britain and Australia, 375
Puberulic acid, **5.88**
  biosynthesis, 114–115
  structure, 112
Puberulonic acid, **5.90**
  biosynthesis, 114–115
Pulcherrimic acid, **8.63**
  biosynthesis, 326
  and pulcherrimin, 325
Pulcherrimin, 325
Pulvilloric acid, **5.177**
  biosynthesis, 141
Pulvinic acid, *see* pulvinic acid lactone
Pulvinic acid lactone, **3.36a**
  biosynthesis, 50
  isolation from mycobiont, 50
Punicoskyrin, **5.247**
Purpurogenone, **5.192**
α-Pycolic acid, **8.35**
Pyrenophorin, **5.18a**, 86
Pyrenophorol, **5.18b**
Pyrethrin I, 71
Pyriculol, **5.173**
Pyridine derivatives, 311–315
Pyrogallol, **3.24**
ε-Pyrromycinone, 190
2-Pyruvoylaminobenzamide, **8.27**

Q

Quadrilineatin, **5.58**
Questin, *see* emodin-5-methyl ether
Questiomycin A, 311
Questiomycin B, *see* o-aminophenol

R

Radicicol, **5.288**
Radicinin, **5.24**, 86
  biosynthesis, 361
Ramulosin, **5.101a**
Ramycin, *see* fusidic acid
Rangiformic acid, **7.18**
Ravenelin, **5.278**
Resistomycin, **5.124**
  biosynthesis, 189
Resorcylic acid lactones, 174, 177
Reticulol, **5.109**
Retinal, *see* Vitamin A